£11-00

A Specialist Periodical Report

Nuclear Magnetic Resonance
Volume 2

A Review of the Literature published between
July 1971 and May 1972

Senior Reporter
R. K. Harris, School of Chemical Sciences, University of East Anglia

Reporters
N. Boden, University of Leeds
W. Derbyshire, University of Nottingham
M. I. Foreman, University of Strathclyde
D. G. Gillies, Royal Holloway College, University of London
R. Grinter, University of East Anglia
R. G. Jones, University of Essex
W. T. Raynes, Sheffield University
I. D. Robb, Unilever Research Ltd., Port Sunlight, Cheshire
D. Shaw, Varian Associates Ltd., Walton-on-Thames, Surrey
G. J. T. Tiddy Unilever Research Ltd., Port Sunlight, Cheshire

© Copyright 1973

The Chemical Society
Burlington House, London, W1V 0BN

ISBN: 0 85186 262 4

Library of Congress Catalog Card No. 72-78527

Printed in Great Britain by
Alden & Mowbray Ltd
at the Alden Press, Oxford

Foreword

This second volume of Specialist Periodical Reports on Nuclear Magnetic Resonance follows the pattern set by Volume 1. Thus most of the chapter headings and most of the reporters remain the same; it is hoped that this practice will continue and will provide helpful continuity for the reader. A separate chapter on The Solid State has been instituted, drawing partly from topics covered in Volume 1 by chapters on Relaxation and on Macromolecules and partly from solid-state bandshape studies (not covered in Volume 1); it is anticipated that n.m.r. work on solids will become of increasing importance to chemists. The chapter on Medium Effects has been extended to include effects which are of a physical rather than of a chemical nature; the former tended to be neglected in Volume 1. This year, however, there is no chapter on Oriented Molecules — this topic will be covered biennially. Fourier Transform N.M.R. Spectroscopy has not been treated separately, but its growing importance makes itself felt in several chapters (especially those on Relaxation and on Experimental Techniques). The present volume contains little mention of Chemically Induced Dynamic Nuclear Polarization or of ENDOR (the interested reader should refer to the Specialist Periodical Report on Electron Spin Resonance). As with last year's volume some areas of n.m.r. that are of more concern to physicists than to chemists have not been reviewed (*e.g.* many n.m.r. studies of metals).

The aim of the Report is still to be phenomenon-oriented (rather than compound-oriented) and critical, though the latter is a difficult aim to pursue because of the large number of n.m.r. papers under review. Trivial applications of n.m.r. are not in general mentioned, though this policy sometimes conflicts with the objective of securing comprehensive coverage of the literature.

The literature coverage for the present volume is in general from July 1971 to May 1972 inclusive, though it will be appreciated that a precise division is not feasible because of late arrival of journals (especially for translated versions) and the over-riding desirability of having the shortest possible delay between the end of the review period and publication date. Attempts have been made to refer in Volume 2 to any important papers inadvertently not mentioned in Volume 1. Literature survey is of course a difficult task and it is easy to accidentally miss relevant papers. It is worth commenting at this point that many of the Reporters use the computer-based listing of n.m.r. research publications provided by the United Kingdom Chemical Information Service 'Macroprofile. N.M.R. — Chemical Aspects'. This service utilizes a search profile operating from Chemical Abstracts magnetic tapes; thus it relies heavily on phrases such as Nuclear Magnetic Resonance in the title or key-words of a paper. This procedure may explain the occasional lack of reference to an important article herein. It is, in fact, vital that authors of research papers give considerable thought to the wording of their titles.

The policy of the Senior Reporter on units is to encourage gradually a greater uniformity among the various reports, and to move towards adoption of SI, though leaving a great deal of discretion in the hands of the individual reporters. It is regretted that in Volume 1, Chapter 1, certain equations omitted a factor of 4π; the corrected SI versions are to be found in the present Chapter 1. It is hoped that at each point in the present volume the usage is properly defined. Several general points where uniformity is attempted are as follows:

(1) Energies are invariably listed in joules rather than calories (occasionally both values may be quoted).
(2) Magnetic fields are quoted in tesla (T) rather than Gauss (1 T = 10^4 G), except in Chapter 9.
(3) There is a moratorium on casual use of the phrases high-field and low-field, which seem to the Senior Reporter to be absurd when frequency-sweep spectrometers are used, especially when the units mentioned are Hz (*e.g.* 'a high-field shift of 6 Hz'). When spectral features are being discussed the phrases low-frequency and high-frequency will be used except in cases of field-sweep instrumentation where some special purpose is served by mentioning field shifts. However, in some theoretical work it is clearly correct and desirable to refer to shielding and deshielding, provided p.p.m. are quoted and never Hz (see Chapter 1).

In this context it is pertinent to point to the 'Recommendations for the Presentation of N.M.R. Data for Publication in Chemical Journals', produced by the IUPAC Commission on Molecular Structure and Spectroscopy (*Pure Appl. Chem.*, 1972, **29**, 627), which, *inter alia*, strongly recommends the δ scale of ^{1}H n.m.r. in preference to the τ scale. Although the Commission have not, as yet, formulated policy for other nuclei the logical outcome is the use of high frequency as positive in every case, and some groups have certainly suggested this (*e.g.* the British N.M.R. Discussion Group).

A minor point concerns the abbreviation for nuclear magnetic resonance of particular isotopes. The use of PMR for proton magnetic resonance has been widespread, and it appears that CMR for carbon magnetic resonance is now of frequent occurrence. This practice has, however, been strongly deplored (E. Lustig, private communication) in terms with which this Senior Reporter is in agreement. It is pointed out that, if CMR is allowed, then NMR takes on the meaning 'nitrogen magnetic resonance'!! In any case it is desirable that

Foreword v

the isotopic mass number be given, and the abbreviation which commends itself is, *e.g.*, ^{13}C N.M.R. or ^{13}C n.m.r.

Two new features, which it is hoped will prove helpful, are to be found in this volume. One is a table of symbols used in the reports, together with their meanings. The second is a list of recent review articles on n.m.r., including their titles and page lengths.

Volume 1 was somewhat delayed in publication due to some unexpected industrial problems and related matters. It is hoped that this volume will be published more speedily. If this occurs it will largely be due to the efficiency of the Editorial Staff of the Chemical Society. I am very grateful to them for their help and expertise.

R. K. HARRIS, *October* 1972

SYMBOLS AND THEIR MEANINGS

This list contains the symbols most frequently used in this volume, but it is not expected to be exhaustive. Some specialized symbols are only defined in the chapters that use them. Trivial use of subscripts is not always mentioned in the list below. Other symbols used in the text, e.g. for physical constants such as h or π, or for thermodynamic quantities such as H or S, are standard (see 'Physico-chemical Quantities and Units', second edition, 1971, by M. L. McGlashan, Royal Institute of Chemistry Monographs for Teachers, No. 15). The spin system notation used is that suggested by Haigh [*J. Chem. Soc.* (*A*), 1970, 1682].

v_i	Larmor precession frequency of nucleus i (in Hz)
ω_i	Larmor precession angular frequency of nucleus i (in rad s^{-1})
v_0, ω_0	Spectrometer operating frequency *or* Larmor precession frequency (general, or of bare nucleus)
v_1, ω_1	Frequency of 'observing' rf magnetic field
v_2, ω_2	Frequency of 'irradiating' rf magnetic field
$\Delta v_{\frac{1}{2}}$	Full width (in Hz) of a resonance line at half-height.
f	Modulation frequency (in Hz)
B_0	Static magnetic induction field (magnetic flux density). (B may occasionally refer to the quadratic electric field parameter for nuclear shielding.)
B_1, B_2	Rf magnetic fields associated with v_1, v_2
σ_i	Shielding constant of nucleus i (used sometimes in tensor form). Usually in p.p.m. (σ may occasionally refer to a collision cross-section or to the spin density matrix.)
$\sigma_\parallel, \sigma_\perp$	Components of σ parallel and perpendicular to a molecular symmetry axis.
$\Delta\sigma$	Anisotropy in σ ($\Delta\sigma = \sigma_\parallel - \sigma_\perp$) *or* difference in σ between two different situations.
δ_X	Chemical shift of nucleus of element X (positive when the sample resonates to high frequency of the reference). Usually in p.p.m. Further information regarding solvent,

	reference, or nucleus of interest may be given by superscripts or subscripts or in brackets. The use of the τ-scale for ^{1}H n.m.r. is to be progressively discouraged.
$\Delta\delta$	Change or difference in δ. In Chapter 10 the symbol Δ is used, rather than $\Delta\delta$.
^nJ	Coupling constant through n bonds (in Hz). Further information may be given by subscripts or in brackets. Normally subscripts are only used for algebraic symbols for nuclei in spectral analysis cases, *e.g.* J_{AX}. Brackets are used for indicating the species of nuclei, *e.g.* $J(^{13}C,^1H)$, and/or, additionally, the coupling path, *e.g.* $J(POCF)$. In general terms the bracket notation will be used, *e.g.* as '($^{29}Si,^1H$) coupling constant'. (Occasionally J may be used for rotational quantum number.)
^nK	Reduced coupling constant (see Chapter 2).
$J^*(P,H)$	Isotopically scaled coupling constant, $= J(P,D) \times \gamma_H/\gamma_D$ (see Chapter 2).
J_R	Residual 'coupling' (*i.e.* splitting) in a double resonance experiment.
T_1^X	Spin–lattice relaxation time of the X nucleus (further subscripts refer to the relaxation mechanism).
T_2^X	Spin–spin relaxation time of the X nucleus (further subscripts refer to the relaxation mechanism).
$T_{1\rho}^X, T_{2\rho}^X$	Spin–lattice and spin–spin relaxation time of the X nucleus in the frame of reference rotating with B_1.
T_2^*	Effective spin–spin relaxation time (to account for magnetic field inhomogeneity).
$\hat{\mathcal{H}}$	Hamiltonian operator (in energy units) — subscripts indicate the nature of the operator.
H_i	Element of matrix representation of $\hat{\mathcal{H}}$.
E_n	Eigenvalue of $\hat{\mathcal{H}}$ (or of a contribution to $\hat{\mathcal{H}}$) (E may occasionally refer to electric field).
$\hat{I}_i$	Nuclear spin operator for nucleus i ($\hat{S}$ is also sometimes used, and may on occasions refer to electron spin).
$\hat{I}_{iy}, \hat{I}_{iy}, \hat{I}_{iz}$	Components of $\hat{I}_i$.
I_i	Magnetic quantum number associated with $\hat{I}_i$. (I may occasionally refer to ionization potential or to moment of inertia.)
m_i	Eigenvalue of $\hat{I}_{iz}$ (magnetic component quantum number). (m may also occasionally refer to electronic mass.)
m_T	Total magnetic quantum number for a spin system (eigenvalue of $\sum_i \hat{I}_{iz}$).
$\hat{F}_G, \hat{F}_{Gx}, \hat{F}_{Gy}, \hat{F}_{Gz}$	Nuclear spin operators for a group, G, of nuclei.
F_G	Magnetic quantum numbers associated with $\hat{F}_G$.
$\hat{A}, \hat{B}$	Symmetrized spin operators (see Chapter 5).

Symbols and their Meanings

$\hat{X}$	Transition moment: $\hat{X} = \sum_i I_{ix}$
α, β	Nuclear spin wavefunctions (eigenfunctions of $\hat{I}_z$) for a spin-$\tfrac{1}{2}$ nucleus (α may also refer to polarizability).
R	Redfield relaxation matrix *or* magnitude of reaction field term.
γ_X	Magnetogyric ratio of nucleus X.
g	Nuclear or electronic g-factor
τ	Pre-exchange lifetime of molecular species *or* delay time (*e.g.* between rf pulses) (*or*, occasionally, proton chemical shift scale).
τ_c	Correlation time.
τ_1, τ_2	Correlation times for the spherical harmonics Y_{1m}, Y_{2m}.
τ_J	Angular momentum correlation time.
τ_t	Translational magnetic relaxation correlation time.
t_{cp}	180° pulse separation in a Carr–Purcell sequence
λ	Spin inversion operator.
η	Nuclear Overhauser enhancement *or* asymmetry factor (*e.g.* in $e^2 Qq$) (*or*, occasionally, refractive index or viscosity).
C_x	Spin–rotation coupling constant of nucleus x (used sometimes in tensor form): $C^2 = \tfrac{1}{3}(C_\parallel^2 + C_\perp^2)$.
$C_\parallel, C_\perp$	Components of C parallel and perpendicular to a molecular symmetry axis.
ΔC	Anisotropy in C ($\Delta C = C_\parallel - C_\perp$).
$\bar{C}$	One third of the trace of C. $\bar{C} = \tfrac{1}{3}(C_\parallel + 2C_\perp)$.
$e^2 qQ$	Nuclear quadrupole coupling constant.
ε_r	Relative permittivity (dielectric constant).
ε_0	Permittivity of a vacuum.
μ	Magnetic dipole moment *or* electric dipole moment.
μ_0	Permeability of a vacuum.
e	Absolute charge on the electron.
G	Magnetic field gradient.
χ	Magnetic susceptibility *or* electronegativity *or* deviation spin density matrix (see Chapter 7).
$\Delta\chi$	Susceptibility anisotropy ($\Delta\chi = \chi_\parallel - \chi_\perp$) *or* difference in electronegativities.
a	Hyperfine (electron–nucleus) interaction constant (for additional comment see Chapter 2).
Q	McConnell's constant relating to a.
δ_{ij}	Kronecker delta ($= 1$ if $i = j$, and is zero otherwise).
Z_A	Atomic number of atom A.
μ_N	Nuclear magneton.
β	Bohr magneton.
s_A	Valence s-orbital of atom A.

p_A	Valence p-orbital of atom A.
$S_A^2(o)$	Electron density in s_A at nucleus A.
$\Pi_{s_A s_B}$	Mutual polarizability of s_A and s_B.
$P_{s_A s_B}$	Molecular orbital bond order between s_A and s_B.
α_A^2	s-Character of hybrid orbital at A.
$\rho_{s_A}^2$	Spin density in s_A.
d_A	Finite perturbation applied at nucleus A.
ϕ	Dihedral angle *or* atomic orbital *or* total spin inversion operator.
D	Self-diffusion coefficient.
$\mathbf{D_r}$	Rotational diffusion tensor.
$D_{\parallel}, D_{\perp}$	Components of $\mathbf{D_r}$ parallel and perpendicular to a molecular symmetry axis.

N.M.R. BOOKS AND REVIEWS

This section lists all books and reviews with n.m.r. as a principal theme, published during 1971 and 1972, which were known to the reporters at the time of going to press. Titles and numbers of pages are given where appropriate. Clearly the selection is somewhat arbitrary since desisions must be made as to when n.m.r. is a 'principal' theme and as to what constitutes a review (as distinct from a research paper). N.m.r. sections of general physical chemistry or spectroscopy textbooks have not been included. The list is sectionalized for the reader's convenience.

(a) *Books*
- R1. 'The Analysis of High Resolution N.M.R. Spectra', R. J. Abraham, Elsevier, 1971, 324 pages.
- R2. 'High Resolution N.M.R. of Macromolecules', F. A. Bovey, Academic Press, 1971.
- R3. 'P.M.R. Spectroscopy in Medical and Biological Chemistry', A. F. Casy, Academic Press, 1971, 426 pages.
- R4. 'Theory and Interpretation of Magnetic Resonance Spectra', W. T. Dixon, Plenum Press, 1972, 164 pages.
- R5. 'Pulse and Fourier Transform N.M.R., Introduction to Theory and Methods', T. C. Farrar and E. D. Becker, Academic Press, 1971, 115 pages.
- R6. '^{13}C N.M.R. for Organic Chemists', G. C. Levy and G. L. Nelson, Wiley-Interscience, 1972, 222 pages.
- R7. 'Magnetic Resonance', K. A. McLaughlan, Clarendon Press, Oxford, 1972, 105 pages.
- R8. 'The Nuclear Overhauser Effect', J. H. Noggle and R. E. Schirmer, Academic Press, 1971, 259 pages.
- R9. 'Nuclear Magnetic Resonance', W. W. Paudler, Allyn and Bacon, Boston, Mass., 1971, 241 pages.
- R10. 'Relaxation in Magnetic Resonance, Dielectric and Mossbauer Applications', C. P. Poole, jun., and H. A. Farach, Academic Press, 1971, 392 pages.

R11. 'The Theory of Magnetic Resonance', C. P. Poole, jun., and H. A. Farach, Wiley-Interscience, 1972, 452 pages.
R12. 'The Sadtler Guide to N.M.R. Spectra', W. W. Simons and M. Zanger, Heyden and Son Ltd., 1972, 600 pages.
R13. 'Magnetic Resonance in Metals', J. Winter, Oxford University Press, 1971, 206 pages.

(b) *Symposium lectures, books of partial relevance, chapters of books, etc.* (Note: If symposium lectures are published in an issue of a normal journal they will not usually be listed at this point.)

R14. 'N.M.R. Studies of Molecular Motion in Solids', P. S. Allen, MTP International Review of Science — Physical Chemistry Section, ed. C. A. McDowell, Butterworths, 1972, **4**, 43—83.
R15. 'Conformational Analysis via N.M.R. Spectroscopy', B. Coxon, *Methods Carbohydrate Chem.*, 1972, **6**, 513—539, ed. R. L. Whistler and J. N. BeMiller, Academic Press.
R16. 'Literature Data for IR, Raman, and N.M.R. Spectroscopy of Silicon, Germanium, Tin and Lead Organic Compounds', K. Light and P. Reich, Deut. Verlag. Wiss., Berlin, 1971, 623 pages.
R17. 'N.M.R. Spectroscopy', J. P., Heeschen, in 'The Characterization of Chemical Purity. Organic Compounds', ed. L. A. K. Staveley, 1971, p. 137—147. [Published as a Supplement to *Pure Appl. Chem.*]
R18. 'N.M.R. Spectroscopy of Annulenes', R. C. Haddon, V. R. Haddon, and L. M. Jackman, *Fortschr. Chem. Forsch.*, 1971, **16**, 103—220.
R19. 'Solvent Effects and N.M.R. Coupling Constants', S. L. Smith, *Fortschr. Chem. Forsch.*, 1972, **27**, 117—187.
R20. 'N.M.R. Spectra', R. F. M. White and H. Williams, in 'Physical Methods in Heterocyclic Chemistry', ed. A. R. Katritzky, 1971, Vol. 4, Chapter 4, pp. 121—235.
R21. 'Pyramidal Nitrogen Inversion', J. B. Lambert, *Topics in Stereochem.*, 1971, **6**, 19—105.
R22. 'Configuration and Conformation by N.M.R.', F. A. Anet and R. Anet, in 'Determination of Organic Structures by Physical Methods', ed. F. C. Nachod and J. J. Zuckerman, 1971, Vol. 3, Chapter 7, pp. 344—420.
R23. 'Applications of High-Field N.M.R. Spectroscopy', W. Naegele, in 'Determination of Organic Structures by Physical Methods', ed. F. C. Nachod and J. J. Zuckerman, 1971, Vol. 4, Chapter 1, pp. 1—49.
R24. 'Pulsed N.M.R. Methods', N. Boden, in 'Determination of Organic Structures by Physical Methods', ed. F. C. Nachod and J. J. Zuckerman, 1971, Vol. 4, Chapter 2, pp. 51—137.
R25. 'Nuclear Magnetic Double Resonance Spectroscopy', W. McFarlane, in 'Determination of Organic Structures by Physical Methods', ed. F. C. Nachod and J. J. Zuckerman, 1971, Vol. 4, Chapter 3, pp. 139—193.

R26. '^{15}N N.M.R.', R. L. Lichter, in 'Determination of Organic Structures by Physical Methods', ed. F. C. Nachod and J. J. Zuckerman, 1971, Vol. 4, Chapter 4, pp. 195—232.

R27. 'N.M.R. Spectra of the Heavier Elements', P. R. Wells, in 'Determination of Organic Structures by Physical Methods', ed. F. C. Nachod and J. J. Zuckerman, 1971, Vol. 4, Chapter 5, pp. 233—262.

R28. '^{13}C N.M.R.', P. S. Pregosin and E. W. Randall, in 'Determination of Organic Structures by Physical Methods', ed. F. C. Nachod and J. J. Zuckerman, 1971, Vol. 4, Chapter 6, pp. 263—322.

R29. '^{31}P N.M.R.', J. R. Van Wazer, in 'Determination of Organic Structures by Physical Methods', ed. F. C. Nachod and J. J. Zuckerman, 1971, Vol. 4, Chapter 7, pp. 323—358.

R30. 'N.M.R. Spectroscopy', J. R. Blackborow and K. D. Crosbie, in 'Spectroscopic Properties of Inorganic and Organometallic compounds', ed. N. N. Greenwood, (Specialist Periodical Reports), The Chemical Society, London, 1971, Vol. 4, pp. 1—186.

R31. 'Significant Progress in ^{19}F N.M.R. Spectroscopy', in 'Fluorocarbon and Related Chemistry', ed. R. E. Banks and M. G. Barlow, (Specialist Periodical Reports), The Chemical Society, London, 1971, Vol. 1, pp. 270—290.

R32. 'Physical Methods — N.M.R. Spectroscopy', J. C. Tebby, in 'Organophosphorus Chemistry', ed. S. Trippett (Specialist Periodical Reports), The Chemical Society, London, 1971, Vol. 2, pp. 236—259.

R33. 'Physical Methods — N.M.R. Spectroscopy', J. C. Tebby, in 'Organophosphorus Chemistry', ed. S. Trippett (Specialist Periodical Reports), The Chemical Society, London, 1972, Vol. 3, pp. 248—269.

R34. Symposium: Wide-line N.M.R. (American Oil Chemists' Society 43rd Fall Meeting, Minneapolis, 1969), Program Chairman, W. A. Bosin, *J. Amer. Oil. Chem. Soc.*, 1971, **48**, 1—17, and 47—69.

R35. 'Conformational Analysis. Scope and Present Limitations', Papers presented at the Brussels International Symposium (Sept. 1969), ed. G. Chiurdoglu, Academic Press, 1971. Many papers have n.m.r. significance, especially: (a) 'N.M.R. studies of the conformations and conformational barriers in cyclic molecules', F. A. L. Anet, pp. 15—29, and (b) 'Recent applications of N.M.R. spectrometry in conformational studies of cyclohexane derivatives', R. D. Stolow, pp. 251-258.

R36. 'Berichtsheft: Molekulare Bewegungen in Flussigkeiten', Herrenalb, 1970. *Ber. Bunsengesellschaft Phys. Chem.*, 1971, **75**, issues 3 and 4, pp. 183—396 (N.M.R. only up to p. 283).

R37. Symposium on Dynamic N.M.R. Spectroscopy. Abstracts of Papers, 161st National Meeting, American Chemical Society (Divisions of Org. Chem. and Phys. Chem.), Los Angeles, 1971.

R38. '^{13}C N.M.R. Spectroscopy', J. D. Roberts, 23rd International Congress of Pure and Applied Chemistry (Special Lectures), Butterworths, 1971, **7**, 71—105. (Supplement to *Pure Appl. Chem.*).

R39. 'Investigation of Conformational Equilibria in small and large Molecules by N.M.R. Spectroscopy', F. A. Bovey, 23rd International Congress of Pure and Applied Chemistry (Special Lectures), Butterworths 1971, **7**, 195—202. (Supplement to *Pure Appl. Chem.*).

R40. Seventh International Conference on Magnetism, 1970. *J. Phys. Paris (Colloq.)* 1971, **32**, C1-1. (Vol. I is pp. C1-1 to C1-584, Vol. II is pp. C1-585 to *ca.* C1-1200).

R41. Eighth Colloquium on N.M.R. Spectroscopy (A N.A.T.O. Advanced Study Institute), Aachen, 1971. *Adv. Mol. Relaxation Processes*, 1972, **3**, 1—367.

R42. The Applications of Computer Techniques in Chemical Research, Proceedings of the Institute of Petroleum (Hydrocarbon Research Group) Conference, Manchester, 1971, ed. P. Hepple, Institute of Petroleum, 1972. The following papers have n.m.r. significance: (a) 'Fourier Transform Spectroscopy', R. R. Ernst, pp. 61—75; (b) 'The Application of F.T. N.M.R. to the Measurement of Nuclear Relaxation Times', D. Shaw, pp. 76—84; and (c) 'A fully Automated N.M.R. Spectroscopy System', G. Michel, H. P. Sauter, and A. Staübli, pp. 101—111.

(c) *Regular Magnetic Resonance Review Series.*

R43. 'Pulsed-Fourier Transform N.M.R. Spectrometer', A. G. Redfield and R. K. Gupta, *Adv. Magn. Resonance*, 1971, **5**, 82—115.

R44. 'Spectrometers for Multiple Pulse N.M.R.', J. D. Ellett, jun., M. G. Gibby, V. Haeberlen, L. M. Huber, M. Mehring, A. Pines, and J. S. Waugh, *Adv. Magn. Resonance*, 1971, **5**, 117—176.

R45. 'N.M.R. and Ultraslow Motions', D. C. Ailion, *Adv. Magn. Resonance*, 1971, **5**, 177—227.

R46. 'N.M.R. in Helium Three', M. G. Richards, *Adv. Magn. Resonance*, 1971, **5**, 305—352.

R47. 'General Review of P.M.R.', G. R. Bedford, *Ann. Reports N.M.R. Spectroscopy*, 1971, **4**, 1—69.

R48. The Investigation of the Kinetics of Conformational Changes by N.M.R. Spectroscopy', I. O. Sutherland, *Ann. Reports N.M.R. Spectroscopy*, 1971, **4**, 71—235.

R49. 'N.M.R. Spectroscopy in Pesticide Chemistry', R. Haque and D. R. Buhler, *Ann. Reports N.M.R. Spectroscopy*. 1971, **4**, 237—309.

R50. 'A Simple Guide to the use of Iterative Computer Programs in the Analysis of N.M.R. Spectra', C. W. Haigh, *Ann. Reports N.M.R. Spectrosopy*, 1971, **4**, 311—362.

R51. 'The N.M.R. Spectra of Polymers', M. E. A. Cudby and H. A. Willis, *Ann. Reports N.M.R. Spectroscopy*, 1971, **4**, 363—389.

R52. '^{19}F N.M.R. Spectroscopy', K. Jones and E. F. Mooney, *Ann. Reports N.M.R. Spectroscopy*, 1971, **4**, 391—495.

R53. 'Static Quadrupole Effects in Disordered Cubic Solids', O. Kanert and M. Mehring, *N.M.R. Basic Principles and Process*, 1971, **3**, 1—81.
R54. 'Nuclear Magnetic Relaxation Spectroscopy', F. Noack, *N.M.R. Basic Principles and Progress*, 1971, **3**, 83—144.
R55. Lectures Presented at the Seventh Colloquium on N.M.R. Spectroscopy, Aachen, April, 1970, ed. P. Diehl, E. Fluck, and R. Kosfeld, *N.M.R. Basic Principles and Progress*, 1971, **4**, 1—309.
R56. 'Analysis of N.M.R. Spectra: A Guide for Chemists', R. A. Hoffman, S. Forsén, and B. Gestblom, *N.M.R. Basic Principles and Progress*, 1971, **5**, 1—165.
R57. 'Computer Assistance in the Analysis of High-Resolution N.M.R. Spectra', P. Diehl, H. Kellerhals, and E. Lustig, *N.M.R. Basic Principles and Progress*, 1971, **6**, 1—96.
R58. 'The Theory of Nuclear Spin–Spin Coupling in High Resolution N.M.R. Spectroscopy', J. N. Murrell, *Progr. N.M.R. Spectroscopy*, 1971, **6**, 1—60.
R59. 'Spin–Spin Coupling between Phosphorus Nuclei', E. G. Finer and R. K. Harris, *Progr. N.M.R. Spectroscopy*, 1971, **6**, 61—118.
R60. 'Nitrogen N.M.R.', E. W. Randall and D. G. Gillies, *Progr. N.M.R. Spectroscopy*, 1971, **6**, 119—174.
R61. 'Fluorine Chemical Shifts', J. W. Emsley and L. Phillips, *Progr. N.M.R. Spectroscopy*, 1971, **7**, 1—526.
R62. 'The Narrowing of N.M.R. Spectra of Solids by High-Speed Specimen Rotation and the Resolution of Chemical Shift and Spin Multiplet Structures for Solids', E. R. Andrew, *Progr. N.M.R. Spectroscopy*, 1971, **8**, 1—39.
R63. 'Pulsed N.M.R. in Solids', P. Mansfield, *Prog. N.M.R. Spectroscopy*, 1971, **8**, 41—101.
R64. '^{13}C Satellite N.M.R. Spectra', J. H. Goldstein, V. S. Watts, and L. S. Rattet, *Progr. N.M.R. Spectroscopy*, 1971, **8**, 103—162.
R65. 'The Stereochemistry of Double Bonds', G. J. Martin and M. L. Martin, *Progr. N.M.R. Spectroscopy*, 1972, **8**, 163—259.

(d) *Review articles in regular journals.*

R66. 'N.M.R. Spectra in Liquid Crystals and Molecular Structure', S. Meiboom and L. C. Snyder, *Accounts Chem. Res.*, 1971, **4**, 81—87.
R67. 'Structural Chemistry in Solution. The R Value', J. B. Lambert, *Accounts Chem. Res.*, 1971, **4**, 87—94.
R68. 'Paramagnetic Probes in Magnetic Resonance Studies of Phosphoryl Transfer Enzymes', M. Cohn and J. Reuben, *Accounts Chem. Res.*, 1971, **4**, 214—222.
R69. 'N.M.R. Studies of Molecular Relaxation Mechanisms in Polymers', D. W. McCall, *Accounts Chem. Res.*, 1971, **4**, 223—232.
R70. 'Conformational Analysis of Tris(Ethylenediamine) Complexes', J. K. Beattie, *Accounts Chem. Res.*, 1971, **4**, 253—259.

R71. 'Chemically Induced Dynamic Nuclear Polarization (CIDNP). I The Phenomenon, Examples, and Applications', H. R. Ward, *Accounts Chem. Res.*, 1972, **5**, 18—24.
R72. 'CIDNP. II. The Radical-Pair Model', R. G. Lawler, *Accounts Chem. Res.*, 1972, **5**, 25—33.
R73. 'The Annulenes', F. Sondheimer, *Accounts Chem. Res.*, 1972, **5**, 81—91.
R74. 'Determination of the Solution Conformations of Cyclic Polypeptides', F. A. Bovey, A. I. Brewster, D. J. Patel, A. E. Tonelli, and D. A. Torchia, *Accounts Chem. Res.*, 1972, **5**, 193—200.
R75. 'Proton Chemical Shifts for Solvents and Other Simple Substances', R. A. Fletton and J. E. Page, *Analyst*, 1971, **96**, 370—373.
R76. 'Fourier Transform Approaches to Spectroscopy', G. Horlick, *Analyt. Chem.*, 1972, **43**, 61A—66A.
R77. 'Nuclear Magnetic Resonance Spectrometry', P. L. Corio, S. L. Smith, and J. R. Wasson, *Analyt. Chem.*, 1972, **44**, 407R—438R.
R78. 'Methods and Applications of Nuclear Magnetic Double Resonance', W. von Philipsborn, *Angew. Chem. Internat. Edn.*, 1971, **10**, 472—490.
R79. 'Pulse Fourier Transform ^{13}C N.M.R. Spectroscopy Principles and Applications', E. Breitmaier, G. Jung, and W. Voelter, *Angew. Chem. Internat. Edn.*, 1971, **10**, 673—686.
R80. 'Use of a Computer in N.M.R. Spectroscopy', E. G. Hoffmann, W. Stempfle, G. Schroth, B. Weimann, E. Ziegler, and J. Brandt, *Angew. Chem. Internat. Edn.*, 1972, **11**, 375—386.
R81. 'N.M.R. in Solid-State Chemistry', A. Weiss, *Angew. Chem. Internat. Edn.*, 1972, **11**, 607—619.
R82. 'Shift Reagents in N.M.R. Spectroscopy', R. von Ammon and R. D. Fischer, *Angew. Chem. Internat. Edn.*, 1972, **11**, 675—692.
R83. 'Conformational Analysis in Heterocyclic Systems: Recent Results and Applications', E. L. Eliel, *Angew. Chem. Internat. Edn.*, 1972, **11**, 739—750 (also contained in 23rd Internat. Cong., *Pure Appl. Chem.* 1971, **7**, 219.)
R84. '^{13}C N.M.R. Spectroscopy: A Brief Review', J. B. Stothers, *Appl. Spectroscopy*, 1972, **26**, 1—16.
R85. 'Application of Spectroscopy in the Study of Glassy Solids. Part II. IR, Raman, E.P.R. and N.M.R. Spectral Studies', J. Wong and C. A. Angell, *Appl. Spectroscopy Rev.*, 1971, **4**, 155—232 (pp. 212—229 are on N.M.R.).
R86. 'Chemical Applications of N.M.R. at High Fields', J. K. Becconsall, P. A. Curnuck, and M. C. McIvor, *Appl. Spectroscopy Rev.*, 1971, **4**, 307—356.
R87. 'N.M.R. in Ferromagnets and Antiferromagnets', M. P. Petrov and E. A. Turov, *Appl. Spectroscopy. Rev.*, 1971, **5**, 265—330.
R88. 'Applications of N.M.R. in Enzymology', J. P. Cohen-Addad, *Biochimie*, 1971, **53**, 173.

R89. '^{13}C Magnetic Resonance', E. W. Randall, *Chem. in Britain*, 1971, **7**, 371—378.
R90. 'Lanthanide Shift Reagents in N.M.R. Spectroscopy', M. Holik, *Chem. Listy*, 1972, **66**, 449—457 (in Polish).
R91. 'Applications of the Intramolecular Nuclear Overhauser Effect in Structural Organic Chemistry', G. E. Bachers and T. Schaefer, *Chem. Rev.*, 1971, **71**, 617—626.
R92. 'Water Exchange Kinetics in Labile Aquo and Substituted Aquo Transition Metal ions by Means of Oxygen-17 N.M.R. Studies', J. P. Hunt, *Coord. Chem. Rev.* (*C*), 1971, **7**, 1—10.
R93. 'Magnetic Resonance Methods in the Study of the Electronic Structure of Transition Metal Complexes', D. R. Eaton and K. Zaw, *Coord. Chem. Rev.*, (*C*), 1971, **7**, 197—227.
R94. 'Utilisations des Complexes Paramagnetiques de Lanthanides en R.M.N.', J. Grandjean, *Ind. Chim. Belge*. 1972, **37**, 220—232.
R95. 'N.M.R. Spectrometers. I. Instrument Designs', D. G. Howery, *J. Chem. Educ.*, 1971, **48**, A327—A344.
R96. 'N.M.R. Spectrometers. II. Commercial Spectrometers', D. G. Howery, *J. Chem. Educ.*, 1971, **48**, A389—A398.
R97. 'Solvatation Preferentielle de Cations Diamagnetiques Dans les Melanges Hydro-Organiques par R.M.N.', J. J. Delpuech, A. Peguy, and M. R. Khaddar, *J. Electroanalyt. Chem.*, 1971, **29**, 31—54.
R98. 'Nuclear Magnetic Double Resonance', R. B. Johannesen and T. D. Coyle, *Endeavour*, 1972, **31**, 10—15.
R99. 'Relationship between Liquid Structure and the Temperature Dependence of the Nuclear Magnetic Relaxation Times', R. M. Yul'met'ev, *J. Struct. Chem.*, 1971, **12**, 509—520.
R100. 'N.M.R. Applications in Biochemistry', S. Forsén and B. Lindman, *Kem. Tidsskr.*, 1971, **83**, 64 (in Swedish).
R101. 'The Application of Chlorine, Bromine, and Iodine N.M.R. Spectroscopy to the Study of Physico–Chemical Processes in Liquids', C. Hall, *Quart. Rev.*, 1971, **25**, 87—109.
R102. 'N.M.R. Applications in Organic Chemistry', S. Mager, *Rev. Chim.* (*Roumania*), 1971, **22**, 107.
R103. 'N.M.R. Applications in Organic Chemistry', S. Mager, *Rev. Chim.* (*Roumania*), 1971, **22**, 233.
R104. 'Investigation of Keto–Enol Tautomerism by N.M.R. Spectroscopy', A. I. Kol'tsov and G. M. Kheifets, *Russ. Chem. Rev.*, 1971, **40**, 773—788.
R105. 'I.R. Spectroscopy and N.M.R. in Nucleic Acid Research', H. Fritzsche, *Z. Chem.*, 1972, **12**, 1—18 (in German).

Contents

Chapter 1 Nuclear Shielding
By W. T. Raynes

1 Introduction	1
2 Basic Aspects of Nuclear Shielding	2
A General Theory	2
B Basic Physical Aspects	4
3 Calculations of Nuclear Shielding	7
4 Transmission of Shielding Effects within Molecules	10
A Introduction	10
B Inductive Effects	12
C Resonance Effects	16
D Magnetic Anisotropy Effects	18
E Electric Field Effects	21
F The Ring-current Effect	24
G Van der Waals and Steric Effects	30
H Intramolecular Hydrogen Bonding	31
I Isotope Shifts	33
5 Shieldings of Particular Nuclear Species	35
A Introduction	35
B Carbon Chemical Shifts	36
C Fluorine Chemical Shifts	42
D Phosphorus Chemical Shifts	44
E Nitrogen Chemical Shifts	46
F Chemical Shifts of Nuclei Other than H,C,F,P, and N	47
6 Chemical Shift Anisotropy	49

Chapter 2 Nuclear Spin–Spin Coupling
By R. Grinter

1 Introduction	50

2 Basic Theory: Notation and Abbreviations — 50

- A Sum-Over-States Perturbation Theory (SOS) — 50
- B Finite Perturbation Theory (FPT) — 51
- C Semi-empirical Molecular Orbital Methods — 52
- D Units — 52

3 Theoretical Work — 52

- A Calculations on H_2 or HD — 53
- B *Ab Initio* Calculations of Coupling in Small Molecules — 53
- C Semi-empirical Calculations — 54
- D π-Electron Coupling — 60
- E Nuclear-spin and Electron-spin Coupling — 62
- F Solvent Effects — 62

4 Coupling of Directly Bonded Nuclei — 62

- A Coupling in HD — 62
- B Coupling between ^{13}C and ^{1}H — 62
- C Coupling between ^{15}N and ^{1}H — 64
- D Coupling between ^{31}P and ^{1}H — 65
- E Coupling between ^{1}H and Other Nuclei — 66
- F $^{1}J(^{13}C,^{13}C)$, $^{1}J(^{13}C,^{15}N)$, $^{1}J(^{11}B,^{13}C)$, and $^{1}J(^{11}B,^{11}B)$ — 67
- G Coupling between ^{13}C and ^{31}P — 69
- H Coupling between ^{13}C and ^{19}F — 70
- I Coupling between ^{13}C and Metals — 70
- J Coupling to ^{19}F, except $^{1}J(^{19}F,^{31}P)$ and $^{1}J(^{19}F,^{13}C)$ — 70
- K Coupling between ^{31}P and ^{19}F — 72
- L Other One-Bond Couplings involving ^{31}P — 72

5 Coupling between Nuclei Separated by Two Chemical Bonds, ^{2}J — 74

- A Geminal Proton–Proton Coupling across Carbon — 74
- B Geminal Proton–Proton Coupling across Other Elements — 76
- C $^{2}J(^{13}CC^{1}H)$ — 76
- D Coupling of Other Nuclei to Protons *via* Carbon, $^{2}J(XC^{1}H)$ — 77
  - (i) X = ^{199}Hg — 77
  - (ii) X = ^{11}B and ^{27}Al — 78
  - (iii) X = ^{73}Ge, $^{119,117}Sn$, and ^{207}Pb — 78
  - (iv) X = $^{15,14}N$ and ^{31}P — 78
  - (v) X = ^{77}Se — 80

	(vi) $X = {}^{19}F$	80
	(vii) $X = {}^{183}W$ and ${}^{195}Pt$	80
E	${}^{2}J({}^{19}FC{}^{19}F)$	80
F	${}^{2}J({}^{13}CC{}^{13}C)$, ${}^{2}J({}^{31}PC{}^{13}C)$, and ${}^{2}J({}^{15}NC{}^{13}C)$	80
G	${}^{2}J({}^{31}PN{}^{13}C)$ and ${}^{2}J({}^{31}PN{}^{31}P)$	81
H	${}^{2}J({}^{31}PM{}^{13}C)$, where M is a metal	82
I	${}^{2}J({}^{31}PM{}^{31}P)$, where M is a Metal	82
J	${}^{2}J({}^{13}CO{}^{13}C)$ and ${}^{2}J({}^{31}PO{}^{13}C)$	83
K	Other Two-bond Coupling Constants	83

6 Coupling between Nuclei Separated by Three Chemical Bonds, ${}^{3}J$ 83

A	Coupling of Protons *via* Two Carbon Atoms	83
B	Three-bond Coupling of Protons *via* Other Atoms	88
C	Three-bond Coupling of Carbon to Protons	88
D	Three-bond Coupling of Other Nuclei to Protons *via* Two Carbon Atoms, ${}^{3}J(XCC{}^{1}H)$	89
	(i) $X = {}^{15,14}N$	89
	(ii) $X = {}^{11}B$ and ${}^{27}Al$	90
	(iii) $X = {}^{31}P$, ${}^{119,117}Sn$, ${}^{199}Hg$, and ${}^{207}Pb$	90
E	Three-bond Coupling to Fluorine *via* Two Carbon Atoms, ${}^{3}J(XCC{}^{19}F)$	91
	(i) $X = {}^{1}H$	91
	(ii) $X = {}^{19}F$	92
	(iii) $X = {}^{15}N$ and ${}^{199}Hg$	93
F	Three-bond Couplings between Phosphorus and Hydrogen, ${}^{3}J({}^{31}PXC{}^{1}H)$	93
	(i) $X = O$	93
	(ii) $X = N$	94
	(iii) $X = S$	94
	(iv) $X = $ a Group IVB Element	94
	(v) $X = $ Au and Pt	94
G	${}^{3}J({}^{13}CCC{}^{31}P)$ and ${}^{3}J({}^{13}CCO{}^{31}P)$	94
H	${}^{3}J({}^{13}CCC{}^{13}C)$	95
I	${}^{3}J({}^{199}HgXC{}^{1}H)$	95
J	${}^{3}J({}^{195}PtXC{}^{1}H)$, $X = N, P$, and As	95
K	Other Three-bond Coupling Constants	95

7 Coupling between Nuclei Separated by Four Chemical Bonds, ${}^{4}J$ 96

A	${}^{4}J({}^{1}HCCC{}^{1}H)$	96
B	Other Four-bond (H,H) Couplings	97
C	Four-bond Coupling of ${}^{13}C$, ${}^{14}N$, and ${}^{31}P$ to ${}^{1}H$ *via* Carbon Atoms	98

Contents

D	Four-bond Coupling of Metals to Protons	98
E	$^4J(^1HCCC^{19}F)$ and $^4J(^{19}FCCC^{19}F)$	99
F	Other Four-bond Couplings	99

8 Coupling between Nuclei Separated by Five Chemical Bonds, ^5J   99

A	$^5J(^1HCCCC^1H)$	99
B	$^5J(^1HCCXC^1H)$, X = N, O, and S	100
C	Five-bond Coupling of ^1H to ^{19}F	101
D	Other Five-bond Couplings	101

9 Coupling between Nuclei Separated by Six or More Chemical Bonds   102

A	^6J	102
B	^7J and ^8J	102

10 Coupling in Aromatic Systems   102

A	($^1H,^1H$) Coupling in Benzenoid Aromatic Systems	102
B	($^1H,^1H$) Coupling in Heteroaromatic Systems	105
C	Coupling in Metal–Cyclopentadienyl Compounds	107
D	($^1H,^{19}F$) Coupling in Aromatic Systems	107
E	($^1H,^{13}C$) Coupling in Aromatic Systems	107
F	($^1H,^{31}P$) Coupling in Aromatic Systems	108
G	Coupling of Protons to Other Nuclei in Aromatic Systems	108
H	($^{13}C,^{15}N$) Coupling in Aromatic Systems	108
I	Other Couplings in Aromatic Systems	108

11 'Through-space' Coupling   108

12 Isotope Effects   109

13 Experimental Work of Significance for the Measurement or Interpretation of Spin–Spin Coupling Constants   109

A	Multiple-resonance Experiments	109
B	Bandshape Analyses	110
C	Effects of Paramagnetic Materials	110
D	Fourier-transform Experiments	111
E	Other Experimental Techniques	111

Chapter 3 Nuclear Spin Relaxation
By N. Boden

1 Introduction 112

2 Spin Relaxation in Gases 113
   A Introduction 113
   B Theoretical Developments 113
   C Diatomic Molecules 114
   D Polyatomic Molecules 116
   E Spin Diffusion Measurements 117

3 Spin Relaxation in Liquids 117
   A Introduction 117
   B Pure Liquids 118
      Linear Molecules 118
      Spherically Symmetric Molecules 119
      Axially Symmetric Molecules 119
      Asymmetric Molecules 122
   C Internal Rotation of Methyl Groups 124
   D Molecular Motion and Association in Mixtures of Liquids 125
      Molecular Rotation Studies 125
      Molecular Association Studies 126
      Aqueous Solutions of Non-electrolytes 128
      Hydrogen-bonded Systems 129
   E Ionic Solutions 129
      Diamagnetic Solutions 129
      Paramagnetic Solutions 131
   F Relaxation by Translational Diffusion 131
   G Spin-diffusion Measurements in Liquids 134
   H Transverse Relaxation and Spin-echo Studies 135
      Spin-echo Spectroscopy 135
      Selective Measurement of Transverse Relaxation Rates 135
      Chemical Exchange Studies 137
   I Models for Molecular Rotation 137
   J Studies of Critical Phenomena in Fluids 139
   K ^{13}C Relaxation Studies 139
      Small Molecules 139
      Complex Molecules 139
   L ^{15}N Relaxation Studies 141
   M ^{31}P Relaxation Studies 141

N Determination of Spin Interaction Constants from
Spin–Relaxation Measurements 142
Scalar Spin–Spin Coupling Constants 142
Nuclear Quadrupole Coupling Constants 142
Spin–Rotation Interaction Constants and Chemical
Shielding Anisotropies 142
O Miscellaneous Papers 144

4 Spin Relaxation in Liquid Crystals 144

A Nematic Phases 144
B Cholesteric Phases 147
C Isotropic Phases 147
D Spin-diffusion Studies 147

5 Spin Relaxation in Solids 148

A Spin–Lattice Relaxation in the Rotating Frame and
Dipolar Relaxation 148
B Spin–Lattice Relaxation by Anisotropic Chemical
Shielding Interaction 149
C Spin–Lattice Relaxation by Spin–Rotation Interaction 150
D Spin–Lattice Relaxation by Nuclear Quadrupole
Interaction 150
E Spin–Lattice Relaxation due to Paramagnetic
Impurities 151
F Investigations of Molecular Reorientation 152
Molecular Solids 152
Internal Rotation in Molecular Solids 155
Internal Rotation in Ionic Solids 156
Spin–Lattice Relaxation by Quantum Mechanical
Rotation at Low Temperatures 158
G Investigations of Self-diffusion 159
General Developments 159
Metals 160
Metal Hydrides 160
Ionic Crystals 161
Molecular Crystals 161
H Studies of Molecules Bound to Solid Surfaces 162
I Ferroelectric Materials 164

Chapter 4 Experimental Techniques
By D. G. Gillies

1 Introduction 165

2 Probes	165
A Variable Temperature	165
B High Pressure	166
C High Field	167
3 Frequency Generation	168
4 Pulse-sequence Generation	171
5 Continuous-wave Spectroscopy	172
A Double-resonance Techniques	172
B 'Other' Nuclei	173
C Use of Computers	173
D Relaxation Measurements	174
6 High-resolution Fourier-transform Spectroscopy	175
A Introduction	175
B Fourier Spectrometers	176
C Relaxation Measurements	177
T_1	177
Spin Echoes	179
D Gated Double Resonance	179
E Difference Spectroscopy	180
7 General Pulsed Spectroscopy	181
A Introduction	181
B Spectrometers	181
C Relaxation Measurements in Liquids	182
$T_{1\rho}$	182
Spin Echoes	183
T_1	183
D High-resolution Studies in Solids	184

Chapter 5 Spectral Analysis
By R. G. Jones

1 Introduction	186
2 New Methods of Studying Known Spin Systems	186
A Proton Spin-echo N.M.R. Spectra of 2,2-Dichloro-1,1-difluoroethane: Sub-spectral Analysis	186

B Spin Hamiltonian for Twofold Symmetry	189
Matrix Elements of the Hamiltonian with Symmetrized Spin Operators	191
Symmetry Properties of the $\|A,M\rangle$ Basis Functions	192
C Double-quantum Effects in the Spectrum of 2-Bromothiazole	193
D The Direct Method Applied to Nuclear Magnetic Double Resonance	194

3 Known Spin Systems — 194

A AB-Based N.M.R. Systems: Choice of Bounds for θ in Trigonometric Forms of Co-factors in Mixed Wave-functions	194
B Systems involving Protons Only	196
Saturated Acyclic Systems	196
(i) n-Propyl derivatives; $[AB]_2C_3$	196
(ii) 2,2′-Dihalogenodiethyl ethers; $[AB]_2$	197
Saturated Cyclic Systems	197
(i) 1,3-Dioxans; AB, AMX, ABX, $[AX]_2$	197
(ii) D-Aldopentapyranosyl derivatives; ABX	198
(iii) N-Substituted styreneimines; ABX	198
(iv) L-Azetidine-2-carboxylic acid (L-ACA) and N-acetyl-L-azetidine-2-carboxylic acid (N-A-L-ACA); ABCDE	199
(v) 1,2,3-Substituted prolines; ABCDEF	199
Olefinic Compounds: Dichloropentenes; ABCDE	200
Unsaturated Five-membered Ring Compounds	201
(i) 3,5-Disubstituted 1,2-dihydro-furans and -thiophens; ABX, $ABXY_3$, $ABXY_2$, ABC	201
(ii) Dimers of cyclopentadienyl compounds	201
Sesquiterpenoids; ABX{H}	202
Aromatic Systems	202
Aromatic substitution patterns	202
Biologically important aromatic acids	203
Benzocycloheptenes	203
Carbanions and Ion Pairing; $[AM]_2X$, AMRX, $[AX]_2$, AA′X	204
Aryldi-t-butylmethanol	205
Acridine and Five Monomethyl Derivatives; 'ABCDX'	205
Sulphur Heterocycles	206
syn- and anti-2-Furanaldoximes	208
C Compounds containing Fluorine	209

	4-Fluoromethylated 5(and 6)-Methyl-1,3-dioxans	210
	Ethyl 2,3-Difluoropropionate; ABCXY	210
	Fluorinated Benzofurans; ABC, ABCD, ABCDRX$_3$	211
	7,7-Difluorobenzocyclopropene; [AB]$_2$X$_2$	212
D	Compounds containing Phosphorus	212
	1,3,2,-Dithiaphospholans; [AB]$_2$X	212
	1,3,2-Oxazaphospholans; ABXY	213
E	Miscellaneous Compounds	213
	Methyl- and Chlorine-substituted Diboranes	213
	Tetrafuranyl- and Tetrathienyl-lead Compounds	214
	1,3,2-Dioxarsolans; [AX$_3$]$_2$	214
	trans-2-Chloro-4,5-dimethyl-1,3,2-dioxarsolan	214
	cis-2-Chloro-4,5-dimethyl-1,3,2-dioxarsolan	215
	o-Phenylenebis(*p*-ethoxyphenyl telluride); [AB]$_2$	215
4	**General Comments**	**215**
	Books	215

Chapter 6 Bandshape Phenomena for Fluids
By R. K. Harris

1	**General Introduction**	**216**
A	Current Developments	216
B	Coverage of the Report	216
C	Nomenclature, Notation, and Units	218
D	Experimental Procedures	219
E	Miscellaneous Applications	221
2	**Exchange of Magnetic Sites**	**222**
A	Theoretical Work	222
B	Examples of Intramolecular Exchange	226
	Internal Rotation about the C—C Bond	226
	Ring Inversion and Related Processes	229
	C—N Internal Rotation in Amides and Related Compounds	231
	Nitrogen Inversion, C—N Internal Rotation, and Related Processes	235
	Metallotropic Rearrangements and Related Processes	238
	Other Migration Processes	241
C	Examples of Intermolecular Exchange	241
	Studies of Ligand Exchange using the Swift–Connick Approach	242

	Proton Exchange Processes	246
	Other Studies of Intermolecular Exchange	247

3 Effects of Quadrupolar Nuclei — 249

   A Theoretical Work — 249
   B Resonances of Spin-$\frac{1}{2}$ Nuclei Coupled to Quadrupolar Nuclei — 250
   C N.M.R. Spectra of Quadrupolar Nuclei — 252
      Nitrogen-14 — 253
      Oxygen-17 — 253
      Alkali Metals — 254
      Halogen Nuclei — 254
      Germanium-73 — 255
   D Effects of Magnetic Site Exchange on N.M.R. Spectra for Spin Systems containing Quadrupolar Nuclei — 255

4 Relaxation Effects — 258

   A Theoretical Work — 258
   B Paramagnetic Effects — 259
   C Gas-phase Studies — 261
   D Saturation Studies — 261

5 Bandshapes in Multiple Resonance — 261

6 Appendix: Activation Parameters for Intramolecular Exchange — 262

Chapter 7 Multiple Resonance
By D. Shaw

1 Introduction — 269

2 Theory of Multiple Resonance — 269

3 Homonuclear Double Resonance — 270

   A Spin Decoupling and INDOR — 270
   B Nuclear Overhauser Effect — 271

4 Heteronuclear Double Resonance — 272

   A Spin Decoupling and INDOR — 272
   B $^{13}C-\{^{1}H\}$ Off-resonance Decoupling — 274

5	Chemical Shift Referencing in Multinuclear/Multi-resonance Spectra	278
6	Relaxation Effects in Multiple Resonance	280
7	$^{13}C-\{^{1}H\}$ Nuclear Overhauser Studies	281

Chapter 8 Macromolecules
By I. D. Robb and G. J. T. Tiddy

1 Introduction	285
2 Reviews	285
3 Synthetic Macromolecules	286
A ^{1}H High Resolution	286
Tacticity Determinations	286
Structure Determinations	290
Miscellaneous Studies	291
B Wide-Line N.M.R.	293
C Spin-Echo N.M.R.	295
D ^{13}C N.M.R.	297
E ^{19}F N.M.R.	299
4 Biological and Related Macromolecules	299
A New Techniques	299
B Synthetic Polypeptides	300
C Natural Polypeptides	303
D Polynucleotides	306
E Proteins Containing Paramagnetic Ions	307
F Membranes	309
G Polysaccharides	309
H ^{13}C N.M.R.	310
I ^{19}F and ^{31}P N.M.R.	311
5 Small Molecules	312
A Counterions and Ligands	312
B Hydration of Macromolecules	315
C Other Studies	318

Chapter 9 The Solid State
By W. Derbyshire

1 General Introduction	323

2 Nuclei of Spin $\frac{1}{2}$ — 324

- A Introduction — 324
- B Proton N.M.R. — 324
  - Clathrate Hydrates — 324
  - Conformational Motion — 325
  - Carboranes — 326
  - Hydrogen Sulphide and Selenide — 326
  - Hydrates — 326
  - Ring Compounds — 327
  - Methyl-group Rotation — 328
  - Ammonia and Derivatives — 329
  - Diffusion Studies — 330
  - Ferroelectricity — 331
  - Polymers — 331
- C Spin-$\frac{1}{2}$ Nuclei other than ^{1}H — 332
  - ^{19}F N.M.R. — 333
  - ^{31}P N.M.R. — 334

3 Nuclei of Spin $> \frac{1}{2}$ — 335

- A Introduction — 335
- B The Pseudoquadrupole Effect — 336
- C Studies of Specific Nuclei — 336
  - Deuterons — 336
  - Sternheimer Antishielding — 338
  - ^{7}Li N.M.R. — 338
  - ^{9}Be N.M.R. — 339
  - ^{11}B N.M.R. — 339
  - ^{23}Na N.M.R. — 339
  - Other Nuclei — 340

4 Lineshape Calculations — 341

5 Line-narrowing Techniques — 344

- A Introduction — 344
- B Sample Spinning — 344
- C Pulse Sequences — 345

6 Surface Phenomena — 345

- A Introduction — 345
- B Adsorption on to Silica, Alumina, and Titania Surfaces — 346
- C Zeolites and Exchange Resins — 346
- D Cellulose Systems — 348

E	Proteins	349
F	Synthetic Polymers	349

7 Systems with Unpaired Electron Spins: Para-, Ferro-, and Antiferro-magnets, Metals, and Semiconductors 350

A	Introduction	350
B	Transition Elements and Rare Earths	350
C	Semiconductors	352
D	Metals (General)	353

8 Instrumental and Operational Developments 353

9 Applications 354

Chapter 10 Medium Effects
By M. I. Foreman

1 Introduction 355

2 Coupling Constants 355

3 Chemical Shifts 358

A	Gas-to-solution Shifts	358
B	Hydrogen-bonding Effects	362
C	Ionic Solvation and Ion-pairing Effects	369
D	Aromatic Solvent-induced Shifts (ASIS Effects)	373

4 Shift Reagents 378

A	Lanthanide Complexes	378
	The Stoicheiometry of the Shift-reagent–Substrate Complex	379
	The Nature of the Lanthanide-induced Shifts	380
	Calculation of the Dipolar Interaction	382
	Temperature-dependence of the Lanthanide-induced Shifts	383
	Effects of Isotopic Substitution	384
	Lanthanide-induced Shifts in Chiral Systems	385
	Shift Reagent in Aqueous Solution	386
B	d-Series Transition-metal Shift Reagents	386
C	Diamagnetic Shift Reagents	389

Author Index 391

chemical shift data has increased to twenty-six — one more than that for fluorine — with nitrogen not far behind. With this large increase for C, P, and N it has been decided to limit the number of references to proton chemical shift studies (which total about two hundred and fifty) only to those of particular value for illustrating proton shielding mechanisms or to those of compounds of particular interest such as the annulenes, substituted derivatives of benzene, *etc*.

The following two conventions, used in last year's Report on nuclear shielding,[2] have been adopted in subsequent sections of this Report. N.m.r. chemical shifts and shielding constants are occasionally given without the appellation p.p.m. (parts per million). Where substituent effects are considered and the chemical shift is referred to the unsubstituted compound, the shift has been given a positive sign if the nucleus under investigation has a higher shielding constant in the substituted compound than in the unsubstituted compound. During the period in which the present article was being written, a small number of journals were inaccessible to this reporter. Therefore, apologies must be offered to some authors for omission of any reference to their work.

2 Basic Aspects of Nuclear Shielding

A. General Theory.—Placing the origin of co-ordinates at the nucleus of interest and the origin of the vector potential of the uniform external magnetic field at a point having position vector r_0 from the co-ordinate origin, we obtain for the component $\sigma_{\alpha\beta}$ of the nuclear shielding tensor the more general form of Ramsey's equation

$$\sigma_{\alpha\beta} = \sigma_{\alpha\beta}^{d} + \sigma_{\alpha\beta}^{dg} + \sigma_{\alpha\beta}^{p} + \sigma_{\alpha\beta}^{pg} \qquad (1)$$

The terms on the right-hand side of equation (1) are defined as follows:*

$$\sigma_{\alpha\beta}^{d} = \frac{\mu_0}{4\pi} \frac{e^2}{2m} \langle 0 | \sum_k r_k^{-3} (r_k^2 \delta_{\alpha\beta} - r_{k\alpha} r_{k\beta}) | 0 \rangle \qquad (2)$$

$$\sigma_{\alpha\beta}^{dg} = \frac{\mu_0}{4\pi} \frac{e^2}{2m} \langle 0 | \sum_k r_k^{-3} (r_{0\alpha} r_{k\beta} - r_{0\gamma} r_{k\gamma} \delta_{\alpha\beta}) | 0 \rangle \qquad (3)$$

$$\sigma_{\alpha\beta}^{p} = \frac{\mu_0}{4\pi} \frac{e^2}{2m^2} \sum_n{}' \frac{\langle 0 | \sum_k r_k^{-3} l_{k\alpha} | n \rangle \langle n | \sum_k l_{k\beta} | 0 \rangle + \langle 0 | \sum_k l_{k\beta} | n \rangle \langle n | \sum_k r_k^{-3} l_{k\alpha} | 0 \rangle}{W_0 - W_n} \qquad (4)$$

* Chapter 1, Vol. 1 of these Reports contains a number of equations in which the factor 4π has been accidentally omitted from the denominators. This applies to equations (14)—(22), (30) and (31) of the earlier Report; these equations (and the footnote on page 7 of Vol. 1) may be corrected by replacing the symbol Ξ_0 by $\mu_0/4\pi$. In addition, equation (31) should contain the factor e^2. Corrected versions of equations (14)—(17), (30), and (31) of the earlier Report are given as equations (2)—(5), (21), and (22), respectively, of this Report.

1
Nuclear Shielding

BY W. T. RAYNES

1 Introduction

'The most important single parameter to be derived from the n.m.r. spectrum is the chemical shift.' This contention,[1] possibly controversial even when published in 1959, became, one suspects, increasingly less acceptable to the majority of n.m.r. spectroscopists during the course of the nineteen-sixties. However, the introduction of new techniques for signal enhancement in the past few years has meant that shielding data, in the form of proton and fluorine chemical shifts from the peripheral regions of molecules, can now be supplemented by phosphorus, nitrogen and, above all, carbon chemical shifts which yield, in principle, a much more intimate knowledge of electronic distributions in the interior of molecules. This, coupled with the enticing prospect of large amounts of new information on nuclear shielding and nuclear shielding components in solids, provides additional support for those who are inclined to support the above quotation.

The structure of the present Report does not differ in essence from that of Volume 1 of the present series.[2] As before the emphasis will be on reported work in the review period (July 1st 1971 to May 31st 1972) which either leads to, or may lead to, an improved understanding of the phenomenon of nuclear shielding in isolated molecules. Therefore, as with last year's Report, no space has been devoted to any of the following topics: experimental methods of chemical shift measurement, the details of methods for the quantum-mechanical calculation of shielding constants, and the mechanisms by which intermolecular effects can alter shielding constants. This third restriction forces the exclusion of all solution phenomena including the study of contact and pseudocontact shifts and of weak complex formation. These topics, however, are covered in Chapter 10.

The number of papers presenting new carbon chemical shift data during the review period — about eighty — is more than double the number referred to in Volume 1. In addition the number of papers with new phosphorus

---

[1] J. A. Pople, W. G. Schneider, and H. J. Bernstein, 'High Resolution Nuclear Magnetic Resonance', McGraw-Hill, London and New York, 1959, p. 87.
[2] W. T. Raynes, 'Nuclear Magnetic Resonance' ed. R. K. Harris (Specialist Periodical Reports), The Chemical Society, London, 1972, vol. 1, p. 1.

Nuclear Shielding

$$\sigma_{\alpha\beta}^{pg} = \frac{\mu_0}{4\pi} \frac{e^2}{2m^2} \varepsilon_{\beta\gamma\delta} r_{0\gamma} \times$$

$$\sum_n{}' \frac{\langle 0|\sum_k r_k^{-3} l_{k\alpha}|n\rangle\langle n|\sum_k p_{k\delta}|0\rangle + \langle 0|\sum_k p_{k\delta}|n\rangle\langle n|\sum_k r_k^{-3} l_{k\alpha}|0\rangle}{W_0 - W_n} \quad (5)$$

In equations (2)—(5) the symbols μ_0, e, and m denote respectively the permeability of free space, the electronic charge, and the electronic mass; r_k is the position vector of the k'th electron from the nucleus of interest; W_n represents the energy of the n'th excited state; p_k denotes the linear momentum operator of the k'th electron; l_k ($=r_k \times p_k$) is the orbital angular momentum operator of this electron and the convention of summation over repeated suffices is used. The prime on the summations in equations (4) and (5) denotes a summation over all values of n except $n = 0$, including the continuum of excited states. $\delta_{\alpha\beta}$ is the substitution tensor ($= 1$ if $\alpha = \beta$, $= 0$ if $\alpha \neq \beta$) and $\varepsilon_{\beta\gamma\delta}$ is the alternating tensor [$= 1$ if ($\beta\gamma\delta$) is an even permutation of (xyz), -1 if ($\beta\gamma\delta$) is an odd permutation of (xyz) and 0 if any two of ($\beta\gamma\delta$) are identical]. If $r_0 = 0$, equation (1) reduces to the familiar two-term expression $\sigma_{\alpha\beta}^d + \sigma_{\alpha\beta}^p$ with the 'conventional' diamagnetic and paramagnetic terms. The terms $\sigma_{\alpha\beta}^{dg}$ and $\sigma_{\alpha\beta}^{pg}$ are both dependent upon the origin of the vector potential, *i.e.* upon the gauge choice — hence the superscript g.

During the review period it has been pointed out[3] that the nuclear shielding tensor is not, in general, symmetric. This means that the component $\sigma_{\alpha\beta}$ will not, in general, be equal to $\sigma_{\beta\alpha}$. From equation (4) the shielding $\sigma_{\alpha\beta}^p$ is obviously not necessarily equal to $\sigma_{\beta\alpha}^p$. However, equation (2) makes clear that $\sigma_{\alpha\beta}^d$ is always symmetric so that the 'conventional' diamagnetic part of the shielding tensor can be fully specified by six components. For a gauge choice away from the nuclear site of interest we see that $\sigma_{\alpha\beta}^{dg}$ and $\sigma_{\alpha\beta}^{pg}$ are also not necessarily equal to $\sigma_{\beta\alpha}^{dg}$ and $\sigma_{\beta\alpha}^{pg}$ respectively. Buckingham and Malm divide the shielding into the sum of an isotropic part, a traceless symmetric part, and an antisymmetric part. Thus

where

$$\sigma_{\alpha\beta} = \bar{\sigma}\delta_{\alpha\beta} + \sigma_{\alpha\beta}^{(s)} + \sigma_{\alpha\beta}^{(a)} \quad (6)$$

$$\bar{\sigma} = \tfrac{1}{3}\sigma_{\gamma\gamma} \quad (7)$$

$$\sigma_{\alpha\beta}^{(s)} = \tfrac{1}{2}(\sigma_{\alpha\beta} + \sigma_{\beta\alpha}) - \bar{\sigma}\delta_{\alpha\beta} \quad (8)$$

$$\sigma_{\alpha\beta}^{(a)} = \tfrac{1}{2}(\sigma_{\alpha\beta} - \sigma_{\beta\alpha}) \quad (9)$$

They show that in deciding the number of independent components of the shielding tensor one must consider the symmetry appropriate to the particular nuclear site, and they give a table listing the number of independent components for the shielding tensor of nuclei located at sites possessing any of the possible kinds of symmetry. For a hydrogen atom in the presence of a uniform

[3] A. D. Buckingham and S. M. Malm, *Mol. Phys.*, 1971, **22**, 1127.

electric field and a uniform electric field gradient, Buckingham and Malm present expressions for the components of the proton shielding tensor which show the existence of non-zero values for the antisymmetric part $\sigma_{\alpha\beta}^{(a)}$ of some of the components.

Weisenthal and de Graaf[4] have shown that the quantum-mechanical expression for the diamagnetic susceptibility of neutral atoms and molecules can be expressed in a form that is independent of gauge choice ('gaugeless') even when approximate zero-order wavefunctions are used. This involves choosing the origin of co-ordinates for the electrons at the centre of the nuclear charge of the molecule. Comments on this approach have been made by Moss and Perry[5] who show why the theory is only applicable to neutral species. The method of Weisenthal and de Graaf would appear to be applicable to Ramsey's equation for nuclear shielding although this application has yet to be made.

B. **Basic Physical Aspects.**—In this subsection will be discussed the results of a number of experiments which have a direct bearing on fundamental aspects of the nuclear shielding phenomenon.

In the previous Report[2] a best experimental value for the proton shielding constant of liquid water (corrected for the bulk susceptibility effect of the water) was given. This value was

$$\sigma(H_2O, 25\,°C) = 25.97\,(\pm 0.30)\text{ p.p.m.} \qquad (10)$$

A considerable improvement on this value can now be reported following the work of Winkler et al.,[6] who have measured very precisely the ratio of the magnetic moments of the electron and the proton by the simultaneous observation of an electronic and a nuclear magnetic transition in atomic hydrogen. By combining their result with that of Lambe[7] who studied the n.m.r. frequency of protons in a spherical water sample, they give (presumably for 25 °C)

$$\sigma(H_2O) = 25.64(\pm 0.07)\text{ p.p.m.} \qquad (11)$$

which is both substantially different from and more precise than the older value. To obtain the proton shielding constant of molecular hydrogen Winkler et al.,[6] employ a reported[8] chemical shift

$$\sigma(H_2O) - \sigma(H_2) = -0.6(\pm 0.3)\text{ p.p.m.} \qquad (12)$$

thereby obtaining a value of 26.2 (± 0.3) p.p.m. for the H_2 molecule. Con-

---

[4] L. Weisenthal and A. M. de Graaf, *Phys. Rev. Letters*, 1971, **28**, 470.
[5] R. E. Moss and A. J. Perry, *Mol. Phys.*, 1972, **23**, 957.
[6] P. F. Winkler, D. Kleppner, T. Myint, and F. G. Walther, *Phys. Rev.* (*A*), 1972, **5**, 83.
[7] E. B. D. Lambe, 'Polarization, Matière, et Rayonnement', Société Française de Physique, Paris, 1969, p. 441.
[8] W. A. Hardy, personal communication, quoted in S. Liebes and P. A. Franken, *Phys. Rev.*, 1959, **116**, 633.

Nuclear Shielding

siderable improvement in precision is possible if one makes use of more recent results,[9,10] viz.

$$\sigma(H_2O, 1\,q.\,30\,°C) - \sigma(H_2O, \text{gas}) = -4.315 \text{ p.p.m.}^9 \quad (13)$$

$$\sigma(H_2O, \text{gas}) - \sigma(CH_4, \text{gas}) = -0.56(\pm 0.02) \text{ p.p.m.}^9 \quad (14)$$

$$\sigma(CH_4, \text{gas}) - \sigma(H_2, \text{gas}) = 4.35(\pm 0.15) \text{ p.p.m.}^{10} \quad (15)$$

This leads to

$$\sigma(H_2, \text{gas}) = 26.17(\pm 0.17) \text{ p.p.m.} \quad (16)$$

A precise value of the shielding constant of molecular hydrogen is obviously of particular importance for theoretical reasons. As put by Winkler et al.,[6] 'it appears that the topic of chemical shifts in simple molecules is ripe for a more detailed elaboration'. From a purely experimental point of view the important result here is that of equation (11) which shows that precision of measurement of shielding constants is now nearly as high as that for the measurement of chemical shifts. The significance of this is that only a little more increase in precision will mean that nuclear magnetic shielding constants in any molecule can be determined precisely.[2] However, as pointed out by Winkler et al., water is not a very good choice for a primary standard largely because of the high sensitivity of the shielding to slight changes of temperature. A further calculated result obtained by Winkler et al.[6] is that for the shielding of an isolated hydrogen atom. With higher-order corrections included the proton shielding constant is 17.733 p.p.m. which is a little less than the 17.750 p.p.m. often quoted.

Cade and Ramsey[11] have obtained values of the high-frequency part of the nuclear shielding in the lower rotational levels ($J = 1$ and 2) of the ground vibrational and electronic states of HD and D_2 using the molecular-beam magnetic resonance method. The high-frequency contribution (i.e. σ^p) can be calculated from the electronic part of the spin–rotation interaction constant. Again a much increased precision is obtained. For the $J = 1$ state of HD they find for the proton $\sigma^p = -5.65$ (± 0.08) p.p.m. and for the deuteron $\sigma^p = -5.61$ (± 0.08) p.p.m. as compared with earlier values of -5.96 (± 0.30) p.p.m. and -5.91 (± 0.30) p.p.m. respectively. Such results as these provide a valuable check on the results discussed in the previous paragraph once a good enough value for σ^d has been obtained.

In the calculation of changes in molecular properties occurring upon isotopic substitution it is unusual to take into account changes in the static electric quadrupole moment of the substituted nucleus. Thus, for instance, in calculating the proton and deuteron shielding constants of HD one would normally assume them equal for stationary nuclei and ignore any effect of the

---

[9] J. C. Hindman, J. Chem. Phys., 1966, 44, 4582.
[10] D. K. Hindermann and C. D. Cornwell, J. Chem. Phys., 1968, 48, 2017.
[11] R. F. Cade and N. F. Ramsey, Phys. Rev. (A), 1971, 4, 1945.

deuteron quadrupole moment on the electronic wavefunctions. The validity of this assumption — especially for heavier nuclei for which the K-shell wavefunction will be the most easily changed by the presence of an electric quadrupole moment and for which the K-electrons provide a large diamagnetic shielding — is brought into question by an experimental result of Sen et al.[12] who have detected an effect on the angular distribution of K-shell X-rays of one of the nuclear states of ^{169}Tm which they attribute to such an interaction.

The molecular-beam electric resonance method has been used[13,14] to study the molecules CH_3F and PH_3. For methyl fluoride Wofsy et al.[13] obtain values for the parallel and perpendicular components of σ^p for both the proton and fluorine shieldings. Of particular interest here is the observation that $\sigma^p_{\parallel}$ for the fluorine nucleus is of large magnitude (and negative sign). Earlier, Hunt and Meyer[15] observed for CH_3F trapped in clathrates that the anisotropy in the ^{19}F shielding (viz. $\Delta\sigma = \sigma_\perp - \sigma_\parallel$) was of different sign from that predicted from existing theory.[16] Since shielding changes in ^{19}F shielding are dominated by the paramagnetic term (with gauge origin at the ^{19}F nucleus) and the C—F bond is cylindrically symmetrical, the theory predicts that $\sigma^p_\parallel = 0$ and therefore that $\Delta\sigma(\simeq \sigma_\perp)$ is negative. In fact, Hunt and Meyer found that $\Delta\sigma = 66$ (± 8) p.p.m. and suggested that $\sigma^p_\parallel$ is large and negative. This result is confirmed by the work of Wofsy et al. who obtain for $\sigma^p_\parallel$ a value of -63.5 (± 1.5) p.p.m. Thus the simple theory of Karplus and Das,[16] although good for σ^d and its components, does not work very well for σ^p.

For phosphine the measured value of σ^p for the ^{31}P nucleus has enabled Davies et al.[14] to set up for the first time an absolute shielding scale for this nucleus with the aid of the σ^d value calculated by the method of Flygare and Goodisman.[17] They give for $\sigma(^{31}PH_3)$ a value of 594.40 p.p.m. For the proton shielding in phosphine they give for σ^d an experimental value of 126.41 p.p.m., to be compared with 126.02 p.p.m. calculated by the method of Flygare and Goodisman. Part of the discrepancy here may be attributed to the fact that in obtaining a value of σ (from which to subtract the measured σ^p) use had to be made of proton chemical shift data for phosphine measured in the solvent benzene[18] which, as is well known, produces somewhat abnormal effects on the proton chemical shifts of dissolved solutes.

The first attempt to set up an absolute shielding scale for lead has been made by Lutz and Stricker.[19] They obtain for the $^{207}Pb^{2+}$ ion in D_2O extrapolated to infinite dilution of the ion a shielding constant of $-17\,810$ (± 60) p.p.m. The antishielding would appear to be present in a number of lead compounds. A very large solvent isotope effect is present since upon changing from

[12] S. K. Sen, D. L. Salie, and E. Tomchuk, Phys. Rev. Letters, 1972, **28**, 1295.
[13] S. C. Wofsy, J. S. Muenter, and W. Klemperer, J. Chem. Phys., 1971, **55**, 2014.
[14] P. B. Davies, R. M. Neumann, S. C. Wofsy, and W. Klemperer, J. Chem. Phys., 1971, **55**, 3564.
[15] E. Hunt and H. Meyer, J. Chem. Phys., 1964, **41**, 353.
[16] M. Karplus and T. P. Das, J. Chem. Phys., 1961, **34**, 1683.
[17] W. H. Flygare and J. Goodisman, J. Chem. Phys., 1968, **49**, 3122.
[18] E. A. V. Ebsworth and G. M. Sheldrick, Trans. Faraday Soc., 1967, **63**, 1071.
[19] O. Lutz and G. Stricker, Phys. Letters, 1971, **35A**, 397.

the solvent D_2O to H_2O the shielding constant of the $^{207}Pb^{2+}$ ion fell by 30 (± 3) p.p.m.

Although strictly speaking outside the scope of the present Report, we note here three papers [20—22] involving the use of nuclear shielding and shielding anisotropy data for the estimation of the effects of spin–rotation interaction[23] and chemical shift anisotropy upon nuclear spin–lattice relaxation times.

3 Calculations of Nuclear Shielding

In the present section are considered papers presenting calculations of nuclear shielding constants for individual molecules. A number of papers dealing with the calculation of ring-current contributions to shielding constants are discussed in Section 4E and some papers concerned with the calculations of ^{13}C, ^{19}F, and ^{31}P shieldings in a wide range of compounds are discussed later under the section dealing with the particular nucleus.

The results of an *ab initio* calculation of the bond-length dependence of the nuclear shielding constant of the hydrogen molecule have been given by Cook *et al.*[24,25] They used self-consistent perturbation theory[26] adapted to deal with the two perturbations involved, namely the nuclear magnetic moment and the external magnetic field. The basis functions consisted of 1s- and three 2p-orbitals on each nucleus with the orbital exponents varied for each internuclear distance in units of 0.1 a.u. from 0.8 a.u. to 1.8 a.u. The origin of the vector potential was taken at the bond midpoint for all internuclear distances, which leads to a very near cancellation[27] of the high-frequency terms $\sigma_{\alpha\beta}^{p}$ and $\sigma_{\alpha\beta}^{pg}$ of equation (1). The results are given in Table 1 for the

Table 1 *Nuclear magnetic shielding parameters in p.p.m. of the hydrogen molecule at several values of the internuclear distance r (ref. 24)*

r/a.u.	σ_{xx}	σ_{zz}	$\Delta\sigma$	σ
0.8	35.130	37.497	2.367	35.919
0.9	33.201	35.581	2.380	33.994
1.0	31.460	33.833	2.373	32.251
1.1	29.881	32.235	2.354	30.666
1.2	28.448	30.771	2.323	29.222
1.3	27.141	29.424	2.283	27.902
1.4	25.948	28.182	2.234	26.693
1.5	24.858	27.033	2.175	25.583
1.6	23.855	25.967	2.112	24.559
1.7	22.930	24.975	2.045	23.612
1.8	22.075	24.049	1.974	22.733

[20] S. W. Dale and M. E. Hobbs, *J. Phys. Chem.*, 1971, **75**, 3537.
[21] K. T. Gillen, *J. Chem. Phys.*, 1972, **56**, 1573.
[22] N. Liu and J. Jonas, *J. Chem. Phys.*, 1971, **55**, 463.
[23] P. S. Hubbard, *Phys. Rev.*, 1963, **131**, 1155.
[24] D. B. Cook, A. M. Davies, and W. T. Raynes, *Mol. Phys.*, 1971, **21**, 113.
[25] W. T. Raynes, A. M. Davies, and D. B. Cook, *Mol. Phys.*, 1971, **21**, 123.
[26] G. Diercksen and R. McWeeny, *J. Chem. Phys.*, 1966, **44**, 3554.
[27] S. I. Chan and T. P. Das, *J. Chem. Phys.*, 1962, **37**, 1527.

components σ_{xx} ($= \sigma_{yy}$) and σ_{zz} where the z-axis is along the bond direction. Also given are the shielding anisotropy values $\Delta\sigma$, defined here by $\sigma_{zz} - \sigma_{xx}$, at different bond lengths and the mean shielding σ, defined by

$$\sigma = \tfrac{1}{3}(\sigma_{zz} + 2\sigma_{xx}) \tag{17}$$

The shielding is seen to be highly dependent on the bond length as one would expect from the fact that at $r = \infty$ it has a value of 17.733 p.p.m.,[6] characteristic of an isolated hydrogen atom, whereas at $r = 0$ the value becomes 59.94 p.p.m.,[28] characteristic of the helium atom. The shielding anisotropy which, of course, must be zero for both $r = 0$ and $r = \infty$ reaches its maximum at about 0.9 a.u., and has, as would be predicted qualitatively, a positive sign. One point the Reporter wishes to draw to the interested reader's attention is the statement of equation (18) of the paper[24] that, in SCF perturbation theory nomenclature, $R_\alpha^{(10)} = G(R_\alpha^{(10)}) = 0$. This is invalid, as can be seen from equation (11) of the paper, since it would make the wavefunctions independent of the magnitude of the perturbing magnetic moment. A calculation of the additional term, which affects the high-frequency part of the shielding only, carried out by Mr. J. P. Riley of Sheffield University, gives a correction of less than ± 0.1 p.p.m. which affects the absolute value of σ by about 0.06 p.p.m. in Table 1.

Expanding the proton shielding constant in a Taylor series about the equilibrium bond length r_e gives what may be described as the 'proton shielding function' for a diatomic molecule. Thus

$$\sigma = \sigma_e^{(0)} + \sigma_e^{(1)}\xi + \sigma_e^{(2)}\xi^2 + \sigma_e^{(3)}\xi^3 + \ldots \tag{18}$$

where

$$\xi = (r - r_e)/r_e \tag{19}$$

is the relative displacement from equilibrium and the $\sigma_e^{(i)}$ are molecular parameters, $\sigma_e^{(0)}$ being the shielding constant for the equilibrium bond length. Raynes et al.[25] give values for the $\sigma_e^{(i)}$ for the H_2 molecule, including a value for $\sigma_e^{(0)}$ of 26.680 p.p.m. Figure 1 shows the proton shielding function for H_2 together with the potential curve. Upon averaging over the nuclear motion and then over the occupied rotational and vibrational states they find for the shielding constant of hydrogen gas at 300 K (ignoring, of course, intermolecular effects on the shielding).

$$\sigma(H_2, 300 \text{ K}), \text{calc} = 26.298 \text{ p.p.m.} \tag{20}$$

which is in good agreement with the best experimental value given above. Over the temperature range -100 to $+200$ °C the shielding constant of hydrogen gas is predicted to fall by 0.038 p.p.m. — an amount that should be measureable. Raynes et al.[25] also give isotope shifts for the various forms of isotopically substituted H_2. These results are discussed in Section 4H.

[28] F. D. Feicock and W. R. Johnson, *Phys. Rev.*, 1969, **187**, 39.

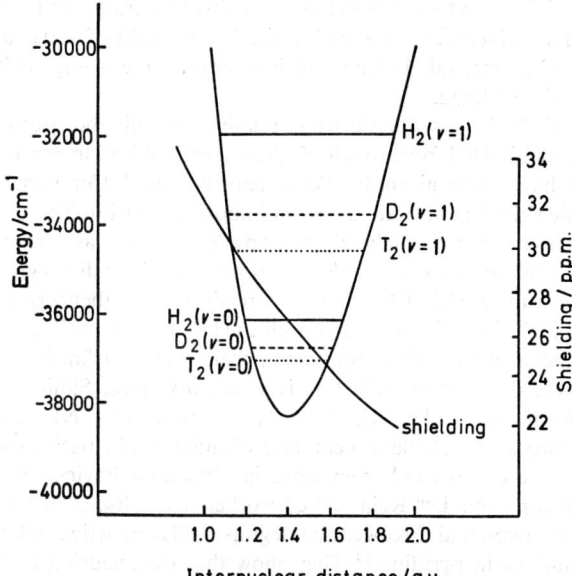

Figure 1 *Bond-length dependence of the nuclear shielding and the potential curve for molecular hydrogen*

Ermler and Kern[29] have calculated zero-point vibrational corrections to a large number of one-electron properties of the water molecule in its ground state using a near-Hartree–Fock potential energy surface. Our concern here is with the diamagnetic contribution to the shielding only. For this Ermler and Kern obtained the results given in Table 2, where the vibrational correction is

Table 2 *Diamagnetic shielding constants σ_e at the equilibrium nuclear geometry and the zero-point vibrational corrections for H_2O, D_2O, and HDO* [29]

Property	At equilib /p.p.m.	Vibrational Corrections/p.p.m.		
		H_2O	D_2O	HDO
$\sigma^d(O)$	416.515	0.019	0.014	0.016
σ^d (^1H or ^2H)	103.983	0.1153	0.0857	$\begin{cases} 0.1123\ (^1H) \\ 0.0895\ (^2H) \end{cases}$

equal to $\sigma^d - \sigma_e^d$. Although these results may be misinterpreted here it seems that, contrary to expectations, the diamagnetic shielding for all nuclei is increased by the vibrational motion. One expects that normally the 'bond-stretching' consequent upon going from the equilibrium geometry to the mean molecular geometry will lead to reduced diamagnetic shielding. The proton isotope shift of HDO relative to H_2O is predicted in Table 2 for the dia-

[29] W. C. Ermler and C. W. Kern, *J. Chem. Phys.*, 1971, **55**, 4851.

magnetic shielding to be -0.0030 p.p.m., again of contrary sign to expectation and also quite different from the observed isotope shift of $+0.030$ p.p.m.,[30] although, of course, this latter result incorporates the isotope effect on the paramagnetic shielding.

Values of $\langle r^{-1} \rangle$ for the electrons relative to both the proton and the chlorine nuclei in HCl from which the diamagnetic shielding constant may be calculated have been given by Petke and Witten.[31] Components of the diamagnetic shielding at both xenon and fluorine nuclei in XeF_2, XeF_4, and XeF_6 have been very accurately calculated by Basch et al.[32] as part of a detailed theoretical study of these molecules. Components of the diamagnetic shielding tensors at the nuclei of NO_2 and O_3 are given by Rothenberg and Schaefer.[33] The inclusion of d-orbitals on all nuclei in these molecules has little effect on the shielding. These authors appear to have defined the shielding such that the diamagnetic shielding is of negative sign. Similar results are available for the molecules NH_3 and CH_3^-,[34] and also for NH_3 and H_2O.[35] In this last work the calculation employed Gaussian lobe basis functions, i.e. p-orbitals were constructed from Gaussian functions having origins in the lobes of each p-orbital.[36] Rein and co-workers have discussed the contributions of one-, two-, and three-centre integrals to diamagnetic shielding for the nitrogen nucleus in pyridine.[37] They show that the inclusion of two-centre overlap densities is essential, with the implication that this is true in general for polyatomic molecules.

4 Transmission of Shielding Effects within Molecules

A. Introduction.—The division of nuclear shielding into a number of physically distinct contributions, which is used in the present section to assist in the interpretation of observed shielding data, is, of course, a very severe approximation made necessary by present-day ignorance of excited-state wavefunctions. However, it does, at least in some cases, render possible an interpretation of trends in shielding in terms of well-established physical and chemical concepts. The total molecular electronic distribution is divided into a 'local' part on the nucleus in question and a part in the remainder of the molecule. The nuclear shielding is then regarded as being composed of three contributions — magnetic fields arising from local diamagnetic currents and giving rise to σ_{loc}^d, magnetic fields arising from local paramagnetic currents and giving rise to σ_{loc}^p and magnetic fields due to induced currents in the distant electrons. Approximate expressions for σ_{loc}^d and σ_{loc}^p are[38]

[30] J. R. Holmes, D. Kivelson, and W. C. Drinkard, *J. Chem. Phys.*, 1962, **37**, 150.
[31] J. D. Petke and J. L. Whitten, *J. Chem. Phys.*, 1972, **56**, 830.
[32] H. Basch, J. W. Moskowitz, C. Hollister, and D. Hankin, *J. Chem. Phys.*, 1971, **55**, 1922.
[33] S. Rothenberg and H. F. Schaefer, *Mol. Phys.*, 1971, **21**, 317.
[34] R. E. Kari and I. G. Csizmadia, *Theor. Chim. Acta*, 1971, **22**, 1.
[35] R. D. Brown, F. R. Burden, and B. T. Hart, *Theor. Chim. Acta*, 1971, **22**, 214.
[36] J. L. Whitten, *J. Chem. Phys.*, 1966, **44**, 359.
[37] R. Rein, G. R. Pack, and J. R. Rabinowitz, *J. Magn. Resonance*, 1972, **6**, 360.
[38] M. Karplus and J. A. Pople, *J. Chem. Phys.*, 1963, **38**, 2803.

$$\sigma_{\text{loc}}^{\text{d}} = \frac{\mu_0}{4\pi} \frac{e^2}{3m} \langle 0|\sum_i r_i^{-1}|0\rangle \tag{21}$$

$$\sigma_{\text{loc}}^{\text{p}} = -\frac{\mu_0}{4\pi} \frac{\hbar^2 e^2}{2m^2} \frac{\langle r^{-3}\rangle_{np}}{\Delta E} \sum_{\text{B}} Q_{\text{AB}} \tag{22}$$

In these equations r_i represents the distance of the i'th electron in the local electron distribution from the nucleus, ΔE represents a mean excitation energy, $_{\text{B}}Q_{\text{AB}}$ accounts for the amount of imbalance in the populations of the orbitals about the nucleus and is zero for spherical symmetry, and $\langle r^{-3}\rangle_{np}$ denotes the mean inverse cube of the distance of the valence-shell p-electrons from the nucleus. Occasionally one must take d-electrons into account.

It is generally believed that differences in $\sigma_{\text{loc}}^{\text{d}}$ together with contributions from distant electrons are wholly responsible for proton chemical shifts, values of $\sigma_{\text{loc}}^{\text{p}}$ and hence differences between them being negligibly small. For all other nuclei it is held that changes in $\sigma_{\text{loc}}^{\text{p}}$ are the dominant cause of the observed chemical shifts and that changes in $\sigma_{\text{loc}}^{\text{d}}$ and in the contributions of distant electrons are small. Recent work discussed earlier[2] suggests that for some series of compounds changes in $\sigma_{\text{loc}}^{\text{d}}$ may be substantial, however. From a reading of the literature it would appear that there is some confusion of these points. Thus one finds some authors claiming for the shielding of, say, ^{19}F and ^{13}C nuclei that the paramagnetic term of Ramsey's equation is dominant over the diamagnetic term. This cannot be true if the shielding constant is positive as it is for these nuclei in most molecules. For the same reason it is not correct to assert for shielding other than that of protons, that $\sigma_{\text{loc}}^{\text{p}}$ is dominant over $\sigma_{\text{loc}}^{\text{d}}$. As stated above, although $\sigma_{\text{loc}}^{\text{d}}$ is generally of greater numerical magnitude than $\sigma_{\text{loc}}^{\text{p}}$, it is $\sigma_{\text{loc}}^{\text{p}}$ which is usually very much more sensitive than $\sigma_{\text{loc}}^{\text{d}}$ to the changing electronic influences occurring in passing along a series of chemical compounds.

The actual mechanisms which contribute to $\sigma_{\text{loc}}^{\text{d}}$ and $\sigma_{\text{loc}}^{\text{p}}$ may be classified[39] into through-bond and through-space effects. The former include the inductive effect, the effect of conjugation, and intramolecular hydrogen bonding. The through-space effects influencing $\sigma_{\text{loc}}^{\text{d}}$ and $\sigma_{\text{loc}}^{\text{p}}$ are the electric field effect, van der Waals interactions, and steric effects. The contributions to the shielding made by the magnetic fields of distant electrons — which are, by definition, through-space effects — arise from group magnetic anisotropy, ring currents, or unpaired electron spins (psuedocontact or dipolar shifts). In addition, unpaired electron spins may be transmitted through bonds *via* spin polarization, thereby causing changes in the local spin distribution giving rise to the so-called contact shift.

When discussing data reported in the literature in terms of the above mechanisms difficulty arises from the fact that often several mechanisms which may be hard to distinguish can be present. For instance, in considering proton chemical shifts produced by polar substituents one must usually

[39] W. T. Raynes, *Mol. Phys.*, 1971, **20**, 321.

consider the magnetic anisotropy and the electric field effect together. Again, the electric field effect and the inductive effect of polar substituents close to the nucleus of interest are very difficult to distinguish from one another. A further complication is provided by the fact that different authors sometimes assume that a different combination of mechanisms is responsible for a given set of data and, with the aid of essentially empirical methods, obtain good agreement for the particular combination of mechanisms they have chosen. One of the purposes of the present section will be to comment on cases where there may be mechanisms present other than those adopted by an author.

In the following discussion the major through-bond effects (induction and conjugation) will be dealt with first, followed by the magnetic anisotropy effect. After that follows a discussion of the electric field and the ring-current contribution to shielding. The remaining effects are then discussed, together with the topic of isotope effects which is viewed largely as a through-bond effect. As mentioned in the Introduction, contact and pseudocontact shifts are discussed in Chapter 10. Much of the discussion in this section is concerned with proton and fluorine shielding. With one or two exceptions carbon, phosphorus, and nitrogen shielding are discussed individually in the next section.

B. Inductive Effects.—In a series of five papers[40,41] Phillips and Wray have discussed the inductive effect in terms of the chemical shifts of a variety of different nuclei in several classes of compounds. Their method of interpreting the observed data is essentially empirical and is based upon the quantitative application of the concept of electronegativity on the Huggins scale. To account fully for observed trends Phillips and Wray find it necessary to allow for the small changes in the electronegativity of a given substituent brought about by the presence of others. For example, consider the series of compounds symbolized by (1), in which X is a halogen, and let us consider the ^{19}F

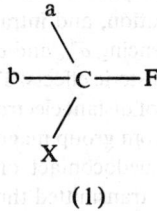

(1)

chemical shifts relative to X = F as X changes along the series Cl, Br, and I. The ^{19}F shielding is written[40a]

$$\sigma = \sigma_0 + k'[\chi'_C - \chi'_F] \tag{23}$$

[40] L. Phillips and V. Wray, *J. Chem. Soc.* (*B*), 1971, (*a*) 2068; (*b*) 2074.
[41] L. Phillips and V. Wray, *J.C.S., Perkin II*, 1972, (*a*) 214; (*b*) 220; (*c*) 223.

Nuclear Shielding

where the σ_0 is the local diamagnetic shielding, which is assumed constant throughout, and the second term is the local paramagnetic shielding. σ_{loc}^p is related to the ionic character of the C—F bond, which is regarded as being proportional to the difference of the effective electronegativities of carbon χ'_C and of fluorine χ'_F in the appropriate compound. The electronegativity χ'_C is found in an additive manner by summing a set of terms for each bond which the carbon atom makes, each term of which sum is proportional to the product of the Huggins electronegativities of two of the atoms. A similar procedure is carried out to find χ'_F. In these expressions terms arise allowing for the contributions to χ'_F and χ'_C of the mutual interaction of the atoms a and b. As may be readily imagined the procedure leads to a very large number of parameters so that in order to make progress some simplification is necessary. Assumptions made are that 'interaction constants between a pair of like or unlike halogens all have the same value' and 'perturbation constants between groups a or b and halogen are assumed to depend upon the nature of a or b but are independent of the halogen concerned'. It appears to this Reporter that some justification for the assumptions should have been made.

Nevertheless, the method does lead to good correlation with experiment. Thus for the series above denoted by (1) they find, using observed data, the general formula for the ^{19}F shift when X = halogen, relative to X = F,

$$\Delta = (\chi_F - \chi_X)[154.12 - 26.63(\chi_a + \chi_b)] \quad (24)$$

The interesting result here is the prediction that for $\chi_a + \chi_b > 5.78$ Huggins units the value of Δ will be negative, whereas for $\chi_a + \chi_b < 5.78$ Huggins units Δ will be positive. The above formula predicts quite well the observed substituent chemical shifts (referred to X = F) for molecules of type (1). This, quoting some of Phillips and Wray's results,[41a] is illustrated in Table 3.

Table 3 *Observed and calculated ^{19}F shifts using the method of Phillips and Wray[40,41]*

Compound	Cl		Br		I	
	Obs.	Calc.	Obs.	Calc.	Obs.	Calc.
CF$_3$X	−29 to −37	−40.22	−48.8	−50.9	−58.3	−67.8
CFH$_2$·CF$_2$X	−12.1	−14.2	−16.9	−18.0	−24.3	−23.7
CF$_3$·CFX·CF$_3$	+8.3	+11.7	+10.8	+14.8	+15.5	+19.5
C$_3$F$_5$X	+14.1	+11.7	+14.8	+14.8	+18.2	+19.5
C$_6$F$_{11}$X	+12	+11.7	+14.2	+14.8		

Another superficially puzzling trend in the data for the series CF$_n$Cl$_{4-n}$ is also accounted for. Thus

	CF$_4$	CF$_3$Cl	CF$_2$Cl$_2$	CFCl$_3$
Δ (obs)	0	−30	−55	−63
Δ (calc)	0	−34	−57	−64

The chemical interpretation of the anomaly is that the increasing ^{19}F shielding expected with chlorine substitution is more than counterbalanced by mutual interactions among the substituents which lead to significant changes in the effective electronegativities.

In their second paper[40a] Phillips and Wray apply the method to octahedrally bonded tin complex dianions of the structure (2). In considering the

$$\left[\begin{array}{c} e \\ b \diagdown | \diagup c \\ Sn \\ a \diagup | \diagdown d \\ f \end{array} \right]^{2-}$$

(2)

^{19}F shifts here it is necessary to distinguish between the non-equivalence of *cis*- and *trans*-relationships. Again, good agreement with experiment is obtained for halogen substituents in the situations a, b, c, d, and e of (2). Extension of the method to geminal-substituent effects on ^{1}H and ^{13}C chemical shifts and directly-bonded-substituent effects on ^{13}C, ^{11}B, ^{31}P, and ^{119}Sn chemical shifts has been made more recently.[41a] In particular, the method predicts good agreements for the anomalous trend of proton shielding in the trihalogeno-substituted methanes, which has the order $CHCl_3 < CHBr_3 < CHF_3 < CHI_3 < CH_4$, and for the carbon shielding in the methyl halides, which also follows anomalous trends. Apparently unaware of an observed value for the ^{13}C shift in CI_4, Phillips and Wray predict $+256$ p.p.m. with respect to methane which is in fair agreement with the observed value[42] of $+290.2$ p.p.m.[43] For CF_4, they predict -251 p.p.m. with respect to CCl_4 which does not compare at all well with the very recent value of -21 p.p.m.[44] Other published work by Phillips and Wray is concerned with the ^{19}F shielding in freely rotating fluoroalkanes[41b] and in fluoro-olefins and *ortho*-substituted benzenes.[41c]

The great need for a more physical understanding of substituent effects on chemical shifts of simple compounds is made evident by the work of Phillips and Wray, which appears to regard the classical inductive effect of organic chemistry as being entirely responsible for these shifts. A few years ago Bucci[45] gave a good correlation between the ^{1}H and ^{13}C shifts of the methyl and ethyl halides on the basis of the electric field model, and many authors have employed the notions of large C—Hal bond magnetic anisotropies (for proton shielding in particular) and of intramolecular dispersion forces to ac-

---

[42] O. W. Howarth and R. J. Lynch, *Mol. Phys.*, 1968, **15**, 431.
[43] Ref. 2, p. 20.
[44] E. L. Motrell and G. E. Maciel, *J. Magn. Resonance*, 1972, **7**, 330.
[45] P. Bucci, *J. Amer. Chem. Soc.*, 1968, **90**, 252.

Nuclear Shielding

count for anomalous shielding values found with bromine and iodine substituents. Very recent work (see below) casts doubt on the classical inductive effect and replaces it by a through-space electric field effect. However, this work is for larger compounds and neither theoretical calculations nor experimental results on methyl halides and similar small compounds make clear to what extent the electric field effect predominates over the inductive effect.

Another case in which mutual interactions involving substituents have been invoked is in substituted derivatives of adamantane studied by van Duersen and Bakker.[46] They measured the shielding of the protons geminal to the substituents in a series of 2,4-disubstituted adamantanes (3), the substituents

(3)

being in either the 2e,4e-positions or the 2e,4a-positions. They find, for instance, that the proton shielding change at the 2a-position occurring upon changing the substituent in either the 4e- or the 4a-position is dependent upon the nature of the substituent in the 2e-position. Possible occurrences not mentioned in ref. 46 are small bond-angle changes involving the proton under investigation, leading to differing carbon electronegativities as well as different values of the through-space field effects.

Hanlan and McCowan[47] have obtained the proton chemical shifts of tetramethyltitanium and some methyltitanium chlorides. Compared with the tetramethyl derivatives of the Group IVB elements the proton shielding of $(CH_3)_4Ti$ is small. However, the increasing replacement of a methyl group by a chlorine atom leads to the further reduction in shielding which one expects on inductive grounds. From their data Hanlan and McCowan given an 'effective electronegativity' of about 3 units for titanium, considerably more than that obtained for silicon, germanium, or lead. As is well known, determinations of electronegativity from proton chemical shift data are suspect for a number of reasons including changes in hybridization involving d-orbitals, as pointed out by these authors. Proton shielding data for a number of n-propyl derivatives $CH_3CH_2CH_2X$ have been given by Schrumpf[48] but an analysis of the data in terms of induction, magnetic anisotropy, electric field effect, etc. has yet to be made. Kawamura et al.[49] have suggested the importance of the inductive effect as well as that of magnetic anisotropy to explain the observed proton shielding in a series of aralkyl hydropolysulphides $R(S_n)H$, where $n = 1—3$.

[46] F. W. van Duersen and J. Bakker, *Tetrahedron*, 1971, **27**, 4593.
[47] J. F. Hanlan and J. D. McCowan, *Canad. J. Chem.*, 1972, **50**, 747.
[48] G. Schrumpf, *J. Magn. Resonance*, 1972, **6**, 243.
[49] S. Kawamura, T. Horii, and J. Tsurugi, *Bull. Chem. Soc. Japan.*, 1971, **44**, 2878.

C. **Resonance Effects.**—A very large number of reports have appeared during the review period presenting experimental chemical shifts, empirical correlations, and molecular orbital calculations concerning aromatic compounds. Here it is possible to do little more than refer to these reports, making the occasional comment. Rodmar and co-workers[50,51] have presented experimental values of ^{1}H and ^{19}F chemical shifts in various substituted fluorothiophens of the types [(4)—(6)]. The data were then fitted[51] to the $\mathscr{F}$ and $\mathscr{R}$

(4)   (5)   (6)

substituents parameters of the theory introduced by Swain and Lupton,[52] which assumes that only electric field and resonance effects are of importance. In the form given by Rodmar et al.,[51] this is

$$\delta = i_k + f_k\mathscr{F} + r_k\mathscr{R} \qquad (25)$$

By a least-squares fit, good agreement was obtained for ^{19}F shifts, but the agreement was poorer for the proton shifts, possibly indicating the existence of magnetic anisotropy effects. CNDO/2 calculations of the ^{19}F shifts were not particularly successful.[53]

Dewar, Golden, and Harris[54] have given an improved treatment of their earlier FM (field-mesomerism) method of interpreting substituent effects on physical and chemical properties. The improvement is required to allow for the 'mesomeric-field effect', *i.e.* the ability of charges produced by the substituent at sites around an aromatic ring by mesomerism to generate a field effect in addition to that produced directly by the substituent. When applied to ^{19}F shifts, however, deviations occur which lead them to conclude that small geometrical and conformational effects can produce significant changes in ^{19}F shielding constants quite apart from any specific electronic effects. This leads Dewar et al.[54] to the challenging conclusion that 'the effects of substituents on chemical properties and on ^{19}F chemical shifts present entirely different problems and that attempts to combine the two will prove fruitless'.

However, several authors have used polar (inductive) and resonance parameters to correlate observed shielding data. Sheppard and Taft[55] have obtained ^{19}F shifts for the *meta-* and *para-*positions in the compounds FC_6H_4R, where R is a sulphur-containing substituent of the form SX, SOX, SF_3, SO_2X, or SF_5, and the results are interpreted in terms of σ_I and σ_R constants,

---

[50] S. Rodmar, L. Moraga, S. Gronowitz, and U. Rosén, *Acta Chem. Scand.*, 1971, **25**, 3309.
[51] S. Rodmar, S. Gronowitz, and U. Rosén, *Acta Chem. Scand.*, 1971, **25**, 3841.
[52] C. G. Swain and E. C. Lupton, *J. Amer. Chem. Soc.*, 1968, **90**, 4328.
[53] S. Rodmar, *Mol. Phys.*, 1971, **22**, 123.
[54] M. J. S. Dewar, R. Golden, and J. M. Harris, *J. Amer. Chem. Soc.*, 1971, **93**, 4187.
[55] W. A. Sheppard and R. W. Taft, *J. Amer. Chem. Soc.*, 1972, **94**, 1919.

permitting conclusions concerning $(p-d)\pi$ and $(p-p)\pi$ interactions for both the Ar—S and S—X bonds to be made. Wiley and Miller[56] have obtained the N-methyl proton chemical shifts of 21 m- or p-substituted NNN-trimethylphenylammonium iodides and found that although a fairly good correlation with the Swain–Lupton parameters was achieved, it appeared both necessary and discouraging to introduce terms to correct for substituent magnetic anisotropy and other effects. Yukawa and co-workers have also discussed the aryl protons in monosubstituted benzenes[57a] and naphthalenes[57b] in terms of substituent parameters, as have Marr and Spoerri[58] for the proton shielding in pyrazines. In the latter case a 'second-order' mesomeric effect is clearly necessary to account for the observed data in addition to the usual field and resonance parameters. Malinowski[59] has given a correlation of ^{13}C, ^{1}H, and ^{19}F chemical shifts in some aromatic compounds involving pairwise corrections to the direct additivity effects of substituents. Wu et al.[60] have obtained the amino-proton chemical shifts of several m- and p-substituted anilinetricarbonylchromium complexes (7) which correlate well with Hammett σ-

R—⟨⟩—NH₂
  |
  Cr(CO)₃

(7)

parameters and indicate direct resonance interactions between the R group and the NH_2 group which are reduced slightly by the presence of the $Cr(CO)_3$ group.

Lavallee and Fleischer[61] have presented a lengthy discussion of the factors influencing the ligand proton chemical shifts in penta-ammineruthenium(II) complexes. Thus, for example, for [2-^{2}H]pyridine-penta-ammineruthenium-(II) (8), where the deuterium atom was present for spectral simplification, the

$$[(NH_3)_5Ru^{II}N\text{⟨⟩}D]^{2+}$$

(8)

[56] G. R. Wiley and S. I. Miller, *J. Org. Chem.*, 1972, **37**, 767.
[57] Y. Yukawa, Y. Tsuno, and N. Shimizu, *Bull. Chem. Soc. Japan*, 1971, **44**, (a) 2843; (b) 3175.
[58] G. S. Marx and P. E. Spoerri, *J. Org. Chem.*, 1972, **37**, 111.
[59] E. R. Malinowski, *J. Phys. Chem.*, 1972, **76**, 1593.
[60] A. Wu, E. R. Biehl, and P. C. Reeves, *J. Organometallic Chem.*, 1971, **33**, 53.
[61] D. K. Lavallee and F. B. Fleischer, *J. Amer. Chem. Soc.*, 1972, **94**, 2583.

4- and 3,5-protons of the pyridine ring are found to be deshielded by 0.15 and 0.3 p.p.m., respectively, relative to the corresponding protons of free pyridine, whereas the 2-proton is more shielded by 0.15 p.p.m. relative to the 2-protons of free pyridine. The latter change is attributed to paramagnetic anisotropy of the ruthenium (*i.e.* the presence of a magnetic moment on the ruthenium atom due to mixing in of low-lying excited states by the external magnetic field, an effect which produces an additional secondary field at nearby positions although is does not of itself produce a 'distortion of the H_2 1s-orbital spatial electron distribution' as stated in the paper).[61] The deshielding observed for pyridine as well as for other ligands indicates a polarization of electron density towards the Ru^{II} ion and rules out the possibility of significant $(d-p)\pi$ back-bonding from the Ru^{II}. However, it is not clear how important is resonance as opposed to induction here. Fischer *et al.*[62] have examined substituent effects on the chemical shift of the hydroxy-proton of 2,6-dimethylphenols and 1-napthhols in dimethyl sulphoxide and have found enhanced resonance effects for $+M$ as well as $-M$ 4-substituents.

Other relevant work reported during the review period includes an application of the Hammett–Streitwieser equation to the observed chemical shifts of non-aromatic protons in methylarenes,[63] the study of the transmission of structural effects by a conjugated chain on proton chemical shifts,[64] the proton chemical shifts of a number of disubstituted anilines,[65] substituent effects on the NH proton chemical shifts of p-substituted phenylureas,[66] and the proton chemical shifts of tetrafuryl- and tetrathienyl-lead.[67]

D. Magnetic Anisotropy Effects.—In this subsection attention is focused on shielding contributions from the magnetic fields of distant chemical bonds or groups. These fields, affecting the shielding directly through space, are present in addition to any effects on shielding caused by induction, resonance, *etc.* However, since the only property upon which they depend (apart from geometry) is the anisotropy in the magnetic susceptibility of the bond or group, it is on proton chemical shifts that they manifest themselves the most readily because of the small range of shifts existing for this nucleus. The simplest quantitative formulation is made by using the point-dipole approximation. For a particular bond or group possessing conical symmetry and having magnetic anisotropy $\Delta\chi$(*e.g.* C—H, C—Cl, or CH_3 but not NO_2 or C=O), the shielding contribution σ_m due to this group is given by equation (26),

$$\sigma_m = \frac{\mu_0}{4\pi} \frac{\Delta\chi}{3R^3}(1 - 3\cos^2\theta) \tag{26}$$

where R is the distance from the nucleus of interest to the site of the point-

[62] A. Fischer, M. C. A. Opie, J. Vaughan, and G. J. Wright, *J.C.S., Perkin II*, 1972, 319.
[63] B. Kamieński and T. M. Krygowski, *Tetrahedron Letters*, 1972, 681.
[64] J. P. Doucet, B. Ancian, and J. E. Dubois, *J. Chim. phys.*, 1972, **69**, 188.
[65] D. Aksnes and F. H. Kronhaug, *Acta Chem. Scand.*, 1971, **25**, 23.
[66] Y. Asabe and Y. Tsuzuki, *Bull. Chem. Soc. Japan*, 1971, **44**, 3482.
[67] G. Barbieri and F. Taddei, *J.C.S. Perkin II*, 1972, 262.

dipole and θ is the angle between the axis of the group and the line joining the site of the point-dipole to the nucleus. It is generally held that for this approximation to be valid R must be fairly large (> 3 nm). A more detailed discussion of the underlying physical ideas involved in the concept of magnetic anisotropy may be found in Volume 1, p. 21.

Schraml and Cawley[68] have discussed the general question of the contribution to the nuclear shielding of a distant, magnetically anisotropic group which can rotate freely within a molecule. If $\Delta\chi^G$ is the group's magnetic anisotropy, the group is rotating rapidly at an angle θ about an axis, the proton is located at a distance R from the site of the magnetic dipole of the 'effective stationary group' and γ is the angle between the vector R and the axis, then the shielding change σ_m^G due to the group is given by

$$\sigma_m^G = \frac{\mu_0}{4\pi} \frac{\Delta\chi^G}{3R^3}(1 - 3\cos^2\gamma)\left(1 - \frac{3}{2}\sin^2\theta\right) \qquad (27)$$

An application of this equation (with R, however, rather on the short side) might be to the shielding of the aldehyde proton in, say, N-ethylformamide by the methyl group as illustrated in (9). The main problem here, as is general

(9)

with magnetic anisotropy calculations, is the choice of location of the point dipole.

Proton chemical shift data of five isomers of 1,2,3,4,5,6-hexachlorocyclohexane have been obtained by Hayamizu and co-workers.[69] Formulae for two of these isomers, α-BHC (10) and δ-BHC (11), are given below (BHC = benzene hexachloride) where the protons are labelled. Use of the point-dipole approximation to calculate the difference in chemical shift between the H^3 and H^4 protons of (10) and a $\Delta\chi$ value for the C—Cl bond obtained by previous

(10)   (11)

[68] J. Schraml and S. Cawley, *Coll. Czech. Chem. Comm.*, 1971, **36**, 2986.
[69] K. Hayamizu, O. Yamamoto, K. Kushida, and S. Satoh, *Tetrahedron*, 1972, **28**, 779.

workers[70] predicts 0.22—0.29 p.p.m., H^3 being the more shielded. However, H^3 is observed to be less shielded than H^4 by about 0.6 p.p.m. On the other hand, the point-dipole approximation explains well the observed (0.01 p.p.m.) difference between the shielding of H^3 and H^4 of (11). The conclusion of Hayamizu et al.,[69] with which the Reporter concurs, is that even if C—H and C—C bond anisotropies and refinements to the point-dipole approximation, such as those of ApSimon et al.,[71] were included it would be unlikely to lead to good agreement between calculated and observed results since other factors such as differential solvent effects, intramolecular van der Waals forces, and electric field effects are probably of importance. Ando and Nishioka[72] have calculated the magnetic anisotropy effects of C—C bonds on the chemical shifts of the terminal methyl and methylene protons of linear hydrocarbons in the form of planar zigzag chains having from three to twenty-two carbon atoms.

Tribble et al.[73] have attempted to account for the proton chemical shifts of a number of hydrocarbons with respect to tetramethylsilane in terms of magnetic anisotropy and van der Waals effects. The particular form used for the anisotropy contribution was that introduced by ApSimon et al.[71] for a finite dipole whereas the van der Waals expression used was the familiar equation

$$\sigma_w = -3B\,\alpha_2 I_2/r_{12}^6 \qquad (28)$$

where B is a constant, α_2 and I_2 are respectively the polarizability and ionization potential of the perturbing atom or group, and r_{12} is the distance from the perturbing atom or group to the proton. By an empirical fit to 46 items of experimental data, values of B and the C—C and C—H bond susceptibilities and anisotropies were obtained. The present Reporter finds it very difficult to accept the physical interpretation of the fit which Tribble et al. obtain, namely the dominance of van der Waals contributions to proton chemical shifts in these hydrocarbons. For instance, the observed proton shielding of ethane relative to tetramethylsilane (-0.856 p.p.m., although -0.882 p.p.m. is a better value[74]) is calculated to be composed of -0.238 p.p.m. due to the change in σ_m and -0.617 p.p.m. (sic) due to the change in σ_w. It is difficult to see why van der Waals forces — always deshielding in effect — are more prominent in H_3C—CH_3 than in H_3C—$SiMe_3$. It seems that inductive effects are not entirely negligible and inductive contributions (including effects of hybridization changes) to proton shielding cannot be regarded as being 'nearly identical for all protons in a hydrocarbon'. Furthermore the fundamental theoretical work required to indicate the superiority of using the finite-dipole approximation rather than the point-dipole approximation as well as

---

[70] J. Homer and D. Callaghan, *J. Chem. Soc.* (*A*), 1968, 518.
[71] J. W. ApSimon, W. G. Craig, P. V. Demarco, D. W. Mathieson, L. Saunders, and W. B. Whalley, *Tetrahedron*, 1967, **23**, 2339.
[72] I. Ando and A. Nishioka, *Makromol. Chem.*, 1972, **152**, 7.
[73] M. T. Tribble, M. A. Miller, and N. L. Allinger, *J. Amer. Chem. Soc.*, 1971, **93**, 3894.
[74] W. T. Raynes and M. A. Raza, *Mol. Phys.*, 1969, **17**, 157.

equation (28) for σ_w has yet to be performed. Magnetic anisotropy effects have also been discussed in attempting to understand proton shielding in substituted acetylenes[75] and in the di-t-butylpropargyl system.[76]

Edge and Sharp[77] have compared proton chemical shifts in *cis*- and *trans*-forms of compounds such as PhN=NMe, PhN=NPr^i, PhN=NBu^i, and PhN=NBu^t. After obtaining values of $\Delta\chi$ for the N=N and N=N groups from single-crystal magnetic susceptibility data, they show that magnetic anisotropy, although of significance, is not sufficient to explain the observed *cis–trans* chemical shifts. Magnetic anisotropy shielding cones for the S=O and N—N=O groups have been discussed by Green and Haller[78] and by ApSimon and Cooney,[79] respectively. Molecular susceptibilities and susceptibility anisotropies obtained by methods other than n.m.r. but relevant to the subject of magnetic anisotropy effects on chemical shifts have been obtained for a number of hydrocarbons,[80] for some aromatic molecules,[81] for benzene, 1,3,5-trifluorobenzene, and hexafluorobenzene,[82] and for acetylene and hydrogen cyanide.[83]

E. **Electric Field Effects.**—In the presence of the electric field, produced, for instance, by a polar substituent, there is a distortion of the electron distribution which leads to changes in the nuclear shielding as compared with the unsubstituted compound. Such shielding changes are usually calculated from the equation given by Buckingham.[84] Thus for a uniform electric field E acting in the vicinity of a conically symmetric bond (*e.g.* X—H), the shielding change is given by

$$\sigma_E = -AE_z - BE^2 \qquad (29)$$

where A and B are bond parameters dependent only on X and which are independent of the electric field and E_z is the component of E along the bond from X to H. In practice, of course, this is an approximation for several reasons: (*a*) the field from a polar group will be highly non-uniform; (*b*) the field may produce changes elsewhere in the molecule which then modify the shielding of the proton in the X—H bond; and (*c*) strictly speaking there should be two quadratic coefficients since the quadratic dependence of the shielding on the field will be different for the components of the field parallel and perpendi-

[75] D. Rosenberg and W. Drenth, *Tetrahedron*, 1971, **27**, 3893.
[76] R. S. Macomber, *J. Org. Chem.*, 1972, **37**, 1205.
[77] S. N. Edge and R. R. Sharp, *J. Chem. Soc. (B)*, 1971, 2014.
[78] C. H. Green and D. G. Hellier, *J. C. S. Perkin II*, 1972, 458.
[79] J. W. ApSimon and J. D. Cooney, *Canad. J. Chem.*, 1971, **49**, 2377.
[80] Z. B. Maksić and J. E. Bloor, *Chem. Phys. Letters*, 1972, **13**, 571.
[81] C. L. Cheng, D. S. N. Murthy, and G. L. D. Ritchie, *Mol. Phys.*, 1971, **22**, 1137.
[82] M. P. Bogaard, A. D. Buckingham, M. G. Corfield, D. A. Dunmur, and A. H. White, *Chem. Phys. Letters*, 1972, **12**, 558.
[83] Y. Kato, Y. Fujimoto, and A. Saika, *Chem. Phys. Letters*, 1972, **13**, 453.
[84] A. D. Buckingham, *Canad. J. Chem.*, 1960, **38**, 300.

cular to the bond. Values of A and B for protons in a number of bonding situations and obtained by a variety of methods may be found in the corresponding section of Volume 1.[2]

Two papers have appeared over the past year concerned with the calculation of electric field parameters. Mukhomorov[85] has extended the variational procedure of Aminova and Gubaidullina[86] for the linear electric field coefficient of the $C(sp^3)$—H bond. This allows for electron exchange and changes the calculated value of A for such a bond from the 3.845×10^{-12} e.s.u. quoted in Volume 1 to 3.615×10^{-12} e.s.u., which is nearer to the generally accepted value of about 3.0×10^{-12} e.s.u. obtained by experiment. Hamer and Reynolds[87] have performed CNDO/2 MO calculations upon the vinyl proton chemical shifts of substituted styrenes relative to styrene. Thus for 4-substituted styrenes (12) the through-bond effects of the X substituent should

H_B
\
C_β — H_C
//
X—⌬—C
\
H_A

(12)

produce identical changes in the shielding of the protons H_B and H_C whereas the through-space effects lead to different changes in the shielding of these two nuclei because of their differing geometrical disposition relative to the axis of the X—C(aryl) bond. Table 4 gives the calculated results of Hamer and Rey-

Table 4 *Calculated changes in the charge densities at the vinylic protons of some 4-substituted styrenes and of the shielding-constant differences between these protons compared with those calculated by the electric field model of equation (29) and those obtained by experiment*

Substituent	Charge-density changes ($\times 10^4$)			Shielding difference/p.p.m. $\sigma_C - \sigma_B$		
	ΔH_B	ΔH_C	$\Delta(H_C-H_B)$	CNDO/2	Field model	Expt.
CMe₃	−11	−12	−1	+0.003	+0.010	+0.009
Me	−8	−9	−1	+0.003	+0.008	+0.006
C≡CH	+2	+7	+5	−0.013	−0.011	−0.017
OMe	−17	−12	+5	−0.013	−0.018	−0.022
F	−2	+16	+18	−0.048	−0.049	−0.066
Cl	+18	+36	+18	−0.048	−0.047	−0.052
CF₃	+26	+48	+22	−0.059	−0.042	−0.047
C≡N	+13	+27	+14	−0.037	−0.042	−0.057
NO₂	+39	+74	+35	−0.093	−0.081	−0.073

[85] V. K. Mukhomorov, *J. Struct. Chem.*, 1971, **12**, 299.
[86] R. M. Aminova and R. Z. Gubaidullina, *J. Struct. Chem.*, 1969, **10**, 236.
[87] G. K. Hamer and W. F. Reynolds, *Chem. Comm.*, 1971, 1218.

nolds for the changes in charge density. As can be seen from the column headed $\Delta(H_C-H_B)$ the possession of increased group dipole by the X substituent in the direction $\overset{-}{X}$—$\overset{+}{C}$(aryl) leads to a greater deshielding of the proton H_C as compared with H_B. This is to be expected on geometrical grounds if the field field effect is of importance, since the component of field along the C_β—H_C bond is large whereas that along the C_β—H_B bond is small. To convert to the shielding difference between H_C and H_B given in the column headed CNDO/2 a multiplicative factor of 26.6 p.p.m. has been used, although no justification for this is given by the authors. However, since σ_{loc}^p is very small for protons the shielding constant of the hydrogen molecule (formerly taken[88] to be 26.6 p.p.m.) is approximately equal to σ_{loc}^d. However, for protons bonded to carbon a value of σ_{loc}^d nearer to 30 p.p.m. would appear to the present Reporter to be a better choice. Under the heading 'Field model' are given the shielding differences $\sigma_C - \sigma_B$ assuming a point electric dipole located at the X-group, an electric field calculated at the C—H bond midpoint, and a value for A of 4.0×10^{-12} e.s.u.

As can be seen from the final column of Table 4, good enough agreement between observed and calculated values is obtained to indicate clearly the existence of intramolecular electric fields. The discrepancies probably reflect the presence of solvent effects and the existence of a contribution to experimental value of $\sigma_C - \sigma_B$ arising from the magnetic anisotropy of the X-group. (An attempt to correct for this was made for the triple bonds in C≡CH and C≡N but no correction was made for the other substituents, some of which have substantial magnetic anisotropies). Hamer and Reynolds provide further theoretical evidence for the existence of an intramolecular electric field effect in a non-conjugated system, namely 4-substituted 1-vinylbicyclo[2,2,2]octane.

Briefly noted here are two very recent studies by methods other than n.m.r. which not only demonstrate the existence of the intramolecular electric fields of polar substituents but question the existence of the inductive effect in such cases.[89,90]

An extensive series of proton chemical shift studies has been made by Hamm and von Philipsborn[91] on the aromatic amines and aromatic N-oxides. The shielding changes at the protons of the latter as compared with the corresponding positions in the former were attributed to the field effects of the NO group, namely the electric field effect and the magnetic anisotropy effect. The larger part of the shielding change appears to be caused by the former. In performing these calculations Hamm and von Philipsborn used several models including those for electric and magnetic dipoles both of finite length. The difficulties, as they point out, are associated with the choice of parameters. In addition the validity of these models at short distances from the centre of

---

[88] J. I. Musher, *Adv. Magn. Resonance*, 1966, **2**, 209.
[89] C. L. Liotta, W. F. Fisher, E. L. Slightom, and C. L. Harris, *J. Amer. Chem. Soc.*, 1972, **94**, 2129.
[90] R. Golden and L. M. Stock, *J. Amer. Chem. Soc.*, 1972, **94**, 3080.
[91] P. Hamm and W. von Philipsborn, *Helv. Chim. Acta*, 1971, **54**, 2363.

the dipole to the nucleus whose resonance is being studied is not yet established.

Homer and Callaghan[92] have extended their work on the ^{19}F shifts of bridgehead-substituted fluorobicyclo[2,2,1]heptanes discussed in Volume 1 to

(13)

similarly substituted fluorobicyclo[2,2,1]heptenes (13). By using the concept of intramolecular electric fields in the form of the equation

$$\delta(^{19}F) = -X\Delta E_z - Y\Delta(E^2) - Z\Delta\langle E^2 \rangle \qquad (30)$$

where the parameters are characteristic of C—F bonds and Δ denotes a difference relative to the bridgehead fluorine shielding in perfluorobicyclo-[2,2,1]heptene, Homer and Callaghan were able to account for the F^7 (*anti*)–F^7 (*syn*) shifts and the F(*exo*)–F(*endo*) shifts for the 5- and 6-positions occurring upon substitution of protons, methyl groups, or halogen atoms at the bridgehead (1- and 4-) positions. The first two terms on the right of equation (30) account for the linear and quadratic electric field effects and the final term is the contribution of intramolecular dispersion forces.

F. **The Ring-current Effect.**—The past year has witnessed a large increase in the number of publications devoted to 'ring current' effects as compared with 1970—71. First, several theoretical papers will be discussed. Roberts[93] has presented improved calculations of ring-current contributions to the proton shielding in benzene, naphthalene, and anthracene as compared with the earlier work of Amos and himself which was discussed in Volume 1. The improvement is brought about by the evaluation of integrals, obtained previously by twice invoking the London Approximation, using now only one application of this approximation which thereby renders them independent of the origin of the vector potential. The results obtained by Roberts for the π-electron (*i.e.* ring-current) contributions to the shielding in benzene and naphthalene are given in Table 5 and include the contributions of individual bonds where the conventional numbering scheme for napthhalene (14) is used. It can be seen that the contributions to the shielding of protons 1 and 2 from bond 1—2 (and from bond 4—10) must be equal on symmetry grounds. For benzene the ring-current contribution to the total shielding is calculated to be -0.72 p.p.m., only about half of the -1.56 p.p.m. empirically obtained

---

[92] J. Homer and D. Callaghan, *J. Chem. Soc.* (*B*), 1971, 2430.
[93] H. G. Ff. Roberts, *Chem. Phys. Letters*, 1971, **11**, 259.

Nuclear Shielding

Table 5 Calculated individual bond contributions (using conventional numbering) and total π-electron (ring-current) contributions to the proton shielding in benzene and naphthalene

Bond	Benzene Proton 1 (p.p.m.)	Naphthalene Proton 1 (p.p.m.)	Proton 2 (p.p.m.)
1—2	−0.919	−0.1020	−0.1020
2—3	0.248	0.258	−0.959
3—4	0.311	0.345	0.275
4—10	—	0.334	0.334
10—5	—	0.162	0.000
5—6	—	0.216	0.122
6—7	—	0.195	0.125
7—8	—	0.035	0·083
8—9	—	−0.734	−0.162
9—1	—	−0.987	0.266
9—10	—	0.000	0.000
Total shielding contribution	−0.720	−1.196	−0.936
Calc. shift from benzene	—	−0.476	−0.216
Obs. shift from benzene	—	−0.46	−0.11

(14)

by the work of Haigh et al.[94] indicating, as suggested by Pople,[95] that local bond anisotropies make a substantial contribution. The experimental values of the proton chemical shifts of naphthalene with respect to benzene which are given in Table 5 are those obtained by Haigh and Mallion.[96] Good agreement was also obtained for the 1-, 2-, and 9-protons of anthracene. Calculations of ring-current contributions to proton shielding in benzene, naphthalene, anthracene, and phenanthrene have been performed by Edwards and McWeeny[97] using self-consistent field perturbation theory. Roberts[98] has presented comparisons of SCF-Hückel and modified-Hückel calculations of the ring-current contributions to the proton shielding in a number of aromatic hydrocarbons which represent an improvement (to take account of gauge dependence) as compared with earlier work. The question of the existence of

[94] C. W. Haigh, R. B. Mallion, and E. A. G. Armour, Mol. Phys., 1970, 18, 751.
[95] J. A. Pople, J. Chem. Phys., 1964, 41, 2559.
[96] C. W. Haigh and R. B. Mallion, Mol. Phys., 1970, 18, 737.
[97] T. G. Edwards and R. McWeeny, Chem. Phys. Letters, 1971, 10, 283.
[98] H. G. Ff. Roberts, Theor. Chim. Acta, 1971, 22, 105.

ring currents in aromatic hydrocarbons has been investigated by Kumanova and Rebane[99] who compared the diamagnetic susceptibilities of several series of cyclic hydrocarbons and deduced that the observed discontinuities in the trends of susceptibilities constitute a positive proof of the presence of ring currents. From this work it appears that about 70% of the experimentally observed anisotropy ($\Delta\chi = -59.7 \times 10^{-6}$ c.g.s. units) comes from the ring current, the remainder being provided by local anisotropic contributions.

Haigh and Mallion[100] have obtained proton shielding data for 3,4-benzophenanthrene (15), pentahelicene (16), and hexahelicene (17) and have

(15)   (16)   (17)

presented[101] a modification of existing theory[102] which takes into account the slight non-planarity of the these benzenoid hydrocarbons. In Table 6

Table 6 *Observed and calculated (using planar and non-planar geometry) τ values of the protons in 3,4-benzophenanthrene (15) and pentahelicene (16)*

	3,4-Benzophenanthrene (15)			Pentahelicene (16)		
Proton	Planar	Non-planar	Obs.	Planar	Non-planar	Obs.
1	1.22	2.11	0.93	—	2.71	1.54
2	2.29	2.35	2.40	2.02	2.57	2.80
3	2.36	2.36	2.47	2.30	2.44	2.57
4	2.13	2.11	2.08	2.08	2.16	2.15
5	2.16	2.16	2.20	2.15	2.18	2.18
6	2.15	2.16	2.26	2.12	2.24	2.22
7	—	—	—	1.99	2.12	2.23

are given comparisons of the observed chemical shifts (τ values) and the calculated values obtained by the assumption of both planar and the actual (*i.e.* non-planar) geometries for (15) and (16). The improvement for (15) is slight, but that for (16) is striking. The poor agreement for the 1-protons of (15) and (16) reflects the presence of steric effects which are highly deshielding and which are not taken into account by the theory. In physical terms the increased

[99] M. D. Kumanova and T. K. Rebane, *J. Struct. Chem.*, 1971, **12**, 507.
[100] C. W. Haigh and R. B. Mallion, *Mol. Phys.*, 1971, **22**, 945.
[101] C. W. Haigh and R. B. Mallion, *Mol. Phys.*, 1971, **22**, 955.
[102] R. McWeeny, *Mol. Phys.*, 1958, **1**, 311.

shielding observed for all protons of (15) and (16), other than the 1-protons, is caused by the non-planarity, so that a given proton is not in the plane of an adjacent ring (to which it does not belong) but is slightly above it, thereby being moved from the highly deshielded positions of the ring plane towards the highly shielded positions above and below the ring. Thus in (16) proton 2 is shielded a little by ring E as compared with the hypothetical planar disposition of rings A and E. Haigh and Mallion[103] have also presented tables of numerical values of the shielding at positions in or out of the plane of a single benzene ring. As compared with the Johnson–Bovey tables,[104] the new tables, based on a large amount of experimental data, show a somewhat smaller deshielding in the ring plane and a considerably smaller shielding above or below the ring plane. Lazzeretti and Taddei[105] have performed uncoupled Hartree–Fock calculations to determine the changes in ring current contributions to the proton shielding of substituted derivatives of benzene. They show that this contribution falls upon substitution with groups such as F, NH_2, OH, Cl, or CH_3 and that the change is generally about the same for all ring positions.

Conveniently discussed at this point is the subject of the proton shielding effects of the cyclopropane ring. Poulter et al.[106] have used the Johnson–Bovey model to estimate the shielding and deshielding effects at various positions near to a cyclopropane ring. In empirical terms this implies a ring current via the σ electrons of the cyclopropane ring although, as pointed out by these authors, it does not constitute a proof of its existence. To obtain agreement with the calculated shielding difference between the protons of cyclopropane and those of 'normal' CH_2 groups, Poulter et al. had to postulate the presence of 4.5 electrons in the σ ring current. This corresponds to a 'barrier' to cyclical electron mobility in cyclopropane. Poulter et al. give proton chemical shifts and estimated contributions thereto from the cyclopropane ring as calculated by the ring-current model for a large number of compounds containing cyclopropyl groups. Two of these results are given below. For (18) the proton shielding difference $\sigma(exo) - \sigma(endo)$ is observed

(18)   (19)

experimentally to be -1.03 p.p.m. whereas the calculated value is -1.20 p.p.m., the difference between exo- and endo-protons being attributable to

[103] C. W. Haigh and R. B. Mallion, Org. Magn. Resonance, 1972, **4**, 203.
[104] C. E. Johnson and F. A. Bovey, J. Chem. Phys., 1958, **29**, 1012.
[105] P. Lazzeretti and F. Taddei, Mol. Phys., 1971, **22**, 941.
[106] C. D. Poulter, R. S. Boikess, J. I. Brauman, and S. Winstein, J. Amer. Chem. Soc., 1972, **94**, 2291.

the 'ring current' in the ring adjacent to that containing the labelled protons. In (19) the observed proton shielding difference $\sigma(5) - \sigma(4)$ is -0.62 p.p.m. whereas that calculated from the ring-current model is -0.53 p.p.m. Of course, in each of these compounds both protons concerned have identical shifts in the absence of the cyclopropyl group. The compilation of results in the paper of Poulter et al.[106] provides an example where confusion could arise from the use of a direction for the chemical shift scale which leads to more positive chemical shift values as the nuclear shielding decreases. This subject was discussed in some detail in Volume 1.[2] In the paper of Poulter et al.[106] values of proton chemical shifts are given on the δ scale (i.e. protons less shielded than tetramethylsilane have positive chemical shifts) whereas the calculated shielding contributions from the cyclopropyl ring anisotropy are properly given positive or negative signs according to whether shielding or deshielding occurs. In other work on the cyclopropyl group Anderson[107,108] has presented data for the shielding and deshielding effects of this group upon substituents and used the method of Anet and Schenk[109] to obtain for cyclopropane an S value of -0.14 p.p.m. This formally indicates antiaromaticity (a paramagnetic ring current) and contrasts with the diamagnetic ring current concept needed to explain the shielding effect of the cyclopropyl ring on adjacent groups. Anderson gives the possible explanations for the negative value of S.

A number of authors have invoked the ring-current concept in organometallic studies. Turbitt and Watts[110] have attempted to obtain proton shielding evidence to support the shielding ($+$) and deshielding ($-$) zones around the ferrocene molecule that are indicated in Figure 2. This was carried

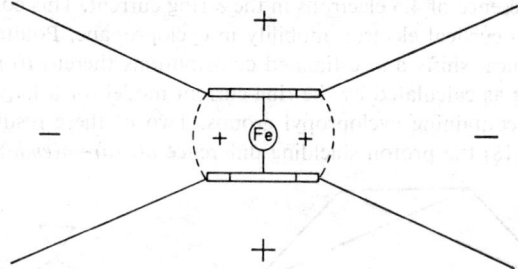

Figure 2 *Shielding ($+$) and deshielding ($-$) zones in the vicinity of the ferrocene molecule*
(Reproduced by permission from *Tetrahedron*, 1972, **28**, 1227)

out by the study of derivatives of [3](1,2)ferrocenophane (20) such as (21), from which clear evidence supporting the shielding scheme of Figure 2 was obtained. Lentzner and Watts[111] have presented proton shielding data for a

---

[107] J. E. Anderson, *Org. Magn. Resonance*, 1971, **3**, 475.
[108] J. E. Anderson, *J. Chem. Soc. (B)*, 1971, 2388.
[109] F. A. L. Anet and G. E. Schenck, *J. Amer. Chem. Soc.*, 1971, **93**, 556, 3310.
[110] T. D. Turbitt and W. E. Watts, *Tetrahedron*, 1972, **28**, 1227.
[111] H. L. Lentzner and W. E. Watts, *Tetrahedron*, 1971, **27**, 4343.

Nuclear Shielding

number of [2]ferrocenophanes (22) and Faller and Jakubowski[112] have discussed the application of the ring-current concept to the subject of organometallic stereochemistry.

(20)   (21)   (22)

We mention here without comment studies of the annulene series for which much new proton shielding data and a general review[113] have appeared during the review period. Studies of the following compounds or classes of compounds have been made: monodehydro[26]annulene, which gives different proton shielding for inner and outer protons and is therefore presumably aromatic and is the largest macrocycle for which ring-current evidence has been observed;[114] methylene-bridged bisdehydroaza[17]annulenes;[115] 1,2-dihalogeno[18]annulene;[116] oxa[17]annulenes;[117] aza[17]annulenes;[118] bridged hetero[11]annulenes;[119] 2-hydroxy-4,9-methane[11]annulenone;[120] the radical anion and dianion of [16]annulene;[121] a perturbed [13]annulenone;[122] and 4:7,10:13-dioxido[15]annulenones.[123] Wilcox et al.[124] have synthesized cyclo-octa[def]biphenylene (23). Both the aromatic protons and the protons of the eight-membered ring show a marked increase in shielding compared with the corresponding protons of 1,8-divinylbiphenylene (24), indicating antiaromaticity in the eight-membered ring. (23) may be the first example of a planar cyclo-octatetraene ring. In a paper on the general question of bicycloaromaticity Grutzner and Winstein[125] have discussed the proton chemical shifts of the bicyclo[3,2,2]nonatrienyl cation and anion.

[112] J. W. Faller and A. Jakubowski, *J. Organometallic Chem.*, 1971, **31**, C75.
[113] F. Sondheimer, *Accounts Chem. Res.*, 1972, **5**, 81.
[114] B. W. Metcalf and F. Sondheimer, *J. Amer. Chem. Soc.*, 1971, **93**, 5271.
[115] P. J. Beeby and F. Sondheimer, *J. Amer. Chem. Soc.*, 1972, **94**, 2128.
[116] G. Schröder, R. Neuberg, and J. F. M. Oth, *Angew. Chem. Internat. Edn.*, 1972, **11**, 51.
[117] G. Schröder, G. Plinke, and J. F. M. Oth, *Angew. Chem. Internat. Edn.*, 1972, **11**, 424.
[118] G. Schröder, G. Heil, H. Röttele, and J. F. M. Oth, *Angew. Chem. Internat. Edn.*, 1972, **11**, 426.
[119] E. Vogel, R. Feldmann, H. Düwel, H. Cremer, and H. Günther, *Angew. Chem. Internat. Edn.*, 1972, **11**, 217.
[120] J. Reisdorff and E. Vogel, *Angew. Chem. Internat. Edn.*, 1972, **11**, 218.
[121] J. F. M. Oth, H. Bauman, J. M. Gilles, and G. Schröder, *J. Amer. Chem. Soc.*, 1972, **94**, 3498.
[122] I. Murata, K. Yamamoto, T. Hirotsu, and M. Morioka, *Tetrahedron Letters*, 1972, 331.
[123] H. Ogawa, M. Yoshida, and H. Saikachi, *Tetrahedron Letters*, 1972, 153.
[124] C. F. Wilcox, J. P. Uetrecht, and K. K. Grohman, *J. Amer. Chem. Soc.*, 1972, **94**, 2532.
[125] J. B. Grutzner and S. Winstein, *J. Amer. Chem. Soc.*, 1972, **94**, 2200.

(23)   (24)

Laarhoven and co-workers[126—128] have carried out proton shielding studies on the arylhexahelicenes and also on substituted hexahelicenes (see also ref. 100). Other papers include a study of the possibility of 'superaromaticity' in 1,2:3,4:7,8:9,10-tetrabenzocoronene,[129] the spatial structure of tetrabenzo[5,7]fulvalene,[130] [7]metacyclophane and its 13-bromo-derivative,[131] the spectra of dibenzo[c,g]phenanthrene and benzo[g,h,i]perylene,[132] and of benzo[b]thiophen and dibenzothiophen.[133]

Anastassiou and Yamamoto[134] have used the method of Anet and Schenck[109] (see also above) to determine the existence or non-existence of ring currents in the heteronins (25) and (26). For cyclononatetraene (25;

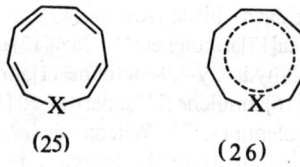

(25)   (26)

$X = CH_2$) and oxonin (25; $X = O$) small and negative S values were obtained. For $1H$-azonine (26; $X = NH$) a large positive S value — greater even than that for benzene — was obtained. These results indicate the absence of aromaticity in cyclononatetraene and oxonin but its presence in azonine. For N-methylazonine the ring current is reduced, probably owing to the need for some non-planarity to avoid large steric strain. One must be careful in interpreting very small S values or small changes in S values with temperature for the reasons given by Anet and Schenck together with the possibility of differential solvent shifts.[135]

G. Van der Waals and Steric Effects.—Intramolecular dispersion forces have been used by a number of authors over the past decade to account in part for

[126] W. H. Laarhoven and R. J. F. Nivard, *Tetrahedron*, 1972, **28**, 1803.
[127] W. H. Laarhoven and R. G. M. Veldhuis, *Tetrahedron*, 1972, **28**, 1811.
[128] W. H. Laarhoven and R. G. M. Veldhuis, *Tetrahedron*, 1972, **28**, 1823.
[129] E. Clar and B. A. McAndrew, *Tetrahedron*, 1972, **28**, 1137.
[130] I. Agranat, M. Rabinowitz, and M. Weissman, *Tetrahedron Letters*, 1972, 273.
[131] S. Fujita, S. Hirano, and H. Nazaki, *Tetrahedron Letters*, 1972, 403.
[132] R. S. Matthews, D. W. Jones, and K. D. Bartle, *Spectrochim. Acta*, 1971, **27A**, 1185.
[133] K. D. Bartle, D. W. Jones, and R. S. Matthews, *Tetrahedron*, 1971, **27**, 5177.
[134] A. G. Anastassiou and H. Yamamoto, *J.C.S. Chem. Comm.*, 1972, 286.
[135] W. T. Raynes and M. A. Raza, *Mol. Phys.*, 1969, **17**, 157.

Nuclear Shielding

observed shielding data. For the purposes of calculation the shielding contribution from these forces σ_w is identified with the expression often used for dealing with intermolecular dispersion forces[136] and given in equation (28), where B is a parameter which is often equated with the quadratic electric field coefficient of equation (29). An alternative view of σ_w, therefore, is that of a fluctuating electric field which averages to zero but the mean square of which does not average to zero. This alternative view is that indicated in equation (30) as used by Homer and Callaghan[92] in the work discussed above. The only other paper during the review period to cite the importance of σ_w is that of Yoshioka et al.[137] who observed significant deshielding between adjacent protons from studies of sesquiterpene lactones. The necessary theoretical work that would make clear the extent to which intramolecular dispersion forces are of importance for the situations where their presence is postulated has yet to be accomplished.

From proton resonance studies of triphenyl-*sym*-triazine (27) and tripyridyl-*sym*-triazine (28), Gil and Pereira[138] have deduced that the latter is more nearly planar than the former and that this indicates that the *ortho–ortho* N···N steric interaction in (28) is less repulsive than a CH···N interaction in

(27)   (28)

(27). To obtain this result it was necessary to make calculations for electric field and magnetic anisotropy effects. A nitrogen chemical shift study here would be informative. Frost and Barzilay[139] have used the different effects of deshielding by alkyl groups *trans* or *cis* to methine, methylene, and methyl groups across a double bond (postulated to be caused by steric deshielding) for the characterization of double bonds.

H. Intramolecular Hydrogen Bonding.

Several papers published in the review period demonstrate the large deshielding effects which can occur for protons involved in intramolecular hydrogen bonds. Gribble and Bousquet[140] have observed the chemical shifts of the 6-proton and the NH proton in *ortho*-

---

[136] W. T. Raynes, A. D. Buckingham, and H. J. Bernstein, *J. Chem. Phys.*, 1962, **36**, 3481.
[137] H. Yoshioka, T. J. Mabry, M. A. Irwin, T. A. Geissman, and Z. Samek, *Tetrahedron* 1971, **27**, 3317.
[138] V. M. S. Gil and A. M. P. Pereira, *Tetrahedron*, 1971, **27**, 5619.
[139] D. J. Frost and J. Barzilay, *Rec. Trav. chim.*, 1971, **90**, 705.
[140] G. W. Gribble and F. B. Bousquet, *Tetrahedron*, 1971, **27**, 3785.

substituted anilides (29) and their thio-analogues. The hydrogen bond formed

(29)

between the NH proton and the X substituent (*e.g.* COMe, OMe, or COPh) is of varying strength and the mean orientation of the RCO group relative to the ring is very sensitive to slight changes in this hydrogen bond. Because of the close proximity of the 6-proton to the carbonyl group (in the predominant conformation of the RCO group) the magnetic anisotropy effect of the carbonyl group is very large at this proton. Consequently, the shielding of the 6-proton can be used to monitor the extent of the NH $\cdots$ X bonding. This was done by Gribble and Bousquet. They also give the NH proton chemical shifts which can be used to measure the hydrogen-bonding ability of X. Other studies on *ortho*-substituted acetanilides made by Andrews *et al.*[141] show steric hindrance to hydrogen bonding with the *ortho*-substituent in some instances and enable conclusions to be made concerning the alternative explanation to that of intramolecular hydrogen bonding, namely, intramolecular dipole–dipole interaction between nearby polar groups.

Proton shielding data for the two stereoisomers of the stable protonated monohalogeno-acetones (30) and (31) are given by Jost *et al.*[142] In the hydrogen-bonded form (30) the C=OH$^+$ proton is less shielded by 0.90 p.p.m. than that of the isomer (31). This seems rather small for a hydrogen

(30) (31)

bond and suggests only weak bonding. The strength of the hydrogen bonding decreases steadily as one passes along the series X = F, Cl, Br, and I, although magnetic anisotropy effects may be important here. It would appear that there are one or two misprints in this communication. Other shielding studies which illustrate the presence of intramolecular hydrogen bonding

[141] B. D. Andrews, A. J. Poynton, and I. D. Rae, *Austral. J. Chem.*, 1972, **25**, 639.
[142] R. Jost, P. Rimmelin, and J. M. Sommer, *Chem. Comm.*, 1971, 1243.

have been performed on halogeno(diethylthiourea)zinc(II) complexes,[143] N-acetoxy-derivatives of N-hydroxyanthines,[144] *ortho*-substituted anilines,[145] and substituted cyclopropanes.[146]

I. **Isotope Shifts.**—The theoretical results of Cook *et al.*[24] for the bond-length dependence of the nuclear shielding of the hydrogen molecule was discussed in Section 3. From these results it is possible to calculate not only the temperature dependence of the observed shielding constant but also the isotope shifts of the various isotopic derivatives of hydrogen. These molecules (HD, HT, DT, D_2, and T_2) for which the word 'isotopomers' is sometimes employed, have nuclei that are substantially more shielded than those of H_2, as one expects for molecules containing isotopes of greater mass.[25] Thus for a particular rotational and vibrational state characterized by the quantum numbers v and J the shielding constant σ_{vJ} can be obtained from equation (18) above, *viz.*

$$\sigma_{vJ} = \sigma_e^{(0)} + \sigma_e^{(1)}\langle\xi\rangle_{vJ} + \sigma_e^{(2)}\langle\xi^2\rangle_{vJ} + \ldots \quad (31)$$

where the expectation values $\langle\xi^n\rangle_{vJ}$ can be readily obtained if the parameters describing the potential curve of the molecule are known. Of course, these quantities have different values for the various isotopomers and one obtains[25] the results given in Table 7 for σ (in p.p.m.) for some of the lower rotational and vibrational levels of these molecules.

Table 7 *Shielding constants* (p.p.m.) *for isotopic derivatives of hydrogen*

v	J	H_2	HD	HT	D_2	DT	T_2
0	0	26.337	26.369	26.384	26.416	26.436	26.459
0	1	26.312	26.351	26.368	26.403	26.426	26.451
0	2	26.263	26.315	26.335	26.379	26.406	26.435
1	0	25.666	25.759	25.800	25.893	25.953	26.020
1	1	25.642	25.741	25.784	25.881	25.943	26.012

It can be seen that the isotope shifts are substantial. To obtain the shielding constant for the compound in the gas (at low densities) one must carry out a statistical average of σ_{vJ} over the available vibrational and rotational states. This leads for H_2 to the value 26.298 p.p.m. given earlier. The isotope shifts are all substantial. The proton isotope shifts $\sigma(HD) - \sigma(H_2)$ and $\sigma(HT) - \sigma(H_2)$ are quite temperature dependent, as is indicated in Figure 3. The only available experimental values are at room temperature for $\sigma(HD) - \sigma(H_2)$ and are 0.036(± 0.002) p.p.m.,[147] 0.04(± 0.01) p.p.m.,[148] and 0.038 p.p.m.[149] and the calculated result is in good agreement with these values.

[143] A. M. Guiliani, *J.C.S. Dalton*, 1972, 497.
[144] N. J. M. Birdsall, T.-C. Lee, and U. Wölcke, *Tetrahedron*, 1971, **27**, 5961.
[145] T. Axenrod and M. J. Wieder, *J. Amer. Chem. Soc.*, 1971, **93**, 3541.
[146] J. Pierre and R. Perraud, *Compt. rend.*, 1972, **274**, C, 205.
[147] D. F. Evans, *Chem. and Ind.*, 1961, 1960.
[148] E. Dayan, G. Widenlocher, and M. Chaigneau, *Compt. rend.*, 1963, **257**, 2455.
[149] L. R. Anders, J. D. Baldeshwieler, and P. C. Lauterbur, quoted by J. I. Musher, *Adv. Magn. Resonance*, 1966, **2**, 200.

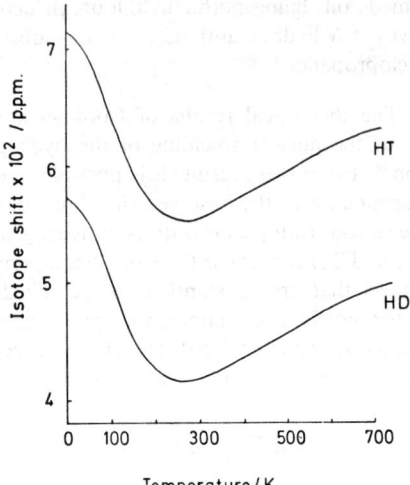

Figure 3 *Temperature dependence of the proton resonance isotope shielding shifts of HD and HT relative to* H_2. *At 100 MHz the scale of isotope shifts is in Hz)*

(Reproduced by permission from *Mol. Phys.*, 1971, **21**, 123)

Isotope-shift studies are usually concerned with the situation when the isotopic substitution occurs for a nucleus other than the one whose resonance is being studied. Hence the term 'secondary isotope effect' employed by Tupčiauskas *et al.*[150] These authors attempted to obtain the 'primary isotope effect' for tin nuclei by obtaining the resonance for ^{117}Sn and ^{119}Sn nuclei in several tin compounds. Although they were not able to obtain directly values for the 'primary isotope effect', it would seem from their work that primary isotope effects are not negligible in all cases. Thus they obtain the following tin chemical shifts in CCl_4 solvent:

$$\sigma(Et_4^{119}Sn) - \sigma(Me_4^{119}Sn) = +5.97(\pm 0.3) \text{ p.p.m.} \quad (32)$$

$$\sigma(Et_4^{117}Sn) - \sigma(Me_4^{117}Sn) = +3.07(\pm 0.3) \text{ p.p.m.} \quad (33)$$

Their results clearly imply that the tin shieldings in each pair of isotopomeric molecules are different. [In the work discussed above on the hydrogen molecule and its isotopomers the calculated shift $\sigma(HD) - \sigma(H_2) \simeq 0.04$ p.p.m. is essentially a primary isotope effect since, within the Born–Oppenheimer approximation and neglecting effects of the deuteron electric quadrupole, there is no difference between the proton and deuteron shieldings in HD.] Sergeev and co-workers have also presented measured values of deuterium isotope effects on ^{13}C chemical shifts[151] and ^{31}P chemical shifts.[152] In the

[150] A. Tupčiauskas, N. M. Sergeev, and Yu. A. Ustynyuk, *Mol. Phys.*, 1971, **21**, 179.
[151] Y. K. Grishin, N. M. Sergeev, and Yu. A. Ustynyuk, *Mol. Phys.*, 1971, **22**, 711.
[152] A. A. Borisenko, N. M. Sergeev, and Yu. A. Ustynyuk, *Mol. Phys.*, 1971, **22**, 715.

latter case a relationship was observed between the ^{31}P isotope shift and the (^{31}P,^{1}H) spin–spin coupling constant which permitted conclusions to be made concerning the existence of small valence-angle changes occurring upon isotopic substitution. Stec et al.[153] have also presented results for deuterium isotope effects on ^{31}P chemical shifts.

In all the above measurements positive isotope shifts were reported. Jensen and Schaumburg[154] have reported very accurate isotope shifts for seven isotopic ethyl fluorides dissolved in deuteriochloroform. Thus for the chemical shifts of the methyl and methylene protons and of the fluorine nucleus relative to those in ethyl fluoride (in p.p.m.) they give in their Table 7:

	methyl	methylene	F
CH$_3$·CHD·F	0.0078	0.0190	0.6453
CH$_2$D·CH$_2$·F	0.0181	0.0029	0.2442
CH$_2$D·CHD·F	0.0239	0.0240	0.7815
CHD$_2$·CHD·F	0.0420	0.0252	—

The Reporter is a little puzzled by the proton isotope shifts which can be deduced from Table 2 of this paper and which appear not only to disagree with those of their Table 7 (given above) but also include some negative isotope shifts as well. Jensen and Schaumburg also give proton and fluorine isotope shifts for ^{13}C substitution at either position in ethyl fluorides and also give some deuterium chemical shifts.

Further isotope shift data have been reported for the ^{19}F shielding in some fluorochloroacetones,[155] and for the ^{13}C shielding of monosubstituted benzenes.[156,157] For benzene Bell et al.[157] find that the ^{13}C isotope shifts for C$_6$H$_5$D with respect to C$_6$H$_6$ are 0.289 p.p.m., for the carbon nucleus bearing the deuterium, 0.110 p.p.m. for the two *ortho* carbon nuclei and 0.011 p.p.m. for the two *meta* carbon nuclei. Very similar isotope shifts were observed for a series of monosubstituted benzenes. They interpret these data as arising from two factors — 'the change in the normal vibrational modes of the benzene ring upon deuterium substitution' and 'the electronic demand of the substituent: an increase in electron density at a given carbon (negative σ) being associated with an increased isotope shift'. However, the second factor is probably almost wholly a consequence of the first.

5 Shieldings of Particular Nuclear Species

A. Introduction.—As stated earlier, the limitations of space and the abundance of new data on the shieldings of other nuclei necessitate the exclusion of further discussion of any of the very large number of papers presenting new proton chemical shift data. However, the paper of Smith et al.[158] which was

---
[153] W. J. Stec, N. Goddard, and J. R. van Wazer, *J. Phys. Chem.*, 1971, **75**, 3547.
[154] H. Jensen and K. Schaumburg, *Mol. Phys.*, 1971, **22**, 1041.
[155] L. H. Sutcliffe and B. Taylor, *Spectrochim. Acta*, 1972, **28A**, 619.
[156] L. R. Womble and G. B. Savitsky, *J. Magn. Resonance*, 1971, **4**, 226.
[157] R. A. Bell, C. L. Chan, and B. G. Sayer, *J.C.S. Chem. Comm.*, 1972, 67.
[158] W. B. Smith, D. L. Deavonport, and A. M. Ihrig, *J. Amer. Chem. Soc.*, 1972, **94**, 1959.

not easily placed in any of the categories of Section 4 should be mentioned. This paper is concerned with the interpretation of substituent effects on the proton chemical shifts in benzene and naphthalene derivatives in terms of the empirical parameter Q [$= P(Ir^3)^{-1}$, where P is the C—X bond polarizability for a substituent X, I is the first ionization potential of X, and r is the C—X bond length].

B. **Carbon Chemical Shifts.**—It is generally agreed that the observed changes in the shielding of carbon nuclei on passing from compound to compound are caused almost entirely by changes in the local paramagnetic term σ_{loc}^p [see equation (22)]. For compounds of a given series RX, the shielding of a carbon nucleus in the group R is usually regarded as being changed largely because of changes in electron density at the nucleus, which occur as X is changed.[159] As the electron density on the carbon nucleus increases the electron cloud expands, the factor $\langle r^{-3} \rangle_{2p}$ in equation (22) falls, and there is an increased shielding. The factors ΔE and ΣQ_{AB} are usually regarded as being little changed along a given series such as that considered here. The mechanisms of charge-density transmission to or from the carbon atom are those discussed in the previous section, with the inductive and mesomeric effects being predominant. The concept of intramolecular electric fields is not usually introduced in explaining carbon shieldings. For some compounds steric effects appear to influence carbon shieldings markedly. Generally speaking shieldings for carbon atoms involved in multiple bonding are significantly lower than those for saturated carbon atoms. This is attributed to a fall in ΔE and hence a higher value of $|\sigma_{loc}^p|$ in the former case as compared with the latter.[160] A 'multiple-bond effect' can also influence the factor ΣQ_{AB} and lead to reduced shieldings.[160]

The picture given above has been challenged recently by the approach of Mason[161] which allows for large changes in σ_{loc}^d for carbon nuclei in some instances (*e.g.* the highly brominated and iodinated methyl halides) and also allows for ΔE changes upon varying a substituent (or the number of atoms of a substituent). None of the papers presenting new carbon shielding data and published during the review period adds in a substantial and quantitative manner to the interpretation outlined above. Below we discuss some of these papers and merely give references to the remainder.

In Volume 1 was given an up-to-date table of the carbon chemical shifts of a number of methyl halides including mixed halogen derivatives. The interesting feature here was that the expected brisk deshielding occurring upon passing along the series from CH_4 to CCl_4 did not occur for the series CH_4 to CI_4. The bromine series was intermediate, the shielding diminishing from CH_4 to CH_2Br_2 but then increasing again from CH_2Br_2 to CBr_4. This anomalous trend was tentatively interpreted in terms of an enhanced σ_{loc}^d contribu-

---

[159] J. M. Sichel and M. A. Whitehead, *Theor. Chim. Acta*, 1966, **5**, 35.
[160] J. A. Pople, *Mol. Phys.*, 1964, **7**, 301.
[161] J. Mason, *J. Chem. Soc. (A)*, 1971, 1038.

Table 8 Carbon relative shielding constants (in p.p.m., relative to CS_2), for chloro- and bromo-substituted ethanes and ethylenes.[162] For non-equivalent carbon nuclei, a superscript '13' has been affixed to denote the carbon nucleus to which the shift refers

Molecule	Cl	Br	Molecule	Cl	Br	Molecule	Cl	Br	Molecule	Cl	Br
CH_3CH_3	188.0	188.0	$^{13}CH_2XCH_3$	153.8	165.4	$^{13}CHX_2CH_3$	124.5	152.4	$^{13}CX_3CH_3$	97.5	155.2
$^{13}CH_3CH_2X$	175.0	173.4	CH_2XCH_2X	148.6	161.3	$^{13}CHX_2CH_2X$	121.4	150.8	$^{13}CX_3CH_2X$	96.1	149.0
$^{13}CH_3CHX_2$	161.2	158.0	$^{13}CH_2XCHX_2$	142.1	152.4	CHX_2CHX_2	118.2	143.4	$^{13}CX_3CHX_2$	92.8	140.3
$^{13}CH_3CX_3$	147.4		$^{13}CH_2XCX_3$	133.0	139.1	$^{13}CHX_2CX_3$	112.5	135.1	CX_3CX_3	87.6	

Molecule	Cl	Br	Molecule	Cl	Br	Molecule	Cl	Br
$CH_2{=}CH_2$			$^{13}CHX{=}CH_2$	67.6	79.0	$^{13}CX_2{=}CH_2$	66.6	96.7
$^{13}CH_2{=}CHX$	76.5	71.3	$^{13}CHX{=}CHX$, cis	74.4	79.1	$^{13}CX_2{=}CHX$	68.6	98.7
			$^{13}CHX{=}CHX$, trans	72.6	84.8			
$^{13}CH_2{=}CX_2$	80.4	66.5	$^{13}CHX{=}CX_2$	76.1	81.3	$CX_2{=}CX_2$	72.4	100.0

tion.[161] The recent result[44] for CF_4 suggests that an anomalous trend is present for the series CH_4 to CF_4 although the shieldings in CH_2F_2 and CHF_3 are as yet unknown. Abnormal results are also found[162] for some of the halogen derivatives of ethane and ethylene as shown in Table 8. For the ethyl chlorides and bromides increased halogen substitution at the adjacent carbon atom (columns of Table 8) always leads to a reduction in the carbon shielding. However, substitution at the carbon atom whose shielding is given (rows of Table 8), although also causing a reduced shielding for chlorine substitution, gives increased shielding as the third and final bromine is substituted. This difference between the trends of chlorine and bromine substitution becomes more marked for the carbon shielding in the ethylene derivatives which are also given in the Table. (The results for the ^{13}C shifts of cis- and trans-dichloroethylene given in Table 7 differ in order from those given by Maciel et al.,[163] who obtained 72.4 p.p.m. and 74.3 p.p.m. respectively relative to CS_2. The ^{13}C shift in ethylene is estimated[164] to be 70.4 p.p.m. above that of CS_2. This implies enhanced shielding for halogen substitution in several instances.) A useful empirical additivity scheme for the ^{13}C shifts of 1,2-disubstituted ethanes (XCH_2CH_2Y) has been proposed by Maciel et al.[165] based on the observed shifts of 54 such compounds. Although good agreement between observed and calculated ^{13}C shifts was obtained (somewhat poorer, however, for the substituents H and I), it was difficult to relate the empirical parameters of the scheme to any theory of ^{13}C shielding. ^{13}C shielding data for the carbon nuclei in acetylenic derivatives X—$C{\equiv}C$—Y have been given by Rosenberg and Drenth[75] and interpreted in terms of electron density changes (including charge polarization of the $C{\equiv}C$ bond) and magnetic anisotropy effects.

The ^{13}C shifts of monosubstituted cyclohexanes are tabulated and discussed by Pehk and Lippmaa[166] for some 36 substituents. All the substituents [apart from —$Sn(C_6H_{11})_3$ in tetracyclohexyltin] lead to a deshielding of the α-carbon atom relative to cyclohexane, with F and NO_2 having the largest effects (62.9 and 57.0 p.p.m. respectively). There appears to be a rough correlation between deshielding at the α-carbon atom and the substituent electronegativity. For the $C{\equiv}N$ group, however, the deshielding is only 0.5 p.p.m. which Pehk and Lippmaa attribute to CN-group magnetic anisotropy. Presumably this effect is present for other substituents such as NO_2. A very good correlation is found between the α-carbon shifts of $C_6H_{11}X$ and the secondary carbon shifts of $(CH_3)_2CHX$ as X changes. The β-carbons of $C_6H_{11}X$ are also deshielded, but by less than the α-carbon; the γ-carbons are generally shielded, as almost always are the δ-carbons. The mechanisms of

[162] G. Miyajima and K. Takahashi, *J. Phys. Chem.*, 1971, **75**, 331, 3766.
[163] G. E. Maciel, P. D. Ellis, J. J. Natterstadt, and G. B. Savitsky, *J. Magn. Resonance*, 1969, **1**, 589.
[164] G. B. Savitsky, P. D. Ellis, K. Namikawa, and G. E. Maciel, *J. Chem. Phys.*, 1968, **49**, 2395.
[165] G. E. Maciel, L. Simeral, R. L. Elliott, B. Kaufman, and K. Cribley, *J. Phys. Chem.*, 1972, **76**, 1466.
[166] T. Pehk and E. Lippmaa, *Org. Magn. Resonance*, 1971, **3**, 679.

Nuclear Shielding 39

these effects are not yet understood but as well as inductive effects (involving possibly alternation of charges) steric effects also appear to be important. Results roughly parallel to those for the ^{13}C shifts of cycylohexane derivatives are found for adamantane derivatives.[167] In particular there is a very close correlation between the α-carbon chemical shifts of the 1-substituted adamantanes (32) and the tertiary carbons of the compounds $(CH_3)_3CX$ as X changes.

(32)

Charge density would appear to be important, although the extended Hückel and Del Re semi-empirical theories for calculating electron densities are not entirely successful in correlating ^{13}C shifts with calculated electron densities for methylcyclohexanes and other compounds.[168] Important new data on the ^{13}C shifts of allene and its substituted derivatives have been given by Stein et al.[169] The only previously reported data were for allene itself[170] (CH_2=C=CH_2), and the extremely low shielding of the central carbon atom (-19.8 p.p.m. with respect to CS_2) has been confirmed. Apart from the substituents CH and CO_2H, monosubstitution leads to an increased shielding of the central carbon.

^{13}C shifts have been measured for the following compounds or classes of aromatic compounds: monohalogenobenzenes,[171] symmetrical *meta*-dihalogenobenzenes,[172] monosubstituted benzenes,[173] substituted bromobenzenes,[174] deuteriobenzonitriles,[175] 1,3,5-trimethyl-2-dichlorobenzene,[176] [^{15}N]quinoline,[177] nitropyrroles and nitroimidazoles,[178a] and anions and cations of nitropyrroles and nitroimidazoles.[178b] For the following organic compounds new ^{13}C data have become available: the C_5—C_8 paraffins and some of the C_9 isomers,[179] methylcyclopentanes, cyclopentanols, and cyclo-

[167] T. Pehk, E. Lippmaa, V. V. Sevostjanova, M. M. Krayuschkin, and A. I. Tarasova, *Org. Magn. Resonance*, 1971, **3**, 783.
[168] N. Cyr, A. S. Perlin, and M. A. Whitehead, *Canad. J. Chem.*, 1972, **50**, 814.
[169] R. Steur, J. P. C. M. Van Dongen, M. J. A. de Bie, W. Drenth, J. W. de Haan, and L. J. M. van de Ven, *Tetrahedron Letters*, 1971, 3307.
[170] R. A. Friedel and H. L. Retcofsky, *J. Amer. Chem. Soc.*, 1963, **85**, 1300.
[171] A. R. Tarpley and J. H. Goldstein, *J. Phys. Chem.*, 1972, **76**, 515.
[172] A. R. Tarpley and J. H. Goldstein, *J. Mol. Spectroscopy*, 1971, **39**, 275.
[173] A. M. Ihrig and J. L. Marshall, *J. Amer. Chem. Soc.*, 1972, **94**, 1756.
[174] J. F. Hinton and B. Layton, *Org. Magn. Resonance*, 1972, **4**, 353.
[175] G. J. Lebel, J. D. Laposa, B. G. Sayer, and R. A. Bell, *Analyt. Chem.*, 1971, **43**, 1500.
[176] R. Price, G. Schilling, L. Ernst, and A. Mannschreck, *Tetrahedron Letters*, 1972, 1689.
[177] P. S. Pregosin, E. W. Randall, and A. I. White, *J.C.S. Perkin II*, 1972, 1.
[178] E. Lippmaa, M. Mägi, S. S. Novikov, L. I. Khmelnitski, B. S. Prihodko, O. V. Lebedev, and L. V. Epishina, *Org. Magn. Resonance*, 1972, **4**, (a) 153; (b) 197.
[179] L. P. Lindemann and J. Q. Adams, *Analyt. Chem.*, 1971, **43**, 1245.

pentyl acetates,[180] uracil, thymine, and 5-halogenouracils,[181] isoborneol,[182] some t-butyl-1,3-dioxans,[183] acyclic alkenes,[184] piperidine and piperazine derivatives,[185] isopropenyl compounds,[186] substituted acrylic acids and their methyl esters,[187] allyl alcohols,[188] cyclopropane derivatives,[189] thiete sulphone,[190] several mesocyclic compounds,[191] isomeric 1,2,5-trimethyl-4-phenylpiperidine-4-ols,[192] $\alpha\beta$-unsaturated oxines,[193] and alcoylthiazoles.[194]

^{13}C shift data for a number of organometallic compounds and organophosphorus and organosilicon compounds have been published. These include studies of Group IVB metal carbonyls,[195] aquocyanocobyric acid and dicyanocobyric acid,[196] π-allylic palladium(II) complexes,[197] the complexes [LM(CO)$_3$] where L = mesitylene, durene, or cycloheptatriene and M = Cr, Mo, or W,[198] methylplatinum compounds,[199] π-cyclopentadienyliron carbonyls,[200] t-phosphine and t-arsine complexes of transition metals,[201] [1-methyl-3-ethyl-(σ-1,3-h^2-π-1,2,3-h^2)-allyl]nonacarbonyltriruthenium,[202] a range of organometallic compounds involving various functional groups,[203] ditertiary phosphines,[204] tri-2- and tri-3-thienylphosphine,[205] a number of tertiary phosphines,[206] 1,3,2-dioxaphosphorinan-2-ones,[207] cis- and trans-

[180] M. Christl, H. J. Reich, and J. D. Roberts, *J. Amer. Chem. Soc.*, 1971, **93**, 3463.
[181] A. R. Tarpley and J. H. Goldstein, *J. Amer. Chem. Soc.*, 1971, **93**, 3573.
[182] O. A. Gansow, M. R. Willcott, and R. E. Lenkinski, *J. Amer. Chem. Soc.*, 1971, **93**, 4295.
[183] A. J. Jones, E. L. Eliel, D. M. Grant, M. C. Knoeber, and W. F. Bailey, *J. Amer. Chem. Soc.*, 1971, **93**, 4772.
[184] D. E. Dorman, M. Jautelat, and J. D. Roberts, *J. Org. Chem.*, 1971, **36**, 2757.
[185] G. Ellis and R. G. Jones, *J.C.S. Perkin II*, 1972, 437.
[186] V. J. Bartuska and G. E. Maciel, *J. Magn. Resonance*, 1972, **7**, 36.
[187] H. Brouwer and J. B. Stothers, *Canad. J. Chem.*, 1972, **50**, 610.
[188] H. Brouwer and J. B. Stothers, *Canad. J. Chem.*, 1972, **50**, 1361.
[189] O. A. Subbotin, A. S. Kozmin, Y. K. Grishin, N. M. Sergeev, and I. G. Bolesov, *Org. Magn. Resonance*, 1972, **4**, 53.
[190] G. C. Levy and D. C. Dittmer, *Org. Magn. Resonance*, 1972, **4**, 107.
[191] T. T. Nakashima and G. E. Maciel, *Org. Magn. Resonance*, 1972, **4**, 321.
[192] A. J. Jones, A. F. Casy, and K. M. J. McErlane, *Tetrahedron Letters*, 1972, 1727.
[193] Z. W. Wolkowski, E. Vauthier, B. Gonbeau, H. Sauvaitre, and J. Musso, *Tetrahedron Letters*, 1972, 565.
[194] R. Garnier, R. Faure, A. Babadjomian, and E. Vincent, *Bull. Soc. chim. France*, 1972, 1040.
[195] O. A. Gansow, B. Y. Kimura, G. R. Dobson, and R. A. Brown, *J. Amer. Chem. Soc.*, 1971, **93**, 5924.
[196] D. Doddrell and A. Allerhand, *Chem. Comm.*, 1971, 728
[197] B. E. Mann, R. Pietropaolo, and B. L. Shaw, *Chem. Comm.*, 1971, 790.
[198] B. E. Mann, *Chem. Comm.*, 1971, 976.
[199] M. H. Chisholm, H. C. Clark, L. E. Manzer, and J. B. Stothers, *Chem. Comm.*, 1971, 1627.
[200] O. A. Gansow, D. A. Schexnayder, and B. Y. Kimura, *J. Amer. Chem. Soc.*, 1972, **94**, 3406.
[201] B. E. Mann, B. L. Shaw, and R. E. Stainbank, *J.C.S. Chem. Comm.*, 1972, 151.
[202] M. Evans, M. Hursthouse, E. W. Randall, E. Rosenberg, L. Milone, and M. Valle, *J.C.S. Chem. Comm.*, 1972, 545.
[203] L. F. Farnell, E. W. Randall, and E. Rosenberg, *Chem. Comm.*, 1971, 1078.
[204] H. Marsmann and H. Horn, *Z. Naturforsch*, 1972, **27b**, 137.
[205] H. J. Jakobsen, T. Bundgaard, and R. S. Hansen, *Mol. Phys.*, 1972, **23**, 197.
[206] B. E. Mann, *J.C.S. Perkin II*, 1972, 30.
[207] A. A. Borisenko, N. M. Sergeev, E. Ye. Nifantiev, and Yu. A. Ustynyuk, *J.C.S. Chem. Comm.*, 1972, 406.

substituted ethylene phosphines,[208] some phosphadiazoles,[209] and some bicyclo[2,2,1]heptyl *o*-trimethylsilyl ethers.[210]

Levy and Cargioli[211] have presented ^{13}C shifts between tetramethylsilane and a number of common standard reference compounds for ^{13}C work. These will be useful in relating literature data obtained with different standards. Solvent isotope effects appear to be significant, if small (1 p.p.m.), as do solvent effects. Thus in CCl_4 carbon disulphide is deshielded by 192.4 p.p.m. relative to $SiMe_4$ whereas in C_6H_{12} the deshielding is 193.0 p.p.m. Ziessow and Carroll[212] have also been concerned with the question of referencing ^{13}C shifts and conclude that cyclohexane is a much better reference than $SiMe_4$. Jensen and Petrakis[213] have given a chart showing the ranges of ^{13}C shifts occurring for several varieties of organic compounds.

^{13}C shielding data have also been reported for the following systems which are of organic or biological interest: branched-chain carbohydrates,[214] porphyrins,[215] carbocations,[216–219] santonin derivatives,[220] some common oligosaccharides,[221] two nitroxide radicals,[222] some alkaloids,[223] some di- and tri-substituted ethylenes,[224] narciclasine tetra-acetate,[225] dimethylcyclohexanes,[226] mutarotating sugars,[227] the ditropylium ion,[228] and tetrahydrocannabinol and its isomers.[229]

An empirical relation between carbon shielding and bond angle for sp^3-hybridized carbon atoms has been used to predict quite well bond angles in some simple compounds[230] — further discussion of the method used is given in Section 5D. The relation between bond angle and carbon shielding in

---

[208] M. Simonnin, R. Lequan, and F. W. Wehrli, *Tetrahedron Letters*, 1972, 1559.
[209] V. V. Negrebetskii, A. V. Kessenikh, A. F. Vasiliev, N. P. Ignatova, N. I. Shvetsov-Shilovskii, and N. M. Melnikov, *J. Struct. Chem.*, 1971, **12**, 731.
[210] H. Schneider, *J. Amer. Chem. Soc.*, 1972, **94**, 3636.
[211] G. C. Levy and J. D. Cargioli, *J. Magn. Resonance*, 1972, **6**, 143.
[212] D. Ziessow and M. Carroll, *Ber. Bunsengesellschaft phys. Chem.*, 1972, **76**, 61.
[213] R. K. Jensen and L. Petrakis, *J. Magn. Resonance*, 1972, **7**, 105.
[214] A. Sepulchre, G. Lukacs, G. Vass, and S. D. Gero, *Angew. Chem. Internat. Edn.*, 1972, **11**, 148.
[215] D. Doddrell and W. S. Caughey, *J. Amer. Chem. Soc.*, 1972, **94**, 2510.
[216] G. A. Olah, P. R. Clifford, Y. Halpern, and R. G. Johanson, *J. Amer. Chem. Soc.*, 1971, **93**, 4219.
[217] G. A. Olah, Y. K. Mo, and Y. Halpern, *J. Amer. Chem. Soc.*, 1972, **94**, 3551.
[218] G. A. Olah, R. A. Schlosberg, R. D. Porter, Y. K. Mo, D. P. Kelly, and G. D. Mateescu, *J. Amer. Chem. Soc.*, 1972, **94**, 2034.
[219] G. A. Olah, R. D. Porter, C. L. Juell, and A. M. White, *J. Amer. Chem. Soc.*, 1972, **94**, 2044.
[220] P. S. Pregosin, E. W. Randall, and T. B. H. McMurry, *J.C.S. Perkin I*, 1972, 299.
[221] D. E. Dorman and J. D. Roberts, *J. Amer. Chem. Soc.*, 1971, **93**, 4463.
[222] G. F. Hatch and R. Kreilick, *Chem. Phys. Letters*, 1971, **10**, 490.
[223] P. W. Sprague, D. Doddrell, and J. D. Roberts, *Tetrahedron*, 1971, **27**, 4857.
[224] J. W. De Haan and L. J. M. van der Ven, *Tetrahedron Letters*, 1971, 3965.
[225] L. Zetta, G. Gatti, and G. Fuganti, *Tetrahedron Letters*, 1971, 4447.
[226] H. Schneider, R. Price, and T. Keller, *Angew. Chem. Internat. Edn.*, 1971, **10**, 730.
[227] W. Voelter, E. Breitmaier, and G. Jung, *Angew. Chem. Internat. Edn.*, 1971, **10**, 935.
[228] H. Volz and M. Volz-de Lecea, *Annalen*, 1971, **750**, 136.
[229] E. Wenkert, D. W. Cochran, F. M. Schell, R. A. Archer, and K. Matsumoto, *Experientia*, 1972, **28**, 250.
[230] D. Purdela, *J. Magn. Res.*, 1971, **5**, 37.

tertiary phosphines has been pointed out by Mann.[206] A good correlation occurs between the ^{13}C shifts of the methyl halides and the inner-shell binding energies of the halogenomethanes,[231] although the exception for CH_3F appears to indicate some double-bond character to the C—F bond. Theoretical calculations of ^{13}C shieldings have been made for a number of methyl compounds,[232] a number of alkanes,[233] and for the CH_5^+ and $C_2H_5^+$ ions.[234] The results for CH_5^+ are of particular interest and are shown in Table 9.

Table 9 Calculated ^{13}C shieldings for CH_5^+ ions

Molecule	σ (p.p.m.), calc.
CH_4	222.1
CH_5^+ (trigonal-bipyramidal), D_{3h}	253.4
CH_5^+ (square-pyramidal), C_{4v}	250.2
CH_5^+ (approx. CH_3^+—H_2), C_s	173.1

Ditchfield and Miller note that charge density at the carbon atom (as measured by gross atomic populations) is insufficient to account for this trend and that one must allow for the larger number of excited states for the C_s structure that can be mixed in by the magnetic field, as compared with the other structures. In effect, ΔE is smaller for the C_s structure so that the paramagnetic shielding is numericaly larger. Although no experimental data are available for CH_5^+ shielding, values for related species are lower than for methane, indicating that $CH_5^+(C_s)$ is the most likely structure.

C. Fluorine Chemical Shifts.—Shielding changes of ^{19}F nuclei in different compounds are generally regarded as being almost entirely caused by changes of σ_{loc}^p. However, unlike ^{13}C shielding, changes of which are most often interpreted as reflecting changes in the $\langle r^{-3} \rangle_{2p}$ factor of equation (22), ^{19}F shifts are usually taken to reflect changes in the $\sum_B Q_{AB}$ factor of this equation. Thus as compared with a conventional single X—F bond, structures such as $X^{\delta+}$—$F^{\delta-}$, in which the electronic distribution about the ^{19}F nucleus approximates more to that of the F^- ion, are interpreted as producing a smaller $\sum_B Q_{AB}$ and hence a more shielded fluorine nucleus. Structures such as $X^{\delta-}=F^{\delta+}$, in which the p-electron distribution is even more unbalanced than in the X—F bond, are interpreted as producing a greater ΣQ_{AB} and hence a less shielded fluorine nucleus.

That the real situation is more complicated than this simple picture allows for is made evident by some recent work. De Marco and Gatti[235] have observed that substitution of bromine for chlorine at X in CF_3CFXCF_2X produces a deshielding of the fluorines. This deshielding is especially marked

---

[231] R. E. Block, *J. Magn. Res.*, 1971, **5**, 155.
[232] R. Radeglia and E. Gey, *J. prakt. Chem.*, 1972, **313**, 1070.
[233] V. N. Solkan, V. M. Mamayev, N. M. Sergeev, and Yu. A. Ustynyuk, *Org. Magn. Resonance*, 1971, 3, 567.
[234] R. Ditchfield and D. P. Miller, *J. Amer. Chem. Soc.*, 1971, **93**, 5287.
[235] A. De Marro and G. Gatti, *J. Magn. Resonance*, 1972, **6**, 200.

Nuclear Shielding

for the CF_2 fluorine nuclei, which may reflect intramolecular dispersion effects. A similar effect has been observed by Sutcliffe and Taylor[155] who find that successive substitution of fluorine for chlorine in the CCl_3 group of $CFCl_2COCCl_3$ leads to an increased shielding of the fluorine nucleus of the $CFCl_2$ group. The importance of the 'through-space' electric field effect on ^{19}F shielding, which may be involved here, has been stressed by Emsley and Phillips in their review.[236]

The work of Rodmar et al. on the ^{19}F shifts of the fluorothiophens has already been mentioned[50,51,53] in Section 4C. ^{19}F chemical shift studies have been made for the following classes of compound: fluoronaphthalene derivatives,[237-239] some fluorinated benzofurans,[240] polyfluoro-1,4-dioxans and -1,4-oxathians,[241] fluoromonosaccharides,[242] deoxyfluoro-D-glucopyranoses,[243] amino-acids,[244] methyl-substituted piperidylfluorophosphoranes,[245] (fluoroaryl)fluorophosphonitriles,[246] some perfluoroalkyl-substituted cyclic phosphines,[247] diazadiphosphetidines and phosphadiazetidinones,[248] arylpolyfluoroalkenylsulphides,[249] chlorofluorosilanes,[250] some fluorosilylamines,[251] some boron trifluoride adducts,[252] mixed tetrahalogenoborate ions,[253] fluorophenylsilanes,[254] trifluoromethylphosphine complexes with molybdenum, chromium, and rhodium,[255] and bis(perfluorovinyl)mercury.[256] ^{19}F chemical shifts have also been calculated for d^{10}- and f^{14}-metal fluorides[257] and have been used to study the *trans* effect in some Co^{III} complexes.[258]

---

[236] J. W. Emsley and L. Phillips, *Progr. N.M.R. Spectroscopy*, 1971, **7**, 1.
[237] W. Adcock, D. G. Matthews, and S. Q. A. Rizvi, *Austral. J. Chem.*, 1971, **24**, 1829.
[238] W. Adcock, S. Q. A. Rizvi, W. Kitching, and A. J. Smith, *J. Amer. Chem. Soc.*, 1972, **94**, 369.
[239] W. Adcock, S. Q. A. Rizvi, and W. Kitching, *J. Amer. Chem. Soc.*, 1972, **94**, 3657.
[240] R. J. Abraham, D. F. Wileman, G. R. Bedford, and D. Greatbanks, *Org. Magn. Resonance*, 1972, **4**, 343.
[241] J. Burdon and I. W. Parsons, *Tetrahedron*, 1971, **27**, 4553.
[242] P. W. Kent, R. A. Dwek, and N. F. Taylor, *Tetrahedron*, 1971, **27**, 3887.
[243] L. Phillips and V. Wray, *J. Chem. Soc. (B)*, 1971, 1618.
[244] E. Bayer, P. Hunziker, M. Mutter, R. E. Sievers, and R. Uhmann, *J. Amer. Chem. Soc.*, 1972, **94**, 265.
[245] M. J. C. Hewson, S. C. Peake, and R. Schmutzler, *Chem. Comm.*, 1971, 1454.
[246] T. Chivers and N. L. Paddock, *Inorg. Chem.*, 1972, **11**, 848.
[247] H. G. Ang, M. E. Redwood, and B. O. West, *Austral. J. Chem.*, 1972, **25**, 493.
[248] R. K. Harris, J. R. Woplin, R. E. Dunmur, M. Murray, and R. Schmutzler, *Ber. Bunsengesellschaft phys. Chem.*, 1972, **76**, 44.
[249] N. M. Sergeev, A. M. Aleksandsov, and L. M. Yagupolskii, *J. Struct. Chem.*, 1971, **12**, 150.
[250] K. Hamada, G. A. Ozin, and E. A. Robinson, *Bull. Chem. Soc. Japan*, 1971, **44**, 2555.
[251] W. Airey, G. M. Sheldrick, B. J. Aylett, and I. A. Ellis, *Spectrochim. Acta*, 1971, **27A**, 1505.
[252] R. C. Stephens, S. D. Lessley, and R. O. Ragsdale, *Inorg. Chem.*, 1971, **10**, 1610.
[253] J. S. Hartman and G. J. Schrobilgen, *Inorg. Chem.*, 1972, **11**, 940.
[254] J. Lipowitz, *J. Amer. Chem. Soc.*, 1972, **94**, 1582.
[255] J. F. Nixon and J. R. Swain, *J.C.S. Dalton*, 1972, 1038, 1044.
[256] R. B. Johannesen and R. W. Duerst, *J. Magn. Resonance*, 1971, **5**, 355.
[257] V. M. Bouznik and L. M. Avkhutsky, *J. Magn. Resonance*, 1971, **5**, 63.
[258] H. A. O. Hill, K. G. Morallee, F. Cernivez, and G. Pellizer, *J. Amer. Chem. Soc.*, 1972, **94**, 277.

Hartman and Schrobilgen[253] stress the importance of a reduction in ΔE occurring when heavier halogens are substituted for some of the fluorine nuclei in the BF_4^- ion in order to explain the reduced shielding of the remaining fluorine nuclei. Thus in dichloromethane solution they observed the following trend in σ relative to that in BF_4^- (in p.p.m.):

BF_3Cl^- $\quad\quad\quad\quad\quad\quad\quad\quad$ BF_3Br^-
-26.7 $\quad\quad\quad\quad\quad\quad\quad\quad$ -37.5

$BF_2Cl_2^-$ $\quad\quad\quad\quad$ BF_2ClBr^- $\quad$ $BF_2Br_2^-$
-47.2 $\quad\quad\quad\quad\quad$ -55.8 $\quad\quad\quad$ -63.1

$BFCl_3^-$ $\quad\quad$ $BFCl_2Br^-$ $\quad\quad\quad\quad\quad\quad$ $BFBr_3^-$
-57.3 $\quad\quad\quad$ -63.5 $\quad\quad\quad\quad\quad\quad\quad\quad$ -72.7

Different degrees of multiple bonding of fluorine to boron may also be important here. Adcock et al.[239] present evidence in favour of a hyperconjugative mechanism for the alternation of ^{19}F shifts by dimethylsilyl substituents $\diagdown$SiMe$_2$,$\diagup$ rather than an inductive effect. The work of Bouznik and Avkhutsky[257] on ^{19}F shielding of ionic fluorides is of particular interest and supports the Kondo–Yamashita model[259,260] which interprets deshielding trends as being caused by overlap forces which increase the importance of the local paramagnetic shielding. This is essentially a steric effect and may be of considerable use in understanding both intermolecular and intramolecular effects on ^{19}F shielding in covalent compounds.

D. Phosphorus Chemical Shifts.—The theory of phosphorus chemical shifts has been treated much more systematically[261] than that of any other comparable nucleus. The treatment of Letcher and van Wazer is concerned solely with σ_{loc}^p and leads to an expression essentially the same as equation (22) but taking into account the contributions of d-electrons as well as p-electrons. Thus,[262] with a slight notation change,

$$\sigma - \sigma_0 = -\frac{\mu_0}{4\pi}\frac{e^2\hbar^2}{3\Delta E m^2}[\langle r^{-3}\rangle_p \zeta_1 + \langle r^{-3}\rangle_d \zeta_2] \quad (34)$$

$$= B(v)\zeta_1 + B(v)f(v)\zeta_2 \quad (35)$$

where the negative quantity $B(v)$ is proportional to $\langle r^{-3}\rangle_p/\Delta E$ and is regarded as being a constant for a given phosphorus co-ordination number v, f is the quantity $\langle r^{-3}\rangle_d/\langle r^{-3}\rangle_p$ (which is always positive) and ζ_1 and ζ_2 are the sums of coefficients obtainable from the charge-density bond-order matrix. A particularly simplifying fact is that for almost all phosphorus compounds the σ-bonds involve only s and p orbitals which therefore contribute via the term

[259] I. Kondo and J. Yamashita, *J. Phys. and Chem. Solids*, 1959, **10**, 245.
[260] Y. Yamagata, *J. Phys. Soc. Japan*, 1964, **19**, 10.
[261] J. H. Letcher and J. R. van Wazer, 'Topics in Phosphorus Chemistry, ^{31}P Nuclear Magnetic Resonance', Interscience, New York, 1967, vol. 5, ch. 2 and 3.
[262] Ref. 261, p. 114.

$B(v)\zeta_1$, whereas the d-orbitals are confined to π-bonds which therefore contribute via the term $B(v)f(v)\zeta_1$. Consequently, the need to invoke the presence of π-bonding occurs whenever the term $B(v)\zeta_1$ is sufficient to account for the observed shift. Since $B(v)$ is negative and $f(v)$ and ζ_2 are positive it is clear that π-bonding produces a deshielding when present. For example, Letcher and van Wazer treated the compounds PH_3 and PMe_3 as being free from π-bonding and thus were able to determine $B(v)$ for PX_3 compounds. From this significant amounts of π-bonding appear to be present in PCl_3, PBr_3, PI_3, P_4S_3, and $P(SMe)_3$ since they were all considerably deshielded relative to the values predicted when the term $B(v)f(v)\zeta_2$ was ignored. (Incidentally, the value for σ_0 of 11 828.5 p.p.m. which is attributed to the ^{31}P nuclei in 85% H_3PO_4 cannot be equated with the shielding constant of the ^{31}P nuclei in this solution which is approximately 350 p.p.m. and can be obtained by subtracting from the 594 p.p.m. obtained for phosphine by Davies et al.,[14] a value of 240 p.p.m., which is the amount by which phosphine is more shielded than the ^{31}P nuclei in the orthophosphoric acid solution.) The value of ζ_1 is determined by the bond polarities of σ-bonds and also by the molecular geometry, which becomes more important as the polarities of the various bonds differ more and more.

During the review period new ^{31}P chemical shift data have been published for the following compounds or classes of compound: P_4 vapour,[263] PBr_3 vapour,[264] compounds of the types $RC{\equiv}C—PR_2^1$ and $RC{\equiv}C—POR_2$,[75] cyclic phosphines,[247] ditertiary phosphines, phosphorous(v) nitride chlorides,[265] some phospholes,[266] salts of O-alkyl alkylphosphonothionoic acids,[267] $[PCl_nBr_{4-n}]^+$ ions,[268] tertiary phosphines, tertiary phosphine complexes with various metals,[269–273] phosphine or phosphite complexes of platinum,[274] cyclopentadienylnickel complexes,[275] substituted diethyl phenylphosphonates,[276] some trialkylphosphine complexes of platinum,[277] some solid-state phosphorous compounds,[278] cyclic derivatives of phosphorus oxyacids,[279] phosphadiazoles, and phosphorus–silicon compounds.[280]

[263] G. Heckmann and E. Fluck, Mol. Phys., 1972, 23, 175.
[264] G. Heckmann, Mol. Phys., 1972, 23, 627.
[265] W. Haubold and E. Fluck, Z. Naturforsch, 1972, 27b, 368.
[266] F. Mathey and R. Mankowski-Favelier, Org. Magn. Resonance, 1972, 4, 171.
[267] M. Mikołajczyk and J. Omelańczuk, Tetrahedron Letters, 1972, 1539.
[268] K. B. Dillon and P. N. Gates, J.C.S. Chem. Comm., 1972, 348.
[269] B. E. Mann, C. Masters, and B. L. Shaw, J. C. S. Dalton, 1972, 704.
[270] C. Masters, B. L. Shaw, and R. E. Stainbank, J. C. S. Dalton, 1972, 664.
[271] B. E. Mann, C. Masters, B. L. Shaw, and R. E. Stainbank, Chem. Comm., 1971, 1103.
[272] B. E. Mann, Chem. Comm., 1971, 1173.
[273] B. E. Mann, B. L. Shaw, and R. M. Slade, J. Chem. Soc. (A), 1971, 2976.
[274] F. H. Allen and S. N. Sze, J. Chem. Soc. (A), 1971, 2054.
[275] J. Thomson, D. Groves, and M. C. Baird, J. Magn. Resonance, 1971, 5, 281.
[276] C. C. Mitsch, L. D. Freedman, and C. G. Moreland, J. Magn. Resonance, 1971, 5, 140.
[277] D. H. Gerlach, A. R. Kane, G. W. Parshall, J. P. Jesson, and E. L. Muetterties, J. Amer. Chem. Soc., 1971, 93, 3543.
[278] K. B. Dillon and T. C. Waddington, Spectrochim. Acta, 1971, 27A, 1381.
[279] G. M. Blackburn, J. S. Cohen, and I. Weatherall, Tetrahedron, 1971, 27, 2903.
[280] D. Purdela, J. Magn. Resonance, 1971, 5, 23.

The treatment of Purdela[280] is an empirical approach which enables X—P—X bond angles to be predicted quite well from ^{31}P chemical shifts. This approach involves approximations which are not only severe but most unusual. For instance the part of the paramagnetic shielding corresponding to the numerator of equation (5) of the present Report averaged over all orientations is taken to be dependent on $\langle 0 | \sum_k r_k^{-1} | 0 \rangle$. Also the procedure whereby the coefficient of this factor (involving the relation between the X—P—X bond angle in YPX$_3$, PX$_4$, and PX$_5$ compounds and an angle ϕ_β) is obtained does not appear to be adequately explained in physical terms. For the ^{31}P chemical shifts of tertiary phosphines Mann suggests[272] the importance of bond-angle changes and the relative lack of importance of substituent electronegativity (although highly electronegative substituents were not considered). Although changes in the ^{31}P shielding of tertiary phosphines on co-ordination to transition metals are not understood there appears to be a fairly linear correlation between the change for a given type of complex and the ^{31}P chemical shift of the free phosphine.[281]

E. Nitrogen Chemical Shifts.—Nitrogen chemical shifts are usually rationalized[282] in terms of changes in σ^p_{loc} as given in equation (22), although it has been suggested that variations in σ^p_{loc} can be significant.[161] New nitrogen chemical shift data have been published for the following: pyridine,[283] aniline derivatives,[284] the enamino-ketone quaternary ammonium salts,[285] a selection of simple organonitrogen compounds,[286] amino-acid derivatives,[287] some six-membered aromatic heterocycles,[288] azidotriphenyl compounds,[289] amines,[290] covalent azides,[291] [^{15}N]quinoline,[177] ring-substituted benzamides and benzonitriles,[292] nitropyrroles and nitroimidazoles and their charged derivatives,[178] azoles and their benzo-derivatives,[293] linear triatomic nitrogen-containing species,[294] and cobalt(III)-penta-ammine complexes.[295]

That the factors influencing nitrogen shielding are the same as those

[281] B. E. Mann, C. Masters, B. L. Shaw, R. M. Slade, and R. E. Stainbank, *Inorg. Nuclear Chem. Letters*, 1971, **7**, 881.
[282] E. W. Randall and D. G. Gillies, *Progr. N.M.R. Spectroscopy*, 1971, **6**, 119.
[283] R. L. Lichter and J. D. Roberts, *J. Amer. Chem. Soc.*, 1971, **93**, 5218.
[284] T. Axenrod, P. S. Pregosin, M. J. Wieder, E. D. Becker, R. B. Bradley, and G. W. A. Milne, *J. Amer. Chem. Soc.*, 1971, **93**, 6536.
[285] J. Dabrowski, K. Kamiénska-Trela, and A. J. Sadlej, *Org. Magn. Resonance*, 1971, **3**, 589.
[286] J. M. Briggs, L. F. Farnell, and E. W. Randall, *Chem. Comm.*, 1971, 680.
[287] P. S. Pregosin, E. W. Randall, and A. I. White, *Chem. Comm.*, 1971, 1602.
[288] M. Witanowski, L. Stefaniak, H. Januszewski, and G. A. Webb, *Tetrahedron*, 1971, **27**, 3129.
[289] P. N. Preston, L. H. Sutcliffe, and B. Taylor, *Spectrochim. Acta*, 1972, **28A**, 197.
[290] R. L. Lichter and J. D. Roberts, *J. Amer. Chem. Soc.*, 1972, **94**, 2495.
[291] W. Beck, W. Becker, K. F. Chew, W. Derbyshire, N. Logan, D. M. Revitt, and D. B. Sowerby, *J.C.S. Dalton*, 1972, 245.
[292] P. S. Pregosin, E. W. Randall, and A. I. White, *J.C.S. Perkin II*, 1972, 513.
[293] M. Witanowski, L. Stefaniak, H. Januszewski, and G. A. Webb, *Tetrahedron*, 1972, **28**, 637.
[294] K. F. Chew, W. Derbyshire, and N. Logan, *J.C.S. Faraday II*, 1972, **68**, 594.
[295] J. W. Lehman and B. M. Fung, *Inorg. Chem.*, 1972, **11**, 214.

influencing carbon shielding is given support by a plot[290] of the ^{15}N chemical shifts of the aliphatic amines RNH$_2$ against the ^{13}C chemical shifts of the corresponding compounds R^{13}CH$_3$. A very close correlation is observed not only for primary but also for secondary amines. It is unclear from the correlation, however, whether it is the $\langle r^{-3} \rangle$ factor or the ΣQ_{AB} factor which is responsible for the correlations. The nitrogen shielding is more sensitive to changes in R in the compounds RNH$_2$ than is the carbon in the compounds RCH$_3$. It is pointed out here that this seems to suggest the importance of the $\langle r^{-3} \rangle$ factor, which may be more sensitive for nitrogen than for carbon on account of the larger nuclear charge. Lichter and Roberts are also able to propose substituent parameters for the ^{15}N chemical shifts of the primary alkylamines.

Chew et al.[294] have attempted to calculate nitrogen chemical shifts in a number of linear triatomic species. This involved calculations of the various factors of equation (22). Fair agreement was obtained, being improved marginally by making allowances for changes in σ^d_{loc} in the manner suggested by Grinter and Mason.[296]

F. Chemical Shifts of Nuclei Other than H, C, F, P, and N.—Only a very small number of papers involving shielding data for nuclei other than H, C, F, P, and N has appeared in the review period. They deal with the following: boron shielding in pentaborane(9),[297] BX$_4^-$ ions,[253] boron trihalides and mixed trihalides,[298] and substituted borazines;[299] tin shielding in some organotin compounds;[300,301] oxygen shielding in nitropyrroles,[178] germanium shielding in a number of compounds;[302] manganese shielding for compounds with manganese–tin bonds;[303] chlorine and bromine shielding in copper monohalides;[304] and niobium shielding in niobium halides.[305]

An interesting discussion of the relations between ^{11}B and ^{13}C chemical shifts of four co-ordinate boron and carbon compounds respectively has appeared.[306] Spielvogel and Purser find a good linear correlation which can be used to predict unknown ^{11}B chemical shifts. Thus

$$(\eta_{i,j})_{^{11}B} = 0.398(\eta_{i,j})_{^{13}C} - 3.12 \quad (36)$$

where the $(\eta_{i,j})$ are pairwise additivity parameters. The factor 0.398 is very

[296] R. Grinter and J. Mason, *J. Chem. Soc.* (A), 1970, 2196.
[297] J. D. Odom, P. D. Ellis, and H. C. Walsh, *J. Amer. Chem. Soc.*, 1971, **93**, 3529.
[298] M. F. Lappert, M. R. Litzow, J. B. Pedley, and A. Tweedale, *J. Chem. Soc.* (A), 1971, 2426.
[299] O. T. Beachley, *J. Amer. Chem. Soc.*, 1971, **93**, 5066.
[300] P. G. Harrison, S. E. Ulrich, and J. J. Zuckerman, *J. Amer. Chem. Soc.*, 1971, **93**, 5398.
[301] A. P. Tupčiauskas, N. M. Sergeyev, and Yu. A. Ustynyuk, *Org. Magn. Resonance*, 1971, **3**, 655.
[302] J. Kaufman, W. Sahm, and A. Schwenk, *Z. Naturforsch*, 1971, **26a**, 1384.
[303] S. Onaka, T. Miyamoto, and Y. Sasaki, *Bull. Chem. Soc. Japan*, 1971, **44**, 1851.
[304] V. M. Bouznik and L. G. Falaleeva, *J. Magn. Resonance*, 1972, **6**, 197.
[305] Y. A. Buslaev, V. D. Kopanev, and V. P. Tarasov, *Chem. Comm.*, 1971, 1175.
[306] B. F. Spielvogel and J. M. Purser, *J. Amer. Chem. Soc.*, 1971, **93**, 4418.

close to an estimate of 0.406 for the ratio of the predicted values of σ^p_{loc}, suggesting that the local paramagnetic term is dominant in determining ^{11}B shieldings. The importance of variations in the local diamagnetic term would not manifest themselves in equation (36) except as a contribution to the constant term of -3.12, provided that they are the same for the boron ion BX_4^- as for CX_4. The close correlation implies that this is so. The ^{11}B shielding data given by Spielvogel and Purser for the halogen derivatives of $Et_3NBH_{3-n}X_n$ run a very close parallel to the ^{13}C shielding in the halogen derivative $CH_{4-n}X_n$ including the reversal in the trend to lower shielding for heavy halogen substitution. A somewhat different anomaly appears in the halogen and mixed-halogen derivatives of $[BH_4]^-$, in which the ^{11}B shielding appears[253] to increase in the order $Cl < F < Br \ll I$. Thus, relative to external $(MeO)_3B$, the shielding constants (in p.p.m.) are shown in Table 10.

Table 10 *Shielding constants for halogenoborate ions*

BF_4^-	BF_3Cl^-	$BF_2Cl_2^-$	$BFCl_3^-$
19.9	16.6	13.8	11.8
BCl_4^-	BCl_3Br^-	$BCl_2Br_2^-$	$BClBr_3^-$
11.6	15.4	22.1	30.9
	BCl_3I^-	$BCl_2I_2^-$	$BClI_3^-$
	30.9	60.8	99.1
BBr_4^-	BBr_3I^-	$BBr_2I_2^-$	$BBrI_3^-$
42.4	62.8	85.8	115.3
BI_4^-			
146.0			

This same anomalous trend is present, although to a smaller extent in the neutral mixed halides.[298] Values, in p.p.m., relative to BF_3,OEt_2, are given in Table 11.

Table 11 *Shielding constants for boron halides*

BF_3	BF_2Cl	$BFCl_2$
-10.0	-19.8	-32.3
BCl_3	BCl_2Br	$BClBr_2$
-46.5	-44.7	-42.3
BBr_3	BBr_2I	$BBrI_2$
-38.7	-26.3	-11.0
BI_3		
$+7.9$		

6 Chemical Shift Anisotropy

As yet there is no complete theory of chemical shift anisotropy which relates the observed parameter (e.g. $\sigma_\| - \sigma_\perp$ for a cylindrically symmetrical situation) to established chemical concepts. The theoretical work of Cook et al.,[24] which gave values of $\sigma_\|$ and $\sigma_\perp$ for the hydrogen molecule, has been discussed earlier. The calculations predict a maximum shielding anisotropy for H_2 of about 2.38 p.p.m. at a bond length of 0.9 a.u. (0.0476 nm). New shielding anisotropy values have been measured for the following: the protons in cyclopentadienyltricarbonylmanganese,[307] cobalt in octacarbonyldicobalt,[308] cobalt and manganese in $CoMn(CO)_9$ (together with the values of the principal components of each shielding tensor),[309] fluorine in C_6F_6 and in $[CF_3CO_2]^-Ag^+$,[310] fluorine in fluoroapatite (together with the principal components of the shielding tensor),[311] fluorine in CH_2F_2,[312] phosphorus in P_4S_3, P_4S_{10}, Zn_3P_2, and Mg_3P_2,[313] protons in the four monohalogeno-derivatives of methane,[314] carbon in benzene,[315,316] and carbon in CS_2.[317] Good agreement is obtained for the two values of ^{13}C shielding anisotropy in benzene, viz. 180 and 190 p.p.m. respectively, the carbon being more shielded when the external magnetic field is perpendicular to the molecular plane.

[307] J. C. Lindon and B. P. Dailey, *Mol. Phys.*, 1971, **22**, 465.
[308] E. S. Mooberry, M. Pupp, J. L. Slater, and R. K. Sheline, *J. Chem. Phys.*, 1971, **55**, 3655.
[309] E. S. Mooberry and R. K. Sheline, *J. Chem. Phys.*, 1972, **56**, 1852.
[310] M. Mehring, R. G. Griffin, and J. S. Waugh, *J. Chem. Phys.*, 1971, **55**, 746.
[311] J. L. Carolan, *Chem. Phys. Letters*, 1971, **12**, 389.
[312] S. G. Kukolich and A. C. Nelson, *J. Chem. Phys.*, 1972, **56**, 4446.
[313] M. G. Gibby, A. Pines, W. K. Rhim, and J. S. Waugh, *J. Chem. Phys.*, 1972, **56**, 991.
[314] K. Hayamizu and O. Yamamoto, *J. Magn. Resonance*, 1971, **5**, 94.
[315] C. S. Yannoni and H. E. Bleich, *J. Chem. Phys.*, 1971, **55**, 5406.
[316] G. Englert, *Z. Naturforsch*, 1972, **27a**, 715.
[317] H. W. Spiess, D. Schweitzer, U. Haeberlin, and K. H. Hausser, *J. Magn. Resonance* 1971, **5**, 101.

2
Nuclear Spin–Spin Coupling

BY R. GRINTER

1 Introduction

In general outline this chapter follows the pattern set in Volume 1 of these Reports, but it does differ in one important respect. At that time there seemed to be good reasons for not attempting to divide couplings over four or more bonds into ^4J, ^5J, ^6J *etc.*, and these were therefore placed in one section—though an attempt was made to divide the field into saturated and unsaturated systems. In retrospect, however, it does appear worthwhile to attempt a separation of the longer-range couplings into groups depending upon the smallest number of bonds which separate the coupled nuclei, and this has been done in the present chapter. Many couplings in aromatic systems remain grouped together in Section 10.

2 Basic Theory: Notation and Abbreviations

The theory on which this chapter is based was outlined in Section 2 of Chapter 2 of Volume 1, and need not be repeated here. However, it is probably worthwhile to summarize a few of the important equations since this will also serve to define some notation and certain abbreviations.

A. Sum-Over-States Perturbation Theory (SOS).—Equations for nuclear spin–spin coupling expressed in the usual formalism of second-order perturbation theory all derive from the basic equation due to Ramsey:[1]

$$J_{AB} = (hI_A \cdot I_B)^{-1} \sum_{n \neq 0} \langle 0|\mathcal{H}'|n\rangle \langle n|\mathcal{H}'|0\rangle /(E_0 - E_n) \quad (1)$$

The nuclear spin–spin coupling between nuclei A and B, J_{AB}, is given in Hz, h is Planck's constant, and I is the nuclear spin angular momentum in units of $\hbar$. $|0\rangle$ represents the wavefunction for the molecular electronic ground state and $|n\rangle$ that of an excited state, their respective energies being E_0 and E_n. $\mathcal{H}'$ is the Hamiltonian operator corresponding to the energy of the electron-coupled spin–spin interaction, which may be split into three parts, *viz.* the orbital terms, the spin-dipolar term, and the Fermi-contact term. Until

[1] N. F. Ramsey, *Phys. Rev.*, 1953, **91**, 303.

quite recently only the last of these contributions to $\mathcal{H}'$ had been widely considered in the interpretation of nuclear spin–spin coupling.

The wavefunctions used in equation (1) may be of either molecular-orbital (MO) or valence-bond (VB) type, and some formulations of equation (1) in these terms have found extensive use, though they consider only the Fermi-contact term. Thus, Pople and Santry,[2] using a linear combination of atomic orbitals (LCAO) MO formalism, derived equation (2), where β is the Bohr

$$J_{AB} = (16h\beta^2/9)\gamma_A\gamma_B S_A^2(0) S_B^2(0)\Pi_{s_A s_B} \quad (2)$$

magneton ($= e\hbar/2mc$), γ is the magnetogyric ratio, $\Pi_{s_A s_B}$ is the mutual polarizability of the valence s-orbitals of atoms A and B, and $S_A(0)$, $S_B^2(0)$ are the electron densities of these same orbitals at the nuclei A and B. By use of the average-energy approximation, equation (2) can be reduced to equation (3), where ΔE is the average excitation energy and $P_{s_A s_B}$ is the MO bond order between the orbitals s_A and s_B. When the wavefunctions are written in VB terms it is often convenient to use hybrid orbitals of the form indicated by equation (4), where α_x^2 is the s-character of the hybrid orbital ϕ_x. J_{AB} is then given by equation (5),[3] where ζ is a normalizing factor for the VB function describing the A—B bond using the hybrid orbitals ϕ_A and ϕ_B and allowing, if necessary, for the polarity of the bond. It should be emphasized that equations (2), (3), and (5) are for the Fermi-contact contribution to the coupling only and that equation (5) is only applicable to directly bonded nuclei A and B.

$$J_{AB} = (16h\beta^2/9\Delta E)\gamma_A\gamma_B S_A^2(0)S_B^2(0)P_{s_A s_B}^2 \quad (3)$$

$$\phi_x = \alpha_x s_x + (1-\alpha_x^2)^{\frac{1}{2}} p_x \quad (4)$$

$$J_{AB} = (64h\beta^2/9\Delta E)\gamma_A\gamma_B \zeta^2 \alpha_A^2 \alpha_B^2 S_A^2(0) S_B^2(0) \quad (5)$$

B. Finite Perturbation Theory (FPT).—More recently, a rather different approach to the calculation of nuclear spin–spin coupling has been proposed by Pople and his co-workers.[4] They have shown, again for the Fermi-contact term only, that:

$$J_{AB} = (16h\beta^2/9)\gamma_A\gamma_B S_A^2(0)S_B^2(0)\left\{\frac{\delta}{\delta d_A}\rho_{sB}^2(d_A)\right\}_{d_A = 0} \quad (6)$$

$\rho_{s_B}^2$ is the diagonal spin-density matrix element calculated for the orbital s_B when the perturbation d_A is applied at atom A. Applications of FPT to the spin-dipolar and orbital terms in $\mathcal{H}'$ have also been described.[5]

---

[2] J. A. Pople and D. P. Santry, *Mol. Phys.*, 1964, **8**, 1.
[3] C. Juan and H. S. Gutowsky, *J. Chem. Phys.*, 1962, **37**, 2198.
[4] J. A. Pople, J. W. McIver, and N. S. Ostlund, *J. Chem. Phys.*, 1968, **49**, 2965.
[5] H. Nakatsuji, K. Hirao, H. Kato, and T. Yonezawa, *Chem. Phys. Letters*, 1970, **6**, 541.

C. Semi-empirical Molecular Orbital Methods.

—In recent years the majority of calculations of nuclear spin–spin coupling have been based upon a set of wavefunctions derived by means of some semi-empirical MO method. The following list gives the most common of these methods, in order of increasing approximation, and the abbreviations which will be used for them in this chapter.

(*a*) Intermediate neglect of differential overlap, (INDO).[6]

In this method, all two-electron repulsion integrals which involve one or more overlap densities are neglected unless the overlap density derives from two orbitals located on the same atomic centre.

(*b*) Complete neglect of differential overlap, (CNDO).[7,8]

Here all two-electron repulsion integrals which involve overlap densities are neglected. There are two parametrizations, CNDO/1[7] and CNDO/2;[8] the latter is the more widely used.

(*c*) Extended Hückel theory, (EHT).[9]

This procedure neglects all electron repulsion integrals; it is an independent-electron theory.

D. Units.

—The equations for J_{AB} above are quoted in their c.g.s. forms. Conversion into the SI form requires that each be multiplied by $\mu_0/4\pi$, where μ_0 is the permeability of a vacuum; this factor is 10^{-7} kg m s^{-2}A^{-2} exactly. It should be noted that, in SI, $\beta = e\hbar/2m$. The units of J are Hz in both the c.g.s. system and in SI, and all the equations for J in this chapter give values in these units.

Equation (7) defines the term 'reduced coupling constant', K_{AB}, as it is used

$$K_{AB} = (2\pi/\hbar\gamma_A\gamma_B)J_{AB} \qquad (7)$$

throughout this chapter.[10,11] In the c.g.s. system its units are cm^{-3} and in SI they are N A^{-2} m^{-3}. Numerically,

$$K_{AB}(\text{c.g.s.}) = K_{AB}(\text{SI}) \times 10 \qquad (8)$$

We now turn to the discussion of work on nuclear spin–spin coupling published during the period under review.

3 Theoretical Work

In this section, only those calculations based on the fundamental theory of nuclear spin–spin coupling will be described. Correlations with other parameters (charge densities, electronegativities, *etc.*) will be discussed under the headings corresponding to the coupling constants involved.

---

[6] J. A. Pople, D. L. Beveridge, and P. A. Dobosh, *J. Chem. Phys.*, 1967, **47**, 2026.
[7] J. A. Pople, D. P. Santry, and G. A. Segal, *J. Chem. Phys.*, 1965, **43**, S129.
[8] J. A. Pople and G. A. Segal, *J. Chem. Phys.*, 1965, **43**, S136; 1966, **44**, 3289.
[9] R. Hoffmann, *J. Chem. Phys.*, 1963, **39**, 1397.
[10] R. M. Lynden-Bell and N. Sheppard, *Proc. Roy. Soc.*, 1962, **A269**, 385.
[11] H. Nakatsuji, H. Kato, I. Morishima, and T. Yonezawa, *Chem. Phys. Letters*, 1970, **4**, 607.

Nuclear Spin–Spin Coupling

A. Calculations on H_2 or HD.—The past year has seen, as far as the Reporter is aware, just two papers devoted exclusively to this important subject. Power and Pitzer,[12] using a coupled Hartree–Fock method, have obtained an approximate wavefunction for HD by minimizing the self-coupling energy, $J(H,H)$. The form of the contact interaction taken from the Dirac equation was used, instead of a delta function, in order to keep $J(H,H)$ finite.[13] Rapid convergence was found and satisfactory results for small molecules have been obtained but not yet published. The Fermi-contact interaction is found to account for almost all the nuclear spin–spin coupling in HD. Moulson and Lowe[14] have discussed the SOS perturbation theory for the calculation of $^1J(H,D)$ and suggest that the principal difficulty of this method, as applied hitherto, has been the use of virtual orbitals of the ground state to construct the excited states needed for the perturbation expansion. They separately optimize the lowest triplet states of u and g symmetry and then form further excited states using virtual orbitals from these. The excited states thus obtained give a much better convergence of the perturbation expansion, and a coupling constant of 42.6 Hz for the Fermi-contact term is found.

An *ab initio* calculation on H_2 by Kowalewski, Vestin, and Roos[15] will be described in the next section.

B. Ab Initio Calculations of Coupling in Small Molecules.—*Ab initio* calculations on H_2, using an extended basis of Gaussian orbitals, have been reported.[15] Although the wavefunction used for the ground state approximated closely to the Hartree–Fock limit, the coupling constant obtained by SOS was *ca.* 114 Hz (experimental value 280 Hz). Inclusion of configuration interaction among the triplet excited states gave a value of 200 Hz. Similar calculations,[15] using the Fermi-contact term only, gave unsatisfactory results for the couplings in ethylene, fluoroethylene, and water; the incorrect sign was obtained for the geminal (H,H) coupling in fluoroethylene. The authors note particularly the lack of convergence in H_2 and H_2O, even when configuration interaction is included, and they suggest that the effects of electron correlation may be very considerable. The importance of electron correlation in the calculation of coupling constants has also been emphasized by Barbier *et al.*[16] They find, using a minimal basis set of LCAO–SCF MO's for CH_4 and the usual second-order perturbation theory, a value of -1.27 Hz for the geminal (H,H) coupling in methane; the experimental value is -12.4 Hz. However, the agreement with experiment is much improved (-15.02 Hz) if third-order contributions which arise from correlation effects are included. The authors note that, since the matrix elements of the contact operator are very small unless both the orbitals involved are localized on the same atom, it can be helpful to interpretation to carry out calculations in terms of localized MO's. Both the impor-

---

[12] J. D. Power and P. M. Pitzer, *Bull. Amer. Phys. Soc.*, 1971, **16**, 309.
[13] T. P. Das and R. Bersohn, *Phys. Rev.*, 1959, **115**, 897.
[14] T. Moulson and J. P. Lowe, *Mol. Phys.*, 1971, **22**, 723.
[15] J. Kowalewski, R. Vestin, and B. Roos, *Chem. Phys. Letters*, 1971, **12**, 25.
[16] C. Barbier, D. Gagnaire, G. Berthier, and B. Levy, *J. Magn. Resonance*, 1971, **5**, 11.

tance of correlation effects and the use of localized orbitals have also been emphasized by Denis and Malrieu[17] in a double-perturbation-theory calculation of the coupling constants of some simple hydrocarbons. These calculations, though rather different in formulation, give results which are quite similar to those obtained with the Pople–Santry formalism [equation (2)] and CNDO/2 MO's, and they tend to fail where such calculations also fail. However, the authors claim that the method is quicker to carry out than those in common use and that it has important conceptual merits. Both Barbier et al.[16] and Denis and Malrieu[17] make extensive use of diagrams in their perturbation theory, and the interesting comparison of these two papers is the justification for placing the Denis–Malrieu work in this section; it really belongs in the next section because the wavefunctions used were semi-empirical ones of the CNDO type.

Ditchfield and Snyder[18] have developed a Hartree–Fock perturbation theory (FPT) including the spin-dipolar, orbital, and contact-spin-dipolar cross-terms, and have applied their results to the couplings in CH_3F using ab initio wavefunctions calculated from a fairly small basis set of Gaussians. The agreement of $^1J(C,F)$, $^1J(C,H)$, and $^2J(H,F)$ with experiment is moderate but the geminal (H,H) coupling is badly overestimated though it does have the correct sign. The conclusion of Nakatsuji et al.[5,11] that the values of the anisotropies of the direct couplings in methyl fluoride estimated by Krugh and Bernheim[19,20] are too high is substantiated.

C. Semi-empirical Calculations.—An underlying trend in the theoretical analysis of coupling constants is the realization that for many nuclei the orbital and spin-dipolar contributions to the coupling cannot be neglected (see below). Thus Blizzard and Santry[21] have described a self-consistent perturbation theory, very similar to that of Ditchfield and Snyder,[18] which it preceded, in which all contributions to the spin–spin coupling are evaluated. The method has much in common with the FPT but is more economical in computer time. Working within the INDO framework, Blizzard and Santry[21] have evaluated a number of couplings involving carbon and fluorine nuclei and have found that the agreement between theory and experiment is on the whole good. They conclude that the frequently neglected orbital and spin-dipolar terms play a very important role in (C,F) and (F,F) coupling and may be larger in magnitude than the Fermi-contact term in the case of (F,F). In (C,F) coupling these terms are generally smaller than the contact term but they are decisive in determining the correct calculation of experimental trends in $^1J(C,F)$ for a series of compounds.

The inclusion of spin-dipolar and orbital terms significantly improves the theoretical values for $^1J(C,C)$ but to a lesser extent than for $^1J(C,F)$ or (F,F)

[17] A. Denis and J.-P. Malrieu, *Mol. Phys.*, 1972, **23**, 581.
[18] R. Ditchfield and L. C. Snyder, *J. Chem. Phys.*, 1972, **56**, 5823.
[19] T. R. Krugh and R. A. Bernheim, *J. Amer. Chem. Soc.*, 1969, **91**, 2385.
[20] T. R. Krugh and R. A. Bernheim, *J. Chem. Phys.*, 1970, **52**, 4942.
[21] A. C. Blizzard and D. P. Santry, *J. Chem. Phys.*, 1971, **55**, 950.

coupling. In many cases the dipolar term could be considered negligible in (C,C) coupling. The theory is not so successful with $^3J(F,F)$ (*cis*), for which the incorrect sign is obtained.

The conclusions of Blizzard and Santry are supported by the work of Nakatsuji *et al.*,[22] who have used SOS with INDO MO's to investigate the anisotropy of indirect nuclear spin–spin coupling in methane, ethane, ethylene, acetylene, and a number of their derivatives. For $^1J(C,X)$ (X = C, N, or F) the anisotropy is of the same order of magnitude as the isotropic coupling, but for non-bonded (C,X) the anisotropies seem negligible. They point out that for (F,F) couplings the anisotropies are particularly large and the orbital and spin-dipolar terms are very important, making decisive contributions which even exceed the Fermi-contact term for the isotropic (F,F) couplings. Earlier work[5,11] concerning the anisotropy of $^1J(C,H)$, suggesting that this was very much smaller than the estimates of Krugh and Bernheim,[19,20] is substantiated (*Cf.* the paper of Ditchfield and Snyder[18] mentioned above).

Schaumburg and his co-workers[23,24] have also included spin-dipolar and orbital terms in semi-empirical calculations, generally in combination with SOS rather than FPT. The effect of including configuration interaction among the triplet excited states was also considered. Their conclusions may be summarized as follows: (*a*) Coupling constants calculated by the INDO–CI–SOS method allow a prediction of trends in magnitude and of signs. (*b*) Spin-dipolar and orbital terms must be included for coupling between nuclei other than hydrogen. (This conclusion is to some extent at variance with that of Blizzard and Santry,[21] see above). (*c*) The discrepancies between theory and experiment are significantly reduced if the *s*-electron density at the nucleus is made a function of the calculated atomic charge. (*d*) On the whole, the results are rather poor, particularly for $^1J(C,F)$ and for geminal couplings in general, and the theory in its present form must be considered unsatisfactory. The fact that the results are uniformly poor for a wide variety of types of calculation[24] points to fundamental problems.

Several other papers using semi-empirical MO's and the contact term only have appeared. Maciel and Bartuska have used the INDO–FPT approach to calculate $^1J(C,C)$ for a variety of configurations of the isopropyl carbonium ion,[25] for isopropenyl compounds and ketones of the forms (1) and (2)

$$H_2C=C\begin{matrix}X\\Me\end{matrix} \qquad O=C\begin{matrix}X\\Me\end{matrix}$$

(1) (2)

---

[22] H. Nakatsuji, I. Morishima, H. Kato, and T. Yonezawa, *Bull. Chem. Soc. Japan*, 1971, **44**, 2010.
[23] A. D. C. Towl and K. Schaumburg, *Mol. Phys.*, 1971, **22**, 49.
[24] H. Jensen and K. Schaumburg, *Mol. Phys.*, 1971, **22**, 1041.
[25] G. E. Maciel, *J. Amer. Chem. Soc.*, 1971, **93**, 4375.

respectively,[26] and for ethyl compounds.[27] The object of the first of the above papers was the investigation of trends rather than the calculation of particular numerical values, and no comparison with experiment was made. It was noted, however,[25] that the value of $^1J(C,C)$ did parallel that of P^2_{ss}, although the fractional variations in the latter still correspond to only about one half of that of the former. In general, for $^1J(C,H)$ and $^2J(CCH)$ there appeared to be no simple relationship between calculated coupling and s–s bond orders, not even a qualitative one. A related observation has been made by Gopinathan and Narasimhan,[28] who have reported calculations of $^1J(C,H)$ using different MO wavefunctions. The calculated couplings were often different although the values of $P^2_{s_C s_H}$ were extremely similar.

In the isopropenyl[26] and ethyl[27] systems, Maciel and Bartuska found evidence for a relationship between $^1J(C,C)$ and the electron-withdrawing power of substituents as measured by their $-I^+$ character. In general, the calculated couplings are found to vary over a wider range than the observed values, and are consistently larger in magnitude.

It is perhaps not surprising, in view of the difficulties encountered in the direct calculation of coupling constants, that a number of laboratories should have turned their attention to the theoretical interpretation of the stereochemical dependence of coupling constants. Within this area vicinal couplings have received considerable attention. Thus Govil[29] has used EHT, CNDO, and INDO MO's with SOS to investigate $^3J(HCXH)$ (X = C, N, or O) as a function of the dihedral angle ϕ. In all cases it is found that the theoretical J values can be fitted to an equation of the form:

$$^3J(HCXH) = A\cos^2\phi + B\cos\phi + C \qquad (9)$$

and values of A, B, and C for the different types of wavefunction and of X are given. For a fixed value of ϕ it is found that $J(HCCH) > J(HCOH) > J(HCNH)$, which is consistent with experimental observations. Better agreement of theory with experiment is found if Burns' rules[30] rather than Slater's are used to determine the orbital exponents. Govil[31] has also examined the stereochemical dependence of $^3J(HCCF)$ in ethyl fluoride using EHT MO's in the Pople–Santry formalism and fitting the calculated J to an equation of the form of equation (9). Again, closer agreement with experiment is found using Burns' rather than Slater's orbital exponents. The effect of substituents on $^3J(H,F)$ is predicted to be greatest when the substituent is on the carbon atom which carries the fluorine and when ϕ is near 0 or 180°.

The dihedral angle dependence of $^3J(HCNH)$ has also been investigated by Gopinathan and Narasimhan.[32] Using methylamine as their model

---

[26] V. J. Bartuska and G. E. Maciel, *J. Magn. Resonance*, 1972, **7**, 36.
[27] V. J. Bartuska and G. E. Maciel, *J. Magn. Resonance*, 1971, **5**, 211.
[28] M. S. Gopinathan and P. T. Narasimhan, *Mol. Phys.*, 1971, **21**, 943.
[29] G. Govil, *Indian J. Chem.*, 1971, **9**, 824.
[30] G. Burns, *J. Chem. Phys.*, 1964, **41**, 1521.
[31] G. Govil, *Mol. Phys.*, 1971, **21**, 953.
[32] M. S. Gopinathan and P. T. Narasimhan, *Mol. Phys.*, 1971, **22**, 473.

and INDO FPT, they find that the (H,H) coupling is significantly influenced by the position of the nitrogen lone-pair, leading to deviations from the Karplus-type equation.[33] They suggest that this equation might be replaced by one of the form of equation (10). They find that calculated values of

$$^3J(\text{HCNH}) = 14.247 \cos^2 \phi - 3.979 \cos \phi + 0.445 - 0.728 \sin \phi + 1.005 \sin 2\phi \quad (10)$$

$^1J(\text{C,H})$ and $^2J(\text{NCH})$ in oximes are also significantly influenced by the position of the nitrogen lone-pair, and it is also worth noting that these couplings are in consistently better agreement with experiment when calculated in the CNDO/2 approximation rather than the INDO. The authors suggest that the neglected orbital and spin-dipolar terms may be important here.

Gopinathan and Narasimhan[34] have also reported theoretical work on $^3J(\text{HCCH})$ and $^3J(\text{HCCF})$ using CNDO/2 and INDO FPT and EHT SOS. Both the theoretical couplings, (H,H) and (H,F), are well reproduced by an equation of the form of equation (9), and values of A, B, and C are given. However, that the couplings depend on factors other than ϕ, e.g. electronegativity, is shown by the fact that $^3J(\text{H,F})$ in ethyl fluoride is best reproduced by equation (11), which the authors compare[32] with equation (10) for

$$^3J(\text{HCCF}) = A \cos^2 \phi + B \cos \phi + C + D \sin \phi \quad (11)$$

$^3J(\text{HCNH})$. Among the other coupling constants calculated,[34] the INDO FPT results for $^1J(\text{C,H})$ and $^2J(\text{CCH})$ were in quite good agreement with experiment and 'superior' to EHT SOS and CNDO/2 FPT calculations. Values for $^2J(\text{HCH})$ and $^2J(\text{HCF})$ were very poorly reproduced both by INDO and CNDO/2 FPT.

Similar results were obtained by the same authors[35] when they used INDO FPT calculations to investigate (H,F) coupling in a number of fluorinated hydrocarbons. Although the variation of $^3J(\text{HCCF})$ with dihedral angle was quite well reproduced, they found that other couplings, particularly $^2J(\text{HCF})$, were very poorly predicted, and they expressed doubt that the inclusion of terms other than the contact will be sufficient to account for the large deviations observed.

Günther, Klose, and Cremer[36] have used CNDO/2 SOS calculations in support of an empirical equation [equation (12)] for the angular dependence

$$^3J(\text{HCCH}) = 4.3 \cos 2\phi - 1.8 \cos \phi + 5.0 \quad (12)$$

of $^3J(\text{HCCH})$ over the single bond between the three-membered ring and the double bond in vinylcyclopropane. They find that the general experimental trend of the coupling is given well by this method.

---

[33] M. Karplus, *J. Chem. Phys.*, 1959, **30**, 11; *J. Amer. Chem. Soc.*, 1963, **85**, 2870.
[34] M. S. Gopinathan and P. T. Narasimhan, *Mol. Phys.*, 1971, **21**, 1141.
[35] M. S. Gopinathan and P. T. Narasimhan, *Mol. Phys.*, 1971, **22**, 543.
[36] H. Günther, H. Klose, and D. Cremer, *Chem. Ber.*, 1971, **104**, 3884.

Gopinathan and Narasimhan[28] have also investigated ^3J(HCCH) in cyclopropane, ethyleneimine, and ethylene oxide, using INDO and CNDO/2 FPT methods and EHT SOS. The *trans* couplings are predicted better than the *cis*-, but the *cis/trans* ratio is very poorly reproduced. Using the same methods, the authors have also examined ^4J(HCCCH) in acetone, isobutene, and propane.[37] The FPT methods predict the conformational dependence of this coupling satisfactorily but the SOS method is inferior. However, this difference may depend on the wavefunctions rather than on the form of the perturbation theory adopted.

Barfield and Sternhell[38] have made an analysis of the conformational dependence of $^5J(H_A,H_B)$, the homoallylic coupling, in systems of the form (3), using a semi-empirical VB method and also INDO FPT. They find that

$$H_A-C_1-C_2{=}C_3-C_4-H_B$$

(3)

both of these methods provide an adequate description of the conformational dependence of the coupling and can be used in the solution of stereochemical problems. In some situations, however, (the homoallylic coupling in *cis*- and *trans*-2-butene, for example), the INDO FPT gives the better results because it includes all valence electrons whereas the VB method considers only the π-system. The results are presented in terms of two dihedral angles, ϕ, measured in a clockwise sense from the C$_1$–C$_2$–C$_3$ plane to H$_A$, and ϕ', which is measured in the same sense from the C$_2$–C$_3$–C$_4$ plane to H$_B$. Both VB and MO calculations then give curves of $^5J(H,H)$ against ϕ which show minima at 0°, 180°, and 360° and maxima at 90° and 270°. The amplitude of the curves is largest for $\phi' = 90°$ and 270° and smallest for $\phi' = 0°$ and 180°. The VB results can be adequately represented by equation (13) but the MO

$$^5J^\pi(H,H) = 4.99 \sin^2 \phi \sin^2 \phi' \qquad (13)$$

results are more complex and the curves show asymmetry and the possibility of negative coupling. On the whole the agreement of theory with experiment is very satisfactory when the difficulties of measuring the couplings and of estimating the dihedral angles and substituent effects are taken into account.

Barfield, Spear, and Sternhell[39] have made a detailed analysis of the dual coupling paths, H—C—C≡C—C—H and H—C—X—C—H, in five-membered rings of the form (4) using the INDO FPT approach. The effects

$$H_5'\diagup\diagdown H_2'$$
$$H_5 \quad X \quad H_2$$

(4)

---

[37] M. S. Gopinathan and P. T. Narasimhan, *J. Magn. Resonance*, 1972, **6**, 147.
[38] M. Barfield and S. Sternhell, *J. Amer. Chem. Soc.*, 1972, **94**, 1905.
[39] M. Barfield, R. J. Spear, and S. Sternhell, *J. Amer. Chem. Soc.*, 1971, **93**, 5322.

of substituents on the homoallylic (^5J) coupling constants are discussed in terms analogous to those used earlier for ^2J(HCH)[40] and ^4J(HCCCH).[41] In this context it is interesting to note that the authors find an SOS analysis more suitable than FPT for providing a simple basis for the qualitative discussion of trends in coupling constants. The dependence of the four-bond H—C—X—C—H coupling on the nature of X is discussed on the basis of Barfield's earlier work on this type of coupling,[42] which, regrettably, was overlooked in the previous Report. Again the authors find it helpful to interpret the dependence on X in terms of an SOS analysis, though the INDO FPT method is used to provide quantitative results. $^4J(trans)$ ($^4J_{25'}$) is predicted to be larger than $^4J(cis)$ ($^4J_{25}$), in reasonable agreement with experiment, and on the basis of the SOS analysis it is concluded that the *cis-* and *trans-*(H,H) couplings are inherently different and that the major differences between the two can be attributed to substituent effects on the homoallylic mechanism.

Calculations of other couplings by semi-empirical methods have been reported by several authors: ^1J(H,C) (refs. 24, 25, 28, 32, 43, and 44); ^1J(C,C) (refs. 25 and 26); ^2J(CCH) (refs. 24, 25, 28, and 44); ^2J(HCH) (refs. 24, 28, 37, and 43); ^2J(NCH) (refs. 32 and 44); ^2J(HCF) (refs. 24 and 35).

In general, the unfortunate conclusion must be that although the calculation of ^1J(C,H) can be quite successful, the results for many other couplings, particularly geminal couplings, are very poor. The inclusion of orbital and spin-dipolar terms, although clearly essential for many couplings, does not at the moment appear to offer a solution to these problems, which must be considered to be fundamental. Nevertheless, Barfield and Sternhell's work[38] shows that very useful results can be obtained theoretically, particularly in the case of the longer-range couplings.

The somewhat gloomy picture presented by the MO theory may be a little relieved by the recent reports of VB calculations of ^3J(HCCH) using SOS PT. For cyclopropane, Chandra and Narasimhan[45] find that VB calculations are more successful than MO, and in particular that they give a much better *cis/trans* ratio, *i.e.* 1.41—1.49, which is to be compared with an experimental value of 1.47—1.73 and an INDO FPT value of approximately 1. Similarly, Duval[46] has reported very good agreement with experiment for ^3J(HCCH) in cyclopropane and ethane using VB theory. It remains to be seen, however, if the VB methods will give equally good results when extended to other nuclei, or indeed when the wavefunctions and the methods of obtaining them are made as sophisticated as the present MO techniques. Nevertheless, the VB theory may well be ripe for renewed application in this difficult field where it had many early successes.

[40] J. A. Pople and A. A. Bothner-By, *J. Chem. Phys.*, 1965, **42**, 1339.
[41] M. Barfield and B. Chakrabarti, *Chem. Rev.*, 1969, **69**, 757.
[42] M. Barfield, *J. Amer. Chem. Soc.*, 1971, **93**, 1066.
[43] R. Radeglia and E. Gey, *J. prakt. Chem.*, 1971, **313**, 1070.
[44] I. N. Bojesen, J. H. Hog, J. T. Nielsen, I. B. Petersen, and K. Schaumburg, *Acta Chem. Scand.*, 1971, **25**, 2739.
[45] P. Chandra and P. T. Narasimhan, *Mol. Phys.*, 1971, **21**, 1067.
[46] E. Duval, *Mol. Phys.*, 1972, **23**, 433.

On a slightly different track, Randić and his co-workers[47] have calculated $^1J(C,H)$ in various unsaturated hydrocarbons, using hybrid orbitals determined by the maximum-overlap method. These are related to the spin–spin coupling constants *via* the relations of Muller and Pritchard[48] [equation (14)] and of Randi´ *et al.*[49] [equation (15)], where α is the coefficient of the

$$^1J(C,H) = 500\alpha^2 \tag{14}$$

$$^1J(C,H) = 1079\alpha^2 (1+S^2) - 54.9 \tag{15}$$

2s-orbital of carbon in the relevant C—H bond hybrid and S is the C—H bond overlap as calculated by the maximum overlap method. These equations are, in effect, derivatives of equation (5) for the special case of (C,H) coupling, and the appearance of the constant term in equation (15) is ascribed[49] to the neglect of ionic structures in the C—H bond. The agreement with experiment is good for both relationships, though the second is somewhat better than the first. The poorest results are obtained for 3,3-dimethylcyclopropene, and reasons for the persistent deviations in small, highly strained rings are discussed.

Dalling and Gutowsky[50] have continued the work of Gutowsky's group on the general and wide-ranging analysis of directly bonded coupling constants. They have studied the Fermi-contact contribution to nuclear spin–spin coupling by determining the relationship between Z, the atomic number, and the electron spin density at the nuclei of isolated atoms, obtaining a measurement of this latter parameter from atomic spectral data. They conclude that the reduced coupling constant, K_{AB},' for a series of isoelectronic and isostructural compounds varies as kZ_A, where k is a relativistic correction which ranges from 1.0 to about 2.5, depending on Z_A. The results are informative and several interesting predictions are made in spite of the fact that equation (16), on which the analysis is based (where a_A is the hyperfine con-

$$K_{AB} = a'_A a'_B / \Delta E \tag{16}$$

stant for atom A *in a molecule* and ΔE is an average excitation energy), would necessarily predict a positive one-bond coupling constant between pairs of identical nuclei, which is not the case for (P,P) coupling.[51] However, the fact that observations in direct contradiction of equation (16) are so rare justifies its use for such a wide-ranging correlative approach to nuclear spin–spin coupling.

D. π-Electron Coupling.—Barfield *et al.*[52] have described theoretical and ex-

[47] M. Randić, Z. Meić, and A. Rubčić, *Tetrahedron*, 1972, **28**, 565.
[48] N. Muller and D. E. Pritchard, *J. Chem. Phys.*, 1959, **31**, 768, 1471.
[49] Z. B. Maksić, M. Eckert-Maksić, and M. Randić, *Theor. Chim. Acta*, 1971, **22**, 70.
[50] D. K. Dalling and H. S. Gutowsky, *J. Chem. Phys.*, 1971, **55**, 4959.
[51] R. V. Emanuel, *J. Chem. Phys.*, 1970, **53**, 856.
[52] M. Barfield, C. J. Macdonald, I. R. Peat, and W. F. Reynolds, *J. Amer. Chem. Soc.*, 1971, **93**, 4195.

perimental studies of long-range (H,H) coupling in ring-substituted styrenes using a semi-empirical VB method and INDO FPT. The conformational dependence of the VB results splits up rather nicely into two mechanisms which show different dependences on ϕ, the dihedral angle between the plane of the benzene ring and the plane of the vinyl group. In the planar conformation the important mechanism is delocalization in the extended π-electron system, and this depends on $\cos^2 \phi$. For out-of-plane conformations a σ–π exchange interaction becomes increasingly important, and this has a $\sin^2 \phi$ dependence. The results of the VB calculations, for all couplings, can therefore be represented very well by means of equations of the form

$$J = A \cos^2 \phi + B \sin^2 \phi \tag{17}$$

where A and B are constants for particular coupled nuclei. The MO results do not permit any such simple analysis but they are quite similar to the VB results, and in general the agreement between theory and experiment is very good. The results show that the π-electron mechanism is the only important one for long-range coupling between the vinyl protons and the *para*-hydrogen of the ring.

Nair and his co-workers[53] have examined the relationship between *ortho*-benzylic coupling [^4J(HCCCH)] and the π-bond order, P, of the aromatic bond concerned. They find a linear relationship between experimental couplings and the square of the bond order, which takes the form

$$^4J(H,H) = 1.21 - 3.67\, P^2 \tag{18}$$

and leads to the conclusion that the σ-electron contribution to this coupling is 1·21 Hz. This last figure would appear to be somewhat large, and this problem and the applications and limitations of the correlation are discussed in detail.

Chuvylkin *et al.*[54] have compared the calculation of π-electron coupling using the equations of Karplus[55] and Ditchfield and Murrell[56] and an equation which they themselves have derived [equation (19)] and which was also

$$^3J(HCCH) = \tfrac{1}{4}Q_{CH}^2 \pi_{HH} \tag{19}$$

given[57] by Van der Hart. π_{HH} is the mutual polarizability of the hydrogen 1s orbitals and Q_{CH} is the spin coupling constant of McConnell.[58] The comparison highlights the differences between the three formulations but does not really explain their origin. Is the difference due to the neglect of certain triplet states in the earlier formulations, and is the simple expression of equation (19) as versatile as that of Ditchfield and Murrell?

[53] P. M. Nair, G. Gopakumar, T. Fairwell, and V. S. Rao, *Indian J. Chem.*, 1971, **9**, 549.
[54] N. D. Chuvylkin, G. M. Zhidomirov, and P. V. Schastnev, *Mol. Phys.*, 1972, **23**, 639.
[55] M. Karplus, *J. Chem. Phys.*, 1960, **33**, 1842.
[56] R. Ditchfield and J. N. Murrell, *Mol. Phys.*, 1968, **15**, 533.
[57] W. J. Van der Hart, *Mol. Phys.*, 1971, **20**, 399.
[58] H. M. McConnell, *J. Chem. Phys.*, 1959, **30**, 126.

E. Nuclear-spin and Electron-spin Coupling.—McIver and Maciel[59] have used INDO FPT methods to investigate the relationship between nuclear spin–spin coupling and electron-spin coupling constants in a few simple hydrocarbons. They show that there is a very close correspondence between these two properties (as might have been expected from their quantum-mechanical formulations), in accord with the available experimental data, in particular the relative constancy of the experimental ratio J_A/a_{AB}, where a_A is the atomic hyperfine coupling constant. The work of Dalling and Gutowsky[50] is relevant here too. On the experimental side, Jakobsen et al.[60] have carried out n.m.r. and e.s.r. investigations of fluorinated phenyl-t-butyl nitroxide radicals and find that the signs of the hyperfine splitting constants are in agreement with the signs of the π-electron contributions to the corresponding long-range coupling constants.

Kato and Kato[61] have shown that to order S^2 (S is the spin of the paramagnetic species), the contact nuclear spin–spin coupling in paramagnetic transition-metal complexes is the same as that of the free ligand.

F. Solvent Effects.—Johnston and Barfield[62,63] have continued their work on the effect of solvents on coupling constants; see Chapter 10.

4 Coupling of Directly Bonded Nuclei

A. Coupling in HD.—The elegant molecular-beam resonance studies of Code and Ramsey[64] have confirmed that the sign of $^1J(H,D)$ is positive; its magnitude is found to be 47 ± 7 Hz.

B. Coupling between ^{13}C and ^1H.—Theoretical calculations on $^1J(C,H)$ have already been noted under references 24, 25, 28, 34, 44, and 47. Attention should be drawn here to the work of Radeglia and Gey,[43] who have calculated $^1J(C,H)$ in CH_3X, using the CNDO/2 SOS method. The agreement of calculated and experimental coupling is quite good, but when the one is plotted against the other two lines are found, depending on whether X is in the first or second row of the Periodic Table. In general, it should be noted again that although calculations reproduce $^1J(C,H)$ better than any other one-bond coupling constant, they are far from satisfactory in many respects.

Correlations of $^1J(C,H)$ with electronegativities and additivity rules have been reported by several workers. Yamamoto et al.[65] find that the $^1J(C,H)$ values in β-1,2,3,4,5,6-hexachlorocyclohexane are given by Malinowski's

---

[59] J. W. McIver and G. E. Maciel, *J. Amer. Chem. Soc.*, 1971, **93**, 4641.
[60] H. J. Jakobsen, T. E. Petersen, and K. Torssell, *Tetrahedron Letters*, 1971, 2913.
[61] H. Kato and H. Kato, *Bull. Chem. Soc. Japan*, 1971, **44**, 1734.
[62] M. D. Johnston and M. Barfield, *Mol. Phys.*, 1971, **22**, 831.
[63] M. D. Johnston and M. Barfield, *J. Chem. Phys.*, 1971, **55**, 3483.
[64] R. F. Code and N. F. Ramsey, *Phys. Rev. (A)*, 1971, **4**, 1945.
[65] O. Yamamoto, K. Hayamizu, S. Satoh, and K. Kushida, *J. Magn. Resonance*, 1971, **5**, 429.

additivity rule for $^1J(C,H)$ in methane derivatives[66] if one takes 68.6 Hz for ζ_{-Cl} and 44.6 Hz for ζ_{-CHCl_2}, so in spite of the cyclic structure and the many chlorine atoms, sp^3 hybridization and electronegativity seem to be the controlling features for $^1J(C,H)$, even in this molecule. Similarly, Tarpley and Goldstein[67] have found good correlations of the directly bonded (C,H) couplings obtained from analyses of the ^{13}C n.m.r. spectra of the halogenobenzenes with additivity relationships. There are some anomalies, however. They also found[68] good correlations of all (C,H) couplings with halogen electronegativity, χ_X, in the m-dichloro-, m-dibromo-, and m-di-iodo-benzenes, and in m-difluorobenzene where the appropriate result is available. If the carbon atoms are designated as shown in (5) then it is found that (C,H)

$$X-C_s=C_{oo}-C_s-X$$
$$|\quad\quad\quad|$$
$$C_o-C_m-C_o$$

(5)

couplings over the same number of bonds involving C_s and C_m behave in a similar manner in their trends with halogen electronegativity. Couplings over the same number of bonds involving C_o and C_{oo} behave similarly but in the opposite manner to couplings over the same number of bonds involving C_s and C_m. Furthermore, considering all couplings involving a given set of 'similar' carbons (C_s and C_m; C_o and C_{oo}), one-bond, three-bond, and four-bond couplings all exhibit the same trend with χ_X, which is opposite to that observed for the two-bond coupling constants.

Hess et al.[69] have interpreted the changes in $^1J(C,H)$ in a methyl group attached to Si, Ge, P, As, or S in terms of the electronegativity of these atoms and of Bent's rules,[70] and Jensen and Schaumburg[24] found that $^1J(C,H)$ in ethyl fluoride is also predicted well by the empirical rules of Hoboken and Malinowski.[66]

Yoder and Schenck[71] have correlated directly bonded (C,H) couplings for some methoxy- and thiomethoxy-derivatives of the Group IV elements with relative basicities determined from i.r. spectral shifts. Rock and Hammaker[72] have examined the relationship between $^1J(C,H)$ and $\nu(CH)$ (the i.r. C—H stretching frequency) for 28 substituted aldehydes. The good correlation of these quantities and the similarly good correlation of $^1J(C,H)$ with the substituent constants σ^* and σ_I is presented as evidence that the effective

[66] N. J. Hoboken and E. R. Malinowski, *J. Amer. Chem. Soc.*, 1961, **83**, 4479.
[67] A. R. Tarpley and J. H. Goldstein, *J. Phys. Chem.* 1972, **76**, 515,
[68] A. R. Tarpley and J. H. Goldstein, *J. Mol. Spectroscopy*, 1971, **39**, 275.
[69] R. E. Hess, C. K. Haas, B. A. Kaduk, C. D. Schaeffer, and C. H. Yoder, *Inorg. Chim. Acta*, 1971, **5**, 161.
[70] H. A. Bent, *Chem. Rev.*, 1961, **61**, 275.
[71] C. H. Yoder and R. Schenck, *J. Inorg. Nuclear Chem.*, 1971, **33**, 2697.
[72] S. L. Rock and R. M. Hammaker, *Spectrochim. Acta*, 1971, **27A**, 1899.

nuclear charge on the carbon atom may play an important part in the variation of $^1J(C,H)$. The correlation of directly bonded (C,H) couplings with Hammett σ constants has been reported for substituted silanes[73,74] and for *ortho*-substituted toluenes, methoxybenzenes, and benzaldehydes.[75] Rosenberg and Drenth[76] have shown that:

$$^1J(C,H) = 6.69\delta(HC\equiv) + 235 \qquad (20)$$

with a correlation coefficient of 0.975 for 30 acetylenes provided that the chemical shift, $\delta(HC\equiv)$, is corrected for diamagnetic anisotropy. Contrary to their earlier suggestion, Neuman and Jonas[77] found no relationship between $^1J(C,H)$ for the NMe_2 group of some amides, thioamides, and amidinium ions and the rotational barriers in these systems. However, the amount of positive charge on NMe_2 should depend on the π-bond order for the central C—N bond, and values of $^1J(C,H)$ are found to parallel this parameter, which provides some support for the suggestion of Haake et al.[78] concerning the relationship between $^1J(C,H)$ and the charge on the atom attached to the methyl group.

Directly bonded (C,H) coupling constants have also been reported in references 79—91.

C. **Coupling between ^{15}N and ^1H.**—The majority of papers on this subject have concerned themselves with the effects of substituents and solvents on the coupling. Thus Axenrod and his co-workers[92,93] have recorded $^1J(^{15}N,H)$ values for a wide variety of substituted anilines. They find that electron-withdrawing substituents cause an increase in the magnitude of $^1J(^{15}N,H)$ which they interpret as being due to the favouring of sp^2 hybridization at the

[73] F. K. Cartledge and K. H. Riedel, *J. Organometallic Chem.*, 1972, **34**, 11.
[74] Y. Nagal, M. A. Ohtsuki, T. Nakano, and H. Watanabe, *J. Organometallic Chem.*, 1972, **35**, 81.
[75] R. E. Hess, C. D. Schaeffer, and C. H. Yoder, *J. Org. Chem.*, 1971, **36**, 2201.
[76] D. Rosenberg and W. Drenth, *Tetrahedron*, 1971, **27**, 3893.
[77] R. C. Neuman and V. Jonas, *J. Phys. Chem.*, 1971, **75**, 3532.
[78] P. Haake, W. B. Miller, and D. A. Tyssee, *J. Amer. Chem. Soc.*, 1964, **86**, 3577.
[79] J. P. Albrand, A. Cogne, D. Gagnaire, and J. B. Robert, *Tetrahedron*, 1971, **27**, 2453
[80] D. Ziessow, *J. Chem. Phys.*, 1971, **55**, 984.
[81] J. P. Jacobsen, J. T. Nielsen, and K. Schaumburg, *Acta. Chem. Scand.*, 1971, **25**, 2785.
[82] G. Miyajima and K. Takahashi, *J. Phys. Chem.*, 1971, **75**, 3766.
[83] P. J. Collin, *Austral. J. Chem.*, 1972, **25**, 425.
[84] M. Aritomi, Y. Kawasaki, and R. Okawara, *Inorg. Nuclear Chem. Letters*, 1972, **8**, 69.
[85] J. F. Hanlan and J. D. McCowan, *Canad. J. Chem.*, 1972, **50**, 747.
[86] R. Garnier, R. Faure, A. Babadjamian, and E.-J. Vincent, *Bull. Soc. chim. France*, 1972, 1040.
[87] H. Hüther and H. A. Brune, *Org. Magn. Resonance*, 1971, **3**, 737.
[88] G. C. Levy and D. C. Dittmer, *Org. Magn. Resonance*, 1972, **4**, 107.
[89] J. Dabrowski and L. J. Kozerski, *Org. Magn. Resonance*, 1972, **4**, 137.
[90] E. Lipmaa, M. Mägi, S. S. Novikov, L. I. Khmelnitski, A. S. Prihodko, O. V. Lebedev, and L. V. Epishna, *Org. Magn. Resonance*, 1972, **4**, 153.
[91] U. Svanholm, *Acta Chem. Scand.*, 1972, **26**, 459.
[92] T. Axenrod and M. J. Wieder, *J. Amer. Chem. Soc.*, 1971, **93**, 3541.
[93] T. Axenrod, P. S. Pregosin, M. J. Wieder, E. D. Becker, R. B. Bradley, and G. W. A. Milne, *J. Amer. Chem. Soc.*, 1971, **93**, 6536.

Nuclear Spin–Spin Coupling 65

nitrogen atom by such substituents. Good correlations of the coupling with Hammett substituent constants are found in both dimethyl sulphoxide (DMSO) and $CDCl_3$. Solvent effects are treated similarly;[92] the ability of a solvent to foster delocalization of the amino-group lone pair and to enhance the sp^2 hybridization at nitrogen increases the magnitude of $^1J(^{15}N,H)$. This point is also taken up by Wasylishen and Schaefer,[94] who point out that because of the solvent-dependence, care is necessary in determining nitrogen hybridization from one-bond ($^{15}N,H$) couplings. Liler[95] has measured $^1J(^{15}N,H)$ for [^{15}N]acetamide in water, DMSO, and acetone. He notes that solvent effects at the carbonyl group, *e.g.* hydrogen-bonding, also affect the ($^{15}N,H$) coupling by the mechanism outlined above.

A detailed analysis of the ^1H n.m.r. spectrum of [^{15}N]pyrrole using homonuclear INDOR has shown that all ($^{15}N,H$) couplings have the same sign.[96] Stilbs[97] has recorded values of $^1J(^{15}N,H)$ in adducts of BF_3 and $SbCl_5$ with ^{15}N-labelled 1,3-dimethylurea.

D. Coupling between ^{31}P and ^1H.—Stec *et al.*[98] have examined the effect of solvent protonic activity on the n.m.r. parameters of some phosphoryl and thiophosphoryl compounds. Increasing solvent acidity is accompanied by a small increase in the value of $^1J(P,H)$ and this is interpreted in terms of the formation of an adduct between the solute and the solvent. The authors were able to interpret these changes in terms of some MO calculations at the CNDO level.

Isotope effects in $^1J(P,H)$ have been reported by two groups. Stec *et al.*[99] found that the ratio $^1J(P,H)/^1J(P,D)$ for various dialkyl phosphates undergoes small but real solvent variations, particularly on going to hydrogen-bonded solvents. These differences are ascribed to the differences in the hydrogen-bonding capacity of the deuteriated and non-deuteriated species. Borisenko and his associates[100] found a deuterium isotope effect for $^1J(P,H)$ in eight three- and four-co-ordinate phosphorus compounds. If $^1J(P,D)$ is converted into $^1J^*(P,H)$ by multiplying by γ_H/γ_D ($= 6.51440$), then $\Delta J [= {}^1J^*(P,H) - {}^1J(P,H)]$ varies from -5.0 Hz to $+4.7$ Hz, being positive for three-co-ordinate and negative for four-co-ordinate phosphorus. Previously reported values of such a primary isotope effect have been negative in sign, so the positive result for the three-co-ordinate molecules suggests that the sign of $^1J(P,H)$ in these compounds might be opposite to that in the four-coordinate, where it is known to be positive.[101] However, triple-resonance experiments showed that $^1J(P,H) > 0$ in the three-co-ordinate compounds, as had already been assumed. The authors doubt that solvent effects can cause the different

[94] R. Wasylishen and T. Schaefer, *Canad. J. Chem.*, 1971, **49**, 3627.
[95] M. Liler, *J. Magn. Resonance*, 1971, **5**, 333.
[96] E. Rahkamaa, *Z. Naturforsch.*, 1971, **26a**, 1187.
[97] P. Stilbs, *Tetrahedron Letters*, 1972, 227.
[98] W. J. Stec, J. R. Van Wazer, and N. Goddard, *J. C. S. Perkin II*, 1972, 463.
[99] W. J. Stec, N. Goddard, and J. R. Van Wazer, *J. Phys. Chem.*, 1971, **75**, 3547.
[100] A. A. Borisenko, N. M. Sergeyev, and Yu. A. Ustynyuk, *Mol. Phys.*, 1971, **22**, 715.
[101] W. McFarlane, *J. Chem. Soc. (A)*, 1967, 1148.

sign of ΔJ. The isotope chemical shift $\Delta\delta_P$ is found to be an approximately linear function of $^1J(P,H)$, as shown in equation (21). The compounds

$$\Delta\delta_P = 1.50 - 1.80 \times 10^{-3} \times {}^1J(P,H) \qquad (21)$$

$C_6H_5PH_2$ and $H_2P(O)OH$ show a secondary isotope effect in that $^1J(P,H)$ for the system H—P—D differs from that for H—P—H in these molecules.[100] $^1J(P,H)$ values are also reported in references 102—107.

E. **Coupling between ^1H and Other Nuclei.**—Two groups have compared the behaviour of $^1J(^{29}Si,H)$ with that of $^1J(^{13}C,H)$. Cartledge and Riedel[73] have examined a range of benzyldimethylsilanes and phenyltetramethyldisilanes and find a good correlation with Hammett σ constants, the slopes of these plots being much greater for $^1J(Si,H)$ than for $^1J(C,H)$. Nagal et al.[74] find very similar results for molecules of general formula $YC_6H_4SiHR^1R^2$ (R^1 and R^2 = H or Me, Y is one of a variety of substituents). If Y is an electron-attracting substituent it increases the magnitude of $^1J(Si,H)$, and the authors tentatively suggest that the similarity of behaviour of $^1J(C,H)$ and $^1J(Si,H)$ may indicate that the latter depends on the s-character of the Si hybrid orbital bonded to H.

Cowley and Damasco[106] have measured $^1J(B,H)$ values in a series of phosphine–borane complexes and find an increase in the magnitude of the coupling as the π-acceptor strength of the uncomplexed phosphine increases. This could be interpreted in terms of the borane hyperconjugative model,[108] but since the trend is also one of increasing $^1J(B,H)$ with increasing electronegativity of substituents on phosphorus, the results might be interpreted equally well in terms of Bent's hypothesis,[70] i.e., as the electronegativity of the phosphorus substituents increases, more s-character is diverted into the B—H bonds, thus increasing $^1J(B,H)$. Jouany et al.[109] have measured $^1J(B,H)$ in adducts of the form $Cl_n(C_2H_5O)_{3-n}P \rightarrow BH_3$ and find that the coupling increases slightly in magnitude as n increases from 0 to 3.

Proton n.m.r. studies of diborane (6) and five methyl-substituted derivatives have been made by Leach, Ungermann, and Onak;[110] they find $^1J(B,H_T)$ =

$$\begin{array}{c} H \diagdown \quad /H\mu \diagdown \quad /H_T \\ \quad B \quad \quad B \\ H \diagup \quad \diagdown H \diagup \quad \diagdown H \end{array}$$

(6)

[102] J. Davis and J. E. Drake, *J. Chem. Soc. (A)*, 1971, 2094.
[103] P. G. Harrison, S. E. Ulrich, and J. J. Zuckerman, *Inorg. Nuclear Chem. Letters*, 1971, **7**, 865.
[104] P. G. Harrison, S. E. Ulrich, and J. J. Zuckerman, *Inorg. Chem.*, 1972, **11**, 25.
[105] G. A. Olah and C. W. McFarland, *Inorg. Chem.*, 1972, **11**, 845.
[106] A. H. Cowley and M. C. Damasco, *J. Amer. Chem. Soc.*, 1971, **93**, 6815.
[107] J. F. Nixon and J. R. Swain, *J.C.S. Dalton*, 1972, 1038.
[108] A. B. Burg, *Rec. Chem. Progr.*, 1954, **15**, 159.
[109] C. Jouany, G. Jugie, and J.-P. Laurent, *Bull. Soc. chim. France*, 1972, 880.
[110] J. B. Leach, C. B. Ungermann, and T. P. Onak, *J. Magn. Resonance*, 1972, **6**, 74.

129—133 Hz and $^1J(B,H_\mu)$ = 40—46 Hz. There are as yet insufficient data to correlate these couplings with structural parameters. One-bond (B,H) couplings have also been measured in derivatives of tetracarba-*nido*-hexaborane (7),[111] the values obtained being $^1J(B_1,H)$ = 202—206 Hz and $^1J(B_6,H)$ =

```
       H\         /H
         C ──H── C
        /  \  |  / \
    H─B₆───B₁───C─H
        \    |    /
         C
         |
         H
```

(7)

141—143 Hz, showing the marked dependence of $^1J(B,H)$ on the structure at the boron atom. Other directly bonded (B,H) couplings have been reported by Davis and Drake[102] and by Young *et al.*[112]

Martin and Fujiwara[113] have measured the (F,H) coupling in FHF⁻ generated in an aprotic solvent. Their value, and that for HF, are compared in Table 1 with the values calculated by Azman *et al.*[114] using CNDO and INDO SOS methods. As the Table shows, the ratio of the calculated couplings is in quite good agreement with the experimental ratio, though the absolute values are poor.

Table 1 *Calculated^a and experimental^b values of $^1J(F,H)/Hz$ in HF and HF$_2^-$*

	$^1J(F,H)$		
Molecule	Exp.	Calc. (*CNDO*)	Calc. (*INDO*)
HF	521	−384.4	−330.9
HF$_2^-$	120.5±0.3	−97.8	−82.8
Ratio	3.93	3.99	4.32

^{a} A. Azman, B. Borstnik, and J. Koller, *Theor. Chim. Acta*, 1969, **13**, 262.
^{b} J. S. Martin and F. Y. Fujiwara, *Canad. J. Chem.*, 1971, **49**, 3071.

F. $^1J(^{13}C,^{13}C)$, $^1J(^{13}C,^{15}N)$, $^1J(^{11}B,^{13}C)$, **and** $^1J(^{11}B,^{11}B)$.—The results of Maciel and his co-workers[25—27] on the calculation of $^1J(C,C)$ have been described in Section 3C. On the practical side, this work has been accompanied by the measurement of $^1J(C,C)$ in labelled ethyl[27] and isopropenyl[26] systems. The ranges of the couplings are 33.0—39.0 Hz in the former and 70.0—84.2 and 41.8—51.8 Hz in the latter, the larger and smaller magnitudes in the isopro-

[111] V. R. Miller and R. N. Grimes, *Inorg. Chem.*, 1972, **11**, 862.
[112] D. A. T. Young, R. J. Wiersema, and M. F. Hawthorne, *J. Amer. Chem. Soc.*, 1971, **93**, 5687.
[113] J. S. Martin and F. Y. Fujiwara, *Canad. J. Chem.*, 1971, **49**, 3071.
[114] A. Azman, B. Borstnik, and J. Koller, *Theor. Chim. Acta*, 1969, **13**, 262.

penyl case being for the formal double and single bonds, respectively. On the whole the couplings in the isopropenyl systems parallel each other quite closely and for both types of molecule the calculations indicate a dependence of the coupling on the $-I^+$ character of a substituent.

Ihrig and Marshall[115] have measured $^1J(C,C)$ values for eight substituted benzenes in which the substituent has a ^{13}C atom directly bonded to a carbon atom of the benzene ring. Substituents having sp^3, sp^2, and sp hybridization of the exocyclic carbon atom are included. The value of the one-bond coupling shows two very pronounced trends:

(a) Increasing the s-character of C-7 (the exocyclic carbon atom) causes $^1J_{17}$ to increase from 44·19 Hz in the sp^2–sp^3 system of toluene, through values in the region of 70 Hz for sp^2–sp^2, to 80·40 Hz for the sp^2–sp system of benzonitrile. This trend is quite consistent with the concept of the dependence of $^1J(C,C)$ on the s-character at the coupled carbons.

(b) The coupling also increases as the electronegativity of the substituent containing C-7 increases. This effect is much larger in the sp^2–sp^2 systems, which all involve a carbonyl group, than in the sp^2–sp^3 systems.

$^1J(C,C)$ has been determined from the ^{13}C n.m.r. spectra of ^{13}C-enriched chlorophylls a and b.[116]

Lichter and Roberts[117] have reported a detailed analysis of ^1H, ^{13}C, and ^{15}N n.m.r. spectra of ^{15}N-enriched pyridine and its hydrochloride. In connection with this work they have discussed (^{13}C, ^{15}N) coupling in some detail. They note that quite a number of $^1J(N,C)$ values from the literature are in agreement with equation (22), proposed by Binsch et al.,[118] which relates

$$^1J(N,C) = \alpha'_N \alpha'_C/80 \qquad (22)$$

$^1J(C,N)$ to the percentage s-character, α', in each of the orbitals forming the C—N bond. However, the very small value of $^1J(N,C)$ for pyridine [0.45 Hz in neat liquid, 0.7 Hz in 30% v/v methanol] can only mean that one or more of the assumptions made in equation (22) do not hold for this molecule, and the authors suggest that the average-energy approximation may be the cause. From a consideration of the $^1J(C,N)$ values above, and of that of the pyridinium ion (12.0 Hz), the authors suggest that the one-bond (C,N) couplings in neat pyridine and methanol solution have different signs. $^1J(^{13}C,^{15}N)$ values have also been reported by Coxon and Johnson.[119]

Odom, Ellis, and their co-workers have measured $^1J(B,B)$ in pentaborane(9)[120] (8) and $^1J(B,C)$ in its 1-methyl derivative;[121] the values are 19.4 ± 0.2 Hz and 72.6 ± 0.5 Hz respectively.

[115] A. M. Ihrig and J. L. Marshall, J. Amer. Chem. Soc., 1972, 94, 1756.
[116] C. E. Strouse, V. H. Kollman, and N. A. Matwiyoff, Biochem. Biophys. Res. Comm., 1972, 46, 328.
[117] R. L. Lichter and J. D. Roberts, J. Amer. Chem. Soc., 1971, 93, 5218.
[118] G. Binsch, J. B. Lambert, B. W. Roberts, and J. D. Roberts, J. Amer. Chem. Soc., 1964, 86, 5564.
[119] B. Coxon and L. F. Johnson, Carbohydrate Res., 1971, 20, 105.
[120] J. D. Odom, P. D. Ellis, and H. C. Walsh, J. Amer. Chem. Soc., 1971, 93, 3529.

(8)

G. Coupling between ^{13}**C and** ^{31}**P.**—Mann[122] has reported the $^1J(C,P)$ values of a series of tertiary phosphines and discussed them in terms of the s-character of the phosphorus and carbon bonding orbitals, a factor which he estimates from the molecular geometry. A plot of $^1J(C,P)/\alpha_C^2$ versus α^2 for the compounds PMe$_3$, PPh$_3$, and PMe$_4^+$ gives a very good straight line, from which a prediction of the CPC bond angle in PBu$_3^t$ may be made. The result is in excellent agreement with that estimated from chemical-shift data. This gives support to the idea that one-bond (C,P) couplings can be interpreted in terms of the Fermi-contact term only, but it should be noted that Simonnin, Lequan, and Wehrli[123] found no simple correlation between $^1J(C,P)$ and $^2J(PCH)$ in a series of substituted ethylene phosphines.

Jakobsen and his co-workers have measured $^1J(C,P)$ values, and other (C,P) couplings, in tri-2- and tri-3-thienylphosphine,[124] where they note, as do others,[122,123] that there is little difference between the magnitudes of $^1J(C,P)$ and $^2J(CCP)$, and indeed the two-bond coupling is usually the larger. In methyl-substituted trithienylphosphine derivatives containing either P^{III} or P^V they find evidence that $^1J(C,P)$ is positive for P^V and negative for P^{III}.[125]

Strongly stereospecific $^1J(C,P)$ values have been found by Gray and Cremer[126] in cis and trans pairs of phosphetan oxides and phosphetanium salts having structures (9) and (10) (X, Y = O, MePh, Bz, or OMe). $^1J(C,P)$ has been measured in four ditertiary phosphines by Marsmann and Horn.[127]

(9)   (10)

---

[121] P. D. Ellis, J. D. Odom, P. W. Lowman, and A. D. Cardin, *J. Amer. Chem. Soc.*, 1971, **93**, 6704.
[122] B. E. Mann, *J.C.S. Perkin II*, 1972, 30.
[123] M.-P. Simonnin, R.-M. Lequan, and F. W. Wehrli, *Tetrahedron Letters*, 1972, 1559.
[124] H. J. Jakobsen, T. Bundgaard, and R. S. Hansen, *Mol. Phys.*, 1972, **23**, 197.
[125] H. J. Jakobsen and M. Begtrup, *J. Mol. Spectroscopy*, 1971, **40**, 276.
[126] G. A. Gray and S. E. Cremer, *Tetrahedron Letters*, 1971, 3061.
[127] H. Marsmann and H.-G. Horn, *Z. Naturforsch.*, 1972, **27b**, 137.

H. Coupling between ^{13}C **and** ^{19}F.—A consistent set of data for seven isotopically substituted ethyl fluorides has been obtained from their ^1H, ^2H, and ^{19}F n.m.r. spectra.[24] The results are discussed in terms of CNDO/2 and INDO SOS calculations including orbital and spin-dipolar terms. Generally speaking, the agreement is poor. $^1J(C,F)$ values have been reported for some fluorochloroacetones[128] and for fluorobenzene.[67]

I. Coupling between ^{13}C **and Metals.**—Mann has measured $^1J(C,^{183}W)$, which is 189 ± 2 Hz in $[(durene)W(CO)_3]$,[129] and $^1J(C,^{57}Fe)$ in $[Fe(CO)_5]$, whose value is 23.4 ± 0.4 Hz.[130] The reduced (C,Fe) coupling (23.8×10^{20} N $A^{-2}m^{-3}$) is therefore intermediate between $^1K(C,^{51}V)$ and $^1K(C,^{59}Co)$, which are 14.6×10^{20} and 40.2×10^{20} N $A^{-2}m^{-3}$, respectively. Carbon-tungsten one-bond couplings have also been reported for some tungsten carbonyl derivatives by Gansow et al.[131] The couplings show little change on going from $W(CO)_6$ (126 Hz) to (cyclohexylamine)$W(CO)_5$ (132 Hz), and the authors tentatively propose that this might be the result of two opposing effects, the increase in $S_w^2(0)$ on the one hand and the decrease of the W—C bond order on the other, when going from CO to cyclohexylamine.

$^1J(C,^{195}Pt)$ values have been obtained from the ^{13}C n.m.r. spectra of a number of methylplatinum compounds.[132] Values range from 470 to 689 Hz. In the complexes trans-$[PtMe(L)(AsMe_3)_2]^+$, where L = CO or C(OMe)Me, a comparison of $^1J(C,Pt)$ for approximately sp^3-, sp^2-, and sp- hybridized carbon atoms can be made. The magnitudes decrease in the order CO > C(OMe)Me > Me, which suggests that the carbon s-character might be important in these couplings. Variations of bond strength cannot, however, be ruled out.

The directly bonded ($^{13}C,^{73}Ge$) coupling in tetramethylgermane has been measured;[50] its magnitude is 18.7 ± 0.9 Hz.

J. Coupling to ^{19}F, **except** $^1J(^{19}F,^{31}P)$ **and** $^1J(^{19}F,^{13}C)$.—(B,F) Couplings in adducts of BF_3 with methylhydrazines[133] and with pyridine, quinoline, and acridine N-oxides[134] have been reported. For the N-oxide adducts the values range from 3.3 to 5.5 Hz, the small magnitudes being due to two opposing contributions, as had been suggested previously.[135] From the manner in which the couplings change with substituent the authors conclude[134] that $^1J(B,F)$ is negative. Lappert et al.[136] have measured $^1J(B,F)$ values for eight

[128] L. H. Sutcliffe and B. Taylor, *Spectrochim. Acta*, 1972, **28A**, 619.
[129] B. E. Mann, *Chem. Comm.*, 1971, 976.
[130] B. E. Mann, *Chem. Comm.*, 1971, 1173.
[131] O. A. Gansow, B. Y. Kimura, G. R. Dobson, and R. A. Brown, *J. Amer. Chem. Soc.*, 1971, **93**, 5922.
[132] M. H. Chisholm, H. C. Clark, L. E. Manzer, and J. B. Stothers, *Chem. Comm.*, 1971, 1627.
[133] L. K. Peterson and G. L. Wilson, *Canad. J. Chem.*, 1971, **49**, 3171.
[134] R. S. Stephens, S. D. Lessley, and R. O. Ragsdale, *Inorg. Chem.*, 1971, **10**, 1610; R. Kuhlman and D. M. Grant, *J. Phys. Chem.*, 1964, **68**, 3208.
[135] R. J. Gillespie, J. S. Hartman, and M. Parekh, *Canad. J. Chem.*, 1969, **46**, 1601.
[136] M. F. Lappert, M. R. Litzow, J. B. Pedley, and A. Tweedale, *J. Chem. Soc. (A)*, 1971, 2426.

boron trihalides and mixed trihalides and have found a good correlation of the coupling with an electronegativity parameter $I_{F(yz)}$ defined, for a molecule of general formula BFYZ, by equation (23), in which IP denotes the first

$$I_{F(YZ)} = IP(BY_3) - IP(BF_3) + IP(BZ_3) - IP(BF_3) \quad (23)$$

ionization potential of the appropriate trihalide. Attempts to rationalize the coupling constants using EHT and the Pople–Santry equation [equation (2)] gave no correlation at all with experiment.[136] Using a modification of the EHT, somewhat better results were obtained, but in general terms it would appear doubtful that much sense can be made of these couplings using the contact term only; this conclusion is supported by the work of Hartman and Schrobilgen.[137] They have measured $^1J(B,F)$ in eight mixed tetrahalogenoborate ions, and find couplings ranging from 1.0 Hz in BF_4^- to 111.3 Hz in $BFBr_3^-$. The coupling constants, and the ^{11}B and ^{19}F chemical shifts, are well reproduced by a set of pair-wise additivity constants, *i.e.*

$$J = \sum_{\text{pairs}} \zeta_{ij} \quad (24)$$

where ζ_{ij} is a parameter associated with the substituents i and j and independent of all other substituents.[138] The fact that the coupling constants are well predicted by this additivity method suggests, according to Vladimiroff and Malinowski,[138] that the Fermi-contact term is not dominant in $^1J(B,F)$ and that orbital and spin-dipolar terms are also important.

Kidd and Matthews[139] have found $^1J(^{121}Sb,F) = 1934 \pm 15$ Hz and $^1J(^{123}Sb,F) = 1047 \pm 25$ Hz from the ^{19}F n.m.r. spectrum of silver hexafluoroantimonate(v) in acetonitrile. Fukushima[140] has measured $^1J(^{209}Bi,F)$ in $KBiF_6$ powder, *i.e.* BiF_6^-; it is 2700 ± 300 Hz. Fukushima notes[140] that $^1J(M,F)$ in MF_6^- ions, where M is a Group VA element, is proportional to the square of the atomic number of M. The regular way in which the coupling constants are related suggests that all these MF pairs have reduced coupling constants of the same sign, namely negative.

Howell and Moss[141] have determined $^1J(^{51}V,F)$ in the VF_6^- ion obtained from $AgVF_6$ in acetonitrile at $-20\,°C$; its magnitude is 88 Hz. $^1J(^{29}Si,F)$ values ranging from 202 to 219 Hz have been reported[142] for seven F_3Si— and —SiF_2—amines.

Gillespie *et al.*[143] have measured the ^{19}F spectra of solutions of mixtures of XeF_4 and XeF_2 in SbF_5. In these solutions the XeF_3^+ ion can be detected; it has C_{2v} symmetry, with two axial fluorine atoms and one fluorine and two

[137] J. S. Hartman and G. J. Schrobilgen, *Inorg. Chem.*, 1972, **11**, 940.
[138] T. Vladimiroff and E. R. Malinowski, *J. Chem. Phys.*, 1967, **46**, 1830.
[139] R. G. Kidd and R. W. Matthews, *Inorg. Chem.*, 1972, **11**, 1156.
[140] E. Fukushima, *J. Chem. Phys.*, 1971, **55**, 2463.
[141] J. A. S. Howell and K. C. Moss, *J. Chem. Soc. (A)*, 1971, 2483.
[142] W. Airey, G. M. Sheldrick, B. J. Aylett, and I. A. Ellis, *Spectrochim. Acta*, 1971, **27A**, 1505.
[143] R. J. Gillespie, B. Landa, and G. J. Schrobilgen, *Chem. Comm.*, 1971, 1543.

lone pairs in the equatorial plane. The coupling constants are found to be $^1J(^{129}Xe,F_{ax}) = 2620$ Hz, $^1J(^{129}Xe,F_{eq}) = 2440$ Hz. XeF^+ ions are also present in the solution and for these a (^{129}Xe, F) coupling of 7260 Hz was measured.

K. Coupling between ^{31}P and ^{19}F.

—Harris et al.[144] have measured $^1J(P,F)$ in a number of diazadiphosphetidines (11) and phosphadiazetidinones (12). The values of $^1J(P,F)$ lie in the normal range. A slight increase in the average value of the one-bond (P,F) coupling is noticeable on going from the open com-

```
        Me                    Me
        |                     |
        N                     N
       / \                   / \
  YF₂P   PF₂Y          O=C     PF₂Y
       \ /                   \ /
        N                     N
        |                     |
        Me                    R
       (11)                  (12)
```

pounds YPF_3NMe_2 (Y = Me, Et, NEt_2, or Ph) to the cyclic systems (11, Y = Me, Et, NEt_2, or Ph) and a further increase on going to (12; R = Me; Y = F, Me, Et, or Ph). This may be related to the decrease in the axial–equatorial bond angle from 90° in the open compounds to 82° in (11) and probably less in (12). There is no discernible trend in $^1J(P,F)$ with substituent parameters, but differences in the coupling between *gauche*- and *trans*-isomers are considerable (ca. 40 Hz).

Schmutzler and his co-workers have reported values of $^1J(P,F)$ for piperidylfluorophosphoranes,[145] dioxabenzofluorophospholes,[146] and aryloxy-substituted fluorophosphoranes.[147] Binder[148] has given values for pyrocatechyl phosphorus trihalides and Olah and McFarland[105] for fluorophosphorus compounds in fluorosulphonic acid solution.

L. Other One-bond Couplings involving ^{31}P.

—McFarlane and McFarlane[149] have investigated the dependence of $^1J(P,P)$ on the internal rotation in biphosphines and have observed a difference of 19 Hz in the coupling for the two diastereomers of 1,2-dimethyl-1,2-diphenylbiphosphine, (13) and (14).

---

[144] R. K. Harris, J. R. Woplin, R. E. Dunmur, M. Murray, and R. Schmutzler, *Ber. Bunsengesellschaft phys. Chem.*, 1972, **76**, 44.
[145] M. J. C. Hewson, S. C. Peake, and R. Schmutzler, *Chem. Comm.*, 1971, 1454.
[146] M. Eisenhut and R. Schmutzler, *Chem. Comm.*, 1971, 1452.
[147] S. C. Peake, M. Fild, M. J. C. Hewson, and R. Schmutzler, *Inorg. Chem.*, 1971, **10**, 2723.
[148] H. Binder, *Z. anorg. Chem.*, 1971, **384**, 193.
[149] H. C. E. McFarlane and W. McFarlane, *Chem. Comm.*, 1971, 1589.

Me–P̈–P̈–Me   Me–P̈–P̈–Ph
 /   \   /   \
Ph   Ph   Ph   Me

   (13)   (14)

The sign of each coupling is found, by means of $^1H-\{^{31}P\}$ tickling experiments, to be negative relative to positive $[^2J(P,H) + {}^3J(P,H)]$. The authors point out that inversion at tervalent phosphorus in biphosphines at ordinary temperatures is slow whereas rotation about the P—P bond is rapid, and owing to differences in steric interaction the rotamer populations of the two diastereomers will be different. Thus any difference in $^1J(P,P)$ between different rotamers will show up as a difference between the two diastereomers, and they therefore suggest that their measurements provide experimental evidence for a dependence of $^1J(P,P)$ upon rotation about the P—P bond. This seems to be the first example of this phenomenon for directly bonded nuclei. The effect of solvent acidity on (P,P) coupling in tetraethylmonothiohypophosphate has been examined and the results have been compared with those from CNDO calculations.[98]

Workers in several laboratories[106,109,150] have correlated the value of $^1J(B,P)$ in adducts of substituted phosphines and borane with the strength of the Lewis base. Rudolph and Schultz[150] have made a careful investigation of this correlation and have summarized many experimental results. By means of $^1H-\{^{31}P\}$ and $^{19}F-\{^{31}P\}$ tickling experiments they have established that $^1J(B,P)$ in $Me_3P \rightarrow BH_3$, $Me_3P \rightarrow BF_3$, $F_2HP \rightarrow BH_3$, and $(Me_2N)F_2P \rightarrow BH_3$ is positive if $^1J(B,H) > 0$ and $^1J(B,F) < 0$. It appears that attempts to rationalize $^1J(B,P)$ values in terms of s-electron density at phosphorus will prove difficult in view of the relatively small values of this coupling for compounds such as $F_3P \rightarrow BH_3$ and $ClF_2P \rightarrow BH_3$.

Zuckerman et al.[103] have used tickling experiments to determine the sign of $^1J(^{119}Sn,P)$ in $Me_3SnPHPh$. It is found to be positive relative to $^1J(P,H) > 0$ and may therefore be considered to be absolutely positive, its magnitude being 538 ± 3 Hz. The corresponding coupling in Me_3SnPPh_2 is 586 ± 10 Hz.[103] The authors conclude that these and other results support the view that s-electron density at phosphorus is concentrated in the bonding orbitals directed towards Sn > H > Ph in that order, which is the opposite of the order of the electronegativities as is required by Bent's rules.[70] This analysis presupposes and to some extent supports the idea that the value of $^1J(Sn,P)$ is Fermicontact dominated.

$^1J(^{183}W,P)$ values have been reported for complexes of the form $W(CO)_5L$, [131,151] and Fischer et al.[151] have summarized the available data on these couplings and have shown that they correlate well with the electronegativity

---

[150] R. W. Rudolph and C. W. Schultz, *J. Amer. Chem. Soc.*, 1971, **93**, 6821.
[151] E. O. Fischer, L. Knauss, R. L. Keiter, and J. G. Verkade, *J. Organometallic Chem.*, 1972, **37**, C7.

of the ligand L. Other one-bond couplings of phosphorus to transition metals have been reported, for ^{103}Rh^{152,153} and ^{57}Fe.^{130}

5 Coupling between Nuclei Separated by Two Chemical Bonds, ^2J

A. Geminal Proton–Proton Coupling across Carbon.—A theoretical investigation of this coupling[28] has already been mentioned in Section 3. The results are very poor and the incorrect sign is obtained in the case of cyclopropane. The effects of substituents on ^2J(HCH) have been examined in several recent papers. Cookson and Crabb[154] have discussed the effect of the electronegativities of X and Y on the geminal coupling in X—CH$_2$—Y in considerable detail, and they have provided a useful, if brief, bibliography of papers in this field, which has been the subject of some dispute. They particularly emphasize the importance of a consideration of ring size and of the effect of lone pairs, and they conclude their paper with two warnings: (a) in 1,3-heterosystems the electronegativity effects are not simply additive, the effect of the two heteroatoms being more than twice that of one, so that the mutual interaction of the heteroatoms cannot be ignored; (b) only values from related systems should be compared, since molecules in which lone-pair and electronegativity effects might appear to be the same can show different ^2J(HCH) values.

Davies and Hudec[155] have investigated the variation of geminal (H,H) coupling with respect to the orientation of the lone pairs on adjacent sulphur and oxygen atoms. Empirical relationships are derived and presented in graphical form. For oxygen, sulphur, and the sulphone group the curves are similar in shape and correspond quite closely to a theoretical curve for methanol.[156] However, the curve for the sulphoxide group is rather different and shows the asymmetry of the group in that it, unlike the others, is not symmetrical about 180°. Cyclic systems can be included in the correlations if appropriate allowances are made.

Fraser and Raby[157] have reported values for the geminal coupling of the allylic methylene protons in a series of selectively deuteriated cyclohexene derivatives. The effect of substituents β to the methylene group is found to agree with theoretical predictions.[40] The correlation of the (H,H) coupling with a number of substituent parameters is investigated but the results are poor, and the best correlation is with the chemical shifts of the vinyl protons. Reasons for this are suggested.

Bailey et al.[158] have measured and discussed geminal couplings in the side-chain protons of substituted ethylbenzenes. Brink and Larsson[159] have

---

[152] B. E. Mann, C. Masters, B. L. Shaw, and R. E. Stainbank, *Chem. Comm.*, 1971, 1103.
[153] B. E. Mann, C. Masters, and B. L. Shaw, *J.C.S. Dalton*, 1972, 704.
[154] R. C. Cookson and T. A. Crabb, *Tetrahedron*, 1972, **28**, 2139.
[155] R. Davies and J. Hudec, *Chem. Comm.*, 1972, 124.
[156] G. E. Maciel, J. W. McIver, N. S. Ostlund, and J. A. Pople, *J. Amer. Chem. Soc.*, 1970, **92**, 4151.
[157] R. R. Fraser and B. F. Raby, *J. Amer. Chem. Soc.*, 1972, **94**, 3458.
[158] K. Bailey, A. W. By, K. C. Graham, and D. Verner, *Canad. J. Chem.*, 1971, **49**, 3143.
[159] M. Brink and E. Larsson, *Tetrahedron*, 1971, **27**, 3875.

determined ^2J(HCH) values for a series of monobenzylmercapto- and *gem*-dibenzylmercapto-compounds. They have discussed the influence of structure and solvent on these couplings, dividing the compounds up into four groups which have characteristic values of ^2J(HCH). The effect of solvent on ^2J(HCH) has also been investigated by Cox and Harrison[160] and by Ihrig and Smith.[161] The former interpret their results in terms of the analysis of this coupling by Pople and Bothner-By.[40] Ihrig and Smith[161] are concerned with the case of the three difluoroethylenes and conclude that their data, in conjunction with previous results, confirm that the orientation of the solute dipole affects the magnitude of ^2J(HCH) and ^2J(HCF). Dipole orientation is not, however, a factor in determining solvent effects on vicinal (H,H) and (F,F) couplings.

The spectrum of 1,1-difluoroethylene has been measured[162] in two nematic solvents. ^2J(HCH) is found to be negative in sign and to change but little in magnitude on going from the isotropic to the nematic phase of the liquid crystal. Signs and magnitudes of couplings, including ^2J(HCH), have been obtained from detailed analyses of the high-resolution ^{1}H n.m.r. spectra of 1,4-dibromobutane[163] and of 2,4-dichloropent-1-ene and 2,4-dichloro-[1-^{2}H]pent-1-ene.[164] Tickling experiments have been used to determine the sign of the (H_b, H_c) coupling in (15).[165] It is found to be negative (-9.5 Hz) relative to a positive value of ^5J(H—C=C=C=C—H), which is positive relative to positive ^3J(H—C—C—H) in (16).

$$\begin{array}{cc}
\text{EtO}\diagdown\diagup H_b & \text{Bu}^t\text{O}\diagdown\diagup H_b \\
C\!=\!C\!=\!C\!=\!C & C\!=\!C\!=\!C\!=\!C \\
H_a\diagup\diagdown H_c & H_a\diagup\diagdown \text{Me} \\
(15) & (16)
\end{array}$$

Fraser, Petit, and Miskow[166] have examined the effect of chiral shift reagents on the coupling of benzylic protons, and they find that this enables them to measure ^2J(HCH) directly. These results can then be compared with the corresponding values calculated from measurements of ^2J(HCD). For benzyl methyl sulphoxide in acetone they find the directly measured (H,H) coupling to be 12.97 ± 0.02 Hz while ^2J(HCD) = 1.92 ± 0.01 Hz, giving an (H,H) coupling of 12.51 Hz. The difference of 0.46 ± 0.2 Hz represents a significant isotope effect; the possible effect of deuterium nuclear quadrupole relaxation was shown to be zero by making measurements at 34 and 45 °C.

---

[160] R. H. Cox and L. W. Harrison, *J. Magn. Resonance*, 1972, **6**, 84.
[161] A. M. Ihrig and S. L. Smith, *J. Amer. Chem. Soc.*, 1972, **94**, 34.
[162] J. Gerritsen and C. MacLean, *J. Magn. Resonance*, 1971, **5**, 44.
[163] D. Aksnes, *Acta Chem. Scand.*, 1972, **26**, 164.
[164] G. Gurato and A. Rigo, *Org. Magn. Resonance*, 1971, **3**, 433.
[165] M. L. Martin, F. Lefevre, and R. Mantione, *J. Chem. Soc. (B)*, 1971, 2049.
[166] R. R. Fraser, M. A. Petit, and M. Miskow, *J. Amer. Chem. Soc.*, 1972, **94**, 3253.

Values of ^2J(HCH) have also been reported in references 24, 82, and 167—193.

B. Geminal Proton–Proton Coupling across Other Elements.—$^2J(\mathrm{H_T BH}_\mu)$ values have been reported for diborane (6) and five methyl-substituted derivatives;[110] the values range from 7.5 to 8.7 Hz but the data were insufficient to correlate couplings with structure. ^2J(HNH) values have been given for 2-chloroaniline,[94] and for [^{15}N]formamide[95] and [^{15}N]acetamide.[95] From a consideration of the magnitudes and signs of ^2J(COC) (-2.8 ± 0.2 Hz) and ^2J(COH) (-2.6 Hz) in CH_3OCOCl and CH_3OH respectively, Ziessow[80] has suggested that ^2J(HOH) in water is negative, in agreement with a calculation by Pople et al.[4]

Tentative values of ^2J(HMoH) have been reported.[194]

C. $^2J(^{13}CC^1H)$. Calculations[25,28,34] of this coupling have already been described (Section 3); it is perhaps useful to note here that, insofar as suitable experimental data are available for comparison, INDO FPT calculations can predict these couplings reasonably well but EHT SOS and VB SOS methods are less successful.

[167] P. L. Durette and D. Horton, *Org. Magn. Resonance*, 1971, **3**, 417.
[168] B. Pedersen, P. Klaeboe, and T. Torgrimsen, *Acta Chem. Scand.*, 1971, **25**, 2367.
[169] L. A. Sternson, D. A. Coviello, and R. S. Egan, *J. Amer. Chem. Soc.*, 1971, **93**, 6529.
[170] M. Brink and E. Larsson, *Tetrahedron*, 1971, **27**, 5713.
[171] M. Revel, M. Bon, and J. Navech, *Compt. rend.*, 1972, **274**, C, 430.
[172] J. L. Sudmeier, G. L. Blackmer, C. H. Bradley, and F. A. L. Anet, *J. Amer. Chem. Soc.*, 1972, **94**, 757.
[173] K. Pihlaja, G. M. Kellie, and F. G. Riddell, *J.C.S. Perkin II*, 1972, 252.
[174] R. G. Jones, P. Partington, W. J. Rennie, and R. M. G. Roberts, *J. Organometallic Chem.*, 1972, **35**, 291.
[175] A. Geens, M. Anteunis, F. de-Pessemier, J. Fransen, and G. Verhegghe, *Tetrahedron*, 1972, **28**, 1097.
[176] M. Anteunis, G. Swaelens, F. Anteunis-de Ketelaere, and P. Dirinck, *Bull. Soc. chim. belges*, 1971, **80**, 409.
[177] P. Dirinck and M. Anteunis, *Canad. J. Chem.*, 1972, **50**, 412.
[178] M. Anteunis and P. Dirinck, *Canad. J. Chem.*, 1972, **50**, 423.
[179] M. Anteunis, F. Anteunis-de Ketelaere, and F. Borremans, *Bull. Soc. chim. belges*, 1971, **80**, 701.
[180] P. Maroni and J.-P. Gorrichon, *Bull. Soc. chim. France*, 1972, 785.
[181] H. Brouwer and J. B. Stothers, *Canad. J. Chem.*, 1972, **50**, 601.
[182] I. Stassinopoulou and C. Zioudrou, *Tetrahedron*, 1972, **28**, 1257.
[183] D. Gagnaire and P. Vottero, *Bull. Soc. chim. France*, 1972, 873.
[184] M. Anteunis, G. Verhegghe, and T. Rosseel, *Org. Magn. Resonance*, 1971, **3**, 693.
[185] N. M. Sergeyev, G. I. Auramenko, V. A. Korenevsky, and Yu.A. Ustynyuk, *Org. Magn. Resonance*, 1972, **4**, 39.
[186] R. Cahill, and T. A. Crabb, *Org. Magn. Resonance*, 1972, **4**, 259.
[187] M. Anteunis and M. Vandewalle, *Spectrochim. Acta*, 1971, **27A**, 2119.
[188] H. Günther and H. Klose, *Chem. Ber.*, 1971, **104**, 3898.
[189] K. L. Williams, S. Mosser, and D. E. Stedman, *J. Amer. Chem. Soc.*, 1971, **93**, 7208.
[190] E. Dradi, G. Tosolini, and G. Gatti, *J. Magn. Resonance*, 1972, **6**, 565.
[191] K. D. Bartle, D. W. Jones, and R. L'Amie, *J.C.S. Perkin II*, 1972, 646.
[192] K. D. Bartle, D. W. Jones, and R. L'Amie, *J.C.S. Perkin II*, 1972, 650.
[193] R. H. Cox, B. S. Campbell, and M. G. Newton, *J. Org. Chem.*, 1972, **37**, 1557.
[194] J. P. Jesson, E. L. Muetterties, and P. Meakin, *J. Amer. Chem. Soc.*, 1971, **93**, 5261.

Tarpley and Goldstein[68] have analysed the ^{13}C n.m.r. spectra of *m*-dichloro-, *m*-dibromo-, and *m*-di-iodo-benzene and have found a positive sign for $^2J(CCH)$ [relative to $^1J(CH) > 0$] and $|^2J(CCH)| < |^3J(CCCH)|$ in all cases. Linear correlation with halogen electronegativity is found for all the (C,H) couplings. Similar observations have been made by the same authors for uracil, thymine, and 5-halogenouracils.[195] The value of $^2J(CCH)$ has been measured in β-1,2,3,4,5,6-hexachlorocyclohexane (all the chlorine atoms are equatorial) by means of Fourier-transform spectroscopy.[65] The coupling is rather large and negative (-5.72 ± 0.04 Hz) and the authors suggest that this could be due to a dependence of the coupling on the dihedral angle between the C—H bond and that of the second carbon to a chlorine atom.

Detailed analyses of the spectra of isotopically substituted ethyl fluorides[24] and thiazoles[44] have been reported. In the case of the thiazoles, signs were determined by tickling experiments and all (C,H) couplings were found to be positive. In both cases, attempts to rationalize the couplings by means of MO calculations gave only moderate satisfaction. All the (C,H) coupling constants

$$\begin{array}{c} H \quad\quad H \\ \big| \quad\quad \big| \\ H - \!\!\!-\!\!\!- SO_2 \\ \big| \\ H \end{array}$$

(17)

for thiete sulphone (17) have been measured;[88] they indicate strain and rehybridization of the ring.

Other reports of $^2J(CCH)$ can be found in references 79, 86, 87, 91, 167, and 179.

D. Coupling of Other Nuclei to Protons *via* Carbon, $^2J(XC^1H)$.—We divide up these couplings according to the Group in the Periodic Table to which X belongs, the transition metals being dealt with collectively at the end.

(*i*) X = ^{199}Hg. Johannesen and Duerst[196] have measured $^2J(HgCH)$ in methyl(perfluorovinyl)mercury, $(CF_2{=}CF)HgMe$, for which a (Hg,H) coupling of -139.39 Hz was determined by means of $^{19}F-\{^{199}Hg\}$ and $^1H-\{^{199}Hg\}$ tickling experiments. The sign is relative to a negative $^3J(FCCF)$. $^2J(HgCH)$ in MeHgR has been compared with $^3J(HgSiCH)$ in Me$_2$SiHgR,[197] and the linear relationship between these two couplings for changing R is taken to indicate that the coupling mechanism is essentially the same in both. Similar conclusions have been drawn with respect to the parallel behaviour of $^2J(HgCH)$ and $^3J(HgPCH)$ in complex Hg^{II} cations.[198]

$^2J(HgCH)$ values are also reported in references 199–201.

[195] A. R. Tarpley and J. H. Goldstein, *J. Amer. Chem. Soc.*, 1971, **93**, 3573.
[196] R. B. Johannesen and R. W. Duerst, *J. Magn. Resonance*, 1971, **5**, 355.
[197] T. F. Schaff and J. P. Oliver, *Inorg. Chem.*, 1971, **10**, 1521.
[198] P. L. Goggin, R. J. Goodfellow, S. R. Haddock, and J. G. Eary, *J.C.S. Dalton*, 1972, 647.

(ii) X = ^{11}B and ^{27}Al. Bogdanov et al.[202] have measured ^2J(BCH) in some alkoxyvinylboranes and triallylborane complexes with amines, finding values ranging from 0.0 to 6.4 Hz. They have determined these couplings from the line broadening in the ^{11}B and ^{1}H spectra by application of the theory of Soloman and Bloembergen.[203,204] Westmoreland and his co-workers[205] have measured the spectrum of NaAlEt$_4$ in dimethoxyethane at 35 °C. Under these conditions the methylene protons show nine lines split (fortuitously) equally by the three equidistant protons on the adjacent methyl group, ^3J(HCCH), and by the ^{27}Al nucleus, ^2J(AlCH), which has $I = ^5/_2$. The magnitude of the splitting is 7.3 ±0.1 Hz, which is therefore the value of ^2J(AlCH).

(iii) X = ^{73}Ge,[119,117] Sn, and ^{207}Pb. The value of ^2J(GeCH) = 2.99±0.03 Hz has been determined from the spectrum of GeMe$_4$ by a curve-fitting technique.[206]

Gielen and his associates[207,208] have measured ^2J(SnCH) values for a large number of organotins. They find[208] that J is not related to $\Sigma\sigma^*$, the sum of the Taft σ^* constants for the substituents bound to tin, but Malinowski's additivity rule[209] can successfully be used to predict the couplings. $^2J(^{119}$Sn CH) in Me$_3$SnPHPh has been found[103] to be +52.7±0.3 Hz relative to ^1J(P,H) > 0, and the positive sign is also inferred for Me$_3$SnPPh$_2$, where the (^{119}Sn,H) coupling is 52.2±0.2 Hz.

Campbell and Green[200] have studied the variation of ^2J(HgCH) and ^2J(SnCH) with temperature. Aritomi et al.[84] have measured ^2J(SnCH) and ^2J(PbCH) in bis(acetylacetonato)dimethyl-tin and -lead in a variety of solvents. They observe that whereas the coupling to lead increases markedly with the donor strength of the solvents, the coupling to tin remains almost constant. The former appears to be very sensitive to the amount of charge on the lead atom.

(iv) X = ^{15,14}N and ^{31}P. McFarlane and McFarlane[210] have examined the spectra of five quaternary ammonium salts derived from pyridine. ^{1}H-{^{14}N} tickling combined with simultaneous decoupling of substituent protons enabled the authors to show that $^2J(^{14}$NCH) and $^3J(^{14}$NCCH) are both positive relative to a positive ^3J(HCCH) in the aromatic ring. The $^2J(^{14}$NCH)

[199] F. G. Thorpe, T. N. Huckerby, P. H. Lindsay, and S. W. Beuer, *Tetrahedron Letters*, 1971, 2821.
[200] C. H. Campbell and M. L. H. Green, *J. Chem. Soc.* (A), 1971, 3282.
[201] W. Kitching, M. L. Bullpitt, P. D. Sleezer, S. Winstein, and W. G. Young, *J. Organometallic Chem.*, 1972, **34**, 233.
[202] V. S. Bogdanov, A. V. Kessenikh, and V. V. Negrebetsky, *J. Magn. Resonance*, 1971, **5**, 145.
[203] I. Soloman and N. Bloembergen, *J. Chem. Phys.*, 1956, **25**, 261.
[204] N. Bloembergen, *J. Chem. Phys.*, 1957, **27**, 572.
[205] T. D. Westmoreland, N. S. Bhacca, J. D. Wander, and M. C. Day, *J. Organometallic Chem.*, 1972, **38**, 1.
[206] J. Kaufmann, W. Sahm, and A. Schwenk, *Z. Naturforsch.*, 1971, **26a**, 1384.
[207] M. Gielen, N. Goffin, and J. Topart, *J. Organometallic Chem.*, 1971, **32**, C38.
[208] M. Gielen, M. De Clercq, and B. De Poorter, *J. Organometallic Chem.*, 1972, **34**, 305.
[209] T. Vladimiroff and E. R. Malinowski, *J. Chem. Phys.*, 1965, **42**, 440.
[210] H. C. E. McFarlane and W. McFarlane, *Org. Magn. Resonance*, 1972, **4**, 161.

values range from 0.8 to 1.05 ± 0.2 Hz and the positive sign may be a little surprising in view of the fact that the reduced couplings ^2K(XCH), (X = H, Si, Sn, or P^V) are negative when the carbon is saturated, *i.e.* for a bond angle of *ca.* 109°. This might be explained by the fact that the NCH bond angle is close to 120°, and in thiazole, where the angle should be similar, tickling experiments have given a negative sign for both $^2J(^{15}$NCH) and $^3J(^{15}$NCCH) relative to the three (H,H) couplings, which were all found to be positive.[44] Homonuclear INDOR methods have been used to show that $^1J(^{15}$N,H), $^2J(^{15}$NCH), and $^3J(^{15}$NCCH) have the same sign in [^{15}N]pyrrole.[96] The whole subject of (^{15}N,C) and (^{15}N,H) coupling has been discussed in some detail by Lichter and Roberts,[117] who have compiled a useful table of representative values. They show how $^2J(^{15}$NCH) and $^3J(^{15}$NCCH) can be rationalized by a close inspection of the coupling path and the orientation of the nitrogen lone pair, and they emphasize the relationship of these couplings to the corresponding ones involving carbon and phosphorus. The effect of solvent and protonation on $^2J(^{15}$NCH) is discussed, and it is suggested that the decrease in the magnitude of the coupling on protonation of the lone pair (−10.06 to −3.01 Hz for pyridine) is due to the increase of the $n \rightarrow \pi^*$ excitation energy which this causes. This type of argument is familiar in the qualitative discussion of chemical shifts, and a similar theory has been advanced by Price[211] to explain the large negative value of $^2J(^{15}$NCH) in pyridine as compared with ^2J(CCH) in benzene. Coxon and Johnson[119] have also reported some values of $^2J(^{15}$NCH).

Like ^2J(NCH), ^2J(PCH) can also be found with either sign. Lequan and Simonnin[212] have measured the ^{1}H n.m.r. spectra of six compounds of the form R_2^1P—CH=CH—R^2 and have determined that in all but one of them the signs of ^2J(PCH) and ^3J(PCCH) are the same. Since the sign of the latter (P,H) coupling is generally believed to be positive,[213] this implies a change of sign for ^2J(PCH). Schmidbaur *et al.*[214] have studied the system Me$_3$PAuMe-Me$_3$P as a function of the concentration of the reactants and of temperature. The dependence of the absolute value of ^2J(PCH) on these factors then gives information about the relative signs of the coupling in the complex and the free ligand. Since ^2J(PCH) in Me$_3$P is known to be +2·6 Hz, a value of the coupling in the complex can be evaluated; it is found to be −8.9 Hz, showing that the sign of ^2J(PCH) changes on co-ordination at phosphorus. A similar experiment has been carried out by Grobe and Möller,[215] who have measured ^2J(PCH) for mixtures of Me$_3$P and BF$_3$. They find a coupling of −11.3 Hz for the Me$_3$P→BF$_3$ complex. On the basis of INDOR experiments it has been suggested[216] that ^2J(PCH) is probably negative in some iridium complexes.

[211] R. Price, Ph.D. Dissertation, University of London, 1969.
[212] R.-M. Lequan and M.-P. Simonnin, *Tetrahedron Letters*, 1972, 145.
[213] R. K. Harris and E. G. Finer, *Bull. Soc. chim. France*, 1968, 2805.
[214] H. Schmidbaur, A. Shiotani, and H.-F. Klein, *Chem. Ber.*, 1971, **104**, 2831.
[215] J. Grobe and U. Möller, *Z. Naturforsch.*, 1971, **26b**, 639.
[216] B. E. Mann, C. Masters, and B. L. Shaw, *J.C.S. Dalton*, 1972, 48.

Values of $^2J(\text{PCH})$ have also been reported in references 102, 123, 146, 194, 198, and 217—225.

(v) X = ^{77}Se. A few values of $^2J(\text{SeCH})$ have been reported.[91,226,227]

(vi) X = ^{19}F. Ihrig and Smith[161] have examined the spectra of the three difluoroethylenes in a large number of solvents. They conclude that solute dipole orientation affects the magnitude of the solvent effect on geminal (H,F) coupling. Fourier-transform proton spin-echo spectroscopy has been used to obtain a value of 55.24 ± 0.08 Hz for $^2J(\text{FCH})$ in 1,1-difluoro-2,2-dichloroethane.[228] Values of $^2J(\text{FCH})$ have also been reported in references 24, 168, 177, 178, and 229—231.

(vii) X = ^{183}W and ^{195}Pt. A value of 3.0 Hz for $^2J(\text{WCH})$ in WMe$_6$ has been measured by Shortland and Wilkinson[232] and some values of $^2J(\text{PtCH})$ have been reported.[132, 233—235]

E. $^2J(^{19}\text{FC}^{19}\text{F})$.—The geminal (F,F) coupling in 1,1-difluoroethylene has been measured in the nematic phase of a liquid crystal;[162] its value of $+32.5$ Hz changes very little on going to the isotropic phase of the solvent. Other $^2J(\text{FCF})$ values are reported in references 128, 161, 196, 230, 231, 236, and 237.

F. $^2J(^{13}\text{CC}^{13}\text{C})$, $^2J(^{31}\text{PC}^{13}\text{C})$, and $^2J(^{15}\text{NC}^{13}\text{C})$.—$^2J(\text{CCC})$ values have been measured for a series of 7-^{13}C-labelled monosubstituted benzene derivatives.[115] The couplings show a sensitivity to the electronegativity of the substituent grouping but very little to its hybridization, unlike the one-, three-, and four-bond (C,C) couplings which are sensitive to both. In all cases $|^2J(\text{CCC})| < |^3J(\text{CCCC})|$.

[217] M. Murray, R. Schmutzler, E. Gründeman, and H. Teichmann, *J. Chem. Soc. (B)*, 1971, 1714.
[218] H. C. Clark and L. E. Manzer, *J. Organometallic Chem.*, 1971, **30**, C89.
[219] B. E. Mann, B. L. Shaw, and R. M. Slade, *J. Chem. Soc. (A)*, 1971, 2976.
[220] C. Masters, B. L. Shaw, and R. E. Stainbank, *J.C.S. Dalton*, 1972, 664.
[221] M.-P. Simonnin, C. Charrier, and R. Burgade, *Org. Magn. Resonance*, 1972, **4**, 113.
[222] F. Mathey and R. Mankowski-Favelier, *Org. Magn. Resonance*, 1972, **4**, 171.
[223] P. Dembech, D. Seconi, P. Vivarelli, L. Schenetti, and F. Taddei, *J. Magn. Resonance*, 1972, **4**, 185.
[224] J.-P. Tuchagnes and J.-P. Laurent, *Bull. Soc. chim. France*, 1971, 4246.
[225] A. V. Dogadina, Yu. D. Nechaev, B. I. Ionin, and A. A. Petrov, *Zhur. obshchei Khim.*, 1971, **41**, 1662.
[226] M. Dräger and G. Gatow, *Spectrochim. Acta*, 1972, **28A**, 425.
[227] M. Garreau and G. Martin, *Bull. Soc. chim. France*, 1971, 4497.
[228] R. L. Vold and R. R. Shoup, *J. Chem. Phys.*, 1972, **56**, 4787.
[229] L. Phillips and V. Wray, *J. Chem. Soc. (B)*, 1971, 1618.
[230] J. Burdon and I. W. Parsons, *Tetrahedron*, 1971, **27**, 4553.
[231] K. Schaumburg, *J. Magn. Resonance*, 1972, **7**, 177.
[232] A. Shortland and G. Wilkinson, *Chem. Comm.*, 1972, 318.
[233] G. M. Whitesides and J. F. Gaasch, *J. Organometallic Chem.*, 1971, **33**, 241.
[234] R. Lazzaroni and C. A. Veracini, *J. Organometallic Chem.*, 1971, **33**, 131.
[235] M. Tsutsui, M. Ori, and J. Francis, *J. Amer. Chem. Soc.*, 1972, **94**, 1414.
[236] E. A. Noe and J. D. Roberts, *J. Amer. Chem. Soc.*, 1972, **94**, 2020.
[237] L. Cavalli, *J. Magn. Resonance*, 1972, **6**, 298.

Gray and Cremer[238] have found that ^2J(PCC) in 2,2,3,4,4-pentamethylphosphetans is stereospecific with respect to the exocyclic phosphorus substituent (see Table 2). The coupling is large when the coupled carbon nucleus is *trans* to this substituent and small when it is *cis*.

Table 2 *Stereospecific values of* ^2J(PCC)/Hz *in 2,2,3,4,4-pentamethylphosphetans.*[238]

(18)   (19)

Methyl orientation with respect to exocyclic P substituent		trans	cis	cis	trans
Substituent	X = Ph	31.8	2.5	4.9	27.8
	X = Me	30.5	2.1	4.3	26.9
	X = Cl	37.1	0.0	2.5	33.5

Jakobsen *et al.*[124] have made definite assignments of ^2J(PCC) in tri-2- and tri-3-thienylphosphine. The values range from 17.50 to 27.08 Hz and are smaller in magnitude than the ^1J(C,P) couplings. A similar observation has been made for *cis*- and *trans*-substituted ethylene phosphines[123] and Mann[122] has noted that in a series of tertiary phosphines ^1J(P,C), ^2J(PCC), and ^3J(PCCC) are all of approximately the same magnitude. ^2J(PCC) in triphenylphosphine has been found to be +19.65 Hz.[124] A value of $^2J(^{15}$NCC) has been reported.[119]

G. $^2J(^{31}$PN^{13}C) **and** $^2J(^{31}$PN^{31}P).—Some values of ^2J(PNC) have been measured by Simonnin *et al.*[123]

(20)

The sensitivity of ^2J(PNP) to nearby substituents has been reported from several laboratories. Allen,[239] for example, found that the (P^1,P^5) coupling in

---
[238] G. A. Gray and S. E. Cremer, *Chem. Comm.*, 1972, 367.
[239] C. W. Allen, *J. Magn. Resonance*, 1971, **5**, 435.

phenyl-substituted phosphonitrilic fluoride trimers (20) was greatly affected by changing the substituents X at P^3 from F_2 (46 Hz) to Ph_2 (33 Hz). This is not in accord with the additivity rule proposed by Finer,[240] though it can be rationalized in terms of Bent's hypothesis[70] or the work of Grant and Lichtman.[241] Nuretdinov et al.[242] have used $^1H-\{^{31}P\}$ and $^1H-\{^{13}C\}$ double resonance to determine the signs of some ^2J(PNP) couplings, which were found to be positive relative to a positive ^3J(POCH). The values ranged between 74 and 126 Hz and were very dependent upon the substituents attached to the phosphorus atoms, decreasing with the introduction of electron-donor substituents. The same observation has been made for ^2J(PNP) in phosphorus(v) nitride chlorides and phosphorus(v) sulphur(vi) nitride chlorides by Haubold and Fluck.[243] A strong dependence of ^2J(PNP) on substituent electronegativity in diazaphosphetidines (11) has been noted.[144] Niecke and Nixon[244] have observed that the temperature-dependence of ^2J(PNP) in $EtN(PF_2)_2$ is much larger than that of the corresponding methyl analogue. They suggest that this implies that steric interactions may have an important effect on the temperature-dependence of the coupling, which has also been shown to be positive.[245]

H. $^2J(^{31}PM^{13}C)$, where M is a Metal.—Gansow et al.[131] have reported cis- and trans-^2J(PWC) values for five monophosphine-substituted tungsten carbonyls. The former range between 7 and 10·5 Hz, the latter between 22 and 45.4 Hz. Mann, Shaw, and Stainbank[246] have determined ^2J(PMC) with M = Rh, Pd, and Ru by means of the ^{13}C n.m.r. spectra of some tertiary phosphine complexes.

I. $^2J(^{31}PM^{31}P)$, where M is a Metal.—Shaw and his co-workers[152,153,216] have made numerous measurements of ^2J(PMP), where M = Rh or Ir. For complexes of the type $IrCl_2X(PMe_2Ph)(Ph_2PCH_2CH_2PPh_2)$,[216] $^1H-\{^{31}P\}$ INDOR experiments have enabled them to determine both the magnitude and the relative sign of ^2J(PIrP) (trans). The former varies between 320 and 450 Hz depending on X, and the absolute sign is most likely to be positive. In mer-$[MX_3L_3]$ (M = Rh or Ir; X = halogen; L = tertiary phosphine) it also seems likely that ^2J(PMP) (cis) is positive.[153] However, Lynden-Bell et al.[245] have shown, using heteronuclear tickling experiments, that ^2J(PMoP) is opposite in sign to $[^1J(P,H) + ^3J(P,H')]$ in cis-$(PH_3)_2Mo(CO)_4$ and has the same sign as $[^1J(P,F) + ^3J(P,F')]$ in cis-$(CCl_3PF_2)_2Mo(CO)_4$. Therefore, since

---

[240] E. G. Finer, J. Mol. Spectroscopy, 1967, **23**, 104.
[241] D. M. Grant and W. M. Lichtman, J. Amer. Chem. Soc., 1965, **87**, 3994.
[242] I. A. Nuretdinov, V. V. Negrebetskii, A. Z. Yankelevich, A. V. Kessenikh, E. I. Loginova, L. K. Nikonorova, and N. P. Grechkin, Doklady Akad. Nauk. S.S.S.R., 1971, **196**, 1369.
[243] W. Haubold and E. Fluck, Z. Naturforsch., 1972, **27b**, 368.
[244] E. Niecke and J. F. Nixon, Z. Naturforsch., 1972, **27b**, 467.
[245] R. M. Lynden-Bell, J. F. Nixon, J. Roberts, J. R. S. Swain, and W. McFarlane, Inorg. Nuclear Chem. Letters, 1971, **7**, 1187.
[246] B. E. Mann, B. L. Shaw, and R. E. Stainbank, Chem. Comm., 1972, 151.

$^1J(P,H)$ and $^1J(P,F)$ have been shown to be positive and negative respectively, and since $^3J(P,H')$ and $^3J(P,F')$ are very small in comparison with the one-bond couplings, the authors conclude that $^2J(PMoP)$ is negative in both cases. But $^2J(PMoP)$ is found to be positive in cis-$EtN(PF_2)_2Mo(CO)_4$.[245] Nixon and Swain[107] have found that $^2J(PMoP)$ increases as the electronegativity of the groups attached to phosphorus increases. Values of $^2J(PMoP)$ have also been reported by Jesson et al.[194] and of $^2J(PPtP)$ by Whitesides and Gaasch.[233]

J. $^2J(^{13}CO^{13}C)$ and $^2J(^{31}PO^{13}C)$.—Ziessow[80] has used $^1H-\{^{13}C\}$ Torrey oscillations to determine the sign and magnitude (-2.8 ± 0.2 Hz) of $^2J(COC)$ from natural-abundance $^{13}CH_3O^{13}COCl$. Ihrig and Marshall[115] found a similar magnitude for the same coupling in $^{13}CH_3O^{13}COPh$ (2.63 Hz).

$$\begin{array}{c} S \\ \| \\ (MeO)_2P-N-PCl_2 \\ | \\ Me \\ (21) \end{array} \qquad \begin{array}{c} S \\ \| \\ (MeO)_2P-N-P(SC_4H_9)_2 \\ | \\ Me \\ (22) \end{array}$$

$$\begin{array}{c} S \\ \| \\ (MeO)_2P-N-P(OMe)_2 \\ | \\ Me \\ (23) \end{array}$$

The value of $^2J(POC)$ and its sign relative to $^3J(POCH)$ has been determined by $^1H-\{^{13}C\}$ INDOR spectroscopy for the compounds (21), (22), and (23).[242] Assuming that $^3J(POCH)$ is positive, $^2J(P^VOC)$ is found to be -6.0 Hz for all three compounds and $^2J(P^{III}OC)$ is $+19.0$ for (23).

K. **Other Two-bond Coupling Constants.**—$^2J(FXeF)$ has been measured[143] in the XeF_3^+ ion (see Section 4J) and found to be 174 Hz for the axial–equatorial coupling. A value of $^2J(BeOP)$ (6.1 Hz at 50 °C) has been measured in $[(NMe_2)_3P=O]_4 \cdots Be^{2+} 2Cl^-$.[247] Other measurements include $^2J(^{119}Sn-PH)$ (ref. 103), $^2J(FPF)$ (refs. 145, 147), $^2J(PBH)$ (refs. 102, 106, 109), $^2J(PCF)$ (refs. 107, 133), $^2J(CCF)$ (ref. 24), $^2J(HgCF)$ (ref. 196), $^2J(PtAsC)$ (ref. 132), $^2J(PMoH)$ (ref. 194), $^2J(HPF)$ (ref. 105), $^2J(HRuC)$ (ref. 248), and $^2J(HCrP)$ (ref. 249).

6 Coupling between Nuclei Separated by Three Chemical Bonds, ^3J

A. **Coupling of Protons via Two Carbon Atoms.**—Vicinal couplings in extensively conjugated and aromatic ring systems, where other couplings have also been measured, will be discussed in Section 10. We deal here with

[247] J. J. Delpuech, A. Peguy, and M. R. Khaddar, *J. Magn. Resonance*, 1972, **6**, 325.

saturated molecules and those having limited conjugation, or with papers which confine their attention to ^3J(HCCH).

Theoretical work in this field, much of it devoted to the interpretation of the stereochemical dependence of ^3J(HCCH), has been quite extensive during the period under review. It is discussed in Section 3C. It will suffice here to reiterate that the general trends, such as are embodied in Karplus' relation,[33] are well reproduced, but that individual theoretical values are frequently unreliable.

The functional form of the Karplus equation[33] has been carefully investigated by Whipple.[250] The equation predicts that $\frac{3}{2}N+\frac{1}{2}L = 2J+J'$ (J and J' are the two vicinal coupling constants) in the [AX]$_2$ spectrum of a symmetrically disubstituted ethane should remain constant with changes of rotamer population, but this is not found to be true experimentally. The sign of the deviation between theory and experiment suggests that a source of the discrepancy might be the opening of the projected HCH angle, but the small number of data available suggest that this angle is decreased rather than increased. Slessor and Tracey[251] have used a form of the Karplus equation to estimate the three-bond coupling constants between a proton and an adjacent methylene group. They show that the coupling to the two methylene protons depends on the ratio of k_1 to k_2 in the Karplus equations (25), where

$$\left.\begin{array}{ll} J_1 = k_1 \cos^2 \phi_1 - C & 0° \leqslant \phi_1 \leqslant 90° \\ J_2 = k_2 \cos^2 \phi_2 - C & 90° \leqslant \phi_2 \leqslant 180° \end{array}\right\} \quad (25)$$

ϕ is the dihedral angle. If this ratio is assumed to be constant, and is set equal to 0.9 on the basis of theoretical parameters calculated by Karplus,[33] then the coupling constants are found to be related to the dihedral angles by equations which do not require a knowledge of k_1 or k_2. The dihedral angles computed in this way from ratios of J_1 to J_2 appear to be uninfluenced by the effects of ring strain and substituent electronegativity, which implies that these factors change k_1 and k_2 but not their ratio.

A modified Karplus curve for the angular dependence of ^3J(HCCH) over the single bond between the three-membered ring and the double bond in vinylcyclopropane has been proposed.[36] (See also Section 3C). Durette and Horton[167] have investigated ^3J(HCCH) in pyranoid sugar derivatives. They correlate experimental results with a Karplus equation modified for the difference, $\Delta\chi$, in electronegativity between a hydrogen atom and a substituent at a carbon atom in the H—C—C—H system. They find:

$$^3J(\text{HCCH}) = \{7.8 - 1.0 \cos \phi + 5.6 \cos 2\phi\}\{1 - 0.1 \Delta\chi\} \quad (26)$$

and this equation reproduces trends quite well, but there are a number of

[248] M. Evans, M. Husthouse, E. W. Randall, E. Rosenberg, L. Milone, and M. Valle, *Chem. Comm.*, 1972, 545.
[249] D. N. Kursanov, V. N. Setkina, P. V. Petrovskii, V. I. Kdanovich, N. K. Baranetskaya, and I. D. Rubin, *J. Organometallic Chem.*, 1972, **37**, 339.
[250] E. B. Whipple, *J. Magn. Resonance*, 1971, **5**, 163.
[251] K. N. Slessor and A. S. Tracey, *Canad. J. Chem.*, 1971, **49**, 2874.

anomalies and the authors council caution in its use for determining stereochemistry. From measurement of ^2J(HCCH) for several partially deuteriated benzocycloheptene derivatives at $-120\,°C$, St. Jacques and Vaziri[252] have concluded that the Karplus relationship is also applicable to seven-membered rings.

^3J(HCCH) in vinylcyclopropane has been investigated by Günther and Wendisch,[253] who have used the Walsh model of cyclopropane to interpret substituent effects on this coupling. The three-membered ring is shown to be a less effective transmitter of long-range couplings than the double bond. The effect of substituents on ^3J(HCCH) in cyclopropanes has been found to be intermediate between that in saturated and unsaturated compounds.[189]

^3J(HCCH) in sp^2–sp^2 and sp^2–sp^3 systems has been discussed in some detail by Rummens,[254] and a relationship between vicinal coupling and bond length has been reported by Ammon and Wheeler.[255] They show that good linear correlations exist between ^3J(HCCH) and C—C bond length in cyclopentadiene and fulvenes. If J_a is the coupling across the formal double bond and J_b that across the formal single bond, then the bond length d is given by equation (27), where $\bar{J}$ is the coupling across the bond in question and

$$d/\text{Å} = 1\cdot395 + 0\cdot0636\,(J-\bar{J}) - 0\cdot0359\,\{(J-\bar{J})/|J-\bar{J}|\} \quad (27)$$

$\bar{J}$ is the mean of J_a and J_b. This equation reproduces bond lengths with a maximum deviation of 0.011 Å and an average deviation of 0.006 Å. This result is of interest in connection with the suggestion[256] that the well-known relationship between vicinal (H,H) coupling and π-bond order is based upon the relationship of the latter to bond length.

Peake et al.[257] have shown that all ^3J(HCCH) values in some 1,3,2-dithiaphospholans (24) are positive, and they note that $^3J(H_a,H_a) \neq {}^3J(H_b,H_b)$ for (24a; R = Ph), a difference which appears to have been undetected

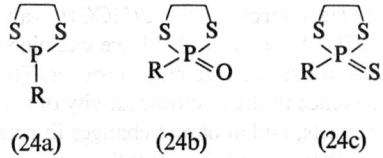

(24a)   (24b)   (24c)

previously. (The protons H_a and H_b are the methylene protons cis and trans to the substituent at phosphorus; their particular orientation was not determined.)

A number of detailed spectral analyses which include values of ^3J(HCCH)

---

[252] M. St-Jacques and C. Vaziri, Org. Magn. Resonance, 1972, **4**, 77.
[253] H. Günther and D. Wendisch, Chem. Ber., 1971, **104**, 3914.
[254] F. H. A. Rummens, J. Magn. Resonance, 1972, **6**, 550.
[255] H. L. Ammon and G. L. Wheeler, Chem. Comm., 1971, 1032.
[256] M. A. Cooper and S. L. Manatt, J. Amer. Chem. Soc., 1969, **91**, 6325.
[257] S. C. Peake, M. Fild, R. Schmutzler, R. K. Harris, J. M. Nichols, and R. G. Rees, J.C.S. Perkin II, 1972, 380.

have been reported. They include the spectra of substituted oxetans,[258] 1,4-dibromobutane,[163] 1,4-dihydro-1,4-ethenoisoquinolin-3(2H)-ones,[259] 2,4-

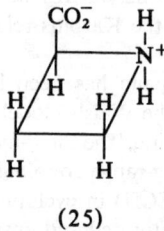

(25)

dichloropent-1-ene,[164] 2,4-dichloro[1-^{2}H]pent-1-ene,[164] and azetidine-2-carboxylic acid (25),[260] for which the vicinal (H,H) couplings are particularly discussed. An accurate value of ^3J(HCCH) (3.471 ± 0.011 Hz) has been obtained for 1,1-dichloro-2,2-difluoroethane by Vold and Shoup[228] using Fourier-transform proton spin-echo spectroscopy.

Considerable interest in the coupling constants of conformationally labile molecules, particularly ethane derivatives, continues. Thus Dawson and Reynolds[261] have examined the effect of solvent on coupling in eight 1,2-dihalogenoalkanes, attributing variations in the geminal and vicinal couplings to conformational changes. Phillips and Wray[262] have reinvestigated the data of Abraham and Gatti[263] on vicinal coupling in XCH_2—CH_2Y systems. They give empirical equations which enable the coupling constants to be calculated, to within ±0.18 to ±0.27 Hz, from the Huggins electronegativities of X and Y. The results show clearly that a substituent exerts a maximum effect when it is *trans*-coplanar to either of the coupled nuclei. Trends which were not explicable in terms of earlier treatments can now be understood.

The general features of the ^{1}H n.m.r. spectra of n-propyl derivatives have been discussed in some detail by Schrumpf,[264] who reports the analysis of several such spectra. He interprets the ^3J(HCCH) values in terms of substituent electronegativity. Bailey et al.[158] have examined the spectra of the side-chain protons of some substituted ethylbenzenes. The coupling constants are discussed with reference to the electronegativity of the substituent groups, steric and electronic effects, and apparent changes in rotamer populations.

Among the many papers describing coupling constants in saturated six-membered rings, the following are of interest. De-Mey et al.[265] have measured vicinal couplings in a 1,2,3,4,5,6-hexachlorocyclohexane of which the

[258] J. Jokisaari, E. Rahkamaa, and H. Malo, *Z. Naturforsch.*, 1971, **26a**, 973.
[259] L. Bauer, C. L. Bell, G. C. Brophy, W. A. Bubb, E. B. Sheinin, S. Sternhell, and G. E. Wright, *Austral. J. Chem.*, 1971, **24**, 2319.
[260] W. A. Thomas and M. K. Williams, *Org. Magn. Resonance*, 1972, **4**, 145.
[261] D. A. Dawson and W. F. Reynolds, *Canad. J. Chem.*, 1971, **49**, 3438.
[262] L. Phillips and V. Wray, *J.C.S. Perkin II*, 1972, 536.
[263] R. J. Abraham and G. Gatti, *J. Chem. Soc.* (B), 1969, 961.
[264] G. Schrumpf, *J. Magn. Resonance*, 1972, **6**, 243.
[265] C. A. De-Mey, A. J. De-Kok, J. Lugtenburg, and C. Romers, *Rec. Trav. chim.*, 1972, **91**, 383.

X-ray structure has been determined. Pihlaja, Kellie, and Riddell[173] have measured 60 and 220 MHz spectra of several 1,3-dioxans known to exist in non-chair conformations and have found ^3J(HCCH) values larger than 10 Hz in a number of cases. All the vicinal (H,H) couplings are almost independent of temperature. ^3J(HCCH) values in severely flattened six-membered rings have been reported by Lambert and Keong.[266] Vicinal (H,H) couplings in five-membered ring systems have been discussed[169,267] and Raza and Reeves[268] have determined ^3J(HCCH) in β-propiolactone in nematic solvents. The spectra of some cyclopentadienylides (26) have been measured,[269] and a moderate correlation of ^3J(HCCH)

$$(26) \quad X = \overset{+}{S}Me_2, \overset{+}{As}Ph_3, \text{or } \overset{+}{P}Ph_3$$

with π-bond orders calculated by the ω technique has been found. All the ^3J(HCCH) values in some isotopically substituted thiazoles were found to be positive.[44] Attempts to calculate these and other couplings using CNDO/2 SOS were only moderately satisfactory.

Hüther and Brune[87] have measured ^3J(HCCH) values in twenty substituted cyclobutene derivatives and have discussed them with respect to the bonding in complex-stabilized cyclobutadiene.

Martin et al.[270] have shown that in alcohols such as $R^1CH=CH-CHOHR^2$ the cis- and trans-stereoisomers behave differently in the presence of rare-earth shift reagents. The spectra of the trans-compounds remain strongly coupled but ^3J(HCCH) becomes easy to measure in the cis-compounds.

Other reports of ^3J(HCCH) are contained in references 65, 81, 89, 168, 171, 174, 175, 177, 180—183, 185, 187, 188, 190—193, 212, 223, 231, and 271—285.

[266] J. B. Lambert and F. R. Koeng, Org. Magn. Resonance, 1971, 3, 389.
[267] S. Wawzonek and W. E. Bennett, Org. Magn. Resonance, 1972, 4, 73.
[268] M. A. Raza and L. W. Reeves, Mol. Phys., 1972, 23, 1007.
[269] E. E. Ernstbrunner and D. Lloyd, Chem. and Ind., 1971, 1332.
[270] G. J. Martin, N. Naulet, F. Lefevre, and M. L. Martin, Org. Magn. Resonance, 1972, 4, 121.
[271] D. Tavernier and M. Anteunis, Bull. Soc. chim. belges, 1971, 80, 219.
[272] C. P. Lillya and R. A. Sahatjian, J. Organometallic Chem., 1971, 32, 371.
[273] W. E. Heyd and C. A. Cupas, J. Amer. Chem. Soc., 1971, 93, 6086.
[274] K. N. Fang, N. S. Kondo, P. S. Miller, and P. O. P. Ts'o, J. Amer. Chem. Soc., 1971, 93, 6647.
[275] K. Hayamizu, O. Yamamoto, K. Kushida, and S. Satoh, Tetrahedron, 1972, 28, 779.
[276] J. P. Albrand, A. Cogne, D. Gagnaire, and J. B. Robert, Tetrahedron, 1972, 28, 819.
[277] H. Booth and T. Masamune, J. C. S. Perkin II, 1972, 354.
[278] G. Wood, G. W. Buchanan, and M. H. Miskow, Canad. J. Chem., 1972, 50, 521.
[279] A. De Marco and G. Gatti, Org. Magn. Resonance, 1971, 3, 599.
[280] C. Gallina, M. Paci, and P. Viglino, Org. Magn. Resonance, 1972, 4, 31.

B. Three-bond Coupling of Protons via Other Atoms.

—Ramachandran et al.[286] have used ^{1}H n.m.r. data and conformational calculations to generate a ^3J(HNCH)–dihedral angle relationship for the peptide unit. Using eight examples in which the number of theoretical assumptions is least, they find that the best fit, as determined by a least-squares procedure, is given by equation (28). The relationship gives reasonable values for the dihedral angle in

$$^3J(\text{HNCH}) = 7.9 \cos^2 \phi - 1.55 \cos \phi + 1.35 \sin^2 \phi \quad (28)$$

cyclic oligopeptide structures but not for all peptides. The disagreement is considered to be the result of some interactions not being taken into account in the usual conformational calculations. Solvent effects on the coupling are also briefly discussed. Liler[287] has measured ^3J(HNCH) in N-alkylformamides, N-alkylacetamides, and their O-protonated cations. He finds, as is usually the case, that the coupling is increased by O-protonation, the cis-couplings being enhanced by factors about twice those of the trans-couplings. Other measurements of ^3J(HNCH) have been reported by Giuliani[288] and by Bichlmeir and West.[289]

^3J(HCOH) has been measured in the six-membered rings formed by dimers of glycolaldehyde,[182] and some values of ^3J(HCSiH) have been reported.[283,290]

Onak[291] has demonstrated that the use of a sufficiently broad band makes it possible to noise-decouple all the borons in pentaborane(9) [see structure (8)], and that with sufficient power, fine structure emerges on the peak attributed to the apical hydrogen atom. Seven peaks are observed, which indicates coupling to both bridge and base-terminal hydrogen atoms with a ^3J(HBBH) value of 5.7 Hz. The coupling to the two types of proton must be the same within ± 0.1 Hz, and since these protons are observed as two separate peaks there can be no tautomeric mechanism which makes them equivalent, as there is in $B_3H_8^-$ and B_6H_{10}. Onak and his co-workers[110] have also used ^{11}B-decoupled proton spectra to measure ^3J(HCBH) in methyl-substituted diboranes (6). They find $^3J(\text{HCBH}_\mu) = 2.4$—3.5 Hz and $^3J(\text{HCBH}_T) = 5.0$—5.3 Hz.

C. Three-bond Coupling of Carbon to Protons.

—^3J(CCCH) values have been obtained by Tarpley and Goldstein from the analysis of the ^{13}C n.m.r. spectra

[281] R. Keat, D. S. Ross, and D. W. A. Sharp, *Spectrochim. Acta*, 1971, **27A**, 2219.
[282] G. W. Burton, M. D. Carr, P. B. D. de la Mare, and M. J. Rosser, *J.C.S. Perkin II*, 1972, 710.
[283] R. J. Ouellette, D. Baron, J. Stolfo, A. Rosenblum, and P. Weber, *Tetrahedron*, 1972, **28**, 2163.
[284] L. Lunazzi and D. Macciantelli, *Chem. Comm.*, 1971, 933.
[285] J. C. Richer and N. Baskevitch, *Bull. Soc. chim. belges*, 1971, **80**, 459.
[286] G. N. Ramachandran, R. Chandrasekraran, and K. D. Kopple, *Biopolymers*, 1971, **10**, 2113.
[287] M. Liler, *J.C.S. Perkin II*, 1972, 720.
[288] A. M. Giuliani, *J.C.S. Dalton*, 1972, 492
[289] B. Bichlmeir and R. West, *J. Organometallic Chem.*, 1971, **32**, 35.
[290] C. Michel and M. Brini, *Bull. Soc. chim. France*, 1971, 4107.
[291] T. Onak, *Chem. Comm.*, 1972, 351.

of symmetrical *m*-dihalogenobenzenes[68] and of uracil, thymine, and the 5-halogenouracils.[195] In all cases $|^3J(\text{CCCH})| > |^2J(\text{CCH})|$. The effect of substituent electronegativity on the (C,H) couplings is fully discussed (see also Section 4B). The same authors have used[13]C spectra to measure $^3J(\text{CCCH})$ for acetone both as the neat liquid (1.46 ± 0.02 Hz) and in solution in benzene (1.50 ± 0.05 Hz).[292]

Ihrig and Marshall[293] have measured the relative signs of long-range (C,H) couplings in methyl [*carboxy*-^{13}C]benzoate. Three-, four-, and five-bond couplings all have the same sign, most probably positive, which would be in accord with other evidence on the signs of (C,H) coupling. $^3J(\text{CCCH})$ values have also been given for β-1,2,3,4,5,6-hexachlorocyclohexane[65] and for thiete sulphone (17).[88]

Values of $^3J(\text{CNCH})$ and $^3J(\text{CSCH})$ have been reported for thiazoles;[44,86] they were all found to be positive.[44]

Ziessow[80] has measured $^3J(\text{COCH})$ in CH_3OCOCl (+4.50 ± 0.04 Hz), using the $^1H-\{^{13}C\}$ Torrey-oscillation technique.

D. Three-bond Coupling of Other Nuclei to Protons *via* Two Carbon Atoms, $^3J(\text{XCC}^1\text{H})$.—These couplings will be grouped under the nature of X.

(*i*) X = [15,14]N. [14]N Tickling experiments combined with simultaneous decoupling of substituent protons have enabled McFarlane and McFarlane[210] to show that $^3J(^{14}\text{NCCH})$ is positive in a series of quaternary ammonium salts derived from pyridine. The values do not appear to be greatly affected by the incorporation of the nitrogen atom in an aromatic ring, though they are somewhat smaller than that which has been suggested for coupling between a *trans* nitrogen and a proton in an unsaturated system. The authors suggest that this may indicate a negative π-contribution or that the availability of the nitrogen *s*-electron in the σ framework is reduced. Lichter and Roberts[117] have discussed coupling of nitrogen to hydrogen *via* carbon atoms and have summarized some of the generalizations concerning the signs and magnitudes of these couplings. They have also measured all the (^{15}N,[13]C) and (^{15}N,H) coupling constants in ^{15}N-enriched pyridine and its hydrochloride. They point out that, taking account of the negative magnetogyric ratio of ^{15}N, the effects on $^2J(^{15}\text{NCH})$ and $^3J(^{15}\text{NCCH})$ of protonation of the nitrogen parallel those shown by $^2J(\text{CCH})$ and $^3J(\text{CCCH})$ in X[13]CCH and X[13]CCCH fragments as the electronegativity of X is increased, and also those shown by $^2J(\text{PCH})$ and $^3J(\text{PCCH})$ in PCH and PCCH fragments, respectively. Rahkamaa[96] has shown that $^1J(^{15}\text{N,H})$, $^2J(^{15}\text{NCH})$, and $^3J(^{15}\text{NCCH})$ in [^{15}N]pyrrole all have the same sign.

$^3J(\text{NCCH})$ values have also been reported for substituted anilines,[93] [^{15}N]acetamide,[95] derivatives of 6-[^{15}N]amino-6-deoxy-D-glucose,[119] and isotopically substituted thiazoles.[44]

---

[292] A. R. Tarpley and J. H. Goldstein, *Mol. Phys.*, 1971, **21**, 549.
[293] A. M. Ihrig and J. L. Marshall, *J. Amer. Chem. Soc.*, 1972, **94**, 3268.

(ii) $X = {}^{11}B$ and ${}^{27}Al$. Bogdanov et al.[202] have analysed the line-broadening in the ${}^{1}H$ and ${}^{11}B$ n.m.r. spectra of organoboron compounds to obtain ${}^{3}J({}^{11}BCCH)$ values. They find that cis-couplings are larger than trans- in magnitude. ${}^{2}J({}^{27}AlCCH)$ has been measured in $NaAlEt_4$; it is found to be 5.8 ± 0.3 Hz.[205]

(iii) $X = {}^{31}P$, ${}^{119,117}Sn$, ${}^{199}Hg$, and ${}^{207}Pb$. Extensive data on the coupling of a ring proton to X in the systems (27) and (28) have recently become available.

(27)   (28)

(a) $X = {}^{31}P$; $Y = O$ or S; $n = 3$
(b) $X = {}^{117,119}Sn$; $Y = O$ or S; $n = 4$
(c) $X = {}^{199}Hg$; $Y = O$ or S; $n = 2$
(d) $X = {}^{207}Pb$; $Y = O$ or S; $n = 4$

They have been analysed in detail by Barbieri and Taddei,[294,295] who have shown that the long-range (X,H) couplings in these molecules bear a very close relationship to each other and to the corresponding (H,H) couplings in furan and thiophen. These relationships may be expressed in the forms of equations (29)—(32). Furthermore, it is found[294,295] that the ratios of the

$$J(P,H) = 0.90 \pm (0.19) J(H,H) + 0.18(\pm 0.54) \quad (29)$$

$$J({}^{117}Sn,H) = 5.51(\pm 0.62) J(H,H) + 2.20(\pm 1.73) \quad (30)$$

$$J(Hg,H) = 14.35(\pm 0.63) J(H,H) - 1.33(\pm 1.73) \quad (31)$$

$$J(Pb,H) = 9.69(\pm 1.29) J(H,H) + 2.05(\pm 3.60) \quad (32)$$

slopes of the graphs of equations (29)—(32) are very similar in magnitude to the ratios of the s-electron density at the nuclei of the appropriate atoms X (after allowing for hybridization), and these facts led Barbieri and Taddei to suggest that all these couplings have a similar mechanism which is dominated by the Fermi-contact term. All the (X,H) couplings are also found to have the same relative sign for any given X,[294,295] and this result has been substantiated for $X = Pb$ and Hg by Ebdon, Huckerby, and Thorpe,[296] who also draw the conclusion that the mechanism of (Hg,H) and (Pb,H) coupling is the same in these systems.

Jakobsen and Begtrup[125] have measured a number of ${}^{3}J(PCCH)$ values in methyl-substituted trithienylphosphine derivatives [i.e. methyl-substituted

---
[294] G. Barbieri and F. Taddei, J. Chem. Soc. (B), 1971, 1903.
[295] G. Barbieri and F. Taddei, J.C.S. Perkin II, 1972, 262.
[296] A. P. Ebdon, T. N. Huckerby, and F. G. Thorpe, Tetrahedron Letters, 1971, 2921.

(27a) and (27b) in which $n = 3$, $Y = S$, and X is either P or $P{=}S$]. The couplings are all found to be positive and dependent upon steric hindrance and the orientation of the phosphorus lone pair. The effect of the methyl group on the (H,P) couplings parallels that noted for (H,H) couplings in the methylthiophens when compared with thiophen.

Coupling of phosphorus to hydrogen through two carbon atoms has also been reported for phosphine-substituted ethylenes,[123,212] alkylphosphine-boranes,[102] complexes of alkylphosphines with alkylfluoroborates,[224] tri-3-furylphosphines,[223] t-butyl complexes of iridium(III),[220] and t-butylphosphines and methiodides.[219]

$^3J(SnCCH)$ values in t-butyltins have been correlated[208] with an additivity rule of the type proposed by Malinowski.[209] Measurements of this coupling have also been reported for some methylcyclopentadienyl–metal compounds.[200]

E. **Three-bond Coupling to Fluorine *via* Two Carbon Atoms, $^3J(XCC^{19}F)$.**—
The important categories here are $X = {}^1H$ and $X = {}^{19}F$.

(*i*) $X = {}^1H$. Calculations on the variation of $^3J(HCCF)$ with dihedral angle have been reported.[31,34] (See also Section 3C.) The results reflect general trends in the coupling, but other calculations gave poor results even though orbital and spin-dipolar terms were included.[24]

Phillips and Wray[229] have examined the vicinal (H,F) coupling in a series of deoxy-fluoro-D-glucopyranoses and have developed a novel empirical additivity rule for predicting the $^3J(HCCF)$ *gauche* coupling with good accuracy. They have determined a number of contributions to the coupling, each associated with a particular structural element, and the sum of these, plus the value of 16.0 Hz measured in the unperturbed system fluoroethane, gives the observed coupling within 0.5 Hz. They also demonstrate that $^3J(HCCF)$ has a stereochemical dependence upon electronegative substituents. Hägele, Harris, and Sartori[297] have discussed the variation of (H,F) coupling with substituent electronegativity in compounds having the general formulae $CF_3CH(XY)$ and $CH_3CF(XY)$. They find that the data may be reasonably fitted with equation (33), where χ_j is the electronegativity of substituent j.

$$\langle ^3J(HCCF)\rangle = 59.39\,(\pm 0.24) - 3.23\,(\pm 0.02)\sum_j \chi_j \quad (33)$$

For the *trans*-coupling in a single rotamer they find that equation (34) applies.

$$^3J^t(HCCF) = 111.4\,(\pm 0.6) - 7.28\,(\pm 0.05)\sum_j \chi_j \quad (34)$$

Williams *et al.*[189] have studied the spectra of mono- and di-fluorocyclopropanes of general formula (29), particularly with regard to $^3J(HCCF)$. The signs of the couplings are determined by the criterion of best fit obtained

[297] G. Hägele, R. K. Harris, and P. Sartori, *Org. Magn. Resonance*, 1971, **3**, 463.

$$\begin{array}{c} F_1 \\ F_2 \\ H_B \diagup \diagdown H_X \\ H_A \quad Y \end{array}$$

(29)

using LAOCON III, and the authors express confidence that their sign combination is unique. They find that $^3J(F_1,H_X)$ and $^3J(F_2,H_X)$ decrease as the electronegativity of Y increases while $^3J(F_2,H_A)$, $^3J(F_2,H_B)$, $^3J(F_1,H_A)$, and $^3J(F_2,H_B)$ increase. This difference of behaviour of the three-bond (F,H) couplings is rationalized with reference to some recent theoretical work[298] which suggests that a substituent introduces an alternation in charge density along a saturated hydrocarbon chain. The geometrical relationship of the fluorine, rather than the hydrogen, to the substituent is also found to be important.

Jensen and Schaumburg[24] find that $^3J(HCCF)$ in ethyl fluorides is predicted quite well by the equation proposed by Ihrig and Smith[299] relating coupling to dihedral angle and substituent electronegativity. Three-bond (H,F) couplings have been measured for the three difluoroethylenes in a large number of solvents and it is found that they all have correlation coefficients of 0.9 or better when plotted against each other, even if they are taken from different molecules.[161] The authors conclude that this indicates that the same kinds of changes in electron distribution, rotational or vibrational states, *etc.*, are responsible for the observed changes with solvent of all the vicinal (H,F) couplings. The solvent effects are closely proportional to $[^3J(H,F)]^{\ddagger}$.

Other reports of $^3J(HCCF)$ may be found in references 168, 177, 228, 236, 279, 300, and 301.

(*ii*) X = ^{19}F. Cavalli[237] has analysed the ^{19}F spectra of three 1,2-disubstituted tetrafluoroethanes and has determined all vicinal and geminal coupling constants. There is a good correlation between $\langle ^3J(FCCF) \rangle$ and the sum of the Huggins electronegativities of the substituents in systems of the form CF_3CF_2-, $CF_3CF\langle$, and $-CF_2-CF_2-$ if the substituents are halogens, which suggests, Cavalli proposes, that electronegativities but *not* dihedral angles are important here. If the substituents are hydrogens, *i.e.* in CF_3CF_2H, CF_3CFH_2, and HCF_2CF_2H, angularly dependent electronegativity factors or dihedral angle changes seem to affect the $^3J(FCCF)$ values.

Abraham *et al.*[301] have completely analysed the ^{19}F spectra of some fluorinated benzofurans and they find that all $^3J(FCCF)$ values are negative. They were unable to represent the (F,F) couplings in terms of a sum of

---

[298] J. A. Pople and M. Gordon, *J. Amer. Chem. Soc.*, 1968, **89**, 4253.
[299] A. M. Ihrig and S. L. Smith, *J. Amer. Chem. Soc.*, 1970, **92**, 759.
[300] R. J. Abraham and W. L. Oliver, *Org. Magn. Resonance*, 1971, **3**, 725.
[301] R. J. Abraham, D. F. Wileman, G. R. Bedford, and D. Greatbanks, *Org. Magn. Resonance*, 1972, **4**, 343.

substituent effects. Smith and Ihrig[161] have found that solute dipole orientation is not a factor in determining the solvent effects on vicinal (F,F) coupling, and Gerritsen and MacLean[302] have found from measurements on o-difluorobenzene in nematic solvents that the indirect (F,F) coupling has a very small anisotropy.

The temperature dependence of ^3J(FCCF) in 1,2-dihalogenoperfluoropropanes has been investigated[303] and (F,F) couplings for perfluorovinylmercury compounds have been reported.[196]

(iii) X = ^{15}N and ^{199}Hg. Values of $^3J(^{15}$NCCF)[119] and ^3J(HgCCF)[196] have been published.

F. Three-bond Coupling between Phosphorus and Hydrogen, $^3J(^{31}$PXC^{1}H).—
^3J(PCCH) has been discussed in Section 6D (iii).

(i) X = O. The variation of ^3J(P^{III}OCH) with dihedral angle has been investigated by Bergesen and Albriktsen,[304] who have completely analysed the spectra of 2-chloro-, 2-methoxy-, and 2-phenoxy-1,3,2-dioxaphosphorinans (30a—c). They are thus able to add some more points to the graph of

$$\diagdown\!\!\!\!-\!\!\!\!\diagup\!\!\!\!-\!\!\!\!O\diagdown\!\!\!\!\diagdown P\!\!\!\!-\!\!\!\!\diagup\!\!\!\!-\!\!\!\!O\diagup\quad | \atop R$$

(30)
(a) R = Cl
(b) R = OMe
(c) R = OPh

^3J(P^{III}OCH) versus dihedral angle, and they find, in agreement with earlier work,[305] that the best correlation is obtained by assuming that the coupling is positive. Increasing the electronegativity of the substituent at P causes the coupling to become more positive. ^3J(P^{V}OCH) has been found to decrease when electron-donor substituents are introduced at phosphorus,[242] but Revel et al.,[171] who have measured ^3J(P^{III}OCH) in seven dioxaphospholans (31),

$$\begin{array}{cc} R\diagdown\!\!\!\!\diagdown P\diagup\!\!\!\!\diagup O\!\!-\!\!CH_2 \\ Y\diagup\quad \diagdown O\!\!-\!\!CH_2 \end{array}\qquad\begin{array}{cc} R\diagdown\!\!\!\!\diagdown P\diagup\!\!\!\!\diagup O\!\!-\!\!CH_2\diagdown\\ O\diagup\quad\diagdown O\!\!-\!\!CH_2\diagup \end{array}CH_2$$

(31) (32)

report that 'it is difficult to find a simple relationship between the (P,H) coupling and the groups R and Y on the phosphorus.' White et al.[306] have given ^3J(P^{V}OCH) values for many dioxaphosphorinan-2-ones (32).

[302] J. Gerritsen and C. MacLean, Spectrochim. Acta, 1971, 27A, 1495.
[303] A. De Marco and G. Gatti, J. Magn. Resonance, 1972, 6, 200.
[304] K. Bergesen and P. Albriktsen, Acta. Chem. Scand., 1971, 25, 2257.
[305] D. Gagnaire, J. B. Robert, and J. Vernier, Bull Soc. chim. France, 1968, 2392.
[306] D. W. White, G. K. McEwen, R. D. Bertrand, and J. G. Verkade, J. Chem. Soc. (B), 1971, 1454.

Other values of $^3J(\text{P}^{\text{III}}\text{OCH})^{116,193,242}$ and $^3J(\text{P}^{\text{V}}\text{OCH})^{99,307}$ have been reported.

(ii) X = N. Simonnin et al.[221] have measured the n.m.r. spectra of some aminophosphines as functions of temperature. They conclude that the sign of $^3J(\text{P}^{\text{III}},\text{NCH})$ is independent of rotation about the P—N bond. The magnitude of the coupling depends upon substituent electronegativity and also probably upon the valence angles at N; it appears to conform to the ideas of Goldwhite and Rowsell[308] and of Bent.[70] Albrand and his co-workers[276] have examined the spectra of six 1,2-dimethyl-1,3,2-diazaphospholans substituted at phosphorus (33). All the $^3J(\text{P}^{\text{III}}\text{NCH})$ values are positive

$$\text{R}-\overset{\displaystyle \overset{\text{Me}}{|}}{\text{P}}\overset{\displaystyle \text{N}-\text{CH}_2}{\underset{\displaystyle \underset{\text{Me}}{|}}{\text{N}-\text{CH}_2}}$$

(33)

and they are also stereospecific. They fall into two groups lying in the ranges 2.2—3.5 Hz and 5.9—10.9 Hz while the (P,H) couplings to the methyl hydrogens are larger, 12.4—15.6 Hz.

Some values of $^3J(\text{P}^{\text{V}}\text{NCH})$ and $^3J(\text{P}^{\text{III}}\text{NCH})$ in P^{III}—N—P^{V} systems have been reported by Nuretdinov et al.[242] The former increases with the introduction of electron-donor substituents at P^{III} while the latter decreases, but not in quite the same order.

Other reports of $^3J(\text{PNCH})$ include P^{III} (refs. 106, 133) and P^{V} (refs. 147, 307).

(iii) X = S. The 1,3,2-dithiaphospholan (24a; R = Ph) shows two small $^3J(\text{P}^{\text{III}}\text{SCH})$ values of opposite sign.[257] By analogy with $^3J(\text{P}^{\text{III}}\text{OCH})$, this may indicate a dihedral angle near 90°.

(iv) X = a Group IVB Element. Zuckerman and his co-workers[103,104] have reported values of $^3J(\text{P}^{\text{III}},\text{X},\text{C},\text{H})$ for Group IV phenylphosphines of general formula PhPHXMe$_3$, where X = C, Si, Ge, or Sn.

(v) X = Au and Pt. Values of $^3J(\text{P}^{\text{III}}\text{AuCH})^{309,310}$ and of $^3J(\text{P}^{\text{III}}\text{PtCH})^{233}$ have been reported.

G. $^3J(^{13}\text{CCC}^{31}\text{P})$ and $^3J(^{13}\text{CCO}^{31}\text{P})$.—Gray and Cremer[126] have found that $^3J(\text{CCCP})$ in phosphetan oxides and phosphetanium salts are stereospecific. The signs of (C,P) couplings in tri-2- and tri-3-thienylphosphines have been determined by Jakobsen et al.[124] The three-bond coupling is positive, and

[307] K. E. De Bruin, A. G. Padilla, and D. M. Johnson, *Tetrahedron Letters*, 1971, 4279.
[308] H. Goldwhite and D. G. Rowsell, *Chem. Comm.*, 1969, 713.
[309] G. C. Stocco and R. S. Tobias, *J. Amer. Chem. Soc.*, 1971, **93**, 5057.
[310] H. Schmidbaur and A. Shiotani, *Chem. Ber.*, 1971, **104**, 2821.

in general the long-range (C,P) couplings are of the same sign as the (C,H) couplings in thiophen [*cf.* Section 6D (iii)] and the (C,F) couplings in monofluorobenzenes. Mann[122] has found that one-, two-, and three-bond (C,P) couplings *via* carbon atoms in a series of tertiary phosphines are approximately equal in magnitude. Some ^3J(CCCP) values have been given by Simonnin *et al.*[123]

Borisenko *et al.*[311] have measured ^3J(CCOP) values in 1,3,2-dioxaphosphorinan-2-ones (32; R = H). They find that the coupling depends on the CCOP dihedral angle and also upon the orientation of the P—O bond. The angular dependence is rather like that of ^3J(POCH),[312] being large when ϕ is near 180° and small when it is *ca.* 60°. Some values of ^3J(CCOP) have also been reported by Mantsch and Smith.[313]

H. $^3J(^{13}CCC^{13}C)$.—Ihrig and Marshall[115] have made the interesting observation that $|^3J(CCCC)| > |^2J(CCC)|$ in some mono-substituted benzene derivatives.

I. $^3J(^{199}HgXC^1H)$.—Schaaf and Oliver[107,314] have measured several couplings of this type with X = Si and Ge. They find[197] that ^3J(HgSiCH) increases as the electronegativity of the substituents on the mercury increases, and they suggest that the linear relationship observed between ^2J(HgCH) and ^3J(HgSiCH) in MeHgR and Me$_3$SiHgR, respectively, indicates that the coupling mechanism is basically the same in both cases. Goggin *et al.*[198] have made an analogous suggestion with respect to the similarity in the behaviour of ^2J(HgCH) and ^3J(HgPCH) in complex Hg^{II} cations.

J. $^3J(^{195}PtXC^1H)$, X = N, P and As.—Appleton and Hall[315—317] have continued their work on the ^3J(PtNCH) coupling and the factors which determine its value. They find,[315] for simple diamine complexes of Pt^{II} and Pt^{IV}, that the following factors are important: the state of oxidation of the platinum; the nature of the ligand *trans* to the chelate ring; and the dihedral angle between the PtNC and the NCH planes. The nature of the diamine and of *cis*-ligands in Pt^{IV} complexes appears to be less important.

Clark and his co-workers have reported some values of ^3J(PtPCH)[218,318] and of ^3J(PtAsCH).[132]

K. **Other Three-bond Coupling Constants.**—Harris *et al.*[144] have measured the signs and magnitudes of $^3J(P^{IV}NPF)$ in diazadiphosphetidines (11). The

[311] A. A. Borisenko, N. M. Sergeyev, E. Ye. Nifant'ev, and Yu. A. Ustynyuk, *Chem. Comm.*, 1972, 406.
[312] D. W. White and J. G. Verkade, *J. Magn. Resonance*, 1970, **3**, 111.
[313] H. H. Mantsch and I. C. P. Smith, *Biochem. Biophys. Res. Comm.*, 1972, **46**, 808.
[314] T. F. Schaaf and J. P. Oliver, *J. Organometallic Chem.*, 1971, **32**, 307.
[315] T. G. Appleton and J. R. Hall, *Inorg. Chem.*, 1971, **10**, 1717.
[316] T. G. Appleton and J. R. Hall, *Inorg. Chem.*, 1972, **11**, 117.
[317] T. G. Appleton and J. R. Hall, *Inorg. Chem.*, 1972, **11**, 124.
[318] H. C. Clark and K. Itoh, *Inorg. Chem.*, 1971, **10**, 1707.

couplings show an appreciable variation which is not readily understood. The following three-bond couplings have also been reported: $^3J(\text{SeCCH})$ (refs. 91, 227), $^3J(\text{SeNCH})$ (ref. 91), $^3J(\text{HPCF})$ (ref. 107), $^3J(\text{HCPF})$ (refs. 146, 225), $^3J(\text{GeOCH})$ (ref. 206), $^3J(\text{CSCP})$ (ref. 124), and $^3J(\text{PMoPH})$ (ref. 107).

7 Coupling between Nuclei Separated by Four Chemical Bonds, ^4J

Theoretical work relevant to this type of coupling has been described in Section 3.

A. $^4J(^1\text{HCCC}^1\text{H})$.

Bauer et al.[259] have reported some rather large allylic couplings along a **W** path in fragments of the form (34) and have discussed the effect of substituents on them. When the electronegativity of the substituent

$$\underset{\diagdown\text{C}_1\diagup}{\text{H}}\overset{\mid}{\underset{}{\text{C}_2}}\underset{}{\text{C}_3}\diagdown^{\text{H}}$$

(34)

at C-2 is increased, there is a marked increase in the magnitude of $^4J(\text{H,H})$. A plot of the coupling against the analogous *meta*-coupling in benzene with the same substituents at the central carbon atom appears to approximate to a straight line passing through the origin. This leads the authors to suggest that the effect of substituents on this coupling may be similar to that of substituents on ^4J (*meta*) in mono-substituted benzenes. Jost, Rimmelin, and Somer[319] have noted that the four-bond (H,H) coupling in protonated halogenoacetones is related to the electronegativity of the halogen substituent (it increases as the electronegativity increases) and is much larger than the corresponding coupling in the unprotonated molecules. They suggest that this marked increase in coupling is to be associated with an increasing contribution from the structure which carries a formal positive charge at the carbonyl carbon atom rather than on the oxygen. A large $^4J(\text{HCCCH})$ value (1.7 Hz) has been detected in a substituted acetone.[320] The coupling does not follow a zig-zag path, the protons concerned being *trans*-related in a non-planar system.

Hüther and Brune[87] have measured and discussed the allylic couplings in 20 substituted cyclobutenes, and the dependence of $^4J(\text{HCCCH})$ on torsional angle in vinylcyclopropenes has been examined.[188]

Detailed analyses of the spectra of three halogen-substituted aliphatic hydrocarbons have been reported. The spectra of 1-chloro-2-bromo-2-fluoropropane in CCl$_4$ and acetone have been completely analysed by Abraham and Oliver.[300] The $^4J(\text{HCCCH})$ values differ considerably and both are positive; this can be rationalized in terms of substituent electronegativity effects. Other literature values of $^4J(\text{HCCCH})$ are examined, and for molecules of general

[319] R. Jost, P. Rimmelin, and J. M. Sommer, *Chem. Comm.*, 1971, 1243.
[320] M. Anteunis, M. Wandewalle, and L. Von Wijnsberghe, *Bull. Soc. chim. belges*, 1971, **80**, 423.

formula $CH_3CXYCHUV$ the following empirical expression is suggested:[300]

$$^4J(HCCCH) = 0.12(\chi_X + \chi_Y) - A \sum_j \chi_j \qquad (35)$$

where χ_j is the electronegativity of substituent j and the sum runs over all substituents. The constant A lies in the range 0.05—0.1; the data are not sufficient to state a firmer value. The ^1H n.m.r. spectra of 1,4-dibromobutane[163] and 2,4-dichloropent-1-ene and 2,4-dichloro[1-^2H]pent-1-ene[164] have been analysed.

As usual, the study of six-membered saturated rings continues to constitute a large proportion of n.m.r. work on long-range coupling. $^4J(HCCCH)$ between two axial hydrogens has been observed[321] in halogenocyclohexanones and halogenodecalones, in apparent contradiction of the observations of Jefford and Waegell.[322] No four-bond couplings between axial and equatorial hydrogen atoms adjacent to the carbonyl group could be observed in any of the molecules studied. The authors conclude[321] that if $^4J(HCCCH)$ is separated into σ and π contributions[41] then $^4J(H_{ax},H_{ax})$ is probably a π coupling. $^4J(HCCCH)$ values in the spectra of five isomers of 1,2,3,4,5,6-hexachlorocyclohexane have been found[275] to be in agreement with Barfield's calculations,[42] and the coupling has also been measured in an isomer of this molecule, of which the X-ray structure is known.[265]

Other $^4J(HCCCH)$ values have been reported in references 65, 174, 181, 185, 187, 222, 223, 285, 323, and 324.

B. Other Four-bond (H,H) Couplings.—Anteunis *et al.*[179] have measured some $^4J(HCOCH)$ values in 1,3-dioxolans (35) substituted with a $(CH_2)_n$ unit of varying size. They find that the coupling between H_c and H_x, involving a non-planar path, varies between 0 and 0.5 Hz depending on the value of

$$H_{x'} \overset{(CH_2)_n}{\underset{\overset{O}{\underset{H_c}{\diagdown}}\overset{C}{\diagup}\underset{H_t}{O}}{\diagup}} H_x$$

(35; $n=5-12$)

n, but in an irregular manner. Some $^4J(HCOCH)$ values in tetra- and dihydrofuran rings have been reported,[81,187] and Brouwer and Stothers[181] have noted a $^4J(HCOCH)$ coupling which will be discussed in Section 8.

[321] C. Cuvelier, R. Ottinger, and J. Reisse, *Tetrahedron Letters*, 1972, 277.
[322] C. W. Jefford and B. Waegell, *Bull. Soc. chim. belges*, 1970, **79**, 427.
[323] C. H. Green and D. G. Hellier, *J.C.S. Perkin II*, 1972, 458.
[324] P. Calinaud and J. Gelas, *Compt. rend.*, 1972, **274**, C, 1001.

Values of ^4J(HCSCH) (ref. 325), ^4J(HCPtCH) (ref. 233), and ^4J(HCNNH) (ref. 289) have been published.

C. Four-bond Coupling of ^{13}C, ^{14}N, and ^{31}P to ^1H via Carbon Atoms.—Tarpley and Goldstein[68] have measured the signs and magnitudes of ^4J(CCCCH) in o-dihalogenobenzenes and have discussed the variation of the latter with halogen electronegativity. They find that the coupling is negative [relative to ^1J(C,H) > 0], in agreement with earlier results for benzenes. In contrast, Ihrig and Marshall[293] find that the ^4J(CCCCH) coupling in methyl benzoate is most probably positive, but one is dealing here with a coupling to an exocyclic carbon atom. Yamamoto et al.[65] have found a negative ^4J(CCCCH) value in β-1,2,3,4,5,6-hexachlorocyclohexane.

Some values of ^4J(^{14}NCCCH), apparently the first reported, have been estimated from linewidths.[210] The couplings, which were measured in quaternary pyridinium salts, have values in the region 0.1—0.01 Hz.

^4J(PCCCH) values have been reported for phospholes[222] and for tri-3-furylphosphines derivatives.[223] Jakobsen and Begtrup[125] have measured ^4J(PCCCH) couplings in methyl-substituted trithienylphosphine derivatives [see Section 6D(iii)]. They have also determined the signs of these couplings and find that all the (H₃C,P) values are positive. Some ^4J(PCSCH) values are negative for P^{III} and positive for P^V.

D. Four-bond Coupling of Metals to Protons.—A value of ^4J(HgCCCH) of 13.2 Hz has been measured in $Me_3SiHg-C\equiv C-Me$,[314] and in (36) the following couplings have been recorded:[201] $(Hg,H_c) \approx (Hg,H_b) = 95$ Hz, $(Hg,H_a) = 104$ Hz. The authors suggest that the fact that the four- and three-

$$H_b \underset{H_c}{\overset{H_a}{\underset{|}{C}}}=C\underset{H}{\overset{H}{\underset{|}{C}}}-HgOAc$$

(36)

bond couplings are so similar in magnitude may be due to the predominance of conformations in which the mercury atom is located quite near the double bond, or to a π-electron contribution.

Four-bond couplings of Sn and Hg to H through carbon atoms have been measured in methylcyclopentadienyl–metal compounds by Campbell and Green,[200] and Ebdon et al.[296] have given values of ^4J(HCOCHg), ^4J(HCSCHg), ^4J(HCOCPb), and ^4J(HCSCPb) in furyl- and thienyl-lead and -mercury, (27c, d) and (28c, d). The authors deduce[296] that all these couplings are of the same sign and conclude that the coupling mechanism is

[325] M. Brink, *Org. Magn. Resonance*, 1972, **4**, 195.

identical for Hg and Pb, and contact-dominated. Very similar results and conclusions have been presented by Barbieri and Taddei,[294,295] who have also discussed the corresponding couplings to Sn [see also Section 6D (iii)].

E. $^4J(^1HCCC^{19}F)$ and $^4J(^{19}FCCC^{19}F)$.—Anteunis and Dirinck[177,178] have reported ^4J(HCCCF) values for dioxan derivatives. Abraham et al.[301] have completely analysed the ^{19}F spectra of some fluorinated benzofurans; the ^4J(HCCCF) values are probably positive and the ^4J(FCCCF) negative. Values of ^4J(FCCCF) are also included in references 128 and 303.

F. Other Four-bond Couplings.—Simonnin et al.[221] have observed that the sign of ^4J(FPNCH) in aminophosphines is independent of rotation about the P—N bond, but a large stereochemical effect on the sums and differences of ^4J(FPNPF) values has been reported.[144] Stereospecific ^4J(HCNSF) couplings have been resolved in $Me_2NS(O)F$ at -110 °C.[281] The sign of ^4J(PMoPCF) is probably positive.[107] The following four-bond couplings have also been reported: ^{4}(PtNCCH) (ref. 315), ^4J(POSnCH) (ref. 104), ^4J(FCHgCF) (ref. 196), ^4J(CCCCP) (refs. 123, 124), ^4J(PNPNP) (ref. 243), ^4J(PIrPCH) (ref. 216), ^4J(FCNCH) (ref. 231), and ^4J(PCSCH) (ref. 125).

8 Coupling between Nuclei Separated by Five Chemical Bonds, ^5J

A. $^5J(^1HCCCC^1H)$.—The work of Barfield and Sternhell[38] on the conformational dependence of homoallylic (H,H) coupling has been described in Section 3C. Here we might note that a comparison of the VB and MO results indicates that the major factor responsible for the differences between cisoid and transoid homoallylic coupling constants is a direct mechanism of negative sign which arises for cis-2-butenes because of the proximity of the bonds containing the coupled nuclei. Barfield and Sternhell[38] also discuss the effect of substituents.

(37)

Santelli[326] has investigated ^5J(HCCCCH) in substituted allenes, i.e. in the fragment (37). If ϕ is the angle between the planes $H_aC_1C_2$ and $H_bC_1C_2$, then Santelli finds that ^5J(H,H) is given approximately by equation (36).

$$^5J(H,H) = 2.25 \sin^2 \phi + 1.18 \qquad (36)$$

This interesting relationship requires more experimental data for its firm establishment. ^5J(H—C≡C≡C—H) has been measured in a number of cumulenes.[165] The cis and trans values differ only slightly and are both positive

[326] M. Santelli, Chem. Comm., 1971, 938.

relative to positive ^3J(HCCH). The authors suggest that the differences between the *cis*- and *trans*-couplings are due to the substituents.

Marshall and Folsom[327] have discussed the homoallylic coupling in 1,4-dihydro-1-naphthoic acid (3.93 Hz) and the corresponding coupling in 1,4-dihydrobenzoic acid (8—9 Hz) and 9,10-dihydroanthracenes (negligibly small). In view of the fact that the benzoic acid is more nearly planar than the naphthoic, the homoallylic coupling does not apparently depend on the planarity of the ring, and the authors propose that homoallylic spin–spin coupling can be transmitted by an olefinic bond but cannot be transmitted significantly through an aromatic bond.

^5J(HCCCCH) couplings have also been measured in dibenzothiophen derivatives[328] and hexachlorocyclohexanes.[65,275]

B. $^5J(^1HCCXC^1H)$, X = N, O, and S.—^5J(HCCNCH) values of 0.3 Hz (*cis*) and 1.0 Hz (*trans*) have been found in complexes of *NN*-dimethylacetamide with SbCl$_5$ and HCl–SbCl$_5$.[329] $|^5J(H_m,H_x)|$ in (38) is 0.06 Hz.[330]

$$H_b \quad H_a$$
$$H_m \quad O$$
$$H_m \quad NMe_x$$
$$O$$

(38)

Brouwer and Stothers[181,331] have reported a number of values of ^5J(HCCOCH). In methyl norbornene-2-carboxylates (39) and methyl norbornane-2-carboxylates (40), they find[331] couplings ranging between less than 0·1 and 0·24 Hz, and the dependence of these couplings on the nature and

(39)   (40)

orientation of the 2- and 3-substituents is discussed. They conclude that their results are qualitatively consistent with the analysis of Barfield and Chakrabarti[41] as far as the dependence of the coupling on molecular geometry is

[327] J. L. Marshall and T. K. Folsom, *J. Org. Chem.*, 1971, **36**, 2011.
[328] F. Balkan and M. L. Heffernan, *Austral. J. Chem.*, 1971, **24**, 2305.
[329] Q. Appleton, L. Bernander, and G. Olofsson, *Tetrahedron*, 1971, **27**, 5921.
[330] C. D. Poulter and R. B. Dickson, *Tetrahedron Letters*, 1971, 2255.
[331] H. Brouwer and J. B. Stothers, *Canad. J. Chem.*, 1971, **49**, 2152.

Nuclear Spin–Spin Coupling

concerned. In some substituted acrylic acids and their methyl esters, R(X)C=C(Y)CO$_2$Me, they have observed[181] some interesting ^5J(HCCOCH) and ^4J(HCOCH) couplings. When R = MeO and X = Y = H they find that the (E)-isomers (i.e. those having the OMe and CO$_2$Me groups *trans*) exhibit readily measurable four- and five-bond (H,H) couplings, whereas in the (Z)-isomers these are not resolved. Brouwer and Stothers suggest that these couplings depend for their magnitude on the configuration of the methyl of the methoxy-group, as Feeney and Sutcliffe proposed.[332] Similar observations were made for the (E)- and (Z)-isomers of 1-methoxyprop-1-ene.[181]

Lunazzi and Macciantelli[284] have measured ^5J(HCCSCH) in *ortho*-substituted thioanisoles. They find that the ^5J coupling between the methyl and the *ortho*-protons takes the values 0.13, 0.17, 0.25, 0.31, 0.36, and 0.43 Hz for *ortho*-substituents Me, Cl, Bu^t, Br, I, and NO$_2$ respectively. Thus the magnitude of the coupling rises as the size of the *ortho*-substituent increases, and the authors conclude that the change in coupling reflects the population of the *anti*-configuration.

C. Five-bond Coupling of ^{1}H to ^{19}F.—De Marco and Gatti[279] have reported complete analyses of the ^{1}H and ^{19}F spectra of 2,2,4,4,4-pentafluoro-n-butane. The long-range ^5J(HCCCCF) coupling is sensitive to solvent, and values $^5J(trans)$ = 0.16 Hz and $^5J(gauche)$ = 0.97 Hz were estimated. They suggest that the difference in the *trans*- and *gauche*-couplings may be rationalized in 'through-space' terms. A large ^5J(HCCCCF) value of 2.68 Hz has been reported in a fluorinated benzofuran derivative.[301]

Some ^5J(FCCNCH)[231] and ^5J(FCPNCH)[133] values have been published. Schuetz and Nilles[333] have measured a value of ^5J(FCSCCH) = 4.2 Hz and ^6J(FCSCCCH) = 0.45 Hz in (41a) and (41b), respectively. They suggest that a **W** path is important for coupling through sulphur.

(41)(a) R = H
(b) R = Me

D. Other Five-bond Couplings.—^5J(CCCCCH) has been shown to have the same sign as the three- and four-bond (C,H) couplings in methyl [*carboxy*-^{13}C]benzoate;[293] it is probably positive. Abraham *et al.*[301] have measured the *para* (C,C) coupling in a series of substituted benzofurans. The magnitude of the coupling ranges between 15 and 20 Hz, much larger than in benzenes,

---
[332] J. Feeney and L. H. Sutcliffe, *Spectrochim. Acta*, 1968, **24A**, 1135.
[333] R. D. Schuetz and G. P. Nilles, *J. Org. Chem.*, 1971, **36**, 2188.

and the influence of the hetero-ring cannot be predicted as the sum of two substituent effects, in contrast to the ^{19}F chemical shifts.

A few other five-bond couplings have been detected, namely ^5J(PNPOCH) (ref. 242), ^5J(FCCHgCF) (ref. 196), ^5J(HCHgCCF) (ref. 196), and ^5J(CCCCCP) (ref. 123).

9 Coupling between Nuclei Separated by Six or More Chemical Bonds

A. ^6J.—Two values of ^6J(H—C≡C—C≡C—C—H) have been measured:[165] they are found to be negative and have values of -0.9 and -1.0 Hz. (There appears to be an error in the numbering of the protons in structure III of this paper).

A stereospecific six-bond (H,H) coupling has been reported[330] for (38). A distinctly measurable coupling of 0.37 Hz is detected between H_b and H_x whereas the estimated upper limit of $^6J(H_a,H_x)$ is 0·05 Hz. 'Through-space' contributions to (H_b,H_x) are thought to be very unlikely; the orientation of the amide system appears important, and there could be a π-contribution to the coupling.

The following six-bond couplings have also been reported: ^6J(FCSCCCH) (ref. 333), ^6J(FCCHgCCF) (ref. 196), and ^6J(CCCCCCP) (ref. 123).

B. ^7J and ^8J.—Abraham et al.[301] have measured seven-bond (H_3C,H) and (H_3C,F) couplings of 0.18 and 0.56 Hz, respectively, in some fluorinated benzofurans. An eight-bond (H_3C,F) coupling of 0.5 Hz in a similar molecule was thought to be a 'through-space' coupling.

10 Coupling in Aromatic Systems

Theoretical work on coupling in conjugated molecules has been discussed in Section 3. In what follows, theoretical correlations will be included with the description of the general content of the paper.

A. (^{1}H,^{1}H) Coupling in Benzenoid Aromatic Systems.—Jakobsen et al.[60] have compared n.m.r. and e.s.r. measurements of proton hyperfine splitting constants in aromatic t-butyl nitroxides. The results are in good agreement with each other, and the signs of the couplings for protons and fluorine are in accord with the signs of the π-electron contributions to (H_3C,H), (H_3C,F), and (F_3C,H) couplings determined by Blears, Danyluk, and Schaefer in substituted benzenes.[334]

A very detailed analysis of the ^{1}H n.m.r. spectra of [1-^{13}C]benzene in the isotropic and nematic phases has been made by Englert, Diehl, and Niederberger;[335] all (H,H) and (^{13}C,H) indirect couplings and their signs have been determined. Matthews, Jones, and Bartle[336] have carried out complete

[334] D. J. Blears, S. S. Danyluk, and T. Schaefer, Canad. J. Chem., 1968, **46**, 652.
[335] G. Englert, P. Diehl, and W. Niederberger, Z. Naturforsch., 1971, **26a**, 1829.
[336] R. S. Matthews, D. W. Jones, and K. D. Bartle, Spectrochim. Acta, 1971, **27A**, 1185.

iterative analyses of the 60 MHz spectrum of dibenzo[c,g]phenanthrene and the 100 MHz spectrum of benzo[ghi]perylene. Assignments were assisted by the use of double resonance, solvent effects, and tickling experiments. The *ortho* coupling constants of unhindered protons were well predicted by Hückel MO π-electron bond orders *via* empirical relationships, and the authors conclude that C—C—H bond distortions can account for the high values for overcrowded protons. Haigh and Mallion[337] have also examined the ^3J(HCCH) couplings of some non-planar condensed benzenoid hydrocarbons and they too find that these couplings give as good a correlation with π-bond order as planar hydrocarbons, provided that the protons concerned are entirely unhindered. Furthermore, the ^3J(HCCH) values for protons one of which is sterically overcrowded are larger than those calculated theoretically by *ca.* 0.6 Hz, as was observed in planar molecules. However, the same reasons cannot be advanced to explain this discrepancy. In the planar compounds the steric pressure is likely to reduce the CCH angle, decrease the H—H distance, and so increase the coupling. In the non-planar hydrocarbons the primary effect of the steric constraint would be expected to be the increase of the HCCH dihedral angle, causing a decrease in the coupling. This interesting phenomenon therefore remains unexplained, but the authors draw attention to a paper[338] in which an INDO FPT calculation is found to give a surprisingly large increase in an olefinic ^3J(HCCH) coupling for a small out-of-plane twist. An analysis of the 220 MHz spectra of benzo[a]pyrene and 6-methylbenzo[a]pyrene has been reported.[339] Hansen and Berg[340] have determined (H,H) and (C,H) couplings from an analysis of the ^{13}C–H satellite spectrum of pyrene. The (H,H) results are in good agreement with a CNDO SOS calculation with the addition of π-electron contributions calculated by McConnell's method,[341] but the calculated (C,H) couplings are less than half of the observed values.

Complete analyses of the ^{1}H n.m.r. spectra of four condensed hydrocarbons containing a methylene group have been reported.[342] The couplings are all positive and of the same order of magnitude as those in the pure aromatics. The effects of solvent and dilution on the couplings have been investigated, and the calculation of coupling constants between ring protons and between ring and methylene protons by means of bond-order and other simple relationships is discussed. The relationship of the four-bond coupling between benzylic and *ortho*-protons to the π-bond order of the intervening aromatic bond has been investigated by Nair *et al.*[53]

A complete analysis of the spectra of benzoic and toluic acids has been reported by Diez and Rico,[343] and the n.m.r. spectra of anilines have been

[337] C. W. Haigh and R. B. Mallion, *Mol. Phys.*, 1971, **22**, 945.
[338] M. Bacon and G. E. Maciel, *Mol. Phys.*, 1971, **21**, 257.
[339] E. Cavalieri and M. Calvin, *J.C.S. Perkin I*, 1972, 1253.
[340] P. E. Hansen and A. Berg, *Acta Chem. Scand.*, 1971, **25**, 3377.
[341] H. M. McConnell, *J. Mol. Spectroscopy*, 1958, **1**, 11.
[342] J. Douris and A. Mathieu, *Bull. Soc. chim. France*, 1971, 3365.
[343] E. Diez and M. Rico, *Anales de Fis.*, 1971, **67**, 91.

analysed by Wasylishen and Schaefer[94] and by Aksnes and Kronhaug.[344] The latter find that the (H,H) couplings of the aromatic ring can be predicted quite well by assuming additivity of substituent effects. They also report couplings between exocyclic and ring protons which are in agreement with earlier results. Wasylishen and Schaefer[345] have also published precise analyses of the ^{1}H and ^{19}F spectra of some halogen-substituted benzaldehydes. The measured couplings are compared with CNDO and INDO FPT calculations and the results are satisfactory, showing no clear superiority for either method, though the effect of the π-contribution is noticeable in INDO but not in CNDO.

Hutton et al.[346] have reported (H,H) couplings in a variety of substituted benzal chlorides. Almqvist and his co-workers[347] have investigated couplings between methyl protons in substituted o-, m-, and p-xylenes, and they compare their results with calculations by Acrivos[348] and by Barfield and Chakrabarti.[349] The agreement between theory and experiment is good in view of the difficulties involved.

Barfield et al.[52] have carried out an extensive experimental and theoretical investigation of the long-range (H,H) coupling in four halogen-substituted styrenes. The signs of all coupling constants having magnitudes greater than 0.1 Hz were determined and in all cases these were found to be in agreement with VB SOS and INDO FPT calculations. On the question of magnitude, agreement between theory and experiment also proved to be very gratifying. The case of each type of coupling, four-bond, five-bond, etc., is discussed in detail, and in the case of the seven-bond couplings the calculations indicate the complete absence of a σ-electron contribution.

Additivity rules have been applied to coupling constants in trisubstituted benzenes[350] and in the phenyl moiety of aryl-lithium and aryl Grignard compounds.[351] On the whole, quite good agreement of predicted with observed coupling is found, but Stephens et al.[350] find that, if a resonance interaction between substituents is possible, the observed couplings are larger than those calculated. Ladd and Parker[351] also find a linear dependence of coupling on substituent electronegativity.

Grutzner et al.[352] have measured (H,H) coupling in the triphenylmethanide, fluorenide, and indenide carbanions in a variety of solvents and with Li$^+$ and Na$^+$ as counter-ions. The variation of the coupling constants with solvent and

[344] D. Aksnes and F. H. Kronhaug, Acta Chem. Scand., 1971, **25**, 1871.
[345] R. Wasylishen and T. Schaefer, Canad. J. Chem., 1971, **49**, 3216.
[346] H. M. Hutton, J. B. Rowbotham, B. H. Barber, and T. Schaefer, Canad. J. Chem., 1971, **49**, 2033.
[347] S. O. Almqvist, A. J. Aasen, B. Kimland, and C. R. Enzell, Acta Chem. Scand., 1971, **25**, 3186.
[348] J. V. Acrivos, Mol. Phys., 1962, **5**, 1.
[349] M. Barfield and B. Chakrabarti, J. Amer. Chem. Soc., 1969, **91**, 4346.
[350] M. D. Stephens, J. D. Reinheimer, and A. H. Kappleman, Canad. J. Chem., 1971, **49**, 3759.
[351] J. A. Ladd and J. Parker, J.C.S. Dalton, 1972, 930.
[352] J. B. Grutzner, J. M. Lawlor, and L. M. Jackman, J. Amer. Chem. Soc., 1972, **94**, 2306.

cation appears to correlate with the distance between the cation and the anion, so that as this separation increases the vicinal coupling decreases. Similar observations have been made by others, and the authors suggest that the couplings are determined, at least in part, by local electric fields. The ^3J(HCCH) values also correlate with π-bond orders using the equation for neutral molecules. ^3J(HCCH) values in the phenyl rings of 1,1-diphenyl-n-butyl-lithium cations in a variety of solvents have been measured by Okamoto and Yuki.[353]

The following references also contain data on (H,H) coupling in aromatic systems: 258, 284, and 354—360.

B. (^{1}H,^{1}H) Coupling in Heteroaromatic Systems.—Bartle *et al.*[361,362] have given full analyses of the proton spectra of four thiophen derivatives. Good agreement between bond orders calculated from *ortho*-coupling constants and simple MO theory is found.[361] Ewing and Scrowston[363] have also presented detailed analyses of the spectra of sulphur heterocycles and have examined solvent effects. They find that use of the program LAOCON III leads to accurate chemical shifts but that substantial errors remain in the coupling constants. The signs of the (H,H) couplings in thieno[2,3-*b*]thiophene have been determined from n.m.r. spectra in the nematic phase.[364]

Heffernan and his associates have made detailed analyses of the ^{1}H n.m.r. spectra of benzothiophen,[365] dibenzothiophen and its *S*-oxides,[328] dibenzothiophen and 9,9'-dicarbazyl,[366] and some isoquinolines.[367] The spectrum of benzothiophen is close to being deceptively simple, but under very high resolution and using several solvents, sufficient lines could be resolved to give an unambiguous analysis. The inter-ring long-range coupling constants and their signs were determined.[365] Very similar problems were encountered in the ABMX spectra of dibenzothiophen and 9,9'-dicarbazyl,[366] but the use of solvent shifts made accurate analysis possible, and the relative signs of the coupling constants of dibenzothiophen were found by means of tickling experiments. In the isoquinoline spectra the magnitudes and signs of the

[353] Y. Okamoto and H. Yuki, *J. Organometallic Chem.*, 1971, **32**, 1.
[354] R. E. Gall, D. Landman, G. P. Newsoroff, and S. Sternhell, *Austral. J. Chem.*, 1972, **25**, 109.
[355] P. N. Preston, L. H. Sutcliffe, and B. Taylor, *Spectrochim. Acta*, 1972, **28A**, 197.
[356] G. A. Olah, R. D. Porter, C. L. Jeuell, and A. M. White, *J. Amer. Chem. Soc.*, 1972, **94**, 2044.
[357] W. B. Smith, D. L. Deavenport, and A. M. Ihrig, *J. Amer. Chem. Soc.*, 1972, **94**, 1959.
[358] J. D. Reinheimer and A. H. Kappelman, *J. Org. Chem.*, 1972, **37**, 326.
[359] D. Welti and D. Sissons, *Org. Magn. Resonance*, 1972, **4**, 309.
[360] K. N. Scott, *J. Magn. Resonance*, 1972, **6**, 55.
[361] K. D. Bartle, D. W. Jones, R. S. Matthews, A. Birch, and D. A. Crombie, *J. Chem. Soc. (B)*, 1971, 2092.
[362] K. D. Bartle, D. W. Jones, and R. S. Matthews, *Tetrahedron*, 1971, **27**, 5177.
[363] D. F. Ewing and R. M. Scrowston, *Org. Magn. Resonance*, 1971, **3**, 405.
[364] C. A. Boicelli, A. Mangini, L. Lunazzi, and M. Tiecco, *J.C.S. Perkin II*, 1972, 599.
[365] F. Balkan and M. L. Heffernan, *Austral. J. Chem.*, 1972, **25**, 327.
[366] F. Balkan, M. W. Fuller, and M. L. Heffernan, *Austral. J. Chem.*, 1971, **24**, 2293.
[367] F. Balkan and M. L. Heffernan, *Austral. J. Chem.*, 1971, **24**, 2311.

long-range couplings were determined with the aid of ^{14}N decoupling. The coupling constants show a number of interesting features.[367] The spectra of isotopically substituted ^{2}H, ^{13}C, and ^{15}N thiazoles have been examined[44] and the relative signs of the couplings determined by tickling experiments. All (H,H) and (^{13}C,H) couplings were found to be positive while all (^{15}N,H) couplings are negative.

Fukui et al.[368] have reported a complete analysis of the spectrum of 3-methylpyrrole. All couplings between ring protons and between ring protons and the N-proton are positive. Complete analyses of the ^{1}H n.m.r. spectra of 1- and 2-methylindazole have been given by Elguero et al.[369] Dembech and his co-workers[370] have reported many couplings in substituted 2-chloro-1-methylbenzimidazoles. They find quite good correlations with the additivity rules of Castellano and Kostelnik[371] and of Hayamizu and Yamamoto,[372] but the correlation of ortho-couplings with Hückel MO bond orders was poor.

The 100 and 200 MHz spectra of 2-pyridone and some of its derivatives have been analysed by von Ostwalden and Roberts.[373] Dorie et al.[374] have reported coupling constants for 35 substituted pyridines and have discussed them with respect to substituent electronegativity and additivity relationships. The couplings are also compared with the mutual polarizability, Π, of the appropriate hydrogen 1s orbitals, calculated from CNDO/2 wavefunctions, and the plot of J versus Π shows a correlation coefficient of 0.945. An investigation of the effect of including some two-centre integrals in the evaluation of $\Pi_{s_A s_B}$ in equation (2) indicated that use of one-centre terms only may lead to an error of the order of 8%.

Elias, Moritz, and Paul[375] have measured (H,H) couplings in N-methyldiazinium iodides. Fairly satisfactory agreement was obtained between observed couplings and those calculated by an unspecified semi-empirical method. Wasylishen and Schaefer[376] have measured three-, four-, and five-bond couplings in syn- and anti-furanaldoximes. The experimental trends in the coupling constants are quite well reproduced by CNDO and INDO FPT calculations.

Other measurements of (H,H) coupling in heterocyclic systems may be found in references 125, 294, 295, and 377—381.

[368] H. Fukui, S. Shimokawa, J. Sohma, T. Iwadore, and N. Esumi, *J. Mol. Spectroscopy*, 1971, **39**, 521.
[369] J. Elguero, A. Fruchier, R. Jacquier, and U. Scheidegger, *J. Chim. phys.*, 1971, **68**, 1113.
[370] P. Dembech, G. Seconi, P. Vivarelli, L. Schenetti, and F. Taddei, *J. Chem. Soc. (B)*, 1971, 1670.
[371] S. Castellano and R. Kostelnik, *Tetrahedron Letters*, 1967, 5211.
[372] K. Hayamizu and O. Yamamoto, *J. Mol. Spectroscopy*, 1968, **25**, 422.
[373] P. W. Von Ostwalden and J. D. Roberts, *J. Org. Chem.*, 1971, **36**, 3792.
[374] J. P. Dorie, M. L. Martin, S. Barnier, M. Blain, and S. Odiot, *Org. Magn. Resonance*, 1971, **3**, 661.
[375] D. J. Elias, A. G. Moritz, and D. B. Paul, *Austral. J. Chem.*, 1972, **25**, 427.
[376] R. Wasylishen and T. Schaefer, *Canad. J. Chem.*, 1972, **50**, 274.
[377] V. M. S. Gil and A. M. P. Pereira, *Tetrahedron*, 1971, **27**, 5619.
[378] E. V. Blackburn, T. J. Cholerton, and C. J. Timmons, *J.C.S. Perkin II*, 1972, 101.
[379] A. Taurins and R. K. C. Hsia, *Canad. J. Chem.*, 1971, **49**, 4054.

C. **Coupling in Metal–Cyclopentadienyl Compounds.**—Sergeyev and his co-workers have measured the ^{13}C satellites in the ^{1}H spectrum of cyclopentadienyltrimethyltin and have discussed it in connection with a comprehensive summary of literature data on the spectra of metal–cyclopentadienyl systems.[382] They have also described the spectra of some (trimethylsilyl)cyclopentadienes and have noted some of the regularities in these spectra.[383]

D. **(^{1}H, ^{19}F) Coupling in Aromatic Systems.**—Aksnes and Kronhaug[344] have measured (H,F) coupling constants within the aromatic ring and between exocyclic CF$_3$ groups and ring protons in several substituted anilines. Schaefer and his associates have measured (H,F) coupling between the aldehydic proton and ring fluorine atoms in fluorobenzaldehydes[345] and between the benzal proton and ring fluorine atoms in some fluorochlorobenzal chlorides.[346] The results have been compared with CNDO and INDO FPT calculations.[345]

Rodmar et al.[384] have analysed the spectra of 46 substituted fluorothiophens, finding many relative signs for the (H,F) couplings by means of double resonance. In a subsequent paper[385] the results are correlated with reactivity constants, electronegativities and, in the case of ^3J, bond orders. The empirical correlations are quite successful but the theoretical calculations based on either McConnell's expression [equation (37)] or on equation (3) are not satisfactory.

$$J^\pi(H,F) = (\beta^2 Q_{CH} Q_{CF}/h\Delta E)(P^\pi_{C_H C_F})^2 \qquad (37)$$

Other references quoting (H,F) couplings for aromatic molecules are 258 and 351.

E. **(^{1}H, ^{13}C) Coupling in Aromatic Systems.**—Englert et al.[335] have measured all the (H,C) coupling constants in [1-^{13}C]benzene; all are positive except the ^4J, which is negative. Tarpley and Goldstein have reported (H,C) couplings in chloro-, bromo- and iodo-benzene,[67] in the corresponding symmetrical m-dihalogenobenzenes,[68] and in uracil, thymine, and the 5-halogenouracils.[195] The dependence of the couplings on substituent electronegativity is discussed in detail (see Sections 4B, 5C, 6C, and 7C) and it is also shown, for the monohalogenobenzenes,[67] that they may be predicted by means of a set of additivity parameters, though there are some anomalies. Hansen and Berg[340] have

[380] T. Sone and K. Takahashi, *Org. Magn. Resonance*, 1971, **3**, 527.
[381] R. E. Rondeau, M. A. Berwick, and H. M. Rosenberg, *J. Heterocyclic Chem.*, 1972, **9**, 427.
[382] Yu. K. Grishin, N. M. Sergeyev, and Yu. A. Ustynyuk, *J. Organometallic Chem.*, 1972, **34**, 105.
[383] N. M. Sergeyev, G. I. Auramenko, A. V. Kisin, V. A. Korenevsky, and Yu. A. Ustynyuk, *J. Organometallic Chem.*, 1971, **32**, 55.
[384] S. Rodmar, L. Moraga, S. Gronowitz, and U. Rosén, *Acta Chem. Scand.*, 1971, **25**, 3309.
[385] S. Rodmar, S. Gronowitz, and U. Rosén *Acta Chem. Scand.*, 1971, **25**, 3841.

reported (H,C) couplings for pyrene; there is very poor agreement with a CNDO SOS calculation.

F. (^{1}H,^{31}P) Coupling in Aromatic Systems.—These coupling constants have aroused considerable interest in the past year, and a particularly detailed analysis of the long-range spin–spin coupling between protons and phosphorus in the six isomeric methyl-substituted trithienylphosphines and their sulphides [see Section 6D (iii)] has been published.[125] The signs of the couplings have been obtained and all observed (H$_3$C,P) couplings are found to be positive in the phosphines. In the phosphine sulphides a change of sign is usually observed, all signs being similar to those of the corresponding ring-proton–methyl-proton couplings in the methylthiophens. A sizeable dependence of the ring-proton–phosphorus coupling on the orientation of the phosphorus lone pair with respect to the ring planes is indicated by the experimental results for the *ortho*-methyl-substituted phosphines. The work of Barbieri and Taddei[295] on similar systems has been described in more detail in Section 6D (iii).

G. Coupling of Protons to Other Nuclei in Aromatic Systems.—Comparisons of (H,H) coupling with couplings over the same paths involving ^{205}Tl (ref. 386), ^{119,117}Sn (ref. 294), and ^{207}Pb (ref. 295) have been made. They suggest that the coupling mechanisms in all cases are very similar and dominated by the Fermi-contact term.

H. (^{13}C,^{15}N) Coupling in Aromatic Systems.—Randall and his co-workers[387] have measured the ^{13}C and ^{15}N n.m.r. spectra of [^{15}N]quinoline in neutral and protonated form; they have obtained all the (C,^{15}N) coupling constants. In CCl$_4$, ^1J(C,N) < ^2J(CCN) < ^3J(CCCN), and the magnitude of ^1J(C,N) (1.4 and 0.6 Hz) is much smaller than the value of *ca* 8.7 Hz which is estimated from the relationship due to Binsch *et al.*[118] The anomalously high two-bond coupling to the carbon in the *peri* position to the nitrogen is interpreted as being due to the effect of the nitrogen lone pair. This idea is supported by the notable solvent dependence of this coupling. The (C,^{15}N) couplings in pyridine have been measured and discussed by Lichter and Roberts.[117]

I. Other Couplings in Aromatic Systems.—Wasylishen and Schaefer[94] have observed coupling between ^{15}N and ring protons in 2-chloro[^{15}N]aniline. Barboiu *et al.*[388] have determined the relative signs of all the (F,F) couplings in pentafluoropyridine; the broadening effect of the ^{14}N was removed by strong irradiation. The couplings are solvent-dependent.

11 'Through-space' Coupling

There has been no systematic study of this topic in the past year, though

[386] J. P. Maher, M. Evans, and M. Harrison, *J.C.S. Dalton*, 1972, 188.
[387] P. S. Pregosin, E. W. Randall, and A. I. White, *J.C.S. Perkin II*, 1972, 1.
[388] V. Barboiu, J. W. Emsley, and J. C. Lindon, *J.C.S. Faraday II*, 1972, **68**, 241.

a number of authors have suggested that their experimental results might be rationalized in terms of this concept. For (H,F) coupling these include data on fluorine-substituted benzal chlorides,[346] 3-acetyl-2-fluoronaphthalene and o-fluoroacetophenone,[389] trifluoromethyl-quinoxalines,[390] 2,2,4,4,4-pentafluoro-n-butane,[279] and fluorinated benzofurans.[301]

Marsmann and Horn[127] have tentatively suggested that a through-space contribution to a (P,C) coupling might explain some rather strange results for ditertiary phosphines.

12 Isotope Effects

Isotope effects on $^1J(P,H)$ have been observed by Stec et al.[99] and by Borisenko and his associates.[100] These results are described in Section 4D. Fraser, Petit, and Miskow[166] have reported an isotope effect in $^2J(HCH)$, which is discussed in Section 5A.

13 Experimental Work of Significance for the Measurement or Interpretation of Spin–Spin Coupling Constants

The following is a brief summary of some of the recently reported experimental techniques which seem to show some promise of assisting in the measurement and/or interpretation of nuclear spin–spin coupling. The summary is not expected to be comprehensive and many of the papers noted here will be discussed in detail by more competent authorities elsewhere in this volume.

A. **Multiple-resonance Experiments.**—Emsley and his associates[391] have shown how complex ^{1}H n.m.r. spectra of molecules in the nematic or isotropic phase may be simplified by partial deuteriation and deuterium decoupling. They illustrate the technique with a spectrum of CH_3CD_2OH in the nematic phase. Barboiu, Emsley, and Lindon[388] have demonstrated that the broadening effect of ^{14}N in the spectrum of pentafluoropyridine can be removed by strong irradiation, so that it becomes possible to use double- and triple-resonance experiments to determine the relative signs of the (F,F) couplings. Sudmeier et al.[172] have decoupled ^{59}Co from the 251 MHz spectrum of cobalt tris(ethylenediamine).

The use of the nuclear Overhauser effect in coupled ^{13}C n.m.r. spectra has been described by Freeman and Hill.[392] The nuclear Overhauser enhancement can be maintained in the spectrum, without decoupling of the protons as is usually the case, provided that the second irradiation field, B_2, is suitably pulse-modulated. Under these conditions the proton saturation that

---

[389] W. Adcock, D. G. Matthews, and S. Q. A. Rizvi, *Austral. J. Chem.*, 1971, **24**, 1829.
[390] E. Abushanab, *J. Amer. Chem. Soc.*, 1971, **93**, 6532.
[391] J. W. Emsley, J. C. Lindon, J. M. Tabony, and T. H. Wilmshurst, *Chem. Comm.*, 1971, 1277.
[392] R. Freeman and H. D. W. Hill, *J. Magn. Resonance*, 1971, **5**, 278.

is responsible for the ^{13}C signal enhancement persists for several seconds whereas the radiofrequency coherence effects (multiplet coalescence) disappear as soon as B_2 is extinguished. Coxon[393] has shown how INDOR techniques can be used to detect small, long-range (H,H) coupling constants.

Ziessow[80] has described the application of the ^{1}H–{^{13}C} Torrey-oscillation technique to the determination of $^2J(^{13}CO^{13}C)$ in a natural-abundance sample of $^{13}CH_3O^{13}COCl$.

B. Bandshape Analyses.—Reeves and his co-workers[394] have extended a matrix formulation of the Bloch equations which includes chemical exchange, paying particular attention to the effect of scalar couplings on the lineshapes. They have applied this theory[395] to NN-dimethylacetamide and NN-dimethyltrifluoroacetamide, and show that it is possible to determine the relative signs of the long-range couplings and to measure unresolved interactions. Bogdanov et al.[202] have used the theory of Solomon and Bloembergen[203,204] to obtain (B,H) couplings from line broadening in the ^{11}B and ^{1}H spectra of some organoboron compounds. Harris and Pyper[396] have shown that the relative signs of J_{AX} and J_{BX} for an $^2A^2B^3X$ spin system (where X is undergoing rapid relaxation) can always be measured by studying the double-quantum transition in the AB region. By this means the two (H,^{14}N) scalar couplings in 2-bromothiazole are shown to have the same sign.

C. Effects of Paramagnetic Materials.—(H,H) and (H,F) decoupling by paramagnetic ion co-ordination at nitrogen[397] and at oxygen[398] has been described by Engel and Nathan. Gansow et al.[399,400] have shown how applications of Fourier-transform n.m.r. to carbon atoms having a long relaxation time may be facilitated by the addition of a kinetically inert, paramagnetic, metal complex with a totally symmetric ground state, e.g. tris(acetylacetonato) chromium(III). Spin–lattice relaxation times are drastically reduced and no chemical shift is detected ±0.1 p.p.m. There is little line-broadening for concentrations of Cr(acac)$_3$ less than 0.1 mol l^{-1} Martin et al.[270] have shown that in alcohols such as R^{1}CH=CH—CHOHR^2 the cis- and trans-stereoisomers behave differently in the presence of a rare-earth shift reagent. The spectra of the trans-isomers remain strongly coupled but ^3J(HCCH) becomes easy to measure in the cis-compounds. The use of chiral shift reagents for the determination of the ^2J(HCH) coupling of benzylic protons has been described by Fraser, Petit, and Miskow[166] (see also Section 5A).

[393] B. Coxon, Carbohydrate Res., 1971, **18**, 427.
[394] L. W. Reeves and K. N. Shaw, Canad. J. Chem., 1971, **49**, 3671.
[395] L. W. Reeves, R. C. Shaddick, and K. N. Shaw, Canad. J. Chem., 1971, **49**, 3683.
[396] R. K. Harris and N. C. Pyper, Mol. Phys., 1972, **23**, 277.
[397] R. Engel, J. Chem. Soc. (C), 1971, 3554.
[398] R. Engel and G. Nathan, J. Chem. Soc. (C), 1971, 3844.
[399] O. A. Gansow, A. R. Burke, and W. D. Vernon, J. Amer. Chem. Soc., 1972, **94**, 2550.
[400] O. A. Gansow, A. R. Burke, and G. N. La Mar, Chem. Comm., 1972, 456.

Nuclear Spin–Spin Coupling

D. Fourier-transform Experiments.—Allerhand et al.[401] have shown how partially relaxed Fourier-transform ^{11}B n.m.r. can enhance the resolution of ^{11}B spectra, and Vold and Shoup[228] have used Fourier-transform proton spin-echo spectra to get very accurate values of two- and three-bond couplings in 1,1-difluoro-2,2-dichloroethane.

E. Other Experimental Techniques.—Gansow and his associates have described an alternately pulsed ^{13}C n.m.r. technique for use with Fourier-transform measurements.[402] They claim that the method is superior to the usual off-resonance decoupling or single-resonance irradiation methods for signal assignment. The technique also appears to have many interesting applications when used in conjunction with lanthanide shift reagents.[403]

Harris and Gazzard[404] have demonstrated the use of spectra in the nematic phase for the determination of isotropic scalar spin–spin coupling constants. The method is particularly applicable when these couplings are large. Katz et al.[405] have noted that first-order spectra can easily be obtained from deuteriated alicyclic and aliphatic hydrocarbons containing a small amount of randomly distributed ^{1}H. The method does not, at first sight, appear useful for the determination of coupling constants, but the authors suggest that it might be possible to choose a concentration of ^{1}H large enough to show spin–spin coupling.

Kessemeier and Rhim[406] have shown that linewidths in solids may be narrowed by irradiation with strong continuous rf magnetic fields under magic angle conditions. They term this a 'rotary spin-echo technique'. The dipolar coupling is reduced by this process, and it might be possible to obtain coupling constants.

[401] A. Allerhand, A. O. Clouse, R. R. Rietz, T. Roseberry, and R. Schaeffer, *J. Amer. Chem. Soc.*, 1972, **94**, 2445.
[402] O. A. Gansow and W. Schittenhelm, *J. Amer. Chem. Soc.*, 1971, **93**, 4294.
[403] O. A. Gansow, M. R. Willcott, and R. E. Lenkinski, *J. Amer. Chem. Soc.*, 1971, **93** 4295.
[404] R. K. Harris and V. J. Gazzard, *Org. Magn. Resonance*, 1971, **3**, 495.
[405] J. J. Katz, G. N. McDonald, and A. L. Harkness, *Chem. Comm.*, 1972, 542.
[406] H. Kessemeier and W.-K. Rhim, *Phys. Rev. (B)*, 1972, **5**, 761.

3
Nuclear Spin Relaxation

BY N. BODEN

1 Introduction

There are many interesting physical, chemical, and biological phenomena whose origins are intimately related to the motions of atoms or molecules. With an understanding of these processes as the objective, spectroscopic techniques are being applied at a rapidly increasing rate to the study of dynamic processes in all states of matter. These techniques include, in addition to nuclear spin relaxation, the often complementary ones of inelastic neutron scattering, i.r. bandshape analysis, depolarized light scattering, and dielectric relaxation. This chapter reports the results of nuclear spin relaxation studies, principally by pulse methods, for the period from July 1971 to May 1972 inclusive. The coverage is similar to that given last year;[1] material is presented against the same background, using the same definitions, symbols, *etc*. All references to relaxation in fluids as studied by high-resolution lineshape analysis and double-resonance techniques are excluded as they are reviewed, respectively, in Chapters 6 and 7. Relaxation studied in gases, liquids, and solids is comprehensively covered; the principal omissions include the more physical papers concerned with spin dynamics and metals, and work on biological and macromolecule systems which is covered in Chapter 8. Pulsed n.m.r. in solids, including such subjects as line narrowing by multiple-pulse techniques and the spin-locked double-resonance experiment for ultrasensitive detection of weakly resonant spin systems, is this year reviewed in Chapter 9. New developments in pulse instrumentation and techniques are covered in Chapter 4.

This article is intended as a literature report, not as a general review of magnetic relaxation theory and practice. Those interested in an introduction to the theory of nuclear spin relaxation are referred to the standard texts on this subject by Abragam,[2] Slichter,[3] or the recently published book by Poole and Farach.[4] Useful discussions of relaxation theory are to be found in the

---

[1] N. Boden, in 'Nuclear Magnetic Resonance', ed. R. K. Harris, (Specialist Periodical Reports), The Chemical Society, London, 1972, Vol. 1, Chapter 3.
[2] A. Abragam, 'The Principles of Nuclear Magnetism', Oxford University Press, London, 1961.
[3] C. P. Slichter, 'Principles of Magnetic Resonance', Harper and Row, New York, 1963.
[4] C. P. Poole and H. A. Farach, 'Relaxation in Magnetic Resonance', Academic Press, New York, 1971.

early papers by Bloembergen, Purcell, and Pound,[5] or Kubo and Tomita,[6] or in the more recent reviews by Redfield[7] and Noack;[8] the latter is particularly useful for its summary of relaxation expressions and spectral density functions. The book by Goldman[9] and reviews by Jeener[10] and Ailion[11] provide excellent accounts of relaxation in the strong collision limit as appropriate to solids.

2 Spin Relaxation in Gases

A. Introduction.—The attraction of making spin relaxation measurements in gases is that they can be interpreted to provide quantitative information about the anisotropic part of the intermolecular interaction potential, information which is not available by measurement of conventional gas transport properties. For this reason relaxation in gases has always been a subject of considerable theoretical and experimental investigation; furthermore, gas measurements are more amenable to theoretical interpretation than in condensed phases and are therefore better understood. Because of the relative difficulties in gas-phase work, it is not, however, too surprising that the general increase in the number of groups working in relaxation is not reflected in this field. This year's work is collated under the same sub-section headings as employed in Volume 1.

B. Theoretical Developments.—The relaxation of spin ½ nuclei in polyatomic molecules in dilute gases is determined by the spin–rotation interaction. In 1967, Bloom et al.,[12] using the conventional correlation function approach, derived expressions for T_1 in gases composed of symmetric and spherical molecules. In 1970, McCourt and Hess[13] made a significant step forward by treating relaxation in spherical-top molecules from the point of view of the kinetic theory of transport and relaxation processes based on the Waldman–Snyder equation (Vol. 1, p. 117). The existence of an error in their work has been pointed out by Ozier.[14] McCourt and Hess[15] have shown how this error resulted from an incorrect treatment of the 'anisotropic' part of the spin–rotation interaction Hamiltonian. In the extreme narrowing limit, the spin relaxation times T_1 and T_2 for spherical top molecules are now given by

$$T_1^{-1} = T_2^{-1} = \frac{2}{3}\langle J(J+1)\rangle \left\{\frac{\bar{C}^2}{\omega_{\text{coll}}} + \frac{4}{45}\frac{\Delta C^2}{\omega_{\text{coll}}}\right\} \quad (1)$$

---

[5] N. Bloembergen, E. M. Purcell, and R. V. Pound, *Phys. Rev.*, 1948, **73**, 679.
[6] R. Kubo and K. Tomita, *Proc. Phys. Soc. Japan*, 1954, **9**, 888.
[7] A. G. Redfield, *Adv. Magn. Resonance*, 1965, **1**, 1.
[8] F. Noack, in 'NMR, Basic Principles and Progress', ed. P. Diehl, Springer, New York, 1971, vol. 3, p. 83.
[9] M. Goldman, 'Spin Temperature and Nuclear Magnetic Resonance in Solids', Oxford University Press, Oxford, 1970.
[10] J. Jeener, *Adv. Magn. Resonance*, 1965, **1**, 205.
[11] D. C. Ailion, *Adv. Magn. Resonance*, 1971, **5**, 177.
[12] M. Bloom, F. D. Bridges, and W. N. Hardy, *Canad. J. Phys.*, 1967, **45**, 3584.
[13] F. R. McCourt and S. Hess, *Z. Naturforsch.*, 1970, **25a**, 1169.
[14] I. Ozier, *Z. Naturforsch.*, 1971, **26a**, 1232.
[15] F. R. McCourt and S. Hess, *Z. Naturforsch.*, 1971, **26a**, 1234.

where the quantities $\bar{C} = \frac{1}{3}(C_\parallel + 2C_\perp)$ and $\Delta C = C_\parallel - C_\perp$ describe the spin-rotation tensor, and ω_{coll} and $\tilde{\omega}_{\text{coll}}$ are collision integrals. Notice that this equation applies equally to tetrahedral and octahedral molecules, contradicting their earlier results, and is equivalent to the expression derived by Bloom et al.[12] The conclusion of Dong and Bloom[16] (Vol. 1, p. 117) that the best value for $|\Delta C|$ is 18.2 kHz rather than 21.0 kHz is confirmed. In fact, the best current value for $|\Delta C|$ is (18.5 ± 0.5) kHz, as obtained by Yi et al.[17] McCourt and Hess[15] have also obtained an expression for relaxation in symmetric top molecules and have discussed the conditions under which it approximates to equation (1).

C. Diatomic Molecules.—Molecular hydrogen is an attractive system to study because the very large separation between rotational levels, and the unobservability of para-hydrogen, mean that a limited number of rotational states must be accounted for at reasonable temperatures; in fact, it is the only system for which lattice correlation functions can be calculated from first principles.[18] There have been no new studies in pure H_2 gas, but the H_2-rare-gas systems have been extensively investigated; the only previous investigations are those by Riehl et al.[19] on the H_2-He system and the earlier fragmentary measurements by Williams[20] on the H_2-Ne and H_2-Ar systems. Foster and Rugheimer[21] have been making precise T_1 measurements in H_2-X mixtures (X = Xe, Kr, Ar, Ne, or He) at 20 amagat* over the temperature range 167—265 K for rare gas fractions of 0—60%. Similar measurements have been reported by Lalita and Bloom[22] for the H_2-He system in the temperature interval 77—730 K. Values of $(T_1^H/\rho)^{OX}$, i.e. the value of T_1^H/ρ for an ensemble of ortho-H_2 molecules each of which only interacts with X, were determined by extrapolating the plot of T_1^H/ρ versus concentration to zero concentration of H_2. The assumption that T_1/ρ is a linear function of the gas concentration has been verified by extensive T_1 measurements on H_2-He mixtures.[23] Assuming that the transition rates of the H_2 molecules between J manifolds are small compared to transition rates within each manifold (M transitions), and using perturbation theory to calculate the relative probabilities for the various possible transitions within each J manifold, the calculations of Oppenheim and Bloom[18] indicate that, for adulterated H_2,

$$(T_1/\rho)^{OX} = (25/2)\{A(1,0;1,1)/[1.02 \times 10^{13}\langle D(J)\rangle]\} \quad (2)$$

[16] R. Y. Dong and M. Bloom, *Canad. J. Phys.*, 1970, **48**, 793.
[17] P. N. Yi, I. Ozier, and N. F. Ramsey, *J. Chem. Phys.*, 1971, **55**, 5215.
[18] M. Bloom and I. Oppenheim, *Adv. Chem. Phys.*, 1967, **12**, 549; J. M. Deutch and I. Oppenheim, *Adv. Magn. Resonance*, 1966, **2**, 225.
[19] J. W. Riehl, J. L. Kinsey, J. S. Waugh, and J. H. Rugheimer, *J. Chem. Phys.*, 1968, **49**, 5276.
[20] D. L. Williams, *Canad. J. Phys.*, 1962, **40**, 1027.
[21] K. R. Foster and J. H. Rugheimer, *J. Chem. Phys.*, 1971, **56**, 2632.
[22] K. Lalita and M. Bloom, *Canad. J. Phys.*, 1971, **49**, 1018.
[23] S.-P. Vega, K. R. Foster, and J. H. Rugheimer, *J. Chem. Phys.*, 1972, **56**, 678.

* The dimensionless unit 'amagat' is the ratio of the density under consideration to the density at standard temperature and pressure.

where $A(J,M; J',M')$ is the transition rate between states $|J',M'\rangle$ and $|J,M\rangle$, and

$$D(J) = [J(J+1)(2J-1)(2J+3)/57.4]\{1+[4.74/(4J^2+4J-7)]\}$$

$J = 1,3,5, \cdots$ for ortho-H_2 and $\langle ---\rangle$ indicates thermal averaging over the populated rotational states. The function $D(J)$ accounts for the larger effective rotational correlation time for H_2 in higher J states. At densities which result in the extreme narrowing limit, but are not high enough that three-body effects are important (between one and several hundred amagat)

$$A(J,M; J',M') = \langle v\sigma(J,M; J',M')\rangle \qquad (3)$$

where $\sigma(J,M; J',M')$ is the H_2 reorientation cross-section for H_2–X collisions, and $\langle ---\rangle$ indicates thermal averaging over the Boltzmann distribution of relative speeds v. Thus, given a suitable model for the intermolecular potential, the necessary scattering calculations can be performed and information about the angle-dependent terms of the interaction obtained. Foster and Rugheimer[21] fitted the available He–X data by coupled-channel molecular scattering calculations, assuming infrequent transitions of H_2 between rotational levels, and using a Lennard-Jones (12,6) potential for the isotropic and angular-dependent parts of the intermolecular potential. The calculations indicate that the apparent anisotropy in the attractive part of the H_2–rare-gas interaction is consistently approximately twice as great as that predicted from molecular polarizability measurements. The suggested explanation for this discrepancy is that the radial dependence of the repulsive part of the anisotropic potential is different from that of the attractive part, *i.e.* the Lennard-Jones potential is an oversimplification.

Spin relaxation in F_2 gas at normal temperatures is expected to differ from that in H_2 because higher rotational states will be appreciably occupied and the rotational motion may be treated classically; it is therefore expected to be determined by spin–rotation interaction and Gordon's classical theory (Vol. 1, p. 118) should be valid. Values for T_1^F in pure F_2 gas have been measured as a function of temperature (187—333 K) and density both by Courtney and Armstrong[24] and by Dybowski *et al.*,[25] and also by the latter group of workers in F_2–X mixtures (X = Ar or He) at 200 K. In the pure gas, $T_1^F/\rho \propto T^{-1.7\pm0.1}$ indicating that relaxation is predominantly *via* spin–rotation interaction; the deviation of the exponent from -1.5 is probably not significant. The plots of T_1^F versus ρ were linear and the values obtained from them, using Gordon's equation, for σ_J^{FF} (the cross-section for exchange of angular momentum in a collision) are between 0.246 nm² at 187 K and 0.101 nm² at 333 K. For comparison, the temperature-independent value calculated from kinetic theory is 0.42 nm². The observed temperature dependence, $\sigma_J^{FF} \propto T^{-1}$, is the familiar one occurring in molecules like NH_3, C_2H_2, HCl, BF_3, CF_4, *etc.*, and suggests that in this temperature region the F_2 molecules behave classically

[24] J. Courtney and R. Armstrong, *J, Chem. Phys.*, 1970, **52**, 2158.
[25] C. R. Dybowski, M. Chien, and C. G. Wade, *J. Chem. Phys.*, 1972, **56**, 4229.

with respect to transfer of angular momentum. In the F_2–rare-gas mixtures at 200 K, $\sigma_J^{FAr} = 0.393$ nm^2 and $\sigma_J^{FHe} = 0.305$ nm^2, and in both cases $\sigma_J^{FF} = 0.20$ nm^2, which is similar to the value 0.24 nm^2 determined in the pure gas at the same temperature.

D. Polyatomic Molecules.—The spin–rotation interaction is known to be the dominant relaxation mechanism in gaseous methane. Since the spin–rotation coupling constants for this molecule have been accurately determined ($\bar{C} = 10.4$ kHz and $\Delta C = 18.5$ kHz) from molecular beam measurements,[17] it ought to be possible to check the theory of spin–lattice relaxation for spherical-top molecules by making precise T_1^H versus ρ measurements in the vicinity of the T_1 minimum. This objective has been pursued by Bloom and his co-workers[12,16,26] in recent years. In 1970, Dong and Bloom[16] reported a study of the T_1^H versus ρ minimum in dilute CH_4 gas; the measurements appeared to be adequately fitted by the conventional theory of Bloom et al.[12] and the assumption that the correlation function of the spin–rotation interaction is a simple exponential function of time. Beckmann et al.[26] now report a brilliant piece of experimental work in which T_1^H was measured in gaseous CH_4 as a function of density at room temperature between 0.006 and 7.0 amagat. T_1 passes through a minimum near 0.4 amagat, in agreement with the previous, but less precise measurements of Dong and Bloom. These measurements are not, however, consistent with the assumption of a simple exponential correlation function. The reason for this is that the centrifugal distortion energy of the CH_4 molecule removes the degeneracy of rotational states having the same value of the quantum number J. Since the tensor term, ΔC, of the spin–rotation interaction couples states whose energies differ due to the centrifugal distortion interaction,[27] the effects of these splittings should manifest themselves in the plot of T_1 versus ρ in the density region where the collision frequency is of the same order as the average distortion splitting. Inclusion of this interaction into the theory of Bloom et al.[12] gives the following general expression for T_1

$$T_1^{-1} = \frac{2IkT}{\hbar^2}\left(\bar{C}^2 \frac{\tau_1}{1+\omega_0^2\tau_1^2} + \frac{4}{45}\Delta C^2 \sum_k \frac{F_k\tau_k}{1+(\omega_0-\omega_k)^2\tau_k^2}\right) \quad (4)$$

where F_k is a measure of the sum of the squares of the matrix elements of the tensor part of the spin–rotation interaction between states having energy differences $\hbar\omega_k$; $\Sigma F_k = 1$. The τ_k obtain from the assumption of Lorentzian rotational lines at low densities, i.e.

$$j_k(\omega) = \frac{2\tau_k}{1+\omega_k^2\tau_k^2}, \quad g_k(t) = \exp(-t/\tau_k) \quad (5)$$

It was shown that the T_1 data could be fitted to this equation if the centrifugal

---
[26] P. A. Beckmann, M. Bloom, and E. E. Burnell, Canad. J. Phys., 1972, **50**, 251.
[27] C. H. Anderson and N. F. Ramsey, Phys. Rev., 1966, **149**, 14.

distortion spectrum is replaced by a single 'average' line at frequency of ca. 200 MHz, which is quite close to the observed average distortion frequency at room temperature.

The effects of centrifugal distortion on spin relaxation can be neglected when the associated energy splitting $E_{c.d.}$ is collisionally quenched, i.e. if the condition

$$(1/\hbar)E_{c.d.} \ll \omega_{coll}$$

is satisfied. This condition imposes a lower limit on the pressures for which equation (1) can be applied. Since $\omega_{coll} \approx 10^9$ Hz at 1 atm. at room temperature, and $E_{c.d.} \approx 10^7$ to 10^8 Hz for CH_4, this lower pressure limit is of the order of 0.1—1.0 atm. The measurements of Beckmann et al.[26] indicate that the effect of centrifugal distortion shows up below 0.5 atm.

A further discussion of proton spin–lattice relaxation in paramagnetically adulterated methane (Vol. 1, p. 120) has been given by Lipsicas and Siegel.[28] It is shown how the experimental temperature dependence of T_1^H, as well as the fact that O_2 is about eight times as effective as NO in the relaxation even though the dynamic parameters for CH_4–O_2 and CH_4–NO collisions are practically the same, can be accounted for by consideration of the paramagnetic moment auto-correlations $\langle \mu(0)\, \mu(t) \rangle$ over the time period of a collision. This time correlation is especially dependent on the coupling of the internal orbital and the spin angular momentum of the molecule with molecular rotation. In the case of oxygen, S is a good quantum number and the electron spins are coupled to rotation only via the electron–electron dipolar coupling. NO, by contrast, is a molecule which fits Hund's case A; the paramagnetic moment vector precesses about the J axis during the time period of the collisions, resulting in a partially averaged magnetic moment.

E. **Spin Diffusion Measurements.**—Khoury and Kobayashi[29] have used the spin-echo technique to measure the self-diffusion coefficient D in CF_4 gas for the conditions 243—348 K and 200—440 atm. The data are represented by

$$\rho D/(\rho D)_0 = 1 - 0.041\,7274\rho - 0.265\,870\rho^2 - 0.054\,3426\rho^3$$

where ρ is in g cm^{-3} and D is in cm^2 s^{-1}. The low-density data were used, along with previously published second virial coefficient data, to evaluate intermolecular force constants for the Lennard-Jones (6,12) potential as $\sigma = 0.4706$ nm and $\varepsilon/k = 152.1$ K, and for the modified Buckingham (6-exp) potential for $\alpha = 15$ as $\sigma = 0.4642$ nm and $\varepsilon/k = 170$ K.

3 Spin Relaxation in Liquids

A. **Introduction.**—There has been a significant increase in the number of publications concerned with spin relaxation measurements in liquids, their

---

[28] M. Lipsicas and M. Siegel, *J. Magn. Resonance*, 1972, **6**, 533.
[29] F. Khoury and R. Kobayashi, *J. Chem. Phys.*, 1971, **55**, 2439.

main motivation being the current quest for an understanding of the dynamic structure of liquids. Unfortunately, in most pure liquids the correlation functions for the spin-dependent interactions which couple the spins to the lattice possess characteristic decay times of the order 10^{-11}—10^{-14} s; their spectral densities in the range 1—200 MHz, accessible to n.m.r. are therefore independent of frequency with the consequence that spin relaxation measurements are insensitive to the actual shape of these correlation functions but depend on correlation times which measure the area under these functions. The work reported shows that it is now possible to determine reliable values for the rotational correlation time τ_2 and angular velocity correlation time τJ from T_1 measurements. On the other hand, the efforts to provide a reasonable theoretical interpretation for these parameters are disappointing; in many cases authors are still content to compare their τ_2 values with the Debye–Stokes–Einstein expression sometimes using the microviscosity modification of Gierer and Wirtz (Vol. 1, p. 124). Zeidler[30] has given a nice, but brief, review of nuclear spin relaxation theory in relation to molecular motion in liquids, but the excellent article by Noack[8] will be found more useful to those interested in a comprehensive summary of the general theory of nuclear magnetic relaxation. Noack's article is primarily concerned with the theoretical and experimental aspects of nuclear magnetic relaxation spectrocsopy, *i.e.* the frequency dependence of T_1 and T_2 and its application to the study of molecular dynamics in the liquid state. Although in most pure liquids molecular motion is too fast to be accessible to this technique (the notable exceptions being viscoelastic liquids), in ionic and macromolecular solutions there are relaxation processes which occur on a longer time scale, enabling this technique to be exploited.

The format for the material reported in this section is similar to the one employed last year. The only significant change is the inclusion of liquid-state spin-echo studies, last year reported as Section 4; other changes are minor and merely reflect the change in emphasis of published work.

B. Pure Liquids.—*Linear Molecules.* In the ortho–para mixtures of D_2 in the liquid state, the two spin systems $I = 1$, $J = 1$ (para-D_2) and $I = 2$, $J = 0$ (ortho-D_2) are only weakly coupled, and their relaxation rates can be independently determined.[31,32] At 20 K, $T_1(J = 0) \approx 10^3$ s and $T_1(J = 1) \approx 1$ s. The relaxation of the $(J = 1)$ molecules is *via* the intramolecular spin–spin interactions modulated by the quadrupole interactions and is satisfactorily accounted for by the theory of Deutch and Oppenheim.[33] The relaxation of the $(J = 0)$ molecules is more difficult to explain. The relaxation rate from modulation of the intermolecular dipole interactions by translational motion is too slow by more than an order of magnitude. A direct relaxation mechanism is proposed involving a fluctuating polarization of the $(J = 0)$ molecules

[30] M. D. Ziedler, *Ber. Bunsengesellschaft phys. Chem.*, 1971, **75**, 229.
[31] B. Maraviglia, H. Meyer, and S. M. Myers, *Phys. Rev. (A)*, 1971, **4**, 1679.
[32] R. Wang, M. Smith, and D. White, *J. Chem. Phys.*, 1971, **55**, 2661.
[33] J. M. Deutch and I. Oppenheim, *Adv. Magn. Resonance*, 1966, **2**, 225.

by their ($J = 1$) neighbours. This polarization proceeds through the electric-quadrupole interaction, which produces a small admixture of $J = 2$ into the $J = 0$ state. A calculation by Harris[34] shows that, in solid D_2, $T_1(J = 1)/T_1(J = 0) \propto x(\Gamma/B)^2$, where Γ is the quadrupole coupling constant for the pair interactions, B the rotational constant, and x the concentration of ($J = 1$) molecules. This result has been qualitatively confirmed by Maraviglia et al.[31] for liquid D_2; its verification for the solid is complicated by cross-relaxation.

Krynicki and Powles[35] have remeasured and reinterpreted the T_1^H in liquid solutions of anhydrous HBr in DBr and HCl in DCl along the liquid–vapour curves. The contribution to $(T_1^H)^{-1}$ from intra- and inter-molecular dipole–dipole interactions and spin–rotation interaction were separated using values for τ_2 determined from T_1^D; using known spin–rotation interaction constants, τ_J were obtained from T_{1sr}^H. Neither the magnitude nor the temperature dependence of $\tau_J \tau_2$ corresponds to $I/6kT$. The ratio $\tau_2:\tau_J \approx 8$ at the melting point and unity at the critical temperature in both liquids. In HBr, $\tau_J \propto T$; in contrast, in HCl, τ_J remains constant up to about 220 K and then increases with temperature as in HBr. This anomalous behaviour of τ_J in HCl is attributed to hydrogen bonding at low temperatures.

Spherically Symmetric Molecules. The only study of a spherical-top molecule reported this year is by Boden and Folland[36] of P_4 in liquid phosphorus. The ^{31}P relaxation is dominated by the spin–rotation interaction, and the expression for the temperature dependence of T_{1sr}^{-1} is the same as in the plastic-crystalline phase, indicating that the rotation mechanism is the same in both these phases. Interpreting $(T_{1sr})^{-1}$ in terms of the transient rotation model,[37] the authors conclude that the P_4 molecules are executing 120° jumps, most likely about their C_{3v} axes.

Axially Symmetric Molecules. In recent years, nuclear spin relaxation time measurements have been successfully employed in the determination of the anisotropy in the rotation of symmetric-top molecules in liquids (Vol. 1, p. 125). Measurements of intramolecular spin–rotation, dipole, or quadrupole relaxation rates for two nuclei in geometrically non-equivalent positions in the molecule are required. Hitherto, only quadrupole relaxation has been exploited for this purpose because of the experimental difficulty in determining the other interactions. This year several investigators have demonstrated how ^{13}C relaxation measurements enable the other two interactions to be conveniently utilized. Huntress' theoretical treatment[38] of spin–rotation relaxation for anisotropic rotational diffusion has been questioned by Bender and Zeidler[39] and Wang et al.,[40] who independently showed that it is not

[34] A. B. Harris, *Solid State Comm.*, 1972, **10**, 329.
[35] K. Krynicki and J. G. Powles, *J. Magn. Resonance*, 1972, **6**, 539.
[36] N. Boden and R. Holland, *Mol. Phys.*, 1971, **21**, 1123.
[37] R. J. C. Brown, H. S. Gutowsky, and K. Shimomura, *J. Chem. Phys.*, 1963, **38**, 76.
[38] W. T. Huntress, *J. Chem. Phys.*, 1968, **48**, 3524.
[39] H. J. Bender and M. D. Ziedler, *Ber. Bunsengesellschaft phys. Chem.*, 1971, **75**, 236.
[40] C. H. Wang, D. M. Grant, and J. R. Lyerla, *J. Chem. Phys.*, 1971, **55**, 4674.

possible to calculate the rates separately for each tensor component, as did Huntress — his general expression does not reduce to the correct expression for relaxation in symmetric-top and spherical-top molecules. The correct expression for a spherical-top molecule is

$$(T_{1sr})^{-1} = 2(kT/3\hbar^2)[(I_\| C_\|)^2(\tau_{J,\|}/I_\|) + 2(I_\perp C_\perp)^2(\tau_{J,\perp}/I_\perp)] \quad (6)$$

where $I_\|$ and $I_\perp$ are the moments of inertia about and perpendicular to the symmetry axis, and $C_\|$ and $C_\perp$ are the corresponding components of the spin–rotation coupling constant. Since $I_\| C_\| \approx I_\perp C_\perp \approx$ trace (IC), equation (6) indicates that the relative contributions to $(T_{1sr})^{-1}$ from rotation about and perpendicular to the symmetry axis are determined mainly by the value of $\tau_{J,\|}/I_\|$ with respect to $\tau_{J,\perp}/I_\perp$. Lyerla et al.[41] have shown how equation (6) gives a satisfactory description of T_{1sr}^C in MeCN and MeI. Since, for these molecules, $\tau_{J,\|}/I_\| \gg \tau_{J,\perp}/I_\perp$, $\tau_{J,\|}$ is determined directly from T_{1sr}^C. Comparison of $\tau_{J,\|}$ and $\tau_{J,\perp}$ (the latter was calculated from $\tau_{2,\perp}$ using the Hubbard relation) with $\tau_{f,k} = (I_k/kT)^{\frac{1}{2}}$, shows that for both molecules rotation about the symmetry axis involves large angular diffusive steps whereas the perpendicular rotation occurs by small steps.

The interest shown in chloroform is not too surprising, since it is a good example of a symmetric-top molecule. Previous work by Huntress[42] has shown that over most of the liquid range this molecule undergoes small-step rotational diffusion. From T_1^D and T_1^{Cl} measurements between 238 and 323 K, he obtained values for the components of the rotational diffusion tensor $D_\|$ and $D_\perp$ for which $E_A(D_\|) = 2.8$ kJ mol^{-1} and $E_A(D_\perp) = 6.7$ kJ mol^{-1}. Extensive measurements of T_1^H are independently reported by two groups. Bender and Zeidler[43] have used isotopic dilution measurements to determine the inter- and intra-molecular relaxation rates over the temperature range 255—320 K; the spin–rotation interaction makes the major contribution to the latter. Dinesh and Rogers[44] have made measurements using the pure liquid between 219 and 363 K; they base their separation of the interactions on the observation that $(T_1^H)^{-1}$ shows a minimum at 338 K, permitting $(T_{1sr}^H)^{-1}$ to be estimated; values for $(T_{1d, inter})^{-1}$ were then obtained using computed values of $(T_{1d, intra}^H)^{-1}$. The activation energies are $E_A(T_{1sr}) = 5.24$ kJ mol^{-1}, which is approximately the average of $E_A(D_\|)$ and $E_A(D_\perp)$, and $E_A(T_{1d, inter}) = 10.05$ kJ mol^{-1}, which is to be compared with $E_A(D) = 9.6$ kJ mol^{-1}.

For a spherical-top molecule in the small-step diffusion limit assuming $\tau_{2,k} \tau_{J,k} = I_k/6kT$ and $D_k = 1/6 \tau_{2,k}$, equation (6) becomes

$$T_{sr}^{-1} = 2(4\pi^2/3\hbar)[(I_\| C_\|)^2 D_\| + 2(I_\perp C_\perp)^2 D_\perp] \quad (7)$$

Using in this expression values for $D_\|$ and $D_\perp$, Bender and Zeidler[43] estimate

---

[41] J. R. Lyerla, D. M. Grant, and C. H. Wang, *J. Chem. Phys.*, 1971, **55**, 4676.
[42] W. T. Huntress, *J. Phys. Chem.*, 1969, **73**, 103.
[43] H. J. Bender and M. D. Zeidler, *Ber. Bunsengesellschaft phys. Chem.*, 1971, **75**, 236.
[44] Dinesh and M. T. Rogers, *J. Chem. Phys.*, 1972, **56**, 542.

$|C_\parallel| = 0.31$ kHz and $|C_\perp| = 0.88$ kHz. Although the signs are not determined from the relaxation rates, an estimate of $(C_\parallel + C_\perp)$ from chemical shift data favours the choice $C_\parallel = -0.31$ kHz and $C_\perp = 0.88$ kHz. The temperature dependence of T_1^C has been measured by Shoup and Farrar[45] in 60% enriched chloroform. It is determined by the intramolecular dipole–dipole interaction with the proton, enabling a direct estimation of $\tau_{2\perp}$, and hence of $D_\perp$ from the relation $\tau_{2\perp} = (6D_\perp)^{-1}$, to be made. The value $E_A(D_\perp) = 6.8$ kJ mol^{-1} is in good agreement with that of Huntress.[42] From the difference between $(T_2^C)^{-1}$ and $(T_1^C)^{-1}$, $J(\text{C,Cl}) = 23.3 \pm 0.8$ Hz was obtained. Similarly, from the difference between $(T_2^C)^{-1}$ and $(T_1^C)^{-1}$ in CHCl$_3$, Dinesh and Rogers[46] estimate $J(\text{H,Cl}) = 5.25$ Hz; moreover, this value is seen to be temperature independent in contradiction to the value previously determined in the same way by Dietrich and Kosfeld.[47]

The two symmetric tops PH$_3$ and AsH$_3$ have been studied, but the results have been interpreted assuming isotropic rotation. Measurements of T_1^H, T_1^D, and T_1^P in pure liquid PH$_3$ and PD$_3$ and their mixtures are reported by Sawyer and Powles.[48] The ^{31}P relaxation is dominated by the spin–rotation interaction, and values for τ_J have been estimated using $C_P = 102$ kHz calculated from the ^{31}P chemical shift. Values for τ_2 are estimated from T_1^D using a value of 115 kHz for the deuteron coupling constant. The temperature dependence of τ_2 is Arrhenius, with $E_A = 2.1$ kJ mol^{-1}, but it becomes non-Arrhenius for τ_J at high temperatures; the product $\tau_J \tau_2$ does not agree with the Hubbard relation. The intermolecular dipole interaction makes the major contribution to T_1^H at lower temperatures and values for τ_t, the translational magnetic relaxation correlation time, are estimated; $E_A(\tau_t) = 4.6$ kJ mol^{-1}. Burnett and Zeltmann[49] have measured T_1^H, T_1^D, and T_1^{As} in pure AsH$_3$ and AsD$_3$ over the temperature range 160—298 K. At the high temperatures, the rotation is described by Steele's inertial model $[\tau_2 = (\pi I/12\ kT)^{\frac{1}{2}}]$; the ratio $[T_1^{As}(\text{AsD}_3) : T_1^{As}(\text{AsH}_3)]^{-1} \approx (I_{\text{AsD}_3} : I_{\text{AsH}_3})^{\frac{1}{2}}$. As in PH$_3$, intermolecular dipole interactions determine T_1^H at low temperatures and spin–rotation interaction at high temperatures. Translational diffusion is shown to proceed via large-size jumps of the order of a molecular diameter. From the relaxation data the values $(e^2qQ/h)_D = 94.3$ kHz and $C_H(\text{AsH}_3) = 17.9$ kHz are obtained.

Jordan et al.[50] report a study of T_1^P, T_2^P, and T_2^{Cl} in liquid PCl$_3$, and T_1^P and T_2^P in PBr$_3$. T_2^P is controlled by the scalar coupling interaction, whereas T_1^P is controlled by spin–rotation interaction at high temperatures and a combination of dipolar, scalar, and anisotropic chemical shielding interactions at lower temperatures. Values calculated for $J(^{31}\text{P}, ^{35}\text{Cl})$ and $J(^{31}\text{P}, ^{79}\text{Br})$ are 127 and 296 Hz, respectively; the value for PCl$_3$ agrees with the value repor-

---

[45] R. R. Shoup and T. C. Farrar, *J. Magn. Resonance*, 1972, **7**, 48.
[46] Dinesh and M. T. Rogers, *Chem. Phys. Letters*, 1971, **12**, 352.
[47] W. Dietrich and R. Kosfeld, *Z. Naturforsch.*, 1969, **24a**, 1209.
[48] D. W. Sawyer and J. G. Powles, *Mol. Phys.*, 1971, **21**, 83.
[49] L. J. Burnett and A. H. Zeltmann, *J. Chem. Phys.*, 1972, **56**, 4695.
[50] A. D. Jordan, R. G. Cavell, and R. B. Jordan, *J. Chem. Phys.*, 1972, **56**, 483.

ted[51] in a previous study, whereas that for PBr_3 is believed to be more accurate than the one previously reported.[52]

Theoretically, Mishima[53] has extended Steele's[54] treatment of molecular rotation in liquids, which is based on the rotational Langevin equation, to axially symmetric molecules. Angular correlation functions are given. It is shown that when the motion is nearly 'free' the contribution of the precessional motion is important. The Fourier transforms of these correlation functions can not be evaluated analytically. Assuming that the rotational friction constant is given by the Stokes–Einstein equation for a spherical molecule and that the viscosity depends on temperature exponentially, these functions were calculated numerically and compared with T_1^H values in cis- and trans-decalin;[55] good agreement with the experimental results is obtained, demonstrating the importance of taking into account the precessional motion of the molecules. Freed[56] has discussed the simplification which occurs in the calculation of angular momentum–orientational correlation functions when rotation is about a fixed axis in space (symmetric-top molecule orientated in a liquid crystal). Theoretical studies of the inter-relation of τ_2 and τ_J for spherical-top molecules in the limit where $\tau_2 \approx \tau_J$ are still awaited.

Asymmetric Molecules. One would, in general, expect to find a non-Arrhenius temperature dependence for the intramolecular contributions to T_1^{-1} in non-spherical molecules in the liquid state. That this is not often realized in practice probably reflects the limited liquid range of the compounds usually studied and the tendency of experimentalists to make linear approximations and estimate mean activation energies. Pajak et al.[57] report what might be regarded as typical non-linear log T_1^H versus T^{-1} plots for both CH_3 and ring protons in toluene and the 2-, 3-, and 4-methylpyridines. In illustration of the above argument, Parker et al.[58] assume Arrhenius behaviour for similar measurements in liquid toluene. The suggestion by Pajak et al. that these plots may be interpreted as a sequence of linear Arrhenius temperature regions each of which is associated with a different mode of motion is most likely an oversimplification. O'Reilly and Peterson[59] have reported a non-Arrhenius temperature dependence of T_1^D in CD_3OD which they attribute to the importance of more than one distinct mode of rotation; expressions for τ_2 are given for the case of rotational jumps about both two and three different molecular axes.

The complex character of the rotational motion of non-spherical molecules

[51] J. H. Strange and R. E. Morgan, *J. Phys. (C)*, 1970, **3**, 1999.
[52] M. Rhodes, D. W. Aksnes, and J. H. Strange, *Mol. Phys.*, 1968, **15**, 541.
[53] K. Mishima, *J. Phys. Soc. Japan*, 1971, **31**, 1796.
[54] W. A. Steele, *J. Chem. Phys.*, 1963, **38**, 2404.
[55] K. Mishima and Y. Tanaka, *J. Phys. Soc. Japan*, 1972, **32**, 581.
[56] J. H. Freed, *J. Chem. Phys.*, 1972, **56**, 1407.
[57] Z. Pajak, J. Jurga, and K. Jurga, *Acta Phys. Polon.*, 1971, **A40**, 893.
[58] R. G. Parker and J. Jonas, *J. Magn. Resonance*, 1972, **6**, 106.
[59] D. E. O'Reilly and E. M. Peterson, *J. Chem. Phys.*, 1971, **55**, 2155.

in liquids is nicely illustrated by Kintzinger and Lehn's[60] study of liquid pyridine. Analysis of T_1^D and T_1^H, measured between 253 and 353 K in [2-^{2}H]-pyridine, [3-^{2}H]pyridine, and [4-^{2}H]pyridine, in terms of the anisotropic rotational diffusion theory given by Huntress,[61] shows the molecule behaves essentially as an axially symmetric rotor; the motion is described by the two diffusion coefficients $D_\parallel$ and $D_\perp$ for rotations about perpendicular and in-plane axes respectively. At 303 K, $D_\parallel = 1.3 \times 10^{11}$ s^{-1} and $D_\perp = 8.5 \times 10^{11}$ s^{-1}, with, respectively, activation energies of 4.2 and 13.0 kJ mol^{-1}. At temperatures below 333 K, $D_\parallel > D_\perp$, but at higher temperatures $D_\parallel \approx D_\perp$, demonstrating the expected transition from anisotropic to isotropic rotation at high temperatures. It is also suggested that at temperatures in the neighbourhood of the boiling point, $D_\perp > D_\parallel$, since here the rotations will be determined by inertial effects. Kitchlew and Rao[62] report an experimental study of T_1^H and molecular motion in liquid thiophen in the temperature range 300—550 K. This molecule is similar to pyridine in that the principal axes of the diffusion tensor coincide with those of the inertia tensor and we might therefore expect similar dynamic behaviour. The contribution of the intermolecular dipole interaction is separated from the observed relaxation rate on the basis of Torrey's hard-sphere translational diffusion model using D values determined by the spin-echo technique: D varies from 2.4×10^{-9} m^2 s^{-1} at 300 K to 38.0×10^{-5} m^2 s^{-1} at 550 K with a non-Arrhenius temperature dependence. The remaining contribution is analysed in terms of the intramolecular dipole interactions and the spin-rotation interactions using the theory of anisotropic rotational diffusion and assuming that τ_J is the same for all three principal axes; this assumption relates the three diffusion constants by the ratios of the corresponding moments of inertia since $\tau_{J,k} = D_k I_k / kT$. This assumption would appear contradictory to the results of Kintzinger and Lehn's[60] work using pyridine. Nevertheless, the results obtained show: (i) $D_\parallel$ varies from 0.21×10^{11} s^{-1} at 300 K to 3.5×10^{11} s at 550 K and obeys an Arrhenius equation up to 500 K with an activation energy 13.4 kJ mol^{-1} which is similar to that for $D_\perp$ in pyridine, (ii) τ_J varies from 1.4×10^{-14} s at 300 K to 10.8×10^{-14} s at 550 K. Since the four protons in thiophen are not geometrically equivalent, it is necessary to take cross-relaxation effects into account, with the result that the observed relaxation rate is not a linear superposition of the three interactions.

Jonas and co-workers[63—65] have nicely demonstrated the utility of pressure-dependent T_1 measurements in the study of molecular motion in the liquid state. The effect of pressure on the rotational and translational motions in benzene,[63] chlorobenzene,[63] bromobenzene,[64] and [Me-^{2}H$_3$]toluene[64] has been studied by measuring T_1^H in mixtures of these compounds and their deuteriated

[60] J. P. Kintzinger and J. M. Lehn, *Mol. Phys.*, 1971, **22**, 273.
[61] W. T. Huntress, *J. Chem. Phys.*, 1968, **48**, 3524; *Adv. Magn. Resonance*, 1970, **4**, 1.
[62] A. Kitchlew and B. D. N. Rao, *J. Chem. Phys.*, 1972, **56**, 649.
[63] T. E. Bull and J. Jonas, *J. Chem. Phys.*, 1970, **52**, 4553.
[64] H. J. Parkhurst, Y. Lee, and J. Jonas, *J. Chem. Phys.*, 1971, **55**, 1368.
[65] R. A. Assink, J. DeZwaan, and J. Jonas, *J. Chem. Phys.*, 1972, **56**, 4975.

analogues. The intermolecular dipolar relaxation rates, which represent the translational motion of molecules, are shown to follow closely the changes in viscosity with pressure. This result is not surprising because viscosity measures linear momentum transfer which is also the mechanism for this type of relaxation; this also explains why the hydrodynamic model for the translational correlation time

$$\tau_t = 12\pi a^3 \eta/kT \qquad (8)$$

is usually found to be quite satisfactory. In contrast, the corresponding expression for τ_2

$$\tau_2 = 4\pi a^3 \eta/kT \qquad (9)$$

is generally a poor representation of the observed behaviour; the activation energies for τ_2 and η do not always agree and the proportionality constant between τ_2 and η is usually much smaller than given. The failure of equation (9) originates, as Kivelson et al.[66] have pointed out, in the fact that the rotational and translational motions depend, respectively, on the anisotropic and isotropic parts of the intermolecular potential; for small symmetric molecules these potentials will differ appreciably and τ_2 and η will behave differently, whereas for large asymmetric molecules they tend to be similar and τ_2 and η are related. Such behaviour is demonstrated by the pressure dependence of T_1^H (intra) in benzene and chlorobenzene.[63] In benzene, the relatively unhindered motion about the C_6 axis results in a considerable difference in the activation volumes for η and τ_2. On the other hand, in chlorobenzene the C_6 symmetry axis is removed and the activation volumes for τ_2 and η are similar. Gillen and Noggle[67] have recently proposed that, for small molecules in liquids, rotational motions which reorientate an electric dipole proceed by small-step diffusion, so that τ_2 can be confidently calculated from equation (9). To determine whether it is the change in molecular shape or dipole moment which is responsible for the increased coupling between τ_2 and η in chlorobenzene, Jonas et al.[65] measured the pressure dependence of T_1^D at 303 K in [4-^{2}H]pyridine, fluoro[4-^{2}H]benzene, [4-^{2}H]toluene, chloro[4-^{2}H]-benzene, bromo[4-^{2}H]benzene, and [4-^{2}H]benzyl cyanide. The results show that there there is a correlation between the empirical parameter κ ($\tau_2 = 4\pi a^3 \eta \kappa/3kT$)[66] and the van der Waals volume of the substituent on the benzene ring. It is therefore concluded that the molecular shape plays a decisive role in determining the degree of coupling between the rotational and translational motions in these liquids.

C. **Internal Rotation of Methyl Groups.**—The application of relaxation measurements to the study of internal motions of smaller groups within large molecules in liquids is complicated by the combined appearance of internal and overall anisotropic rotation. The efforts being made in this direction are

[66] D. Kivelson, M. G. Kivelson, and I. Oppenheim, *J. Chem. Phys.*, 1969, **52**, 1810.
[67] K. T. Gillen and J. H. Noggle, *J. Chem. Phys.*, 1970, **53**, 801.

nicely illustrated by this year's work on the CH_3 group which, owing to its simplicity, has become the model system. Analysis[68,69] of the methyl ^1H and ^{13}C spin–rotational relaxation rates in liquid toluene in terms of the Burke and Chan[70] model has shown them to originate from the spin–internal-rotation coupling. This behaviour is not unexpected since it is generally accepted that an isolated methyl group such as that in toluene is essentially a free rotor. Similarly, Kuhlmann and Grant[71] have demonstrated that the significant spin–rotation interaction contribution to the ^{13}C relaxation of the methyl carbons in mesitylene arises through the spin–internal-rotation interaction. On the other hand, in o-xylene, where interactions between the adjacent methyl groups give rise to a three-fold barrier to rotation of the order 10 kJ mol^{-1}, the methyl ^{13}C relaxation is dominated by dipolar interactions; in this case, the rate of internal rotation is comparable to the overall rotation rate.

In 1,2,3-trimethyl benzenes, the CH_3 groups at C-1 and C-3 reorientate in a three-fold potential whereas the one at C-2 rotates in a six-fold potential. Alger et al.[72] have ingeniously shown how ^{13}C relaxation measurements may be used to demonstrate that the six-fold barrier to rotation is lower than the three-fold one. The ^{13}C relaxation rates of the 2-methyl in hemimellitine (1,2,3-trimethylbenzene) and the 2- and 5-methyls in isodurene (1,2,3,5-tetramethylbenzene) have substantial contributions from spin–rotation interactions, whereas the relaxation of the 1- and 3-methyls in both compounds are dominated by the $^{13}C-^1H$ dipolar interactions. Thus, the 2-methyls are undergoing essentially free rotation whereas the rotations of the 1- and 3-methyls are restricted: the barrier is of the order 6.3 kJ mol^{-1}.

Measurements[73] of the pressure dependence of the T_1^D values of the ring and [$^2H_3$]methyl deuterons in [$^2H_8$]toluene show, as expected, that increase of pressure affects the overall rotation more than the internal rotation of the [$^2H_3$]methyl group; at 373 K, the activation volumes for these two types of motion are, respectively, 7.7×10^{-6} and 2.0×10^{-6} m^3 mol^{-1}.

D. Molecular Motion and Association in Mixtures of Liquids.—*Molecular Rotation Studies.* The interest in testing the various models for τ_2 and τ_t for solutes dissolved in magnetically inert solvents continues (Vol. 1, p. 131). Last year the work of Burke and Chan (Vol. 1, p. 132) on the effect of solvent on T_1^F in substituted 1,1,1-trifluoroethanes was reported (Vol. 1, p. 132); it was concluded that Hill's theory for τ_2 provided a consistent interpretation of all their results. Chatterjee et al.[74] have tested this theory further by studying the effect of solvent on T_1^H of the pyridine molecule in [$^2H_5$]-pyridine, [$^2H_6$]benzene, [$^2H_{12}$]cyclohexane, [$^2H_6$]acetone, [^2H]chloroform,

[68] R. G. Parker and J. Jonas, *J. Magn. Resonance*, 1972, **6**, 106.
[69] C. F. Schmidt and S. I. Chan, *J. Magn. Resonance*, 1971, **5**, 151.
[70] T. E. Burke and S. I. Chan, *J. Magn. Resonance*, 1970, **2**, 120.
[71] K. F. Kuhlmann and D. M. Grant, *J. Chem. Phys.*, 1971, **55**, 2998.
[72] S. T. D. Alger, D. M. Grant, and R. K. Harris, *J. Phys. Chem.*, 1972, **76**, 281.
[73] D. J. Wilbur and J. Jonas, *J. Chem. Phys.*, 1971, **55**, 5840.
[74] N. Chatterjee, J. J. Czubryt, E. Bock, E. Tomchuk, H. M. Hutton, and D. W. Kydon, *Ber. Bunsengesellschaft phys. Chem.*, 1971, **75**, 777.

CS_2, and CCl_4 at 298 K. These measurements are also found to be in good agreement with the prediction of the Hill theory. Shepperd et al.[75] have studied the effects of temperature and solvent on τ_2 for methyl nitrate as determined from ^{14}N T_1 measurements. The Debye–Stokes–Einstein model is shown to be applicable to solutions in polar solvents such as dimethylformamide, but not to solutions in non-polar saturated hydrocarbons such as cyclohexane. This interesting result demonstrates that the reorientation motion motion of a polar molecule is intimately related to the polarizability of its environment. Measurements of T_1^H for 1,2-dichloroethane dissolved in CCl_4 by Pislewski et al.[76] show that in dilute solution the molecule rotates on average about an axis with respect to which its mean moment of inertia amounts to $I = \frac{1}{3}(I_x + I_y + I_z)$ where I_x, I_y, I_z are the moments of inertia with respect to the three principal axes of the molecule; apparently, the Hill model is not applicable. This result emphasizes the importance of taking the anisotropy of the motion into account when comparing experiment with theory, a point made by Sato and Nishioka.[77] These authors found that τ_2 and τ_t, determined from T_1^H measurements by the adiabatic rapid passage method, of the three non-equivalent protons in norbornadiene in [^{2}H]chloroform between 298 and 353 K, indicate that the rotational and translational motions are isotropic; the τ_2 values are shown to be in good agreement with the predictions of the Hill model.

Molecular Association Studies. It is well known that binary liquid mixtures exhibit non-ideal behaviour because of intermolecular association. Spin relaxation measurements may be used to probe the nature of the interactions involved through their effect on molecular dynamics; moreover, in strongly interacting systems it ought, in favourable cases, to be possible to determine the lifetime and geometry of the association product.

The $CHCl_3$–C_6H_6 mixture is the best-known example of a weak interaction system. Previous quadrupole relaxation measurements by Huntress[43] on $CHCl_3$ and $CDCl_3$ in, respectively, equimolar mixtures of $CHCl_3$–C_6D_6 and $CDCl_3$–C_6H_6, have shown that the tumbling motion of the $CHCl_3$ molecule is slowed by a factor of four on complexing with benzene. On the other hand, Anderson's[78] T_1^H measurement for benzene in $CDCl_3$ is equal to that in C_6D_6 and CCl_4, which led him to conclude that the complex between benzene and chloroform is a weak one and does not move as a unit. Sato and Nishioka[79] have reported new T_1^H measurements in chloroform–benzene mixtures at 298 K which show that in the mixture the rotation of the benzene molecule is less hindered, presumably owing to the fact that the C_6 axis coincides with the C_3 axis of chloroform. They also show that τ_t for chloroform is indepen-

---

[75] C. M. Shepperd, T. Schaefer, B. W. Goodwin and J. T'Raa, *Canad. J. Chem.*, 1971, **49**, 3158.
[76] N. Pislewski, T. Kulek and P. Pieranski, *Acta Phys. Polon.*, 1972, **A41**, 121.
[77] K. Sato and A. Nishioka, *J. Magn. Resonance*, 1972, **6**, 231.
[78] J. E. Anderson, *J. Chem. Phys.*, 1969, **51**, 3578.
[79] K. Sato and A. Nishioka, *Bull. Chem. Soc. Japan*, 1971, **44**, 1506.

dent of composition, $\sim 3 \times 10^{-10}$ s, and roughly the same as in pure chloroform, $\sim 1 \times 10^{-10}$ s. Thus, the translational motion of the chloroform molecule is not affected by coupling with benzene. Similarly, Rothschild[80] has shown that the reduced root-mean-square jump distance $\langle r^2 \rangle^{\frac{1}{2}}/d$ (where d is the distance of closest approach) of a C_6H_6 molecule [as determined using Torrey's theory in conjunction with T_1^H (inter) measured in $CDCl_3$ and self-diffusion coefficients measured by the spin-echo method] does not vary significantly over the whole concentration range from its value in pure benzene, $\langle r^2 \rangle^{\frac{1}{2}}/d \sim 1$; $\tau_t \sim 0.2 \times 10^{-10}$ s is likewise insensitive to composition. Thus the translational motion appears to be carried by large distance jumps, of the order of a molecular diameter, of the individual molecules and not by translational jumps of the complex as a whole. Rothschild has also obtained Fourier transforms of i.r. bandshapes and shown that both the benzene and chloroform molecules behave essentially as 'random rotors' performing gas-like orientational jumps through average angles of 0.4 to 0.7 rad around their inertial axes. On the basis of the above evidence it may be concluded that in $CHCl_3$–C_6H_6 the constituent molecules of the complex experience a relatively large degree of rotational and translational freedom similar to that in the pure liquids. Rothschild concludes that this picture holds generally for the interactions in weakly non-ideal mixtures of liquids.

Brüssau and Sillescu[81] have demonstrated how a qualitative estimate for the lifetime of a strong solvent–solute interaction in dilute macromolecular solutions can be obtained by measuring intramolecular relaxation times for the solvent. The method is applicable when the lifetime falls within the limits defined by τ_2 of the solvent and macromolecule, *i.e.* typically 10^{-11}—10^{-9} s. The rotational motion of the solvent molecule is described in terms of a simple two-state model, the 'free' state and the 'bound' state; the theory of spin–lattice relaxation in two-state systems with rapid exchange is extended to allow for anisotropic rotational motions of the solvent molecules. This theory is then applied to the results of ^2H and ^{35}Cl relaxation measurements in dilute $CDCl_3$ solutions of polyvinylpyrrolidone. In the bound state, the chloroform molecule rotates rapidly around an active site of the macromolecule, probably the carbonyl group of the pyrrolidone ring as shown in (1).

(1)

The lifetime of this hydrogen bond is long compared with τ_2 for $CDCl_3$ in the pure solvent, but very short compared to the fastest internal motion within

[80] W. G. Rothschild, *J. Chem. Phys.*, 1971, **55**, 1402.
[81] R. G. Brüssau and H. Sillescu, *Ber. Bunsengesellschaft phys. Chem.*, 1972, **76**, 31.

the macromolecule. Similar measurements are reported for dilute D_2O solutions.

Ragozzino and Bartone[82] have shown how in binary liquid mixtures, $A + B$, comparison of the dipolar T_{1AB}^{-1} and T_{1BA}^{-1} with T_{1AA}^{-1} and T_{1BB}^{-1}, respectively, can provide an insight into the A–B interactions, e.g. $T_{1AB}^{-1} > T_{1AA}^{-1}$ indicates a decrease in reciprocal mobility which suggests interaction phenomena such as molecular associations and structure-breaking processes. They demonstrate that for a weakly interacting system, e.g. acetic acid–benzene, these mixed translational contributions can be reliably determined by assuming ideal mixture behaviour and a suitable model for intermolecular relaxation. This approach is, however, inapplicable when strong interactions are present, e.g. acetic acid–water.

Other reported relaxation studies of molecular interactions in binary liquid mixtures include an extension[83] of Sato and Nishioka's T_1^H study (Vol. 1, p. 134) of molecular association through inter-amide hydrogen-bonding in N-methylacetamide and NN-dimethylacetamide to the solvents p-dioxan and chloroform; also Modak et al.[84] have extended their T_1 measurements in CS_2 and CCl_4 (Vol. 1, p. 132) to the solutes aniline and quinoline.

Hartnell and Allerhand[85] have demonstrated the use of ^7Li T_1 measurements in the study of the aggregation of organolithium compounds in solution.

Aqueous Solutions of Non-electrolytes. Czubryt et al.[86] have reported a T_1^H study of the water–dioxan system; T_1^H measurements were made for the dioxan molecule as a function of composition in D_2O over the temperature range 283—353 K. Contrary to a previous report, $[T_1^H \text{(intra)}]^{-1}$ is strongly concentration-dependent in dilute solution and exhibits a similar behaviour to the solution viscosity, density, and free energy of mixing. It is claimed that the motions of the dioxan and water molecules became correlated at ~ 0.14 mole fraction dioxan corresponding to the formation of the weak complex $C_4H_8O_2,6H_2O$. Similar behaviour has been observed in the pyridine–water system by Kasugai et al.,[87] who showed that $(T_1^H)^{-1}$ has a maximum at 0.3 mole fraction pyridine corresponding with the maximum in viscosity; this result agrees with previous unmentioned T_1^H measurements. The occurrence of a maximum in $(T_1^H)^{-1}$ corresponding with maxima or minima in other physical properties appears to be characteristic of aqueous solutions of hydrogen-bonding solutes (Vol. 1, p. 133).

Presing and Noack[88] have demonstrated the power of nuclear magnetic

---

[82] E. Ragozzino and C. Bortone, *Mol. Phys.*, 1971, **22**, 525.
[83] K. Sato and A. Nishioka, *Bull. Chem. Soc. Japan*, 1971, **44**, 2931.
[84] S. G. Modak, P. L. Khane, C. Maude, and V. G. Kher, *Indian J. Pure Appl. Phys.*, 1971, **9**, 735.
[85] G. E. Hartnell and A. Allerhand, *J. Amer. Chem. Soc.*, 1971, **93**, 4415.
[86] J. J. Czubryt, N. Chatterjee, E. Bock, E. Tomchuk, H. M. Hutton, and D. W. Kydon, *Ber. Bunsengesellschaft phys. Chem.*, 1971, **75**, 904.
[87] Y. Kasugai, Y. Arata, and S. Fujiwara, *Bull. Chem. Soc. Japan*, 1971, **44**, 2557.
[88] G. Presing and F. Noack, *Kolloid-Z.*, 1971, **247**, 811.

relaxation spectroscopy in their study of the molecular motions of poly(ethylene glycol) 30 000 in acqueous (D_2O) solution. In the range between 3 kHz and 120 MHz, T_1^H exhibits two well-resolved dispersion steps indicative of two motional mechanisms similar to the ones previously observed in biopolymer solutions (Vol. 1, p. 133).

Hydrogen-bonded Systems. Nuclear spin relaxation measurements may be used to study the effect of solvent on hydrogen-bonded systems, since τ_2 will be dependent on the size of the polymer unit. As an example, Sato and Nishioka[89] have shown that the self-association of acetic acid exerts a pronounced effect of T_1^{-1} of the hydroxyl proton, the relaxation behaviour being dependent upon the concentration of acid and the kind of solvent. T_1^{-1} was found to exhibit a maximum in CCl_4, and two maxima in C_6D_6 and $CDCl_3$, on dilution. The maximum at 0.5 mole fraction in CCl_4 is attributed to a maximum in the length of the polymer chains. In C_6D_6 and $CDCl_3$ the maximum in the high concentration region is similarly explained, but that in the low concentration region is attributed to the relaxation behaviour of the dimer and the dimer–monomer equilibrium.

E. **Ionic Solutions.**—The general increase in activity in the field of spin relaxation measurements is not reflected by the number of publications concerned with structure and dynamics in ionic solutions; this could, of course, be merely an annual fluctuation. The interest of many authors in this subject appears to be stimulated by the importance of ion binding in biological systems (see Chapter 8).

Diamagnetic Solutions. A number of groups have been interested in using T_1 measurements to study interactions in aqueous electrolyte solutions. The advantages of using quadrupolar T_1 measurements for this purpose are obvious. For spin-½ nuclei, relaxation contributions may arise from spin–rotation as well as dipole interactions. Hertz *et al.*[90] have recently shown that the spin–rotation interaction becomes important for the ^{19}F relaxation in aqueous KBF_4 solutions at high temperatures. Liu and Jonas[91] now find that this relaxation mechanism is important for both ^{19}F and ^{31}P in the PF_6^- anion in D_2O solution at temperatures above 300 K. Using spin–rotation constants calculated from chemical shielding values, they show that the relaxation of both these nuclei is consistent with the M- rather than the J-diffusion model (Vol. 1, p. 123); an interesting result, but its significance depends on the reliability of the spin–rotation constant calculation. Jonas[92] and his coworkers have also demonstrated that there is a characteristic difference in the effect of structure-breaking and structure-forming electrolytes on the pressure

[89] K. Sato and A. Nishioka, *Bull. Chem. Soc. Japan*, 1971, **44**, 2042.
[90] H. G. Hertz, G. Keller, and H. Versmold, *Ber. Bunsengesellschaft phys. Chem.*, 1969, **73**, 549.
[91] Nan-I Liu and J. Jonas, *J. Chem. Phys.*, 1971, **55**, 463.
[92] Y. Lee and J. Jonas, *J. Magn. Resonance*, 1971, **5**, 267.

F

dependence of T_1^H in aqueous electrolyte solutions. In 4.5 molar solutions of the structure-breaking electrolytes CsBr, RbBr, and KBr at 298 K, T_1^H values are longer than in pure water and decrease with increasing pressure; at the same temperature and pressure the structure-breaking ability of the cations increases as $K^+ < Rb^+ < Cs^+$. In contrast, in concentrated solutions of the structure-making electrolytes LiCl, CaCl$_2$, and LaCl$_2$, T_1^H values are shorter than in pure water, and first increase to a maximum and then begin to decrease with decreasing pressure; the structure-making ability of these cations increases as $Li^+ < Ca^{2+} < La^{3+}$, which is parallel to the increase in charge-to-radius ratio of the cations. Clearly, pressure-dependent T_1 studies should be helpful in elucidating the structure of ionic solutions. T_1^H and T_1^D measurements have been reported by Lemius and Domngang[93] for aqueous solutions of CaCl$_3$, GaBr$_3$, and GaI$_3$; Ga^{3+} is hydrated by six water molecules. For the ^{2}H in D$_2$O of hydration, (e^2qQ/h) is found to be ca. 180 kHz. The hydration number for the iodide ion is $\approx$ 17 and for the bromide ion $\approx$ 10. Hertz et al.[94] have reported measurements of T_1 for $^7Li^+$ in H$_2$O and D$_2$O solutions of LiCl, LiBr, and LiI. The (^{1}H, ^{7}Li) dipole–dipole relaxation rate is separated from the quadrupole one; from this the rotational correlation time for the radius vector connecting the cation with the first hydration water molecule is determined and found to be different from the correlation time of the H—H vector in H$_2$O in the first hydration sphere, indicating that the water molecule rotates about the Li—O axis. The effect of concentration on the quadrupole contribution to $(T_1^{Li})^{-1}$ shows a breakdown of cubic symmetry in the hydration sphere as the salt concentration increases. Ader and Loewnstein[95] have measured $T_1(^{59}Co)$ and $T_1(^{14}N)$ in aqueous solutions of three Co complexes: M$_3$Co(CN)$_6$ (M = Li, K, Cs, or NMe$_4$), Co(NH$_3$)$_6$Cl$_3$, and Co(en)$_3$Cl$_3$ (en = ethylenediamine). For the bare Co(CN)$_6^{3-}$ ion they obtain $(e^2qQ/h) \approx$ 0.9 MHz, compared with the value 1.5 MHz in solid Co(NH$_3$)$_6^{3+}$. They suggest that in these ions the electric-field gradient at the ^{59}Co nucleus arises from a 'static' distortion of the spherical symmetry of the ion as a result of association with counter-ions, rather than by a 'dynamic' deformation caused by short-lived collisions with solvent molecules or counter-ions (Vol. 1, p. 135). Stungis and Rugheimer[96] have reported T_1 and T_2 measurements for all three species of nuclei in aqueous solutions of HBF$_4$ at room temperature as a function of concentration.

For completeness, some linewidth studies of quadrupole relaxation in ionic solutions will be briefly mentioned here; these are reported in detail in Chapter 6. Larsen[97] has used ^{75}As linewidth measurements to investigate the structure of Me$_4$As$^+$X$^-$ (X$^-$ = Cl$^-$, I$^-$, or OH$^-$) in aqueous solution. The results suggest that the 'single kinetic entiry' which constitutes the aqueous

---

[93] B. Lemius and S. Domngang, *J. Chim. phys.*, 1971, **68**, 1372.
[94] H. G. Hertz, R. Tutsch, and H. Versmold, *Ber. Bunsengesellschaft phys. Chem.*, 1971, **75**, 1177.
[95] R. Ader and A. Lowenstein, *J. Magn. Resonance*, 1971, **5**, 248.
[96] G. H. Stungis and J. H. Rugheimer, *J. Chem. Phys.*, 1971, **55**, 263.
[97] D. W. Larsen, *J. Phys. Chem.*, 1971, **75**, 3880.

Nuclear Spin Relaxation 131

ion-pair includes the counter-ions separated by 0.25—0.30 nm with, at most, a few H_2O molecules held between the counter-ions; the hydration layer surrounding the cations is not included as part of the 'single kinetic entity'. The same author has also reported[98] a ^{14}N linewidth study of complex formation between lithium ions and pyridine. The measurements are explained in terms of the following models for the complexes: linear $Li^+Cl^-(py)$ for LiCl in aqueous pyridine; linear $Li^+(H_2O)(py)$ for LiBr in aqueous pyridine, and tetrahedral $Li^+Cl^-(py)_3$ for $LiCl^-$ in neat pyridine. Wennerström et al.[99] have made linewidth measurements of ^{35}Cl, ^{79}Br, ^{81}Br, and ^{127}I in aqueous solutions of various substituted ammonium, phosphonium, and sulphonium salts chosen as model systems for studying the binding of ions to proteins and other biologically interesting compounds. Lindblom[100] has also discussed the application of $^{23}Na^+$ linewidth measurements to the elucidation of the mechanism of the binding of ions to membrane surfaces.

Paramagnetic Solutions. An appreciable number of publications have been concerned with proton relaxation in aqueous solutions of biological macromolecules containing bound paramagnetic ions. These studies are reported in Chapter 8 and will not be discussed further here except to mention the note by Koenig,[101] which points out that in general in these studies it is not necessary to differentiate between the electron T_1 and T_2 as previously asserted by Reuben et al.,[102] and also the demonstration by Gaber et al.[103] of the utility of nuclear magnetic relaxation spectroscopy in characterizing the frequency dependence of the paramagnetic ion electron relaxation time. This information is required for interpretation of the proton relaxation measurements in solutions where the electron relaxation time is the effective correlation time for the nuclear–electron dipolar interaction (Vol. 1, p. 137). $(T_1^H)^{-1}$, measured as a function of B_0 (5×10^{-4}—12 T) in aqueous solutions of bovine erythrocyte superoxide dismutase, exhibits a peak which indicates that the spin–lattice relaxation time of the Cu^{2+} spin is field-dependent. Using this information, the authors were able to show that there is at least one rapidly exchanging water molecule on each of the two Cu^{2+} per protein molecule. It is speculated that the water-binding sites on the Cu^{2+}, with a Cu—O bond distance ≈ 0.2 nm, are similarly accessible to the superoxide anion and are the loci for the catalytic activity of the enzyme.

F. Relaxation by Translational Diffusion.—In n.m.r. the translational and rotational contributions to the relaxation rate in liquids may be separated experimentally, enabling T_1 measurements to be used to study the details of

---

[98] D. W. Larsen, *J. Phys. Chem.*, 1972, **76**, 53.
[99] H. Wennerström, B. Lindmann, and S. Forsén, *J. Phys. Chem.*, 1971, **75**, 2936.
[100] G. Lindblom, *Acta Chem. Scand.*, 1971, **25**, 2767.
[101] S. H. Koenig, *J. Chem. Phys.*, 1972, **56**, 3188.
[102] J. Rueben, S. H. Read, and M. Cohen, *J. Chem. Phys.*, 1970, **52**, 1617.
[103] B. P. Gaber, R. D. Brown, S. H. Koenig, and J. A. Fee, *Biochim. Biophys. Acta*, 1927, **271**, 1.

the diffusive steps. The standard theoretical treatment for relaxation by isotropic translational diffusion has been given by Torrey[104] who, after Chandrasekhar,[105] considers the translational motion as a random walk process. It is also assumed that the spins under consideration are at the centre of the diffusing molecules. Hubbard[106] has considered the importance of off-centre spin effects in the short correlation time limit. More recently, Harmon and Muller[107] have included the effects of a non-uniform spin density and off-centre spins by using a radial distribution function and the Hubbard correction in the low-frequency limit of the Torrey theory. The Torrey theory, not incorporating these effects, has recently been revised by Kruger[108] and discussed by Fiorito and Meister[109] with respect to the frequency dependence of T_1 in the diffusive and jump limits. In the jump limit, the theory gives

$$(T_1^{-1})_{\text{inter}} = \frac{2}{5}\pi\gamma^4\hbar^2(n/dD)\,[1+(5\langle r^2\rangle^{\frac{1}{2}}/12d^2]\qquad(10)$$

where $\langle r^2\rangle^{\frac{1}{2}}$ is the root-mean-square molecular jump distance, n is the number density of spins, and d is the distance of closest approach of the molecules. Clearly, values for $\langle r^2\rangle^{\frac{1}{2}}/d$ may be estimated from measurements of T_1 and D; furthermore, using this value for $\langle r^2\rangle$, the mean time between translational jumps can be obtained from $\tau_J = \langle r^2\rangle/6D$. Note that in the jump limit $\tau_t = \tau_J/2$. For benzene,[110] methane,[110] ethane,[107,110] HCl,[35] chloroform,[39] and benzene in $CHCl_3-C_6H_6$ mixtures,[80] it has been found that $\langle r^2\rangle^{\frac{1}{2}}/d \approx 1$. Thus, it may be concluded that in small-molecule non-viscous liquids the translational motion occurs by large distance jumps of the order of the molecular diameter. Determinations of τ_J have been reported by a number of authors. Rothschild[80] has shown that τ_J for benzene in $CHCl_3-C_6H_6$ mixtures does not change significantly with composition and is similar to its value in neat benzene, $\tau_J \sim 0.2 \times 10^{-10}$ s. Bender and Zeidler[30] report τ_J in chloroform; it increases from 0.16×10^{-10} s at 320 K to 0.39×10^{-10} s at 255 K. Zeidler[111] has made an interesting comparative study of translational diffusion in liquid acetonitrile by quasielastic neutron scattering and T_1 measurements. Values for τ_J from neutron scattering are roughly a factor of 5 shorter than the corresponding n.m.r. values; at 298 K, τ_J is 4×10^{-12} s and 23×10^{-12} s by the two methods respectively. Zeidler suggests that the reason for this discrepancy is an incorrect assumption, in calculating $(T_1^{-1})_{\text{inter}}$, that the two interacting spins move independently; he argues that, owing to intermolecular interactions, the translational motions of neighbouring molecules are correlated. This argument does not appear to be consistent

[104] H. C. Torrey, Phys. Rev., 1953, **92**, 962.
[105] S. Chandrasekhar, Rev. Mod. Phys., 1943, **15**, 1.
[106] P. S. Hubbard, Phys. Rev., 1963, **131**, 275.
[107] J. Harmon and B. H. Muller, Phys. Rev., 1969, **182**, 400.
[108] G. J. Kruger, Z. Naturforsch., 1969, **24a**, 560.
[109] R. B. Fiorito and R. Meister, J. Chem. Phys., 1972, **56**, 4605.
[110] B. H. Muller, Phys. Letters, 1966, **22**, 123.
[111] M. D. Zeidler, Ber. Bunsengesellschaft phys. Chem., 1971, **75**, 769.

with Rothschild's observation. In conclusion, Zeidler questions whether the jump times determined by n.m.r. are meaningful. Clearly, further comparative studies of this type are called for together with a critical evaluation of the application of the translational relaxation theory.

It is becoming apparent that translational rather than rotational motions are responsible for proton relaxation in viscous liquids. This hypothesis is based on reported T_1^H measurements in glycerol, which has become the model for viscous liquids. In a previous study by Noack and Preissing,[112] the rotational motion was assumed to be the dominant relaxation mechanism, but the values they obtained for τ_2 differed significantly from those obtained by Drake and Meister[113] from T_1^D in [$^2H_8$]glycerol and also from the dielectric values (Vol. 1, p. 130). Noack et al.[114] now report that the self-diffusion coefficient, measured by the pulsed gradient method, has the same activation energy (57 kJ mol^{-1}) as the shift of the T_1^H minimum (10 kHz to 120 MHz; 253 to 243 K). In view of this observation, they have reinterpreted their T_1^H measurements in terms of Torrey's translational relaxation theory: excellent agreement is obtained between the diffusive limit of this theory, i.e. $\langle r^2 \rangle^{\frac{1}{2}}/d \ll 1$, and the T_1^H measurements for temperatures below 300 K. Fiorito and Meister[109] have reported a corresponding analysis of the Noack and Preissing T_1^H measurements. Additionally, these authors have also shown how Torrey's theory explains the temperature dependence of T_1^H and T_2^H which they measured at 30 MHz as a function of pressure (up to 300 MN m^{-2}). Identical results are obtained by the frequency reduction and T_1, T_2 reduction approaches for temperatures above 285 K, but there is some discrepancy at lower temperatures. Fiorito and Meister also show that similar T_1^H and T_2^H measurements in liquid butane-1,3-diol and 2-methyl-pentane-2,4-diol are also successfully interpreted by the Torrey theory. The paper by these authors is well worth reading by anyone interested in applying Torrey's translational relaxation theory to liquids. These studies on glycerol clearly represent a reversal of previous notions of the relative importance of translational and rotational relaxation in viscous liquids. The interpretation given by Leblond, Uebersfeld, and Korringa[115] to their T_1^H and T_2^H measurements in propan-2-ol and tetraethylene glycol is, therefore, questionable: since the T_1/T_2 ratios at the minima are, respectively, 4.8 and 2.3 instead of 1.6 as predicted by the familiar Debye type magnetic intensity spectrum originating from random rotational motion, they analyse their measurements in terms of a log-Gaussian distribution of correlation times. When organic free radicals are added to these liquids, the protons are shown to relax through dipole interaction with the electron spin while their molecule is temporarily adsorbed in the solvation layer of a radical. These relaxation measurements are part of a general study of the dynamic structure of radical solutions using n.m.r., e.s.r., and dynamic nuclear polarization measurements.

[112] F. Noack and G. Preissing, Z. Naturforsch., 1969, 24a, 143.
[113] P. Drake and R. Meister, J. Chem. Phys., 1971, 54, 3046.
[114] G. Preissing, F. Noack, R. Kosfeld, and B. Gross, Z. Physik., 1971, 246, 84.
[115] J. Leblond, J. Uebersfeld, and J. Korringa, Phys. Rev. (A), 1971, 4, 1532, 1539.

Carazza[116] has given expressions for T_1 and T_2 for the case of anisotropic translational diffusion. His treatment is a generalization of that given by Abragam (ref. 2, p. 300) for isotropic diffusion in the diffusive limit. These expressions should be useful in analysing relaxation measurements for solute molecules dissolved in liquid crystals.

G. **Spin-diffusion Measurements in Liquids.**—There has been a marked increase in the use of the spin-echo method for measurement of self-diffusion coefficients in fluids. Two papers[117,118] have reported measurement of D in liquid ^{3}He at temperatures below 25 mK; in one of them,[117] Hahn's stimulated echo is used for the measurement. Kitchlew and Rao[119] have measured the temperature dependence of D (300—380 K) in the following liquids: pyridine, piperidine, dimethylaniline, benzene, chlorobenzene, p-xylene, m-xylene, o-xylene, nitrobenzene, and thiophen. Whereas the variation of D and η seems to follow the temperature dependence predicted by the Stokes–Einstein equation [$D = kT/6\pi a\eta$] and the hard-sphere model of Longuet–Higgins and Pople[120] [$D \eta/T = 3kc/10\pi a'$, where c is the fractional volume occupied by the molecules in the liquid and a' is the hard-sphere radius related to a used above, through $a' = (9c/5)a$] in most cases, the apparent hard-sphere radii do not appear to have any physical significance in either theory. In all cases, the temperature dependence of D is unexpectedly non-Arrhenius. In contrast, the self-diffusion coefficients reported by O'Reilly and Peterson[59,137] for methanol, acetone, methyl nitrate, nitrobenzene, chlorobenzene, and toluene all exhibit normal Arrhenius behaviour. It is interesting to note that the variation in D measured[121] in liquid water between 248 and 303 K by a tritium tracer technique is, as expected, non-Arrhenius; this paper is useful for its comparison of literature values of D at 298 K. Several papers have reported determination of D for use in conjunction with intermolecular T_1 measurements and Torrey's translational relaxation theory to study the diffusive motion of molecules in liquids: these include measurements in glycerol,[114] thiophen,[62] chloroform,[30] acetonitrile,[111] and benzene–chloroform mixtures.[80]

The pulsed field gradient method has been applied by Kaerger[122] to the investigation of the diffusion of water on 13X, 4A, and 5A zeolites, and by Finch et al.[123] to the measurement of the diffusion constant for water in muscle.

Several authors have reported measurements of D for ions in aqueous

[116] B. Carazza, *Lettere Nuovo Cimento*, 1972, **3**, 366.
[117] A. Tyler, *J. Phys. (C)*, 1971, **4**, 1479.
[118] L. R. Corriccini, D. D. Oshenoff, D. M. Lee, and R. C. Richardson, *Phys. Rev. Letters*, 1971, **27**, 650.
[119] A. Kitchlew and B. D. N. Rao, *Mol. Phys.*, 1971, **21**, 1145.
[120] H. C. Longuet-Higgins and J. A. Pople, *J. Chem. Phys.*, 1956, **25**, 884.
[121] H. R. Pruppacher, *J. Chem. Phys.*, 1972, **56**, 101.
[122] J. Kaeger, *Z. phys. Chem. (Leipzig)*, 1971, **248**, 27.
[123] E. D. Finch, J. F. Harmon, and B. H. Muller, *Arch. Biochem. Biophys.*, 1971, **147**, 299.

solutions. Ader and Lowenstein[95] have measured D for ^{59}Co ions in a 1 molar solution of $K_3Co(CN)_6$; $\langle r^2 \rangle^{\frac{1}{2}} \geqslant 0.32$ nm, showing that the jump distance is comparable with the diameter of the $Co(CN)_6{}^{3-}$ ion. Hertz et al.[94] report measurements of D for the Li^+ ion in aqueous solutions of LiCl, LiBr, and LiI and give a structural interpretation to its concentration dependence.

H. Transverse Relaxation and Spin-echo Studies.—*Spin-echo Spectroscopy.*

A coupled homonuclear spin system yields a spin-echo envelope which, in general, is modulated in a manner determined by the spectral parameters and the pulse spacing. The Fourier transform of this decaying envelope, sampled at the midpoints of the pulse intervals, yields a multi-line spectrum in the frequency domain, the 'spin-echo spectrum' (see Vol. 1, pp. 141, 142). The fact that the lines of spin-echo spectra are not subject to broadening by static field inhomogeneity as in conventional slow-passage spectroscopy has stimulated the recent interest in developing spin-echo spectroscopy as an analytical tool. It has been suggested[124] that line frequencies should be measurable to a precision of ± 0.001 Hz, enabling very accurate measurements of coupling constants and chemical shifts. Vold and Chan[125] have shown that, owing to the presence of the rf field in a Carr–Purcell spin-echo experiment, cross relaxation couples the different spectral magnetizations; this has important consequences for spin-echo spectroscopy. First, spin-echo 'linewidths' depend in general upon rf pulse spacing and, at constant pulse spacing, upon the chemical shifts and thus the external field strength. Second, neither spin-echo linewidths nor intensities are determined solely by the behaviour of a single spectral magnetization. Thus, in principle, the effects of cross-relaxation will severely handicap the analysis of complex spin-echo spectra. Vold and Shoup[124] have been developing the techniques of subspectral analysis appropriate to spin-echo spectroscopy. They have shown that the proton spectrum of 2,2-dichloro-1,1-difluoroethane (which is a typical ABX_2 spin system) measured as a function of pulse spacing can be analysed in terms of three distinct subspectra of the ab variety. Fortunately, in this system the effects of cross-relaxation are small and an analysis was possible; $J_{FF} = 3.471 \pm 0.011$ Hz. Tokuhiro and Fraenkel[126] have been directly investigating the spin-echo modulation for typical examples of homonuclear AX, A_3X, and $[AX]_2$, and heteronuclear ABX systems and comparing the results with their previously developed theory. It should be emphasized that much of the information in the Fourier transform spectrum is unavailable to direct observation.

Selective Measurement of Transverse Relaxation Rates. The selective measurement of transverse relaxation rates for each line in a complex spectrum is, both theoretically and experimentally, a far more difficult problem than the selective measurement of T_1; indeed, Vold and Freeman are to be congratu-

---
[124] R. L. Vold and R. R. Shoup, *J. Chem. Phys.*, 1972, **56**, 4787.
[125] R. L. Vold and S. O. Chan, *J. Chem. Phys.*, 1972, **56**, 28.
[126] T. Tokuhiro and G. Fraenkel, *J. Chem. Phys.*, 1971, **55**, 2797.

lated for their contributions to this subject. Vold and Chan[127] (Vol. 1, p. 141) have proposed what appears to be a reasonable procedure; this involves subjecting the spin system to a non-selective Carr–Purcell sequence of rf pulses, Fourier-transforming the free precession signal obtained by stopping the pulse train part way down the echo decay, and measuring the intensities of the lines in the resulting multi-line spectrum as a function of the pulse train duration. It has, however, now been shown, theoretically by Vold[128] and experimentally by Vold and Shoup,[124] that in the presence of homonuclear spin coupling, although the lines in the partially relaxed spectra have the same frequencies as the slow-passage counterparts, their intensities and phases depend in a complex manner upon relaxation rates, chemical shifts and coupling constants, pulse spacing, and preparation time. Thus, the selective determination of transverse relaxation rates from partially relaxed spectra is severely complicated by homonuclear spin coupling. Amplitude- and phase-distortion effects may be avoided by using sufficiently small pulse spacings to effect spin-locking. However, in this experiment, and for systems without scalar coupling, the existence of transverse cross-relaxation may result in non-exponential decays of the intensity of each line of the partially relaxed spectrum. If cross-relaxation is comparable in magnitude with self-relaxation, interpretation of single or non-exponential decays in terms of theoretically well-defined relaxation parameters will be extremely difficult. Freeman and Hill[129] have demonstrated the use of the conventional $T_{1\rho}$ spin-locking pulse sequence for the selective measurement of the ^{13}C transverse relaxation rates in o-dichlorobenzene. The decay of the lines in the partially relaxed spectrum, obtained by Fourier transformation of the free induction decay following turn-off of B_1, were found to be exponential. The values for $T_{1\rho} = T_2$ compared with the corresponding T_1 values are: for $\underline{C}Cl$, $T_{1\rho} = 4.2$ s, $T_1 > 66$ s; for $\underline{C}CCl$, $T_{1\rho} = 7.7$ s, $T_1 = 7.8$ s; for $\underline{C}CCCl$, $T_{1\rho} = 6.4$ s, $T_1 = 6.3$ s. Cross-relaxation effects are apparently not important in this system. During spin-locking, coherent proton decoupling was used (to obtain the Overhauser enhancement), but was switched to incoherent operation (to eliminate residual splitting) during the acquisition of the free induction decay. Incoherent decoupling could not be employed throughout the experiment since destructive interference causes a rapid decay of the transverse magnetization. The need to avoid the condition $\gamma_C B_{1C} = \gamma_H B_{1H}$ is emphasized since cross-relaxation between the two spin systems interferes with the natural transverse relaxation processes.

Haeberlen et al.[130] have reported some interesting, carefully made, conventional, unselective CPSE measurements on ^{13}C, ^2H, ^1H, and ^{19}F in some small molecules in the liquid phase. The ^2H resonance of C_6D_6 and D_2O, the ^1H resonance of CH_3I and of extremely pure H_2O and C_6H_6, and the ^{13}C resonance

---

[127] R. L. Vold and S. O. Chan, J. Magn. Resonance, 1971, 4, 208.
[128] R. L. Vold, J. Chem. Phys., 1972, 56, 3210.
[129] R. Freeman and H. D. W. Hill, J. Chem. Phys., 1971, 55, 1985.
[130] U. Haeberlen, H. W. Spiess, and D. Schweitzer, J. Magn. Resonance, 1972, 6, 39.

of CS_2 yield straight lines when the echo decay rate R_2 is plotted against t_{cp}^2 as predicted by $R_2 = T_2^{-1} + \frac{1}{12} \gamma^2 G^2 D\, t_{cp}^2$; G is the field gradient across the sample and t_{cp} the 180° pulse spacing. In some unexpected cases, $T_2 < T_1$, e.g. in CS_2 at 221 K, $T_1 = 52.5$ s and $T_2 = 35$ s. In a pure, carefully degassed sample of C_6H_6, they also found that $T_2 < T_1$, but in a sample containing dissolved oxygen their observations are as follows: for large t_{cp}, R_2 behaves as expected except that the extrapolation does not lead to $T_1 = T_2$; for small t_{cp}, the measurements fall below the extrapolated line, and in fact, approach T_1^{-1}. The ^1H and ^{19}F resonances in, respectively, pure and impure H_2O and C_6F_6, behave similarly to benzene. The authors were unable to explain these unexpected observations. R_2 for the ^{13}C spins in C_6H_6 and $C_6H_5{}^{13}CH_3$ shows an oscillatory dependence on $1/t_{cp}$ owing to the scalar coupling of the ^{13}C spins to the directly bonded protons; similar effects have previously been reported by Shoup and Vanderhart[131] (Vol. 1, p. 139). The results for the $^{13}CH_3$ resonance are considered in terms of the slow exchange limit of an AX_3 system.

Chemical Exchange Studies. Last year we reported the application of the CPSE experiment to the study of exchange of solvent molecules between the solvation sphere of a paramagnetic ion and the bulk solvent (Vol. 1, p. 136). In this work a closed expression was used for the dependence of R_2 upon t_{cp} for the case of exchange between two unequally populated sites with unequal relaxation times (both T_1 and T_2). Carver and Richards[132] have since reported, as promised, the derivation of this expression; it is essentially an extension of the earlier derivation given by Allerhand and Gutowsky,[133] who assumed equal relaxation times in the two sites. Numerical calculations are given which are illustrative of the conditions appropriate to the interaction of small molecules or ions with macromolecules: two different cases are considered for the bound molecules: (i) $T_1 = T_2$; (ii) $T_2 < T_1$. Fedotov[134] has reported use of the spin-echo method to study chemical exchange in aqueous solutions of glycine ethyl esters.

An elementary account of the spin-echo experiment and its application to the study of spin diffusion and chemical exchange processes is given by Boden[135] in an article concerned generally with pulsed n.m.r. methods.

I. Models for Molecular Rotation.—Although it is now possible to determine reliable values for the correlation times τ_2 and τ_J from T_1 measurements, it is relatively difficult to provide a reasonable theoretical interpretation for them. O'Reilly[136] has given a brief review of some of the currently favoured

[131] R. R. Shoup and D. L. Vanderhart, *J. Amer. Chem. Soc.*, 1971, **93**, 2053.
[132] J. P. Carver and R. E. Richards, *J. Magn. Resonance*, 1972, **6**, 89.
[133] A. Allerhand and H. S. Gutowsky, *J. Chem. Phys.*, 1964, **41**, 2115; 1965, **42**, 1587.
[134] V. D. Fedotov, *Teor. i eksp. Khim.*, 1971, **7**, 347.
[135] N. Boden, in 'Determination of Organic Structures by Physical Methods', ed. F. C. Nachod and J. J. Zuckerman, Academic Press, New York and London, 1971, Vol. 4 p. 51.
[136] D. E. O'Reilly, *Ber. Bunsengesellschaft phys. Chem.*, 1971, **75**, 208.

models for molecular rotation in liquids and the associated expressions for τ_2; he discusses the small-step rotational diffusion model, the conditional inertial rotation model, and the quasi-lattice random flight model. The latter model was formulated by O'Reilly himself; it assumes that rotation occurs by large-angle steps when the molecule is excited to an interstitial site by a translational step, i.e. complete correlation between translational and rotational motions is assumed with the result that τ_2 is inversely proportional to the self-diffusion coefficient. In two interesting papers with Peterson, O'Reilly[137] has shown that this model accounts quantitatively for the temperature dependence of τ_2 in the polar liquids MeNO$_2$, Me$_2$CO, C$_6$H$_5$NO$_2$, and C$_6$H$_5$Cl, and the pressure dependence of the ring proton intramolecular relaxation rates in C$_6$H$_5$Me and C$_6$H$_5$Cl. The authors also show that the microviscosity model of Gierer and Wirtz which is often invoked to explain the departure of observed τ_2 values from the Debye–Stokes–Einstein expression does not apply in most liquids. The quasi-lattice model clearly represents a significant advance over the Debye–Stokes–Einstein continuum model of rotational diffusion and, importantly, τ_2 can be related to molecular properties.

It is now realized that in many simple liquids, molecular rotation occurs by large-angle steps, and that the rotational and translational motions are appreciably decoupled. Both the small-step rotational diffusion and quasi-lattice random flight models will be inapplicable in such liquids. The extended J- and M-diffusion models introduced by Gordon (Vol. 1, p. 123) have given a reasonable interpretation for the behaviour of linear molecules and were used by McClung (Vol. 1, p. 123) in his description of the rotation of spherical-top molecules. O'Reilly[138] has reported a calculation of the angular velocity correlation function for a hindered rotation model which assumes the molecules rotate freely for a small fraction of the total time. For a spherical-top molecule where the mean barrier height $\bar{\varepsilon}$ averaged over the torsional states is not too small compared with kT,

$$\tau_2 \approx \frac{\langle \theta^2 \rangle^{\frac{1}{2}}}{4(1-\lambda_2)} \left(\frac{\pi I}{2kT}\right)^{\frac{1}{2}} \left(\frac{kT}{\bar{\varepsilon}}\right) \exp(\bar{\varepsilon}/kT) \qquad (11)$$

where $\lambda_2 = \sin[\frac{5}{2}\langle\theta^2\rangle^{\frac{1}{2}}]/\{5 \sin[\frac{1}{2}\langle\theta^2\rangle^{\frac{1}{2}}]\}$ and $\langle\theta^2\rangle$ is the mean-square-angular displacement. This model is similar to the conditional inertial rotation model in predicting $\tau_2 \propto I^{\frac{1}{2}}$, as has been observed by Atkins et al.[139] for NH$_3$ and ND$_3$. O'Reilly also obtained a relation between τ_2 and τ_J for a spherical-top molecule:

$$\tau_2 \tau_J = I\langle\theta^2\rangle/6kT(1-\lambda_2) \qquad (12)$$

Since, for small θ, $\lambda_2 \approx 1 - \langle\theta^2\rangle$, equation (12) reduces to the well-known Hubbard relation in the small-step limit. Thus, $\langle\theta^2\rangle$ may be determined from

[137] D. E. O'Reilly and E. M. Peterson, J. Chem. Phys., 1971, 55, 2155; 1972, 56, 2262.
[138] D. E. O'Reilly, J. Chem. Phys., 1972, 56, 2924.
[139] P. W. Atkins, A. Lowenstein, and Y. Margalit, Mol. Phys., 1969, 17, 329.

measurement of $\tau_2 \tau_J$. We might expect $\langle \theta^2 \rangle$ to be determined by the intermolecular potential and therefore be temperature independent, which is not always observed experimentally (Vol. 1, p. 123). McClung[140] has shown that his theoretical model is compatible with the i.r. bandshapes of CH_4 and CD_4 in both the pure liquids and in dilute solutions in liquid noble gases. In the neat liquids, the results are explained in terms of a J-diffusion model where every 'collision' is effective in completely randomizing the angular momentum. In dilute solutions of CH_4 in liquid noble gases on the other hand, the results are consistent with the M-diffusion model where every 'collision' randomizes only the direction of the angular momentum vector in the CH_4 molecules, but never changes its magnitude. According to McClung this result would be expected for collisions involving only two molecules, since angular momentum must be conserved, but it is somewhat surprising in a liquid where collisions involving more than two molecules occur with high frequency.

J. Studies of Critical Phenomena in Fluids.—No new studies have been reported, but Hamann, Hoheisel, and Richtering[141] have given, at a conference on critical phenomena [reported in *Ber. Bunsengesellschaft phys. Chem.*, 1972, **76** (issues 3 and 4), all papers] a short but critical review of n.m.r. studies at critical points in one- and two-component systems. The main part is concerned with determination of self-diffusion coefficients by means of the spin-echo method. They claim that most of the experimental results indicate that there is no anomaly for the self-diffusion coefficient at the critical point within the precision of the reported measurements (± 2—5%) and for the approach to the critical point obtained [$|(T-T_c)/T_c| \approx 10^{-4}$; $|(C-C_c)/C_c| \approx 10^{-2}$, where C is concentration]. The same holds for measurements of T_1.

K. ^{13}C Relaxation Studies.—Interest in establishing the mechanisms for ^{13}C relaxation in organic molecules continues (Vol. 1, p. 138) and is encouraged by the rapidly increasing use of ^{13}C Fourier transform n.m.r. spectroscopy. A knowledge of T_1^C and its mechanisms is important both as a tool for probing molecular dynamics in liquids (such studies are reported elsewhere in this section) and also for spectroscopic purposes. The studies reported this year conveniently fall in two distinct categories as follows.

Small Molecules. T_1^C measurements as a function of temperature and frequency have been reported for a variety of small organic molecules in the liquid state.[142–150] In many cases ^{13}C-enriched samples were used, and nuclear

---

[140] R. E. D. McClung, *J. Chem. Phys.*, 1971, **55**, 3459.
[141] H. Hamann, C. Hoheisel, and H. Richtering, *Ber. Bunsengesellschaft phys. Chem.*, 1972, **76**, 249.
[142] A. Olivson and E. Lippmaa, *Chem. Phys. Letters*, 1972, **11**, 242.
[143] T. D. Alger, S. W. Collins, and D. M. Grant, *J. Chem. Phys.*, 1971, **54**, 2820.
[144] T. D. Alger and D. M. Grant, *J. Phys. Chem.*, 1971, **75**, 2538.
[145] G. C. Levy, *J.C.S. Chem. Comm.*, 1972, 47.

Overhauser enhancement techniques were employed to determine the contribution from the (^{13}C,^{1}H) dipole interaction. The following conclusions are drawn from these studies. T_1^C is generally in the range 10—100 s. With the exception of methyl groups, it is dominated by dipole interaction with the attached protons at room temperature and by spin–rotation interaction at higher temperatures. Isolated methyl groups tend to behave essentially as free rotors and it is found[69,71,72] that the major contribution to the ^{13}C relaxation rate is from the spin–internal-rotation interaction. On the other hand, T_1^C in rotationally restricted methyl groups tend to be dominated by dipole interactions with the attached protons, e.g. in o-xylene[71,72] and Me$_4$Si.[147] For ^{13}C bonded to a ^{79}Br nucleus, the scalar coupling interaction becomes important because of the close proximity of the ^{13}C and ^{79}Br resonance frequencies;[146-148] in the case of ^{81}Br the scalar interactions is less efficient and both dipole–dipole and scalar interactions are important. The presence of nearly equal amounts of ^{79}Br and ^{81}Br isotopes can lead to non-exponential relaxation of the ^{13}C magnetization, as has been demonstrated by Levy[148] for ^{13}C nuclei bonded to bromine in bromobenzene. The anisotropic chemical shielding interaction is not an important mechanism except in CS_2.[150,151] A detailed temperature (167—308 K) and frequency dependence (15, 30, and 62 MHz) study of T_1^C in the latter liquid by Spiess et al.[150] shows that the relaxation is determined by the anisotropic chemical shift interaction at low temperatures and high frequencies and the spin–rotation interaction at the higher temperatures. In neat formic acid the ^{13}C relaxation proceeds predominantly through dipolar interactions with the attached protons; when paramagnetic ions (hexa-aquochromic perchlorate) are added they are found[151] to shorten T_1^C by acting directly on the carbon nuclei rather than via the proton spins, i.e. the 'three-spin effect' is of negligible importance.

Complex Molecules. In complex molecules, the inversion recovery or progressive saturation Fourier transform techniques may be used to determine the spin–lattice relaxation rates of resolved ^{13}C resonances (see Vol. 1, p. 140). The application of these techniques to the assignment of resonances in proton-decoupled spectra and to the study of segmental motion in complex molecules is very promising. An adaptation of the progressive saturation technique in cases where the sensitivity is low has been discussed by Freeman et al.[152] and applied to the determination of the T_1^C in cortisone acetate. In an interesting paper, Allerhand et al.[153] have critically examined the application of these

[146] J. R. Lyerla, D. M. Grant, and R. D. Bertrand, *J. Phys. Chem.*, 1971, **75**, 3967.
[147] T. C. Farrar, S. J. Druck, R. R. Shoup, and E. D. Becker, *J. Amer. Chem. Soc.*, 1972, **94**, 699.
[148] G. C. Levy, *J.C.S. Chem. Comm.*, 1972, 352.
[149] H. W. Spiess, D. Schweitzer, U. Haeberlen, and K. H. Hausser, *J. Magn. Resonance*, 1971, **5**, 101.
[150] C. F. Schmidt and S. I. Chan, *J. Chem. Phys.*, 1971, **55**, 4670.
[151] R. Freeman, K. G. R. Pachler, and G. N. La Mar, *J. Chem. Phys.*, 1971, **55**, 4586.
[152] R. Freeman, H. D. W. Hill, and R. Kaptein, *J. Magn. Resonance*, 1972, **7**, 82.
[153] A. Allerhand, D. Doddrell, and R. Komoroski, *J. Chem. Phys.*, 1971, **55**, 189.

techniques to the determination of ^{13}C spin–lattice relaxation rates in the proton-decoupled spectra of sucrose, cholesteryl chloride, and adenosine 5'-monophosphate, chosen as typical examples of complex organic compounds likely to be studied by the Fourier transform technique. With the exception of a few side-chain groups, all protonated carbons have T_1 values of less than 1 s; this is due to the slower rotation of these molecules than smaller ones. Some side-chain carbons of cholesteryl chloride show evidence of internal reorientation and have relaxation times up to 2 s. Non-protonated carbons have T_1 values in the range 2—8 s. These relaxation times are sufficiently short to make ordinary Fourier transform n.m.r. a very sensitive technique in the study of complex molecules. The ^{13}C nuclei in these complex molecules relax mainly through (^{13}C,H) dipole interaction with the attached protons. Thus, nearly all carbons in complex molecules should have the same nuclear Overhauser enhancement, making it possible to use ^{13}C integrated intensities to make a carbon count in spectra of complex molecules. Furthermore, T_1^C values may be used in the assignment of the spectra of such molecules, allowance being made for methyl groups and other side-chain carbons with appreciable internal motion. For polystyrene solutions in tetrachloroethylene, Allerhand and Hailstone[154] have shown that, for molecular weights of 10^4 and higher, T_1^C is independent of molecular weight because the effective rotational correlation time is determined by segmental motion and not by the rate of overall rotation, thus providing the opportunity to study the motions of the side-chains.

L. ^{15}N Relaxation Studies.—Lippmaa et al.[155] have shown that the ^{15}N spin–lattice relaxation times measured in ^{15}N-enriched small organic molecules in the liquid state are very long: 20—1000 s at room temperature. T_1^N is determined by dipole interactions with protons at room temperature and by spin–rotation interactions at higher temperatures. When no hydrogen atoms are directly attached to the ^{15}N nucleus under study, then T_1 is very long and both dipole and spin–rotation interactions are of comparable magnitude, e.g. in nitrobenzene at 283 K, $T_1^N \approx 800$ s, $T_{1sr} = 800$ s, T_{1dd}(inter) = 1000 s, and T_{1dd}(intra) = 3300 s. There is no evidence for anisotropic chemical shielding interactions. Clearly, the long T_1 times encountered in most organic molecules will make difficult the universal use of Fourier transform spectroscopy but, on the other hand, they will be beneficial in chemical polarization experiments.

M. ^{31}P Relaxation Studies.—Dale and Hobbs[156] have measured T_1^P as a function of temperature and B_0 in some liquid organic and inorganic compounds: PBr_3, P_4O_6, $PO(OH)_3$, $P(OMe)_3$, $P(OEt)_3$, $P(OPr^i)_3$, $P(OBu^n)_3$, $OP(OMe)_3$, $SP(OMe)_3$, $OP(OEt)_3$ $OP(OBu^n)_3$. By making a number of as-

---

[154] A. Allerhand and R. K. Hailstone, *J. Chem. Phys.*, 1972, **56**, 3719.
[155] E. Lippmaa, T. Saluvere, and S. Laisaar, *Chem. Phys. Letters*, 1971, **11**, 120.
[156] S. W. Dale and M. E. Hobbs, *J. Phys. Chem.*, 1971, **75**, 3537.

sumptions such as isotropic rotation and the applicability of the Hubbard and Debye–Stokes–Einstein relations, an approximate separation of the contributions to T_1^{-1} was achieved. In all compounds, contributions from dipolar anisotropic chemical shielding, and spin–rotation interactions are important. A rough parallel is observed between the chemical shift of a particular compound and the absolute value of the chemical shielding anisotropy; a similar relation for the spin–rotation coupling constant is expected but the correlation was poor. These results indicate that the second-order paramagnetic shielding dominates the ^{31}P chemical shifts in these compounds.

N. **Determination of Spin Interaction Constants from Spin Relaxation Measurements.**—Apart from providing information about the dynamics of the spin-bearing molecules, spin relaxation measurements may be used to determine, albeit indirectly, values for spin-dependent interactions which couple the nuclear spins to the lattice. This is an important and alternative route to these data when they cannot be determined directly by steady-state or Fourier transform spectroscopy. In this section we report those measurements which are not covered elsewhere in this review.

Scalar Spin–Spin Coupling Constants. The $T_{1\rho}$ method (Vol. 1, p. 129) has been employed by Sears[157] to determine: $J(^{19}F, ^{35}Cl) = 27.6 \pm 0.5$ Hz, $J(^{19}F, ^{37}Cl) = 23.0 \pm 0.5$ Hz in $CHFCl_2$; $J(^{19}F, ^{35}Cl) = 13.7 \pm 0.5$ Hz, $J(^{19}F, ^{37}Cl) = 11.4 \pm 0.5$ Hz in $CFCl_3$, and by Burnett and Zeltmann[49] to determine $J(^{75}As, ^{1}H) = 92.1 \pm 1$ Hz and $J(^{75}As, ^{2}H) = 14.2 \pm 0.5$ Hz in AsH_3.

Nuclear Quadrupole Coupling Constants. The major difficulty in calculating (e^2qQ/h) values from T_1 measurements is the estimation of the relevant value for τ_2: various approaches have been employed (Vol. 1, p. 128). Dinesh and Rogers[158] have suggested an alternative one which depends on the assumption that the coupling constants are the same in the solid and liquid states. They have combined solid-state n.q.r. measurements of (e^2qQ/h) on 25 nitrogen compounds with T_2^N measurements in the corresponding liquids to obtain experimental values of τ_2. Comparison of τ_2 with values calculated using the Debye–Stokes–Einstein expression gave an empirical correction factor to be used with the latter in the calculation of τ_2 for other liquids. In this way the following new values in MHz for (e^2qQ/h) were estimated from T_2^N measurements: 2-methylpyrazine, 4.774; 2,5-dimethylpyrazine, 5.313; diethylamine, 4.87; triethylamine, 4.90; n-propylamine, 4.56; isobutylamine, 3.57; isoamylamine, 4.32; allylamine, 4.46; benzylamine, 2.83; cyclohexylamine, 4.10; tetraethylurea, 5.75; formamide, 3.67; NN-dimethylformamide, 4.37; NN-dimethylacetamide, 5.21; quinoline, 4.56; the accuracy is *ca.* 10%.

Spin–Rotation Interaction Constants and Chemical Shielding Anisotropies. Spiess *et al.*[149] have given an elegant demonstration of how, for the linear

[157] R. E. J. Sears, *J. Chem. Phys.*, 1971, **56**, 983.
[158] Dinesh and M. T. Rogers, *J. Magn. Resonance*, 1972, **7**, 30.

molecule CS_2, the ^{13}C spin–rotation interaction constant and the chemical shielding anisotropy can be obtained directly from $(T_1^C)^{-1}$ measurements. The method is based on the fact that the spin–rotation interaction is closely related to the paramagnetic part of the chemical shift.[159] For a linear molecule, to a good approximation,

$$\Delta\sigma = \sigma_\| - \sigma_\perp = -\frac{|e|}{2m}\frac{I}{\mu_N g_k}C_k \qquad (13)$$

where $C_k = C_{\perp k}$ is the spin–rotation constant of the kth nucleus and, μ_N and g_k are the nuclear magneton and nuclear g-factor. In the small-step rotational diffusion limit,

$$T_{1sr}^{-1} \cdot T_{1cs}^{-1} = (4/135)(\omega_0^2 I^2/\hbar^2) C^2 (\Delta\sigma)^2 \qquad (14)$$

Equations (13) and (14) may be used to determine absolute values for C and $\Delta\sigma$ if T_{1sr} and T_{1cs} can be measured. In $^{13}CS_2$, T_1^C is caused by chemical shielding anisotropy and spin–rotation interactions; by measuring T_1^C as a function of ω_0 [14, 30, 62 HMz] these two contributions were separated. Over the temperature interval 167 to 308 K, the product $T_{1sr}^{-1}.T_{1cs}^{-1}$ is constant, confirming the assumption of small-step diffusion. Using this value in equations (13) and (14) gave $|C| = 13.8 \pm 1.4$ kHz and $|\Delta\sigma| = 438 \pm 44$ p.p.m. For comparison, the values $C = -12.6$ kHz and $\Delta\sigma = 428$ p.p.m. were calculated from the known values of C and $\Delta\sigma$ for CO and the relative isotropic chemical shifts of ^{13}CO and $^{13}CS_2$. Similar calculations of spin–rotation constants for ^{13}C are reported by other authors.[69,142,150] The value for $\Delta\sigma$ also agrees quite well with 425 ± 15 p.p.m. recently measured[160] in solid $^{13}CS_2$. The close agreement of calculated and observed values demonstrates how the above method of eliminating the correlation times allows the determination of reliable spin–rotation constants and shift anisotropies by relaxation measurements. This is particularly important for symmetric molecules like CS_2, which are unsuitable for microwave or electric molecular-beam resonance experiments.

There are, unfortunately, few molecular systems to which the above approach is applicable; more often, values for spin–rotation interaction constants are estimated from Hubbard's spin–rotation relaxation equation for spherical molecules $[T_{1sr}^{-1} = (2IkT/\hbar^2)C^2\tau_J$, where $C^2 = (C_\|^2 + 2C_\perp^2)/3]$. The accuracy of C is determined by the value of τ_J used; these values are usually calculated from τ_2 via the expression $\tau_2\tau_J = I/6kT$. Deverell[161] has used this approach to calculate C for a number of ^{19}F and ^{31}P compounds and found large discrepancies between these values and corresponding ones determined either directly from molecular-beam studies or calculated from

---

[159] N. F. Ramsey, *Phys. Rev.*, 1950, **78**, 699; W. H. Flygare, *J. Chem. Phys.*, 1964, **41**, 793; W. H. Flygare and J. Goodisman, *ibid.*, 1968, **49**, 3122.
[160] A. Pines, W. K. Rhim, and J. S. Waugh, *J. Chem. Phys.*, 1971, **54**, 5438.
[161] C. Deverell, *Mol. Phys.*, 1970, **18**, 319.

the chemical shift. In a well-argued paper, Gillen[162] has shown that these discrepancies, and similar ones reported by other authors, are due primarily to incorrect estimates of τ_2. Using average values of τ_2, calculated from the complete rotational diffusion tensors of the molecules PCl_3, PBr_3, and $POCl_3$ in Hubbard's isotropic equation, he obtained values for C which agree reasonably well with ones deduced from an absolute shielding scale based on a recent molecular-beam determination[163] of the ^{31}P spin–rotation interaction tensor in PH_3 ($C_\parallel = -116.4$ kHz and $C_\perp = -114.9$ kHz). The values for the spin-rotation constants reported for PCl_3 and PBr_3 by Jordan et al.[50] are inconsistent with Gillen's values.

It should be noted that generally interest is not in the determination of values of C from T_{1sr} measurements, but in getting τ_J; it is now well established that satisfactory values for C may be calculated from the isotropic chemical shift of a nucleus once a reliable absolute shielding scale has been established from a molecular-beam measurement.

O. **Miscellaneous Papers.**—Blicharski and Blicharski[164] report plots showing the frequency dependence of T_1 and T_2 in the presence of a log-Gaussian distribution of rotational correlation times; these should be useful for interpreting measurements in magnetic relaxation spectroscopy. Blicharski[165] has also presented an interesting unified theoretical treatment for relaxation arising out of intramolecular electric and magnetic multiple interactions between a single nucleus of arbitrary spin and the electric cloud. The theory is only valid for $\omega_0 \tau \ll 1$ and may be applied to liquids of low viscosity and, with some restrictions, in dense gases.

4 Spin Relaxation in Liquid Crystals

This section reports papers concerned with the application of spin-relaxation measurements to the elucidation of molecular dynamics in the nematic, cholesteric, and isotropic phases of thermotropic liquid crystals; measurements in lyotropic liquid crystals are reported in Chapter 8.

A. **Nematic Phases.**—There is still considerable confusion amongst authors as to the relative importance of the roles of the long-range collective order fluctuations and short-range fluctuations, such as translational diffusion, in determining the proton relaxation rate. Theory[166] gives the following expression for the relaxation rate due to modulation of the intramolecular dipolar coupling between a pair of protons in the benzene ring of a nematogen

[162] K. T. Gillen, J. Chem. Phys., 1972, 56, 1573.
[163] P. B. Davies, R. M. Newmann, G. C. Wofsy, and W. Klemperer, J. Chem. Phys., 1971, 55, 3564.
[164] B. Blicharski and J. S. Blicharski, Acta Phys. Polon., 1972, A41, 347.
[165] J. S. Blicharski, Acta Phys. Polon., 1972, A41, 353.
[166] P. Pincus, Solid State Comm., 1969, 8, 707; J. W. Doane and D. L. Johnson, Chem. Phys. Letters, 1970, 6, 29; T. C. Lubensky, Phys. Rev. (A), 1970, 2, 2497; C. C. Sung, Chem. Phys. Letters, 1971, 10, 35.

by the long-range order fluctuations and short-range molecular reorientations:

$$\frac{1}{T_1} = \omega_D{}^2 \frac{S^2}{K} \frac{kT}{D+(K/\eta)^{\frac{1}{2}}} \omega_0{}^{-\frac{1}{2}} + B(T) \qquad (15)$$

where ω_D is the dipolar frequency (describing the strength of the dipolar interaction), $S = \frac{1}{2}\langle 3\cos^2\theta - 1\rangle$ is the order parameter (θ is the angle between the molecular axis and the optical axis), D is the diffusion constant, K is the Frank elastic deformation constant, and η is the orientational average of the Leslie viscosity coefficients. The term $B(T)$ has been added to take care of frequency independent contributions arising from fluctuations in the magnitude of S and short distance variations in the direction of preferred order; its exact form is not known. Experimental studies by Dong, Forbes, and Pintar[167] of T_1^H in p-azoxyanisole confirm the frequency dependence predicted by equation (15), but the observed temperature dependence is much stronger than predicted. Similar observations with respect to temperature dependence are reported by Dong[168] for p-ethoxybenzylidene-p-n-butylaniline and p-(p-ethoxyphenylazo)phenyl heptanoate, and by Watkins and Johnson[169] for p-azoxyanisole, butyl p-(p-ethoxyphenoxycarbonyl)phenyl carbonate, p-(p-ethoxyphenylazo)phenyl hexanoate, p-(p-ethoxyphenylazo)phenyl undecylenate, and p-[N-(p-methoxybenzylidene)amino]-n-butylbenzene; in the latter three compounds the temperature dependence is Arrhenius. This behaviour has led several authors to question the relevance of fluctuations in the orientational order for T_1^H and to consider the importance of relaxation by translational diffusion. The importance of intermolecular interactions has been demonstrated[170] for the proton spin–lattice relaxation in the nematic phase of bis-(4-aminofluorobenzene)terephthalate by comparing it with the quadrupole spin–lattice relaxation in the deuteriated counterpart. Johnson and Watkins[169] and Doane and Moroi[170] have suggested that since translational diffusion can, under certain conditions, give rise to a frequency dependence of the form

$$T_1^{-1} = C - D\omega_0{}^{-\frac{1}{2}} \qquad (16)$$

where C and D are constants, the observed frequency dependence does not distinguish between the two mechanisms. A rough calculation shows, however, that this contribution to T_1^H in p-azoxyanisole, while significant, is frequency-independent. Samulski, Dybowski, and Wade[171] suggest that modulation of the intramolecular dipolar interaction of phenyl ring protons by Brownian motion may contribute significantly to the temperature dependence. The fact that several relaxation mechanisms are responsible for proton relaxation has clearly complicated the interpretation of the results; moreover,

[167] R. Y. Dong, W. F. Forbes, and M. M. Pintar, *J. Chem. Phys.*, 1971, **55**, 145.
[168] R. Y. Dong, *J. Magn. Resonance*, 1972, **7**, 60.
[169] C. L. Watkins and C. S. Johnson, *J. Phys. Chem.*, 1971, **75**, 2452.
[170] J. W. Doane and D. S. Moroi, *Chem. Phys. Letters*, 1971, **11**, 339.
[171] E. T. Samulski and C. R. Dybowski, and C. G. Wade, *Chem. Phys. Letters*, 1971, **11**, 113.

since experimentalists have not yet observed multiple exponential relaxation functions, the role of the alkyl protons in determing the observed T_1^H is not understood. The importance of the alkyl substituent in determining the relaxation rate is demonstrated by measurements of T_1^H in the homologous series of 4,4'-bis(alkoxy)azoxybenzenes.[172] Martins[173] has avoided this complication by making measurements in methyl-deuteriated p-azoxyanisole, and has shown that T_1^H for the benzene ring protons is independent of temperature but shows the same frequency dependence as p-azoxyanisole, in accordance with equation (15). He also demonstrates that the temperature dependence of T_1^H in p-azoxyanisole originates from the methyl proton relaxation which is frequency-independent and is caused by short-range motions whose nature is not yet understood; the activation energy for this motion is approximately 38 kJ mol^{-1}. Thus, as far as can be concluded at present, it appears that relaxation of the ring protons is in agreement with theory, but the relaxation mechanisms for the alkyl protons are not well understood although intermolecular relaxation controlled by self-diffusion appears to be important. Measurements in better defined spin-systems, preferably of ^{2}H, are called for in order to elucidate the relaxation mechanisms for protons in the alkyl side-chains and the rate of spin diffusion in these systems.

Theory predicts an anisotropy in T_1^H owing to modulation of the intramolecular interaction between ring protons which is due to collective order fluctuations. The angular dependence of T_1^H as the nematic director n_0 is rotated in B_0 has been discussed theoretically by Doane and Mori:[170]

$$T_1^{-1} = \omega_D^2(1-3\cos^2\phi)^2 \frac{S^2}{K} \frac{kT}{D+(K/\eta)^{\frac{1}{2}}} \cdot \omega_0^{-\frac{1}{2}} F(\theta) \quad (17)$$

where

$$F(\theta) = 1 - \left(\frac{5}{2}-2^{-\frac{1}{2}}\right)\sin^2\theta + (2-2^{-3/2})\sin^4$$

θ is the angle between n_0 and B_0, and ϕ is the orientation of the internuclear vector with respect to the long axis of the molecule. Tarr et al.,[174] using both a.c. and d.c. electric fields to orientate n_0, have observed anisotropy for T_1^H in p-methoxybenzylidine-p'-cyanoaniline and p-methoxybenzylidene-p-n-butyl-aniline, but found that both a.c. and d.c. fields induce turbulence in the liquid crystal which affects T_1. The effects of turbulence must therefore be incorporated in the expression for T_1 or some alternative method used to orientate n_0. The use of selectively deuteriated liquid crystals for these experiments is to be encouraged since the contributions to T_1^H from the alkyl-chain protons whose relaxation is considered to be isotropic will tend to mask any anisotropy in

---

[172] C. R. Dybowski, B. A. Smith, and C. G. Wade, *J. Phys. Chem.*, 1971, **75**, 3834.
[173] A. F. Martins, *Mol. Crystals Liquid Crystals*, 1971, **14**, 85; *Phys. Rev. Letters*, 1972, **28**, 289.
[174] C. E. Tarr, M. A. Nickerson, and C. W. Smith, *Appl. Phys. Letters*, 1970, **17**, 318; C. E. Tarr, A. M. Fuller, and M. A. Nickerson, *ibid.*, 1971, **19**, 179.

T_1. This no doubt explains why Watkins and Johnson[169] find T_1^H in p-[N-(p-methoxybenzylidene)]-n-butylbenzene to be independent of the alignment of n_0 with respect to B_0. The dependence of T_1 on orientation in the magnetic field is, therefore, still to be determined.

B. Cholesteric Phases.—Dong et al.[175] have shown that proton relaxation in the cholesteric phase of cholesteric mixtures of p-azoxyanisole and cholesteryl chloride is determined by the same mechanism as in the nematic phase; only small changes in T_1^H are observed at the cholesteric-to-nematic phase transition. Dybowski and Wade[176] report a T_1 study of cholesterol and its esters (n-propionate, n-decanoate, myristate, and oleate); the results suggest that the relaxation mechanism in the mesophases of all the esters is similar to that in nematic liquid crystals.

C. Isotropic Phases.—Fluctuations in the local order and translational diffusion provide the mechanisms for proton relaxation in the isotropic phase. In methyl-deuteriated p-azoxyanisole, Martins[173] has shown $T_1 \approx T_1^* + \alpha\Theta$ where T_1^* and α are frequency-independent constants and Θ is the temperature. This behaviour is consistent with a model based on fluctuations in the local-order parameter. The temperature dependence of T_1^H differs from that expected for self-diffusion in normal liquids; therefore, it is argued that if self-diffusion is an important mechanism for T_1^H, then it must be a 'correlated' motion. In p-azoxyinisole, T_1^H is given by the same expression, but the coefficient α is now frequency-dependent. The frequency dependence of T_1 must be attributed to the relaxation of the methyl protons, the mechanism for which is not yet understood. In contrast to the behaviour discussed in p-azoxyanisole, Dong[168,177] has found that, in the isotropic phase of p-methoxybenzylidene-p-n-butylaniline, translational diffusion is the predominant relaxation process; this is so because any local order is destroyed by a relatively fast self-diffusion process. T_1^H is frequency-dependent up to approximately 18 K above the transition temperature in a manner consistent with translational relaxation theory; $D = 9.0 \times 10^{-11}$ m² s at 323 K.

D. Spin-diffusion Studies.—Blinc et al.[178] have found, using the neutron scattering method, that in the nematic phases of p-azoxyanisole, anisalazine, and 4,4'-diheptyloxyazoxybenzene, the apparent self-diffusion constant is temperature-independent and $\approx 1 \times 10^{-9}$ m² s. This behaviour is explained by the similar measurements of Otnes et al.,[179] which show that the components of the D tensor have opposite temperature dependence. In the isotropic phases, normal thermally activated diffusion is observed with $E_D \approx 40$ kJ

---

[175] R. Y. Dong, M. M. Pintar, and W. F. Forbes, *J. Chem. Phys.*, 1971, **55**, 2449.
[176] C. R. Dybowski and C. G. Wade, *J. Chem. Phys.*, 1971, **55**, 1576.
[177] R. Y. Dong, W. F. Forbes, and M. Pintar, *Mol. Crystals Liquid Crystals*, 1972, **16**, 213.
[178] R. Blinc, V. Dimic, J. Pirs, M. Vilfan, and I. Zupancic, *Mol. Crystals Liquid Crystals*, 1971, **14**, 97.
[179] K. Otnes, R. Pynn, J. A. Janik, and J. M. Janik, *Phys. Letters (A)*, 1972, **38**, 335.

mol^{-1}; measurements by the pulsed gradient spin-echo method give the same D as neutron scattering in this phase.

5 Spin Relaxation in Solids

Many interesting properties of solids are intimately related to atomic and molecular motions. N.m.r. has played, and continues to play, a decisive part in our understanding of these properties. For a general account of this work the reader is referred to the useful review by Allen.[180] In solids, n.m.r. lineshape studies (see Chapter 9) give information about the mechanism of atomic and molecular motions, as well as their rates in the region 10^4—10^6 Hz. Spin-lattice relaxation measurements are complementary, enabling a more accurate determination of these rates in the extended range 1—10^9 Hz, thereby giving reliable activation energies. In this section we report papers concerned with spin-relaxation mechanisms, atomic and molecular motions, and the mechanisms of phase transitions.

A. **Spin–Lattice Relaxation in the Rotating Frame and Dipolar Relaxation.**—Rotating-frame relaxation measurements are now routinely made under conditions where the weak collision BPP-type theory[5] is applicable; many of these applications will be discussed later in this section. Here we only mention the paper by Blicharski[181] in which $T_{1\rho}$ and $T_{2\rho}$ are calculated in the weak collision case for dipolar, quadrupolar, spin–rotation, and scalar interactions considered separately, *i.e.* interference effects are neglected. It is also assumed that the correlation functions for these interactions decay exponentially in time with time constant τ_c. Expressions are given for $T_{1\rho}$ and $T_{2\rho}$ in the double rotating tilted frame, *i.e.* $\omega \neq \omega_0$; below are summarized the results, many of which have been given previously, for exact resonance conditions:

Dipolar relaxation between two identical and equivalent nuclei of spin I:

$$\frac{1}{T_{1\rho}} = \frac{I(I+1)\hbar^2\gamma^4 r^{-6}}{5}\left[\frac{3\tau_2}{1+\omega_1^2\tau_2^2} + \frac{5\tau_2}{1+\omega_0^2\tau_2^2} + \frac{2\tau_2}{1+4\omega_0^2\tau_2^2}\right] \quad (18)$$

$$\frac{1}{T_{2\rho}} = \frac{I(I+1)\hbar^2\gamma^4 r^{-6}}{20}\left[\frac{3\tau_2}{1+\omega_1^2\tau_2^2} + \frac{14\tau_2}{1+\omega_0^2\tau_2^2} + \frac{20\tau_2}{1+4\omega_0^2\tau_2^2}\right] \quad (19)$$

Quadrupolar relaxation in the case $I = 1$:

$$\frac{1}{T_{1\rho}} = \frac{3}{80}\left(\frac{e^2qQ}{\hbar}\right)\left(1+\frac{\eta^2}{3}\right)\left[\frac{3\tau_2}{1+\omega_1^2\tau_2^2} + \frac{5\tau_2}{1+\omega_0^2\tau_2^2} + \frac{2\tau_2}{1+4\omega_0^2\tau_2^2}\right] \quad (20)$$

$$\frac{1}{T_{2\rho}} = \frac{3}{320}\left(\frac{e^2qQ}{\hbar}\right)\left(1+\frac{\eta^2}{3}\right)\left[\frac{3\tau_2}{1+\omega_1^2\tau_2^2} + \frac{14\tau_2}{1+\omega_0^2\tau_2^2} + \frac{20\tau_2}{1+4\omega_0^2\tau_2^2}\right] \quad (21)$$

---
[180] P. S. Allen, in 'MTP International Review of Science—Physical Chemistry Series One', ed. A. D. Buckingham, 1972, Vol. 4, p. 43.
[181] J. S. Blicharski, *Acta Phys. Polon.*, 1972, **A41**, 223.

Spin–rotation interaction for a nucleus I in a linear molecule with angular momentum J for the condition $\omega_1 \ll |\omega_0 - \omega_J|$, where ω_J is the angular Larmor precession frequency for the vector J:

$$\frac{1}{T_{1\rho}} = \frac{1}{3}C^2 \langle J(J+1)\rangle \left[\frac{\tau_J}{1+\omega_1^2\tau_J^2} + \frac{\tau_J}{1+(\omega_0-\omega_J)^2\tau_J^2}\right] \quad (22)$$

$$\frac{1}{T_{2\rho}} = \frac{1}{6}C^2 \langle J(J+1)\rangle \left[\frac{\tau_J}{1+\omega_1^2\tau_J^2} + \frac{\tau_J}{1+(\omega_0-\omega_J)^2\tau_J^2}\right] \quad (23)$$

Scalar relaxation of a nucleus I coupled to a non-resonant rapidly relaxing nucleus S in the case of $\omega_1 \ll |\omega_0 - \omega_s|$:

$$\frac{1}{T_{1\rho}} = \frac{1}{3}A_{IS}^2 S(S+1)\left[\frac{\tau_1}{1+\omega_1^2\tau_1^2} + \frac{\tau_2}{1+(\omega_0-\omega_S)^2\tau_2^2}\right] \quad (24)$$

$$\frac{1}{T_{2\rho}} = \frac{1}{6}A_{IS}^2 S(S+1)\left[\frac{\tau_1}{1+\omega_1^2\tau_1^2} + \frac{3\tau_2}{1+(\omega_0-\omega_S)^2\tau_2^2}\right] \quad (25)$$

here τ_1 and τ_2 are the longitudinal and transverse relaxation times for spin S. For the anisotropic chemical shielding interaction, as given by Blinc et al.:[182]

$$\frac{1}{T_{1\rho}} = \frac{2}{45}(\sigma_\| - \sigma_\perp)^2 \omega_0^2 \left[\frac{3\tau_2}{1+\omega_1^2\tau_2^2} + \frac{5\tau_2}{1+\omega_0^2\tau_2^2} + \frac{2\tau_2}{1+4\omega_0^2\tau_2^2}\right] \quad (26)$$

It is not normal practice to measure $T_{2\rho}$; Blicharski suggests it could be measured using an additional receiver coil aligned along the direction of the field B_0. The receiver signal in this case should be proportional to the transverse magnetization $M_\perp$ in the rotating frame under conditions of resonance, and for $M_\perp(0) = M_0$:

$$M_\perp(t) = M_0 \exp(-t/T_{2\rho})\cos\omega_1 t \quad (27)$$

Under strong collision conditions, $\tau_c \gg T_{2,RL}^0$ ($T_{2,RL}^0 \equiv$ rigid lattice T_2), $B_1 \approx B_{loc}$, and $T_{1\rho} \approx T_{1D}$, the dipolar relaxation time (Vol. 1, p. 148). In an excellent article, Ailion[11] reviews the existing theories for relaxation under these conditions and also gives an account of the experimental techniques used for measurement of T_{1D}; the weak collision theory of relaxation in the rotating frame is also discussed. The articles by Jeener[10] and Goldman[9] also give useful treatments of relaxation in the strong collision limit.

B. Spin–Lattice Relaxation by Anisotropic Chemical Shielding Interaction.— The T_1^p measurements in the low-temperature crystalline phase of solid white phosphorus, reported by Boden and Folland,[36] provide an interesting demonstration of how measurement of T_1 as a function of B_0 in the neighbourhood of $\omega_0\tau \approx 1$ enables a unique determination of the dipolar frequency, the

[182] R. Blinc, G. Lahajnar, and J. Slivnik, *Chem. Phys. Letters*, 1971, **11**, 344.

absolute value of the chemical shielding anisotropy, and τ_2 when relaxation is due to modulation of intramolecular dipolar and anisotropic chemical shielding interactions by the reorientational motion of the molecule. This possibility arises because of the difference between the expressions for T_{1cs} and T_{1D} with respect to ω_0 in the two limits $\omega_0\tau \gg 1$ and $\omega_0\tau \ll 1$. In P$_4$, $|\sigma_\| - \sigma_\perp| = 286 \pm 15$ p.p.m. Blinc, Lahajnar, and Slivnik[182] have discussed the determination of $|\sigma_\| - \sigma_\perp|$ by measurement of $T_{1\rho}$ as a function of B_0 but at constant B_1 in the region $\omega_1\tau_2 \approx 1$, when both dipolar and anisotropic chemical shielding interactions contribute to the relaxation. The method is demonstrated by application to powdered UF$_6$; apparently the value (not given) for $|\sigma_\| - \sigma_\perp|$ agrees with the one determined from field dependence studies of the steady state absorption line.

C. Spin–Lattice Relaxation by Spin–Rotation Interaction.

—Recent studies (Vol. 1, p. 150) have provided qualitative evidence for the importance of the spin–rotation interaction as a relaxation mechanism for ^{1}H and ^{19}F nuclei in the solid state. Experimental evidence for its importance for the ^{31}P nucleus is now available. Sawyer and Powles[48] have shown that T_1^P is spin–rotation controlled in the plastic phase of PH$_3$. Boden and Folland[183] have been able to determine uniquely the spin–rotation relaxation rate in plastic crystalline white phosphorus: although $\omega_0\tau_2 \ll 1$, τ_2 values were calculated from $(T_{1cs})^{-1}$, separated from the other contributions by its quadratic dependence on ω_0, using $|\sigma_\| - \sigma_\perp| = 286$ p.p.m.: $\tau_2 = (4.7 \pm 0.1) \times 10^{-13}$ s exp$(5.7 \pm 0.4$ kJ mol$^{-1}/RT)$. Using this τ_2, the intramolecular dipolar relaxation rate was calculated; the diffusion-modulated intermolecular dipolar relaxation was calculated from the self-diffusion coefficient. Subtraction of these contributions from $(T_1^P)^{-1}$, gave a frequency-independent relaxation rate, proportional to τ_2^{-1}, characteristic of the spin–rotation interaction. Assuming molecular rotation occurs by fast uncorrelated jumps between discrete potential wells, the approximate expression (28) is obtained

$$(T_{1sr})^{-1} = \frac{1}{3}\left(\frac{IC\langle\theta^2\rangle^{\frac{1}{2}}}{\hbar}\right)^2 \frac{1}{\tau_2} \tag{28}$$

where $\langle\theta^2\rangle^{\frac{1}{2}}$ is the root-mean-square jump angle. Using a value of $(1/2\pi)|C| = 2.11$ kHz calculated from the chemical shift, $(T_{1sr})^{-1}$ is shown to be consistent with large angle reorientational jumps of the P$_4$ molecules by approximately 120°.

D. Spin–Lattice Relaxation by Nuclear Quadrupole Interaction.

—We briefly report in this section papers concerned with the study of pure nuclear quadrupole relaxation and of nuclear magnetic relaxation due to quadrupole interaction. Korchemkin[184] has used the method of quantum kinetic equations to obtain expressions for T_1 and T_2 for quadrupole relaxation of nuclei with

---
[183] N. Boden and R. Folland, *Chem. Phys. Letters*, 1971, **10**, 167.
[184] M. A. Korchemkin, *Soviet Phys. Solid State*, 1972, **13**, 1800.

integral spin in molecular solids when relaxation results from torsional vibrations. Nuclear quadrupole spin–lattice relaxation in the strong collision case in molecular solids is discussed by Ainbinder et al.[185] Measurements of ^{35}Cl quadrupole spin–lattice relaxation have been used[186,187] to study the reorientation of the CCl_3 group about its threefold axis in some chlorinated methylbenzenes (in p-chlorobenzotrichloride, $E_A = 45.6$ kJ mol^{-1}) and to study[188] the rotary lattice modes of the $PtCl_6$ octahedra in R_2PtCl_6 compounds.

The effect on the quadrupole spin–lattice relaxation time of rotating a rigid crystal about an axis making an angle $\cos^{-1}(1/3)^{\frac{1}{2}}$ with B_0 has been calculated by Ramakantha.[189] When the electric-field gradient is axially symmetric, it is found that T_1 for both direct and Raman processes alters. It is clearly important to take this effect into account in the interpretation of line narrowing by crystal rotation. Snyder and Hughes[190] give expressions for the orientation dependence of the allowed quadrupole transition $\Delta m = \pm 1$ and $\Delta m = \pm 2$, for a crystal in a large external magnetic field; the form of these expressions is independent of the relaxation process involved. Hughes and Reed[191] have compared this theory with experimental measurements for ^{23}Na in sodium nitrate at room temperature. Parker and Schmidt[192] have shown that when a hindered rotation does not change the spatial charge distribution, but merely permutes the locations of nuclear magnetic moments, the resulting spin–lattice relaxation of a nearly fixed nucleus is likely to be dipolar even if this nucleus has a quadrupole moment; this effect is demonstrated by T_1 measurements for ^{7}Li in $LiN_2H_5SO_4$ in which the NH_3^+ group reorientates about its threefold axis.

E. Spin–Lattice Relaxation due to Paramagnetic Impurities.—This subject has been under intensive investigation for a twenty-two year period. The basic features of the relaxation process have been well established (Vol. 1, p. 151); current theoretical and experimental studies are concerned with understanding the details. Punkkinen[193] has presented a numerical solution for the differential equation governing nuclear spin–lattice relaxation in the one-paramagnetic-centre approximation. His results suggest that assumption of a random distribution of paramagnetic impurities could explain the difference in the experimental ^{19}F spin-diffusion constant in CaF_2 and the value calculated using a uniform distribution model. Hauval et al.[194] have measured T_1^H in $La_2Mg_3(NO_3)_{12}.24H_2O$ doped with Nd in the range 1.5—40 K and in mag-

---

[185] N. E. Ainbinder, B. F. Amirkhanov, I. V. Izmestjev, A. N. Osipenko, and G. B. Soifer, Fiz. Tverd. Tela, 1971, **13**, 424.
[186] T. Kiichi, N. Nakamura, and H. Chihara, J. Magn. Resonance, 1972, **6**, 516.
[187] E. V. Izmest, Optika i Spektroskopiya, 1971, **30**, 1038.
[188] R. L. Armstrong and D. F. Cook, Canad. J. Phys., 1971, **49**, 2389.
[189] A. Ramakanth, J. Phys. (C), 1971, **4**, 223.
[190] R. E. Snyder and D. G. Hughes, J. Phys. (C), 1971, **4**, 227.
[191] D. G. Hughes and K. Reed, J. Phys. (C), 1972, **4**, 2945.
[192] R. S. Parker and V. H. Schmidt, J. Magn. Resonance, 1972, **6**, 507.
[193] M. Punkkinen, Phys. Kondens. Mater., 1971, **13**, 79.
[194] G. M. van Den Heuvel, T. J. B. Swanenburg, and N. J. Paulis, Physica, 1971, **56**, 365.

netic fields of 0.7—20 T: at temperatures above 10 K the relaxation is in accord with Bloembergen's theory, but at lower temperatures interactions between the paramagnetic ions become important. Heuval et al. invoke the concept of a dipole–dipole reservoir of magnetic ions to explain these results and Buishvili et al.[195] give a theoretical treatment of this problem, comparing their theory with the results of Heuval et al. Buishvili et al. also given an elementary theory of n.m.r. saturation, taking into account saturation transfer from the weaker Zeeman reservoir to the dipole–dipole reservoir of magnetic ions.

Until now, it has not been experimentally possible to study the relaxation of the nuclei near a paramagnetic impurity; King et al.[196] have described a single-crystal technique which enables impurity-shifted resonances to be detected; using the saturation method they were able to study the spin–lattice relaxation of these resonances in $Y(C_2H_5SO_4)_3,9H_2O$ doped with Yb^{3+} at liquid helium temperatures. An unexpected result is that nearly all of the near proton lines are in strong thermal contact with the bulk protons; spin-diffusion processes are dominant for all but the very nearest protons, suggesting that the so-called diffusion barrier between bulk and near protons has little significance in this crystal. Clearly, this technique should be invaluable for studying relaxation in paramagnetic systems.

Other reports include a study of T_1^F by Valentine and Nolle[197] in γ-irradiated LiF at 300, 77, and 4.2—1.7 K: the relaxation in the range 4.2—1.7 K is characterized by a spin-diffusion regime intermediate to the slow- and rapid-diffusion cases; at 300 K, the relaxation is controlled by F-centre spin–lattice relaxation. Mak and Mahendroo[198] have measured $T_{1\rho}^F$ in $BaF_2:Eu^{2+}$ between 4.2 and 300 K; direct relaxation (i.e. not controlled by spin diffusion) is observed from 45 to 55 K.

Kolb and Wolf[199] have reported use of T_1^H measurements as a function of frequency (0.4—44 MHz) to investigate the migration of metastable triplet excitons in naphthalene and anthracene crystals at 300 and 77 K. Assuming scalar proton–electron interaction and statistical incoherent hopping motion of the excitons, they obtain the following exciton–proton correlation times at the two temperatures: 5.5×10^{-12} and 4.1×10^{-12} s in anthracene and 2.1×10^{-11} and 1×10^{-11} s in naphthalene.

F. **Investigations of Molecular Reorientation.**—*Molecular Solids.* Hexagonal close-packed D_2, although structurally a simple solid, represents a complicated system from the point of view of spin–lattice relaxation; there are two nuclear spin systems, ortho molecules ($J = 0, I = 2$) and para molecules ($J = 1, I = 1$), with corresponding Zeeman levels, so the systems are strongly coupled together as well as being independently coupled to the lattice. Wong et al.[32]

---

[195] L. L. Buishvili, G. R. Khutsishvili, and M. D. Zviadaze, *Phys. Status Solidi (B)*, 1971, **48**, 851.
[196] A. R. King, J. P. Wolfe, and R. L. Ballard, *Phys. Rev. Letters*, 1972, **28**, 1099.
[197] K. M. Valentine and A. W. Nolle, *Phys. Rev. (B)*, 1971, **4**, 15.
[198] C. Mak and P. P. Mahendroo, *Phys. Letters (A)*, 1972, **39**, 195.
[199] H. Kolb and H. C. Wolf, *Z. Naturforsch.*, 1972, **27a**, 51.

and Weinhaus et al.[200] have reported independent experimental investigations of these relaxation mechanisms, and the subject has been discussed theoretically by Harris.[34] Studies of two tetrahedral molecules have been reported. Niemelä et al.[201] have made T_1^H measurements for polycrystalline SnH_4 between 61 and 123 K. There is a phase transition at 99.7 K, presumably a plastic–crystal transition; at temperatures below this transition, relaxation is governed by reorientational motion (the temperature dependence of τ_2 is distinctly non-Arrhenius) and above this transition by translational motion ($E_A = 12.6$ kJ mol^{-1}). For solid white phosphorus (P_4), T_1^P measurements by Boden and Folland[36] showed that, below the plastic phase transition at 196.3 K, relaxation is due to anisotropic chemical shielding and intramolecular dipole–dipole interactions modulated by molecular reorientation [$\tau_2 = 7.5 \times 10^{-14}$ s exp(15.7 kJ mol^{-1}/RT)], whereas in the plastic–crystalline phase there are significant contributions from the spin–rotation interaction also [$\tau_2 = 4.7 \times 10^{-13}$ s exp(5.7 kJ mol^{-1}/RT)]; τ_2 changes discontinuously at the transition temperature, demonstrating that the transition is of the isothermal first-order type.

Several groups of workers[202] have previously used T_1^H measurements to characterize the reorientation process in benzene crystal. These measurements have been shown to be consistent with a model which assumes that the molecules undergo fast uncorrelated jumps about the sixfold axis between geometrically equivalent orientations. For such a model, the reorientational correlation function decays exponentially in time with time constant τ_2 corresponding to the mean interval between these jumps. For benzene, $\tau_2 = 9.1 \times 10^{-15}$ s exp(17.6 kJ mol^{-1}/RT). T_1^H Measurements by Kadaba et al.[203] have shown that the motion of the benzene molecule in the Br_2,C_6H_6 solid compound is similar to that in crystalline benzene: $\tau_2 = 4.3 \times 10^{-15}$ s exp(20.9 kJ mol^{-1}/RT). Interestingly, the bromine n.q.r. frequently in the solid compound is similar to that for solid Br_2, demonstrating that the complex is not a 'charge transfer' complex but is bound by van der Waals and electrostatic multipole interactions. Albert, Gutowsky, and Ripmeester[204] report T_1^F measurements in polycrystalline hexafluorobenzene; the log T_1^F versus T^{-1} plot exhibits two distinct minima, which are attributed to in-plane reorientation of the C_6F_6 molecules at two crystallographically different sites. Boden (ref. 135, p. 115) has discussed similar measurements in terms of the yet unpublished crystal structure. Hexafluorobenzene crystallizes with monoclinic symmetry, space group $P2_1/n$ with $Z = 6$; two of these molecules are at sites of symmetry (Wyckoff position c) and four at general positions (Wyckoff position e). For C_6F_6 molecules at type c sites, $\tau_2 = 2.35 \times 10^{-15}$ s exp(31.3

[200] F. Weinhaus, H. Meyer, and S. M. Meyers, *Phys. Letters (A)*, 1971, **37**, 245.
[201] L. Niemelä, M. Hirvonen, and U. Lähteenmäki, *Phys. Letters (A)*, 1972, **39**, 323.
[202] E. R. Andrew and R. G. Eades, *Proc. Roy. Soc.*, 1953, **A218**, 537; J. E. Anderson, *J. Chem. Phys.*, 1965, **43**, 3575; U. Haeberlen and G. Maier, *Z. Naturforsch*, 1967, **22a**, 1236.
[203] P. K. Kadaba, D. E. O'Reilly, E. M. Peterson, and C. E. Scheie, *J. Chem. Phys.*, 1971, **55**, 5289.
[204] S. Albert, H. S. Gutowsky, and J. A. Ripmeester, *J. Chem. Phys.*, 1972, **56**, 2844.

kJ mol$^{-1}/RT$), whereas at type e sites, $\tau_2 = 1.1 \times 10^{-13}$ s exp(12.6 kJ mol$^{-1}/RT$). An interesting situation is produced by asymmetric substitution of the benzene ring: the in-plane reorientation will now be a disordering process and the simple model used above is no longer relevant. Brot and Darmon[205] have discussed this process for the hexasubstituted chloromethylbenzenes, which reorientate relatively easily because the methyl group and chlorine atom exclude very similar volumes. Chezeau et al.[206] have reported the use of T_1^H and $T_{1\rho}^H$ measurements to study the motional disorder in 1,3,5-trichlorotrimethylbenzene, dichlorodurene, and tetrachloro-p-xylene; it was necessary to invoke a Cole–Cole distribution of correlation times to account for the relaxation due to the in-plane reorientation. Interestingly, the polar chloromethylbenzenes, but not the non-polar ones, exhibit a progressive orientational ordering as the temperature is lowered.

(2)

N.m.r. techniques have been used by Fyfe et al.[207] to search for motion in the symmetrical polycyclics coronene, perylene, and triphenylene. Coronene (2) was the only solid in which motion was detected; the T_1^H minimum was observed and $E_A = 24.2$ kJ mol^{-1}. Brookeman and Rushworth[208] have measured T_1^H in cyclohepta-1,3,5-triene (C_7H_8) from 50 K to its melting point, 198 K. There is considerable molecular motion below the λ-like transition at 154 K: in the region 100—130 K, $E_A = 25.5$ kJ mol^{-1}; below 95 K, $E_A \approx 1.0$ kJ mol^{-1}, and T_1^H in this region is probably governed by lattice vibrations and torsional oscillations. Cycloheptatriene is a non-polar partially saturated seven-membered carbon ring, and it is considered that the motion below the λ-transition is a complicated co-operative process involving ring inversion. Above the transition, quasi-isotropic reorientation occurs. Surprisingly, $T_{1\rho}$ and T_{1D} measurements by Lauer et al.[209a] for single crystals (the anisotropy of the relaxation rates was <10%) of biphenyl and naphthalene indicate considerable freedom for molecular motion; the relaxation

[205] C. Brot and I. Darmon, J. Chem. Phys., 1970, 53, 2271.
[206] J. M. Chezeau, J. H. Strange, and C. Brot, J. Chem. Phys., 1972, 56, 1380.
[207] C. A. Fyfe, B. A. Dunell, and J. A. Ripmeester, Canad. J. Chem., 1971, 49, 3332.
[208] J. R. Brookman and F. A. Rushworth, J. Phys. (C), 1971, 4, 1903.
[209] (a) C. Lauer, D. Stehlik, and K. H. Hausser, J. Magn. Resonance, 1972, 6, 524; (b) K. van Putte and J. van Den Enden, J. Phys. (C), 1971, 4, L161; (c) J. L. Page and F. A. Rushworth, J. Phys. (C), 1969, 2, 415; (d) D. W. Aksnes, Acta Chem. Scand., 1971, 25, 631.

measurements are (with respect to their frequency and temperature variation) practically identical for these two substances, no doubt reflecting their similar crystal rather than molecular structures. Reorientation about an axis perpendicular to the molecular plane is suggested to be the most likely motion determining the relaxation rates; for both molecules $E_A \approx 100$ kJ mol^{-1} and $\tau_0 \approx 10^{-18}$ s. Fluorine substitution is shown to slow down the mobility of biphenyl, whereas the opposite is true for naphthalene crystals; these effects probably originate from changes in the crystal structures. The motion in these fluorinated crystals is clearly very complex and is the most likely explanation for the anomalous dipolar relaxation rate compared with T_1^{-1}.

van Putte and van Den Enden[209b] have separated the intra- and intermolecular contributions to $(T_1^H)^{-1}$ in ethylene by dilution with C_2D_4. At 60 MHz, $T_{1\,\text{min}}$(intra) = 2.0 s, as calculated by Page and Rushworth[209c] assuming the ethylene molecules undergo 180° flip reorientation rather than random reorientation about the C—C axes; but $T_{1\,\text{min}}$(inter) = 150 ± 10 ms is a factor of two smaller than that calculated by Page and Rushworth. T_1^ρ measured by Aksnes[209d] in solid PCl_3 and $POCl_3$ down to about 70 K from their melting points are Arrhenius, with respective activation energies of 6.3 and 36.7 kJ mol^{-1} and indicate $\omega_0 \tau_2 \gg 1$. It is argued that both molecules reorientate preferentially about the C_3 axes; in $POCl_3$ this rotation involves breaking weak intermolecular Cl···O bridges, accounting for the higher barrier.

Measurements by Weithase, Noack, and Schutz[210] of $T_{1\rho}^H$ in hexagonal ice single crystals as a function of temperature for various rotating magnetic field strengths reveal for the first time the expected $T_{1\rho}$ minimum. This has enabled the value of the reorientational correlation time to be accurately determined as $\tau_2 = 1.99 \times 10^{-17}$ s exp(58.2 kJ mol$^{-1}/RT$).

Internal Rotation in Molecular Solids. The reorientation of the methyl group about its threefold axis invariably dominates the proton spin–lattice relaxation process in many molecular crystals at low temperatures. van Putte has studied the role of methyl group reorientation in proton relaxation in solid ketones,[211] and symmetrical alkenes and alkynes.[212] In solid $CD_3COCD_3CH_3$, the observed non-exponential relaxation is shown to be in good agreement with the theory of Hilt and Hubbard.[213] In the protonated ketones, spin diffusion breaks down the effects of the cross-correlations and the deviation from exponential relaxation decreases with increasing chain length. Similar effects are observed for proton relaxation in the alkenes and alkynes. In the low-temperature phase of hexamethyldisilane, $Me_3SiSiMe_3$, Albert et al.[214] have shown that the methyl and Me_3Si groups reorientate independently, with

---

[210] M. Weithase, F. Noack, and J. van Schutz, *Z. Physik*, 1971, **246**, 91.
[211] K. van Putte, *J. Magn. Resonance*, 1971, **5**, 367.
[212] K. van Putte, *J.C.S. Faraday II*, 1972, **68**, 357.
[213] R. L. Hilt and P. S. Hubbard, *Phys. Rev. (A)*, 1964, **134**, 392.
[214] S. Albert, H. S. Gutowsky, and J. A. Ripmeester, *J. Chem. Phys.*, 1972, **56**, 1332.

respective activation energies of 6.5 and 31.4 kJ mol^{-1}, whereas in hexamethylethane, Me$_3$CCMe$_3$ the reorientations of the methyl and t-butyl groups are coupled with E_A = 13.0 kJ mol^{-1}. Similarly, as shown by Saffar et al.,[215] in trimethylacetonitrile, Me$_3$CCN, the rotation of the methyl and t-butyl groups are coupled, E_A = 16.3 kJ mol^{-1}. It is interesting that in methylchloroform as studied by McIntyre and Johnson,[216] the MeCCl$_3$ molecule rotates as a unit, with E_A = 18.0 kJ mol^{-1}, while in methyltrichlorosilane the methyl group reorientates independently with E_A = 5.4 kJ mol^{-1}. In trimethyltin fluoride,[217] in which the tin atoms are bridged by fluorines to form a polymer chain, for methyl reorientation $E_A \approx$ 2 kJ mol^{-1} and for reorientation of the trimethyltin group about its chain axis E_A = 73 kJ mol^{-1}. Temperature- and frequency-dependent T_1^H measurements by von Schütz and Wolf[218] demonstrate that the methyl group reorientation kinetics in methylnaphthalene crystals is strongly dependent on the group arrangement; in the β-position, E_A = 3.1 kJ mol^{-1} and is determined by intermolecular interactions, whereas in the case of the α-position and for adjacent methyl groups, E_A = 10 kJ mol^{-1} and arises largely from intramolecular interactions. In all the above studies, except where specified, the observed relaxation was exponential.

In polycrystalline amino-acids the C_3 reorientation of the NH$_3$$^+$ group determines the proton spin–lattice relaxation mechanism, cf. the role of the methyl group. McElroy et al.,[219] using T_1^H and $T_{1\rho}^H$ measurements, have determined the NH$_3$$^+$ reorientation kinetics in various amino-acids; their results are shown in Table 1; included in the Table are Ratovic's[220] data for α-aminoisobutyric acid in which E_A = 8.0 kJ mol^{-1} for methyl rotation.

Table 1 *Kinetic parameters for* —NH$_3$$^+$ *reorientation in some amino-acids*

	τ°/s	E_A (NH$_3$$^+$)/kJ mol^{-1}
l-Penicillamine	1.3 × 10^{-12}	20.0
Homocystine	1.3 × 10^{-14}	39.7
l-Tyrosine	3.3 × 10^{-13}	30.1
Glycine	1.2 × 10^{-14}	28.0
dl-Lysine, HCl	3.0 × 10^{-13}	16.7
3-Aminoisobutyric acid	—	38.4

Internal Rotation in Ionic Solids. Spin–lattice relaxation measurements have frequently been employed to study the H$_2$O motion in salt hydrates. T_1^H Measurements by El Saffar et al.[221] show that, in AlCl$_3$,6H$_2$O, the H$_2$O

[215] Z. M. El. Saffar, P. Schultz, and E. F. Meyer, *J. Chem. Phys.*, 1972, **56**, 1477.
[216] H. M. McIntyre and C. S. Johnson, *J. Chem. Phys.*, 1971, **55**, 345.
[217] S. E. Ulrich and B. A. Dunell, *J.C.S., Faraday II*, 1972, **68**, 680.
[218] J. U. von Schütz and H. C. Wolf, *Z. Naturforsch.*, 1972, **27a**, 42.
[219] R. G. C. McElroy, R. Y. Dong, M. M. Pintar, and W. F. Forbes, *J. Magn. Resonance*, 1971, **5**, 262.
[220] S. Ratkovic, *Phys. Letters (A)*, 1971, **37**, 41.
[221] Z. M. El. Saffar, W. Mulcahy, and G. Rochau, *J. Chem. Phys.*, 1971, **55**, 3307.

molecules undergo 180° flips with $E_A = 46.0$ kJ mol^{-1}. In HAuCl$_4$,H$_2$O the diaquated proton H$_5$O$_2$$^+$ exists in a non-planar *trans* configuration with the bridging proton in off-centre positions. O'Reilly *et al.*[222] have used T_1^H and $T_{1\rho}^H$ measurements to determine the activation energy for interbond jumps of this bridging proton; it is 24 kJ mol^{-1} in solid phase II (218—290 K) and 49 kJ mol^{-1} in solid phase III (below 218 K). Clearly the geometry of the H$_5$O$_2$$^+$ ions is different in the two phases.

The internal motion of the ammonium ion in simple ammonium salts has always attracted interest in n.m.r. This year, Shimomura *et al.*[223] have re-investigated NH$_4$$^+$ reorientation in the ammonium halides. In NH$_4$Cl, at temperatures below the λ-transition, the single T_1^H and $T_{1\rho}^H$ minima are interpreted in terms of simultaneous S_4 and C_3 reorientations, the kinetic data for which are given in Table 2.

Table 2 *Kinetic parameters for* NH$_4$$^+$ *reorientation in* NH$_4$Cl

Mode	τ°/s	E_A/kJ mol^{-1}
S_4	1.88×10^{-14}	17.1
C_3	2.33×10^{-14}	18.5

Two reorientation modes for the NH$_4$$^+$ ion in solid NH$_4$VO$_3$ are invoked by Peternelj *et al.*[224] to explain the occurrence of distinct minima in T_1^H at 120 and 195 K; these minima are attributed to, respectively, C_3 and C_2 reorientations (Table 3). Strange and Terenzi,[225] employing T_1^H and $T_{1\rho}^H$ measurements between 60 and 500 K, have determined the kinetic parameters (Table 4) for the reorientation (No mode is specified; in interpreting the data it is assumed that the T_1 minima originate from a single process) of the NH$_4$$^+$ ion in cubic salts of the type (NH$_4$)$_2$MX$_6$. The motions of the Me$_4$N$^+$ ion in the salts Me$_4$N$^+$X$^-$ [X = Cl or I] have been characterized (see Table 5) by Albert *et al.*[226] using T_1^H measurements. The data exhibit two minima for each compound, indicating that the methyl group reorientates independently. T_1^H Measurements have also been employed by Tegenfeldt and Ödberg[227] to investigate the reorientational motion of the MeNH$_3$$^+$ ion in the three solid phases of methylammonium chloride.

Table 3 *Kinetic parameters for* NH$_4$$^+$ *reorientation in* NH$_4$VO$_3$

Mode	τ°/s	E_A/kJ mol^{-1}
C_2	5×10^{-15}	21.3
C_3	8×10^{-15}	8.0

[222] D. E. O'Reilly, E. M. Peterson, C. E. Scheie, and J. M. Williams, *J. Chem. Phys.*, 1971, **55**, 5629.
[223] K. Shimomura, T. Kodama, and H. Negita, *J. Phys. Soc. Japan*, 1969, **27**, 255; 1971, **31**, 1291.
[224] J. Peternelj, M. I. Valic, and M. M. Pintar, *Physica*, 1971, **54**, 604.
[225] J. H. Strange, and M. Terenzi, *J. Phys. and Chem. Solids*, 1972, **33**, 923.
[226] S. Albert, H. S. Gutowsky, and J. A. Ripmeester, *J. Chem. Phys.*, 1972, **56**, 3672.
[227] J. Tegenfeldt and L. Ödberg, *J. Phys. and Chem. Solids*, 1972, **33**, 215.

Table 4 *Kinetic parameters for* NH_4^+ *reorientation in* $(NH_4)_2MX_6$

Compound	$\tau°/s$	$E_A/kJ\,mol^{-1}$
$(NH_4)_2SiF_6$	1.5×10^{-14}	9.2
$(NH_4)_2SnBr_6{}^a$	3.0×10^{-14}	6.0
$(NH_4)_2SnCl_6$	2.6×10^{-13}	5.2

^{a} Below λ-transition at 145 K

Table 5 *Kinetic parameters for* Me *and* Me_4N^+ *reorientation in* $Me_4N^+X^-$ *salts*

	Methyl group		Me_4N^+	
	$\tau°/10^{-13}\,s$	$E_A/kJ\,mol^{-1}$	$\tau°/10^{-13}\,s$	$E_A/kJ\,mol^{-1}$
Me_4Cl^a	1.4	28.4	1.0	54.5
Me_4Br	1.7	26.8	3.4	48.1
Me_4I	3.0	23.0	0.8	46.0

^{a} Below 418 K

Utton and Tsang[228] have reported an investigation of the ^1H and ^{19}F relaxation due to the internal rotations of the $SiF_6{}^{2-}$ and $Mg(H_2O)_6{}^{2+}$ ions in magnesium fluorosilicate hexahydrate. The laboratory frame spin–lattice relaxation is non-exponential owing to the strong interionic (H,F) dipolar interactions. For the rotation of the $SiF_6{}^{2-}$ ion, $\tau_2 = 10^{-14\pm1}$ s $\exp(31\pm4$ kJ mol$^{-1}/RT)$; the rotation of the $Mg(H_2O)_6{}^{2+}$ ion is relatively slower and was not characterized by the T_1 measurements.

Spin–Lattice Relaxation by Quantum Mechanical Rotation at Low Temperatures.
Spin–lattice relaxation due to internal rotation of methyl groups and ammonium ions at high temperatures as discussed above is successfully accounted for using a classical model for the reorientation. At low temperatures, where only the lowest librational states are occupied, the role of quantum mechanical tunnelling rotation must, however, be considered. For a symmetrical molecule or group in a finite crystal field the degeneracy with respect to the equivalent orientations of the molecule is generally removed by tunnelling, producing a splitting of the torsional states into a number of substates. Thermal transitions between these states produce a relaxation mechanism which has been observed[229] at low temperatures in solids containing methyl groups where the barrier to classical reorientation is <5 kJ mol^{-1}. The mechanism for these transitions determines the symmetry restrictions on the allowed transitions. Clough[230] and Wallach[231] (Vol. 1, p. 150) have given

---

[228] D. B. Utton and T. Tsang, *J. Chem. Phys.*, 1972, **56**, 116.
[229] P. S. Allen and A. Cowking, *J. Chem. Phys.*, 1968, **49**, 789; G. P. Jones, R. G. Eades, K. W. Terry, and J. P. Llewellyn, *J. Phys (C)*, 1968, **1**, 415; J. Haupt and W. Muller-Warmuth, *Z. Naturforsch.*, 1968, **23a**, 208; 1969, **24a**, 1066; P. S. Allen and C. J. Howard, *Mol. Phys.*, 1969, **16**, 311.
[230] S. Clough, *J. Phys. (C)*, 1971, **4**, 2180.
[231] D. Wallach, *J. Chem. Phys.*, 1971, **54**, 4044.

different theoretical treatments and each has spiritedly refuted the other's approach.[232] There is insufficient experimental data to resolve this controversy at present. Both authors agree, however, that the temperature dependence of T_1 at low temperatures is governed by thermal excitations between the ground and first-excited torsional states, not by excitation to the top of the barrier; the apparent activation energy for T_1 should therefore correspond to the separation between these levels. Haupt[233] has also reached this conclusion using a similar approach to that of Wallach. Experimental verification of this result is provided by the T_1^H measurements of Carrolan et al.[234] for 4-methyl-2,6-di-t-butylphenol; T_1 exhibits two minima, a deep one at 100 K and a relatively shallow one at 14 K. The former is assigned to methyl group rotation within the t-butyl groups; tunnelling plays no significant role at this temperature and the classical BPP theory accounts for the measurements ($E_A = 8.4$ kJ mol^{-1}). At 14 K, the relaxation is determined by the ring methyl protons ($E_A = 1.06$ kJ mol^{-1}); after correcting for the effects of cross-relaxation, the T_1 minimum value is only $\frac{1}{3}$ that of the high temperature value, which is consistent with Clough's theoretical model. The value obtained for E_A compares with the value 1.13 kJ mol^{-1} calculated from $3\hbar(V/2I)^{\frac{1}{2}}$. A brief survey of tunnelling effects in n.m.r. at low temperatures is given in the review by Allen.[180]

Other papers concerned with effects due to the existence of nuclear spin symmetry states of molecules at low temperatures in solids are as follows: studies of nuclear spin symmetry species conversion in solid methane and hexamethylbenzene using level crossing are reported, respectively, by Glattli et al.[235] and Jones and Bloom;[236] a note by Haupt[237] reports a fascinating new dynamic proton polarization effect in polycrystalline γ-picoline (which contains a methyl group) produced by a rapid change of temperature from 8 to 30 K; Watton et al.[238] have used n.m.r. lineshapes to study the spin-symmetry states of ammonium ions in ammonium salts at low temperatures.

G. **Investigations of Self-diffusion.**—*General Developments.* In calculating the relaxation times T_1, $T_{1\rho}$, and T_2 due to diffusion in atomic and molecular solids in the weak collision limit, the BPP theory[5] is used with correlation functions deduced by Torrey,[239] assuming the spins perform a random walk to nearest neighbour positions in the lattice. Torrey's calculation neglects the temporal correlation in successive nuclear jumps caused by the same point defect. Eisenstadt and Redfield[240] have discussed the importance of account-

---

[232] S. Clough and F. Poddy, *J. Chem. Phys.*, 1972, **56**, 1790.
[233] J. Haupt., *Z. Naturforsch.*, 1971, **26a**, 1578.
[234] J. L. Carrolan, S. Clough, N. D. McMillan, and B. Mulady, *J. Phys. (C)*, 1972, **5**, 631.
[235] H. Glattli, A. Sentz, and M. Eisenkremer, *Phys. Rev. Letters*, 1972, **28**, 871.
[236] E. P. Jones and M. Bloom, *Phys. Rev. Letters*, 1972, **28**, 1239.
[237] J. Haupt, *Phys. Letters (A)*, 1972, **38**, 389.
[238] A. Watton, A. R. Sharp, H. E. Petch, and M. M. Pintar, *Phys. Rev. (B)*, 1972, **5**, 4281.
[239] H. C. Torrey, *Phys. Rev.*, 1953, **92**, 962; 1954, **96**, 690; H. A. Resing and H. C. Torrey, *ibid.*, 1963, **131**, 1102.
[240] M. Eisenstadt and A. G. Redfield, *Phys. Rev.*, 1963, **132**, 635.

ing for these correlations. Wolf[241] has extended Torrey's theory to include the effects of correlation for a monovacancy mechanism in an isotopically pure f.c.c. crystal. The shape of the Fourier spectrum obtained is similar to that for Torrey's random walk model but the maximum occurs at $\omega\tau \approx 1.3$ instead of 1.7, i.e. $\tau_J = 0.78\ \tau$, where τ is the mean time interval between successive jumps in the Torrey theory.

An excellent account of the use of spin relaxation measurements to study diffusion of atoms and molecules in the ultraslow-motion or strong collision limit can be found in the review article by Ailion.[11] In 1968, Ailion and Ho[242] predicted that in the ultraslow-motion region $T_{1\rho}$, measured for a single crystal, would have an angular dependence which depends on the diffusion mechanism. Considering only dipole interactions, Samuelson and Ailion[243] have recently extended these calculations to the two-spin system TlCl, showing that the angular dependence of $T_{1\rho}$ for ^{35}Cl is in agreement with values calculated assuming the dominant mechanism to be Cl vacancy diffusion ($E_A = 70.6$ kJ mol^{-1}), as is known by other experiments. This technique promises to be useful for the determination of the dominant diffusion mechanism in both ionic and molecular crystals.

Metals. Seeger, Wolf, and Mehrer[244] have compared the tracer and n.m.r. self-diffusion measurements in aluminium. This paper contains a useful yet brief review of spin relaxation methods as applied to the study of self-diffusion in solids. The conclusion drawn is that the monovacancy formation volume (and presumably also the self-diffusion activation volume for a monovacancy mechanism in aluminium) is smaller than one atomic volume; divacancies contribute to diffusion at high temperatures. This conclusion is consistent with the activation volume of 0.71 ± 0.13 in units of atomic volume determined by Engardt and Barnes[245] in the temperature range 673—723 K from pressure-dependent T_2 measurements.

Metal Hydrides. Many metals will dissolve hydrogen to form non-stoicheiometric hydrides. Nuclear spin relaxation methods have been employed by Weaver[246,247] to investigate hydrogen diffusion in scandium hydride and yttrium trihydride. In scandium hydride,[246] E_A is strongly concentration-dependent, with two processes becoming evident as the hydrogen-to-metal ratio (x) approaches 2. The variation of E_A with x is interpreted as an effect of non-rigid-band behaviour for the electron states. Appreciable occupation of octahedral sites, normally vacant in other hydrides, is indicated. In the hydrogen-deficient hexagonal phase of yttrium trihydride,[247] non-exponential

[241] D. Wolf, *Z. Naturforsch.*, 1971, **26a**, 1816.
[242] D. C. Ailion and P. Ho, *Phys. Rev.*, 1968, **167**, 662.
[243] G. L. Samuelson and D. C. Ailion, *Phys. Rev. (B)*, 1972, **5**, 2488.
[244] A. Seeger, D. Wolf, and H. Mehrer, *Phys. Status Solidi (B)*, 1971, **48**, 481.
[245] R. D. Engardt and R. G. Barnes, *Phys. Rev. (B)*, 1971, **3**, 2391.
[246] H. T. Weaver, *Phys. Rev. (B)*, 1972, **5**, 1663.
[247] H. T. Weaver, *J. Chem. Phys.*, 1972, **7**, 3193.

spin-lattice relaxation is observed and attributed to the formation of regions of YH_3 within the hydrogen-deficient structure.

Ionic Crystals. Hoodless, Strange, and Wylde[248] have used ^{23}Na T_1, T_2, and $T_{1\rho}$ measurements (385—900 K) to study the self-diffusion of Na^+ in pure and Cd^{2+}-doped single crystals of NaI. An expression for $T_{1\rho}$ due to dipolar coupling between unlike spins is given. A comparison of the sodium-ion jump frequencies obtained from relaxation and ionic conductivity measurements indicates that, at high temperatures, diffusion occurs predominantly *via* cation single vacancies whereas, at lower temperatures, impurity-vacancy complexes contribute to the relaxation, indicating that the free and complexed vacancies have similar diffusion rates. Riggin, Knispel, and Pintar[249] have employed $T_{1\rho}^H$ measurements to study self-diffusion of the NH_4^+ ion in the high-temperature (398—443 K) phase of NH_4NO_3, which they liken to the plastic phase of a molecular solid; interestingly, at the melting point $D = 2 \times 10^{-12}$ m^2 s^{-1}, which is typical of the corresponding values obtained for plastic crystals.

Molecular Crystals. For ionic crystals and metals the self-diffusion coefficients determined by spin relaxation measurements usually agree with values determined by other methods. In contrast, for organic plastic crystals, there is in many cases considerable disagreement between both absolute values and apparent activation energies of the self-diffusion coefficients derived from relaxation measurements and those from the direct radiotracer technique (Vol. 1, p. 157). The radiotracer results appear to be self-consistent: for all solids investigated the values obtained for D_o and E_D $[D = D_o \exp(-E_D/RT)]$ are consistent with a monovacancy mechanism. For solids with high entropies of fusion (S_f), *e.g.* hexamethylethane, there is good agreement between the tracer and n.m.r. data but there is a gradual divergence between the two sets of data in progressing to lower entropies of fusion. The close agreement between the divergence of the two sets of data and the increasing disorder of the solids suggests that the spin relaxation measurements are sensitive to some process other than simple monovacancy diffusion.

A considerable number of papers have been published this year, but they have not resolved this discrepancy. Baughman and Turnbull[250] and Albert *et al.*[214] have reported, respectively, a linewidth and a $T_{1\rho}^H$ study of diffusion in hexamethylethane; their results agree with previous n.m.r. and tracer measurements. Boden *et al.*[251] have critically evaluated 'self-diffusion' data obtained by T_1, $T_{1\rho}$, and T_2 measurements in a carefully purified sample of the intermediate S_f compound perfluorocyclohexane; the τ valves exhibit an Arrhenius temperature dependence ($E_A = 59.2$ kJ mol^{-1}) in contradiction with previous linewidth measurements by Bladon *et al.* (Vol. 1, p. 157) which

[248] I. M. Hoodless, J. H. Strange, and L. E. Wylde, *J. Phys. (C)*, 1971, **4**, 2742.
[249] M. T. Riggin, R. R. Knispel, and M. M. Pintar, *J. Chem. Phys.*, 1972, **56**, 2911.
[250] R. H. Baughman and D. Turnbull, *J. Phys. and Chem. Solids*, 1972, **33**, 121.
[251] J. Cohen and P. P. Davis, *Mol. Phys.*, 1972, **23**, 819.

show a transition in the apparent activation energy. The low S_f compound pivalic acid has been investigated by Jackson and Strange[252] by $T_{1\rho}^H$ and T^H measurements: typically E_A (63 kJ mol^{-1}) is significantly smaller than the tracer value (91.2),[253] but, interestingly, the diffusion coefficients agree at the transition point. Bladon et al.[254] have shown that, for the non-typical plastic crystals succinonitrile and tetramethylsuccinonitrile, E_A values from linewidth and creep measurements agree. Folland and Strange,[255] studying the pressure dependence of $T_{1\rho}^H$, have determined the activation volume associated with translational relaxation in hexamethylethane (S_f = 20.1 J K^{-1} mol^{-1}), norbornylene (S_f = 10.2,) and cyclohexane (S_f = 9.2); they are, respectively, 1.2, 0.9, and 0.7 in units of molecular volume. These authors suggest that the dominant mechanism is monovacancy diffusion; an anomalous decrease of the activation volume with pressure in the case of norbornylene and cyclohexane is tentatively attributed to the presence of vacancies of a more complex nature in the low S_f plastic crystals. Tanner,[256] using the pulsed-gradient spin-echo method and observing the Hahn stimulated echo following a 90°–90°–90° rf pulse sequence, has determined the absolute diffusion coefficient in cyclohexane just below its melting point: significantly, the value obtained, 2×10^{-13} m^2 s^{-1}, is in agreement with the corresponding value obtained by Roeder and Douglass[257] using $T_{1\rho}$ measurements, but much smaller than the value, 7×10^{-11} m^2 s^{-1}, obtained by extrapolation of the radiotracer measurements of Hood and Sherwood.[258]

In conclusion, so far no clear-cut experimental evidence has been presented which can account for the apparent discrepancy between the n.m.r. and radiotracer self-diffusion data in the low entropy of fusion organic plastic crystals. Clearly, the resolution of this discrepancy could further our understanding of the disorder in this state of matter.

H. **Studies of Molecules Bound to Solid Surfaces.**—Measurements of spin-lattice relaxation rates in adsorbed molecules as a function of frequency, temperature, and coverage (θ) can often yield information about the nature of the adsorption sites in addition to characterizing the surface diffusion process. Current interest is in adsorbates on silica and silicate adsorbents. Winkler et al.[259] have made a detailed analysis of T_1^H of benzene adsorbed on to a commercial silica gel; T_1^{-1} is determined by inter- and intra-molecular proton–proton interactions, and intermolecular interactions of the benzene protons with paramagnetic impurities and with surface hydroxy-groups in about equal importance. The temperature dependence of T_1^H for a coverage of

[252] R. L. Jackson and J. H. Strange, Mol. Phys., 1971, **22**, 313.
[253] H. M. Hawthorne and J. N. Sherwood, Trans. Faraday Soc., 1970, **66**, 1783.
[254] P. Bladon, N. C. Lockhart, and J. N. Sherwood, Mol. Phys., 1971, **22**, 365.
[255] R. Folland and J. H. Strange, J. Phys. (C), 1972, **5**, L50.
[256] J. E. Tanner, J. Chem. Phys., 1972, **56**, 3850.
[257] S. B. W. Roeder and D. C. Douglass, J. Chem. Phys., 1970, **52**, 5525.
[258] G. M. Hood and J. N. Sherwood, Mol. Crystals, 1966, **1**, 97.
[259] H. Winkler, M. Nagel, D. Michel, and H. Pfeifer, Z. phys. Chem. (Leipzig), 1971, **248**, 17.

two statistical monolayers, i.e. $\theta = 2$, and the dependence of T_1^H on θ, is explained by the existence of three types of adsorption site or region with different relaxation rates and rapid exchange of benzene molecules between them. On each high-energy site (region 1) only one benzene molecule can be adsorbed. Region 2 consists of benzene molecules adsorbed at surface OH groups while region 3 is those molecules adsorbed on top of regions 1 and 2. The mean lifetime of benzene molecules at surface hydroxy-groups is found to correspond to the mean jump interval. It is not possible to calculate reliable surface diffusion coefficients from these T_1 measurements. Boddenburg et al.[260] have used the field-gradient spin-echo technique to measure surface diffusion coefficients for benzene absorbed on silica; the diffusion coefficient shows a maximum near $\theta = 1$, in agreement with previous adsorption rate measurements; it cannot be explained why the apparent activation energy for diffusion is larger than for T_2. Freude[261] has also made use of the spin-echo technique to study surface diffusion in the systems n-hexane or diethyl ether on silica gel. At 298 K, the diffusion coefficient for n-hexane is approximately constant at 2×10^{-9} m^2 s^{-1} for θ between 0.03 and 2, whereas for ether it decreases with decreasing surface coverage. T_1^H Measurements show that n-hexane molecules are localized at adsorption centres with lifetime $\sim 10^{-8}$ s only up to $\theta = 0.01$. In contrast, for ether this type of behaviour is observed up to $\theta = 0.5$ because ether molecules are localized at hydroxy-groups more readily.

The proton motion of the hydroxy-group bound to silicon in a decationated Linde Y zeolite has been studied by Mestdagh et al.[262] The Arrhenius temperature dependence of T_1^H and T_2^H observed above 573 K is interpreted in terms of isotropic diffusion of the proton with respect to a paramagnetic continuum of Fe^{3+} impurities: $\tau_J = 3 \times 10^{-11}$ s exp(42 kJ mol^{-1}/RT); $\langle r^2 \rangle^{\frac{1}{2}} \approx 0.75$ nm, which is a factor of two larger than the separation of neighbouring hydroxy-groups. The importance of considering scalar spin–spin interactions between adsorbate protons and paramagnetic impurities in the adsorbent is pointed out by Vucelic et al.[263] The existence of such an interaction for CH$_4$ adsorbed on to a synthetic zeolite Linde 5A containing approximately 1 p.p.m. by weight Fe^{3+} is shown by the ratio $T_1^H/T_2^H \approx 60$. This value is considerably larger than for analogous interactions in liquid solutions and is probably due to two factors: first, the separation between the proton and paramagnetic ion is decreased compared with the liquid state; second, the diffusional correlation time is longer than in the liquid while the rotational correlation time is practically unchanged, allowing an increased time for scalar interaction.

Headley et al.[264] have reported a preliminary investigation into the use of ^{7}Li T_1 measurements to probe the adsorption sites for ions on porous solid

[260] B. Boddenburg, R. Haul, and G. Oppermann, J. Colloid Interface Sci., 1972, **38**, 210.
[261] D. Freude, Z. Phys. Chem. (Leipzig), 1971, **247**, 209.
[262] M. M. Mestdagh, W. E. Stone, and J. J. Fripiat, J. Phys. Chem., 1972, **76**, 1220.
[263] D. Vucelic, M. Susic, I. Zupancic, and M. Hunter, J. Chem. Phys., 1971, **55**, 4152.
[264] L. C. Headley, W. E. Wallace, P. Waldstein, J. Magn. Resonance, 1971, **5**, 168.

materials. In aqueous lithium chloride solution in contact with porous porcelain, for a constant salt concentration, T_1^{-1} is proportional to the specific surface area. This result is explained in terms of chemical exchange between the bulk phase ions and ions in electrical double layers near the solid surface. The application of halide ion relaxation measurements to study the nature of of ion adsorption sites has possibilities. T_2^H Measurements by Belfort[265] on porous powdered glass equilibriated with aqueous sodium chloride solution provide evidence of two environmental states of water; about 80% of the adsorbed water is motionally restricted, while the remaining 20% is most probably bulk water.

I. **Ferroelectric Materials.**—It has recently been shown (Vol. 1, p. 159) that the microscopic critical dynamics which drive the ferroelectric transition can be studied through an anomalous peak in the dipole or quadrupole relaxation rate at the transition temperature T_c. The occurrence of such peaks is now reported for other ferroelectric materials. In the hydrogen-bonded ferroelectric $NaH_3(SeO_3)_2$, Kuroda et al.[266] observed a peak in $(T_1^H)^{-1}$ for a powdered sample at T_c, but T_1^H measurements[267] in a single crystal show no influence of the ferroelectric transition on T_1^H when B_o is parallel to the a axis but there is a contribution with B_o along the b axis. In CsH_2AsO_4, T_1^{-1} for ^{133}Cs exhibits a logarithmic singularity on approaching T_c from above and drops very fast with decreasing temperature below T_c.[268] In a $K_4Fe(CN)_6,3H_2O$ single crystal, T_1^H has a discontinuity at the Curie point, but there is no evidence for a contribution from the ferroelectric model.[269] Bonera et al.[270] have employed ^{23}Na n.m.r. measurements to study the phase transitions in sodium tungsten bronzes, Na_xWO_3; these are non-stoicheiometric compounds which for $0.4 < x < 1.0$ are metallic and have perovskite structure. The temperature dependence of the ^{23}Na quadrupole coupling constant shows that a phase transition from the high-temperature ideal perovskite structure to a distorted structure occurs. The observation of an anomalous rise in the relaxation rate indicates that the phase transition is driven by microscopic critical dynamics. The transition temperatures T_c depend upon the sodium content and were found to be approximately 390 K for $x = 0.517$, 400 K for $x = 0.72$, and 525 K for $x = 0.855$. The transition is a continuous one, but evidently it could not be decided whether it is of second order, with a departure from the Landau behaviour close to T_c, or of first order with a transition temperature close to the critical temperature.

[265] G. Belfort, Nature, Phys. Sci., 1972, **237**, 60.
[266] N. Kuroda, Y. Tabata, and A. Kawamori, J. Phys. Soc. Japan, 1971, **31**, 609.
[267] A. A. Silvidi and D. T. Workman, J. Chem. Phys., 1971, **55**, 4673.
[268] R. Blinc, M. Mali, J. Slak, and S. Zumer, J. Chem. Phys., 1972, **56**, 3566.
[269] A. Avagadro, E. Cavelius, D. Müller, and J. Peterson, Phys. Status Solidi (B), 1971, **48**, 247.
[270] G. Bonera, F. Borsa, M. L. Crippa, and A. Rigamonti, Phys. Rev. (B), 1971, **4**, 52.

4
Experimental Techniques

BY D. G. GILLIES

1 Introduction

As expected, the usage of Fourier-transform techniques has continued to expand during the review period. It has naturally dragged the chemist into the time domain. There has been a growing interest in the measurement of relaxation times in high-resolution spectra with a view to gaining more insight into the nature of molecular motion. In pursuit of this goal the chemist is forced to study the various contributions to the relaxation rate of a particular nucleus. The advantage to be gained by the study of all the nuclei in a system is illustrated by the work on chloroform–benzene where proton, deuterium, carbon, and chlorine relaxation data have all been used to gain a better understanding of the molecular motion.

Noack[1] has indicated the value of measuring relaxation times at different field values as functions of temperature in order to separate the various relaxation mechanisms. If we add pressure as a further variable, an area where Jonas has been active, we put a very demanding requirement on our experimental techniques. The chemist is also being forced to take a greater interest in the solid state, particularly as high-resolution ^{13}C spectra have now been obtained from solid samples. The growing amount of data on anisotropic chemical shifts from these and other methods and the realization that one cannot describe a chemical shift by one number will no doubt upset some of the semi-empirical correlations that have been made to explain isotropic shift data.

With these new developments the apparatus continues to become more complex. Computers are taking a more and more active part in n.m.r. systems and indeed are enabling new experiments to be performed with relative ease.

2 Probes

A. **Variable Temperature.**—Jensen et al.[2] have described modifications to a Varian HR60 probe which allow spectra with good resolution to be obtained

---
[1] F. Noack, in 'N.M.R., Basic Principles and Progress,' Springer-Verlag, Heidelberg, 1971, vol. 3, p. 83.
[2] F. R. Jensen, L. A. Smith, C. H. Bushweller, and B. H. Beck, Rev. Sci. Instr., 1972, 43, 894.

at temperatures as low as −170 °C with an estimated variation through the sample region of ±0.01 °C. The discussion and excellent constructional details will be of great help to anyone contemplating construction of a variable-temperature probe. Heat leaks both to the probe body and to the magnet are minimized by ensuring that the coolant gas is always in contact with the inner surface of a Dewar vessel from the heat exchanger until well clear of the probe. The quartz Dewar tube in the sample region was very straight and being of 8.93 mm internal diameter it allowed a relatively unrestricted flow of gas past the 5 mm external diameter sample tube. This free flow eliminated the gradients common to more restricted designs. The authors estimated that the temperature gradients were less than 0·004 °C cm^{-1}. A further bonus was that the lower working pressure eliminated previous problems of sample ejection. The receiver coil was mounted on the inside wall of the insert and finally fixed in position only after checking that the probe could be balanced.

Krynicki and Powles[3] measured the temperature dependence of T_1 in liquid solutions of HBr in DBr and HCl in DCl. The HCl measurement was a repeat of earlier work which was thought to be in error because of temperature gradients across the sample. Two separate glass Dewars of earlier design[4] were used for above and below room temperature. The former probe incorporated a long heating coil wound non-inductively on the stainless steel probe tube. It was the original low-temperature probe that was modified to reduce thermal gradients. The basic design consisted of a liquid nitrogen reservoir thermally connected to the sample by a thick-walled copper tube inside which a small heater provided the necessary temperature control. By filling the cylindrical cavity surrounding the top third of the copper tube with slowly frozen tetrachloroethane the temperature gradient was much reduced. This was at the expense of thermal efficiency although sample temperature control was always better than ±0.4 °C.

Gillen et al.[5] used a single-coil probe in their pulse studies on methyl iodide. The coil was contained in a glass Dewar. The sample was temperature-controlled by a stream of dry compressed air for studies above room temperature, and at lower temperatures by a stream of nitrogen generated by boiling off gas at a controlled rate from a 25 l liquid nitrogen Dewar.

B. High Pressure.—Jonas[6] has now described his apparatus in excellent detail. He points out that in making variable-temperature studies on relaxation times the density of the fluid changes. Measurements as a function of pressure allow the study of changes in relaxation times due solely to changes in density. The temperature range for the apparatus was −50 to +350 °C and the pressure 0—5 kbar (0—5 × 10^8 N m^{-2}). There was no tendency for

---

[3] K. Krynicki and J. G. Powles, *J. Magn. Resonance*, 1972, **6**, 539.
[4] D. W. Smith, Ph.D. Thesis, University of Kent at Canterbury, England, 1970.
[5] K. T. Gillen, M. Schwartz, and J. H. Noggle, *Mol. Phys.*, 1971, **20**, 899.
[6] J. Jonas, *Rev. Sci. Instr.*, 1972, **43**, 643.

oxygen to enter the apparatus and so invalidate relaxation data. The wide-gap (9.5 cm) Varian V-3800-1 magnet, which produced a field of 1.4092 T, allowed plenty of room for the high-pressure probe and gave acceptable resolution of 4 Hz at 56.4 MHz for an 8 mm diameter sample (non-spinning of course). The pressure-generating equipment was mounted together with the probe on a movable table so that the probe could be removed from the magnet without disconnecting the high-pressure tubes containing the transmitting fluid. Two probes are described; one for operation below 80 °C was made of beryllium–copper alloy and the other was of titanium alloy. Variable-tempature operation was provided by a thermostated liquid bath for the low-temperature probe and a heater coil for the high-temperature probe. The magnet was thermally insulated from the probe by circulating water at magnet temperature. A detailed description of the insert assembly is given. The apparatus has been successfully used to produce high-resolution Fourier-transform spectra at high pressure.[7]

Fiorito and Meister[8] have reported high-pressure studies on hydrogen-bonded liquids. Use of a stainless steel pressure vessel (2″ external diameter, 0.5″ internal diameter) allowed pressures of 0—3500 kg cm^{-2} to be attained, and by immersion in a liquid bath temperatures in the range -30 to $+70$ °C were available. The pressure was transmitted in a manner similar to that used by Jonas.

Vaughan et al.[9] have given a detailed description of their apparatus for measurements on solids at up to 100 kbar (10^{10} N m^{-2}) pressure. The probe fits between the pole caps of a magnet with a 4.45 cm gap. A 40 ton hydraulic press was used to squeeze the sample between the two faces of a Bridgman anvil. This type of measurement is clearly less interesting to the chemist than those described above.

C. High Field.—An apparatus for measuring ^{13}C relaxation times at 62 MHz has been described.[10] The high field of 5.8 T was provided by a Siemens superconducting magnet (SUMA 75/50/280H, maximum field 8.0 T). A special probe assembly 30 mm in diameter was constructed to provide for a cylindrical sample 7.5 mm in diameter and 15 mm in length over whose volume the resolution was 5 p.p.m. A single-coil configuration was adopted in order to maximize the B_1 field over the large sample. The axis of the sample tube was along the static field. The consequent sharp bends in the split receiver coil were held to have negligible effect on the homogeneity of the rf field. The sample coil was matched into a 50 Ω cable by using a simple capacitative divider (see Figure 1). The matching capacitor, C_2, was of fixed value. The variable tuning capacitor, C_1, was an integral part of the construction and consisted of a metal plunger which could be moved in and out of an

[7] D. J. Wilbur and J. Jonas, *J. Chem. Phys.*, 1971, **55**, 5840.
[8] R. B. Fiorito and R. Meister, *J. Chem. Phys.*, 1972, **56**, 4605.
[9] R. W. Vaughan, C. F. Lai, and D. D. Elleman, *Rev. Sci. Instr.*, 1971, **42**, 626.
[10] H. Jaeckle, U. Haeberlen, and D. Schweitzer, *J. Magn. Resonance*, 1971, **4**, 198.

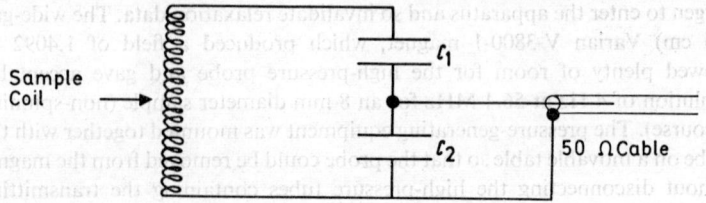

Figure 1 *Impedance matching of a sample coil to a 50 Ω cable*

outer cylinder. Separation (the dielectric) was provided by a thin teflon annular cylinder. This is standard practice for capacitors required to withstand very high voltages. Advanced versions of this probe operating at 14, 30, and 62 MHz were mentioned in a later paper although no further details were given.[11] These later models allow for variable-temperature operation by means of the nitrogen gas-stream method.

A spectrometer operating over the range 270—710 MHz has been described[12] suitable for operating in fields up to 12.5 T. The basic probe construction is similar to that described above. The single coil consisted of two turns of copper ribbon. Unlike the above example the sample (non-spinning) was mounted at 90° to the axis of the solenoid. The tuning and matching circuitry was as in Figure 1 but in this case both capacitors were variable and both of the cylindrical teflon-sleeved type. This allowed adjustment over the wide frequency range. Useful equations are given enabling the appropriate choice of capacitor values to be made for matching into 50 Ω. The coil system formed one arm of a tee network which used a commercial broadband hybrid tee.[13]

A probe used for measuring ^{31}P shielding anisotropies at high fields in solids has been described.[14] The design includes a goniometer to facilitate the orientation of crystals in the magnetic field.

3 Frequency Generation

In this section further examples of the techniques discussed in Volume 1 of this series[15] are taken from the current literature.

Sternlicht and Zuckerman[16] have described modifications to a Varian XL-100 spectrometer to allow Fourier-transform operation. The change from the CW mode, which involves sideband drive and centre-band detection, necessitated the synthesis of some different frequencies. The new transmitter

[11] H. W. Spiess, D. Schweitzer, U. Haeberlen, and K. H. Hausser, *J. Magn. Resonance*, 1971, **5**, 101.
[12] A. M. Gottlieb, V. C. Srivastava, and P. Heller, *Rev. Sci. Instr.*, 1972, **43**, 676.
[13] D. G. Gillies, in 'Nuclear Magnetic Resonance,' ed. R. K. Harris (Specialist Periodical Reports), The Chemical Society, London, 1972, vol. 1, p. 177.
[14] M. G. Gibby, A. Pines, W. K. Rhim, and J. S. Waugh, *J. Chem. Phys.*, 1972, **56**, 991.
[15] Ref. 13, p. 171.
[16] H. Sternlicht and D. M. Zuckermann, *Rev. Sci. Instr.*, 1972, **43**, 525.

frequency was generated by a General Radio (GR) frequency synthesizer whose maximum output frequency was 70 MHz. The synthesizer crystal was phase-locked to the spectrometer master frequency of 15.40096 MHz. Details of this circuit were not given but are available. This circuit would need only slight modification to allow the phase-locking of any synthesizer to the Varian system. The GR frequency is used directly for nuclei other than hydrogen and fluorine. The latter two required the use of a (Relcom) frequency doubler. The synthesized transmitter frequency v_T was split two ways in a PM12-2 hybrid coupler (Electronic Navigation Industries). One path led to the rf gate and power amplifier, the other to a mixer (Hewlett Packard) referenced at the local oscillator frequency $(v_T + 10.70)$ MHz supplied from the Varian console. The output difference frequency was filtered and fed as reference to the second mixer in the receiver.

Redfield and Gupta[17] have described their pulsed Fourier-transform spectrometer. It was specifically designed to observe proton signals in biological systems using a fluorine lock signal. A detailed description is given of the techniques used in frequency synthesis. Extensive use was made of crystal-controlled oscillators which, however, were not mutually phase-locked. Fine adjustment of the proton and fluorine output frequencies was by means of two voltage-controlled oscillators (Wavetek). The authors point out that although their system is not of the highest stability it is more than adequate for protein work. With a view to minimizing the presence of radiation at the proton frequency during the off periods a gated mixer transmitter, followed by a frequency doubler, was used (see Figure 2). This ensures that there is no component present at 100 MHz. The 39.3 and 10.7 MHz signals are supplied from the main synthesizer and generated from 100 MHz and 10.7 MHz crystals $(39.3 = 100/2 - 10.7)$.

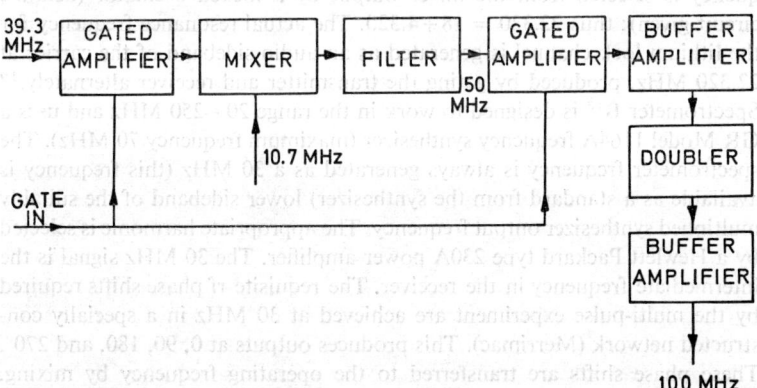

Figure 2 *Simplified block diagram of a low-power mixer/doubler for 100/94 MHz. Gated amplifiers are MC 1545 (Motorola) and the mixer is Type M6A by Relcom* (Reproduced by permission from *Adv. Magn. Resonance*, 1971, **5**, 81)

[17] A. G. Redfield and R. K. Gupta, *Adv. Magn. Resonance*, 1971, **5**, 81.

Ellett et al.[18] have now described at length two spectrometers that have been used in their multi-pulse studies on solids. The techniques of frequency synthesis are of course of general interest. Spectrometer A was solely for observation of protons or fluorine at 54 MHz with locking to a ^{7}Li resonance at 22.308 and 20.987 MHz respectively. The design incorporated one crystal oscillator at 18 MHz. The 54 MHz signal was generated by frequency tripling (detailed circuit given). A simplified block diagram illustrating the method of synthesis of the lithium frequencies is shown in Figure 3. The frequency

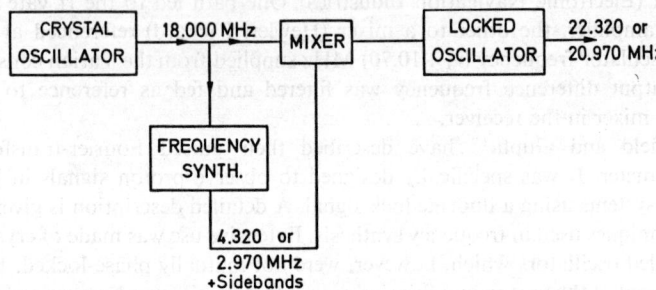

Figure 3 *Simplified block diagram showing method of frequency synthesis* (Reproduced by permission from *Adv. Magn. Resonance*, 1971, **5**, 117)

synthesizer consists of standard TTL integrated logic circuits. Division by factors of two is done by flip-flops and division by five involves SN7490 decade counters. The divided frequencies are added in SN7400 NAND gates. The desired frequency, *e.g.* 4.320 = (18/5 + 18/25), is present in the output together with sidebands separated by 18/25 MHz. The desired carrier frequency is selected from the mixer output by a locked oscillator (detailed circuit given): thus 22.320 = 18 + 4.320. The actual resonance frequency for the lithium lock channel is generated as an audio sideband of the carrier at 22.320 MHz, produced by gating the transmitter and receiver alternately.[19] Spectrometer B[18] is designed to work in the range 20—250 MHz and uses a GR Model 1164A frequency synthesizer (maximum frequency 70 MHz). The spectrometer frequency is always generated as a 30 MHz (this frequency is available as a standard from the synthesizer) lower sideband of the suitably multiplied synthesizer output frequency. The appropriate harmonic is selected by a Hewlett Packard type 230A power amplifier. The 30 MHz signal is the intermediate frequency in the receiver. The requisite rf phase shifts required by the multi-pulse experiment are achieved at 30 MHz in a specially constructed network (Merrimac). This produces outputs at 0, 90, 180, and 270°. These phase shifts are transferred to the operating frequency by mixing. This is an excellent broadband approach to phase shifting. The final mixing is of the single-sideband type. Simple mixing produces upper and lower side-

[18] J. D. Ellett, M. G. Gibby, U. Haeberlen, L. M. Huber, M. Mehring, A. Pines, and J. S. Waugh, *Adv. Magn. Resonance*, 1971, **5**, 117.
[19] Ref. 13, pp. 183 and 184.

bands and although of course only one is effective in the n.m.r. experiment, the presence of the unnecessary component would limit the power capabilities of the following broadband power amplifier. The use of hybrid tees, power dividers, wideband mixers, general hybrid networks, and commercial amplifier modules is amply illustrated in the above spectrometers and the increasing availability of such components has enabled the development of elegant n.m.r. techniques.

4 Pulse-sequence Generation

During the review period papers have appeared with examples of pulse-programming techniques varying from the simplest to the most sophisticated.

Kitchlew and Rao[20] used the simplest system consisting basically of a series of units from the Tetronix 160 series with data acquisition on a storage oscilloscope, the latter being triggered once in each cycle to initiate data acquisition. Ramadan and Tward[21] have described an analogue system consisting of various modules which may be cascaded together to produce any pulse sequence. Digital techniques allow all events to be accurately timed with respect to one master clock (see for example ref. 22). As the experiments become more sophisticated, so the programmer is required to perform functions more complex than merely switching the rf energy and initiating data acquisition. For example, pulses may be required for switching a second rf in double-resonance experiments, clocking data into a digital storage device, producing field inhomogeneity pulses, and for blanking the receiver during rf pulses.

Redfield and Gupta[23] generated their pulse sequence using a shift register and switched diode matrix, derived from a suggestion of Jeener.

Ellet et al.[24] have described two pulse programmers. In the first the sequence is determined by hardwired logic circuits mounted on plug-in cards.[25] The second is much more sophisticated and is interfaced to a PDP-12 computer. No hardware modifications are required to change the pulse sequence, merely software. One can try a new pulse sequence merely by writing a software programme. The interface can utilize manually set analogue delays and multiples thereof by digital counting. Alternatively, accurate digital delays may be selected. These are provided by a 1 MHz clock and six associated decade dividers providing timing intervals from 1 μs to 1 s. These pulses may then be counted in a manually or computer preloaded counter to provide the desired delay.

Cushley et al.[26] have described the interfacing of an IBM 1800 computer to a Bruker HFX-3 spectrometer and proved the importance of synchronizing

---

[20] A. Kitchlew and B. D. N. Rao, *J. Chem. Phys.*, 1972, **56**, 649.
[21] B. Ramadan and E. Tward, *Rev. Sci. Instr.*, 1971, **43**, 1618.
[22] R. L. Vold and S. O. Chan, *J. Chem. Phys.*, 1972, **56**, 28.
[23] Ref. 17, p. 95.
[24] Ref. 18, p. 122.
[25] See also ref. 16.
[26] R. J. Cushley, D. R. Anderson, and S. R. Lipsky, *Analyt. Chem.*, 1971, **43**, 1281.

the data acquisition to the pulse sequence. The timing was computer-controlled.

Two groups have reported the advantages arising from the use of frequency synthesizers to provide the master frequency for the digital timing circuitry.[27,28]

5 Continuous-wave Spectroscopy

A. Double-resonance Techniques.

Moritz[29] has reported an ingenious method for the measurement of accurate line positions. A second servo loop was used in a field–frequency lock system. This consisted of a voltage-controlled audio oscillator which was used to excite a resonance of interest. The signal was phase-detected in dispersion mode and fed back as an error signal to the voltage-controlled oscillator. The frequency was measured as an average over 10^5 periods. The reproducibility was found to be ± 2 mHz.

Jenkins and Phillips[30] have used a scheme which perhaps could be considered as the heteronuclear analogue of Moritz's method. Fluorine and proton spectra may be recorded simultaneously; INDOR experiments are facilitated. Field–frequency lock is set up in the usual way for one nucleus, say on TMS. A second servo loop is formed by bringing a fluorine sideband into resonance by adjusting an audio modulation frequency. The signal is detected in dispersion mode and the error signal fed to a voltage-controlled crystal oscillator which controls the fluorine carrier frequency. This latter frequency is then directly related to the proton resonance frequency since both the proton and fluorine control resonances are simultaneously at resonance in the same magnetic field.

Emsley et al.[31] have overcome the difficulty caused by large deuterium quadrupole splittings in proton spectra in liquid crystal solvents by a combination of coherent and incoherent decoupling irradiation at the deuterium frequency.

Kuhlmann and Grant[32] have measured Overhauser enhancement factors in ^{13}C spectra by a CW method. Proton decoupling was accomplished by pseudo-random phase modulation of the output from a Hewlett Packard frequency synthesizer and subsequent power amplification by a Boonton 230A amplifier. Coupled spectra were obtained by offsetting the proton centre frequency by 1 MHz (300 kHz was insufficient). This method is superior to amplitude switching since constant rf heating effects in the probe are maintained (the system was tuned for equal excitation at the two frequencies). A controller was constructed to synchronize the decoupler to the sweep and also to allow delays between sweeps to enable the system to return to thermal

---

[27] A. Allerhand, D. Doddrell, and R. Komoroski, *J. Chem. Phys.*, 1971, **55**, 189.
[28] U. Haeberlen, H. W. Spiess, and D. Schweitzer, *J. Magn. Resonance*, 1972, **6**, 39.
[29] A. G. Moritz, *Mol. Phys.*, 1971, **20**, 945.
[30] P. N. Jenkins and L. Phillips, *J. Phys.* (*E*), 1971, **4**, 781.
[31] J. W. Emsley, J. C. Lindon, J. M. Tabony, and T. H. Wilmshurst, *Chem. Comm.*, 1971, 1277.
[32] K. F. Kuhlmann and D. M. Grant, *J. Chem. Phys.*, 1971, **55**, 2998.

equilibrium. Switching the decoupler on at the beginning of a sweep allows the intensity build-up of a collapsed multiplet to be studied. The collapse is instantaneous and the increase in intensity (a ^{13}C T_1 process) may be measured as a function of time simply by off-setting the start of the frequency sweep.

An indirect determination of T_1 for ^{29}Si in the molecule $^{29}SiHCl_3$ by proton double-resonance experiments has been reported,[33] although the values obtained are not very accurate.

B. 'Other' Nuclei.—The cases cited here are meant to be illustrative rather than an exhaustive compilation.

Spectra of ^{15}N in natural abundance in some amines have been reported by Lichter and Roberts.[34] The Varian DFS 60 spectrometer was operated at 6.07 MHz and the samples were run as neat liquids in 10 mm external diameter tubes. Shifts were measured relative to a 10M enriched nitric acid sample contained in a concentric 5 mm external diameter tube. Recording conditions were typically: width 250 or 100 Hz; sweep rate 10 or 4 Hz s^{-1}; experimental time 1—2 h. Complications arose from the negative nuclear Overhauser effect caused by incoherent proton irradiation. This arises because of the negative magnetogyric ratio of ^{15}N. A further complication occurs, since in the presence of proton exchange, the n.o.e. may be attenuated, the extent being determined by the exchange rate. In the ammonium ion the two effects were observed to cancel, producing a signal of zero intensity!

The benefit of a multinuclear approach is typified by the study of ^1H, ^2H and ^{14}N resonances in dimethylformamide in a nematic solvent.[35] ^2H and ^{14}N spectra were recorded at 8 and 4 MHz respectively on a Varian wide-line spectrometer. A similar spectrometer was used to observe ^{25}Mg resonances in aqueous phosphate solutions.[36]

Lippmaa et al.[37] measured the ^{14}N, ^{15}N, ^{17}O, and ^{13}C spectra, at 4.32, 6.08, 8.11, and 15.08 MHz respectively, of some nitropyrroles and nitroimidazoles. The 15 mm spinning samples were examined using a frequency-sweep time-sharing method.

An inexpensive modification to the Varian XL-100 has been described which allows observation of deuterium resonances.[38] A ramp derived from a computer of average transients (CAT) is used to produce field-sweep spectra via the deuterium channel which is normally used for field–frequency locking.

C. Use of Computers.—The use of the CAT in CW n.m.r. is well established. The presence of an on-line general purpose computer which is there primarily

---

[33] A. Briguet and A. Erbeia, *J. Phys. (C)*, 1972, **5**, L58.
[34] R. L. Lichter and J. D. Roberts, *J. Amer. Chem. Soc.*, 1972, **94**, 2495.
[35] E. T. Samalski and H. J. C. Berendsen, *J. Chem. Phys.*, 1972, **56**, 3920.
[36] R. G. Bryant, *J. Magn. Resonance*, 1972, **6**, 159.
[37] E. Lippmaa, M. Magi, S. S. Novikov, L. I. Khmelnitski, A. S. Prihodko, O. V. Lebelev, and L. V. Epishina, *Org. Magn. Resonance*, 1972, **4**, 153.
[38] R. E. Santini, *Analyt. Chem.*, 1971, **43**, 801.

for processing data from pulsed studies is of course also beneficial to CW work. Thus multiscan averaging may be readily performed and the data processed in the manner of one's choice. The possibility also exists of producing restricted frequency sweeps at arbitrary offsets under computer control so that time is not wasted in regions of the spectrum which are of no interest.

A detailed description has appeared of an interface between a PDP-8/I computer and a JEOL C60HL spectrometer.[39] Hardware and software suitable for multiscan averaging are described. A procedure for digitization of spectra on to magnetic tape has been reported.[40] Several spectra are recorded and processed on a remote computer. Peaks that do not correspond sufficiently between sweeps are discarded. The computer is used to perform the tedious task of measurement of line positions.

D. Relaxation Measurements.—Spin–lattice relaxation times of ^{13}C have been measured using the adiabatic rapid-passage technique.[32,41] Typically, six forward and reverse sweeps were accumulated in the first and second halves of a CAT memory, with a delay between successive scans to allow the spin system to return to thermal equilibrium. A constant scan period was used so that the time interval between forward and reverse sweeps was adjusted by altering the starting point of the sweep.

Transverse and longitudinal relaxation times have been measured for ^{31}P, and transverse relaxation times for ^{35}Cl, in phosphorus halide systems[42] by CW techniques. Transverse relaxation times for both nuclei were estimated from the linewidths, ^{35}Cl from derivative and ^{31}P from absorption spectra. The linewidths were dominated by quadrupole contributions for both nuclei. One presumes that the inhomogeneity contribution for ^{31}P is negligible in the HA100 spectrometer which was used. The minimum value for T_2^{-1} was approximately 5 s^{-1}, corresponding to a linewidth at half height of about 1.7 Hz. T_1 values for ^{31}P were obtained by the saturation recovery method. The recovery was exponential because $T_2 \ll T_1$ (maximum value for $T_1^{-1} = 0.4$ s^{-1}). The results obtained for phosphorus–halogen coupling constants and spin-rotation constants showed general agreement with previous theoretical and experimental work, indicating the viability of CW measurements in situations of this type.

Lippmaa et al.[43] have reported the temperature dependence of spin–lattice relaxation times of ^{15}N nuclei in some organic molecules. The adiabatic rapid-passage technique was usually used except when $T_2 \ll T_1$ in the presence of scalar interactions. In these cases the saturation recovery technique was used (see above).

[39] R. G. Jones, P. Partington, B. W. Reedy, and T. Trill, *J. Phys. (E)*, 1972, **5**, 44.
[40] R. E. Rondeau and V. L. Donlan, *Analyt. Chem.*, 1971, **43**, 1699.
[41] T. D. Alger, S. W. Collins, and D. M. Grant, *J. Chem. Phys.*, 1971, **54**, 2820.
[42] A. R. Jordan, R. Cavell, and R. B. Jordan, *J. Chem. Phys.*, 1972, **56**, 483.
[43] E. Lippmaa, T. Saluvere, and S. Laisser, *Chem. Phys. Letters*, 1971, **11**, 120.

6 High-resolution Fourier-transform Spectroscopy

A. Introduction.—In Volume 1 of this series Fourier spectroscopy was not covered on the grounds that the subject had been reviewed elsewhere.[44] In fact, the coverage of that review extends into the period for this volume which will concentrate on the newer developments. A general paper[45] entitled 'Fourier transform Approaches to Spectroscopy' has appeared. Some useful pictorial representations are given demonstrating the equivalence of convolution in one domain and multiplication in the other domain. Pickett and Strauss[46] have given a useful account of signal-to-noise ratios in Fourier spectroscopy. A review article on ^{13}C Fourier spectroscopy has appeared.[47] The applications discussed form a useful compilation but the discussion of the principles should be read with caution; for example, the description of the $180°-\tau-90°$ method for measuring T_1 values as a spin-echo experiment is clearly erroneous. Shaw[48] has reviewed the measurement of relaxation times by Fourier methods.

In the important area of ^{13}C Fourier spectroscopy of macromolecules it has been shown that the expected increase in sensitivity at higher magnetic fields may not materialize since the increase in basic sensitivity can be offset by a decrease both in the nuclear Overhauser enhancement and in the $T_2:T_1$ ratio.[49]

An important paper 'Phase and Intensity Anomalies in Fourier Transform N.M.R.' has been published by Freeman and Hill[50] and reviewed elsewhere[51] (see also ref. 52). Anomalies arise when there is residual transverse magnetization at the end of the pulse interval. The anomalies vary in a cyclic manner as a function of the frequency offset from resonance and may be removed by inserting a random delay before the next pulse (see also Section 6C). The quadriga Fourier-transform (QFT) method of Schwenk,[53] which was developed specifically for observation of the resonances of low-abundance insensitive nuclei with T_1 and T_2 values large compared with T_2^*, is formally very similar, although not of proven practicality for high-resolution situations. Similar frequency-dependent effects ensue since transverse components of magnetization are present at the beginning of each pulse because the pulse interval is made less than T_2^* for reasons of sensitivity. Schwenk's solution involved the acquisition of four sets of data taken at four different offset

---

[44] D. G. Gillies and D. Shaw, in 'Annual Review of N.M.R. Spectroscopy,' ed. E. F. Mooney, Academic Press, London, 1972, vol. 5.
[45] G. Horlick, *Analyt. Chem.*, 1971, **43**, 61A.
[46] H. M. Pickett and H. L. Strauss, *Analyt. Chem.*, 1972, **44**, 265.
[47] E. Breitmaier, G. Jung, and W. Voelter, *Angew. Chem. Internat. Edn.*, 1971, **10**, 673.
[48] D. Shaw, in 'The Applications of Computer Techniques in Chemical Research,' Institute of Petroleum, 1972, p. 76.
[49] D. Doddrell, V. Glushko, and A. Allerhand, *J. Chem. Phys.*, 1972, **56**, 3683.
[50] R. Freeman and H. D. W. Hill, *J. Magn. Resonance*, 1971, **4**, 366.
[51] Ref. 44, p. 572.
[52] D. E. Jones and H. Sternlicht, *J. Magn. Resonance*, 1972, **6**, 167.
[53] A. Schwenk, *J. Magn. Resonance*, 1971, **5**, 376.

frequencies, each lower than the lowest Larmor frequency and each separated by one quarter of the repetition rate. Addition of the Fourier-transformed spectra was shown to eliminate many of the 'anomalies'.

Yet another analysis of the possible sensitivity advantages of driven-equilibrium and spin-echo Fourier-transform (DEFT and SEFT) methods over the conventional Fourier experiment has appeared.[54] In the practical situation that ensues in high-resolution spectroscopy (*i.e.* use of a high-resolution magnet) there is no real advantage.[55]

High-resolution Fourier spectra have now been obtained at high pressure[7] (see also Section 2B).

B. Fourier Spectrometers.—The basic approach to realization of a Fourier system is governed by the level of computer involvement. At the very least a computer is required to perform conveniently the Fourier transformation using the now well-known fast Fourier transform (FFT) algorithm.[56] In the system of Redfield and Gupta[17] free induction decays are accumulated in a 1024 channel computer of average transients. The data are read out periodically to a small remote IBM 1130 computer; 1024 data points are sufficient to produce proton spectra of suitable resolution for protein work. In this scheme the computer is used merely for data processing. A more common scheme with 4K data points uses a Fabri-Tek (now Nicolet) signal averager and a PDP-8/I computer[27] a scheme also adopted in many Bruker systems. In this scheme the computer is basically used for processing the data accumulated in the averager. Experimental conditions (*e.g.* pulse length, interval, *etc.*) are set manually in a separate programmer (see Section 4).

The high level of computer involvement is illustrated by the work of Cushley *et al.*,[26] which involves equipment similar in concept to the Varian XL-100 FT system. Their Bruker HFX-3 spectrometer was directly interfaced to an IBM 1800 computer with 24K of memory. This enabled the sampling of 8K data points in double precision (or 16K in single precision), to be performed, giving spectra of higher resolution for a given bandwidth. In this type of system the computer not only acquires data but controls the experiment. A 4K Fabritek 1064 averager fitted with a display oscilloscope and an XY recorder was used only for on-line buffered display and data read-out. The display was used as in Varian systems for on-line interactive phase correction. The IBM computer had facilities for both disk and magnetic tape storage.

In the usual Fourier system the rf carrier is placed at the edge of the frequency range of interest since frequency components from above and below the carrier are not distinguished in frequency.[57] They are, however, distinguished in phase, and Redfield and Gupta[17] have described a method that lifts the restriction on the placement of the rf carrier. The method requires

---

[54] D. E. Jones, *J. Magn. Resonance*, 1972, **6**, 183.
[55] Ref. 44, p. 605.
[56] J. W. Cooley and J. W. Tukey, *Math. Computation*, 1965, **19**, 297.
[57] Ref. 44, p. 597.

two identical phase detectors which must be out of phase by $\pi/2$ to an accuracy of 0.01 rad (see also ref. 18). Detailed circuitry is given, together with the setting-up procedure. By displaying the signals as an xy-presentation on an oscilloscope beautiful pictures are produced which are used to make phase and other adjustments. The outputs of these two phase detectors can be regarded as the real and imaginary parts of a complex signal. They are stored in two halves of the accumulator alternately. Details of the subsequent Fourier analysis are given. Corrections are made to allow for the following factors:

(a) time alternation of the input data, *i.e.* the time samples from one phase detector are always taken one sampling interval after the other;

(b) the use of long pulses, which allows changes in the amplitude and phase of magnetization during a pulse;

(c) instrumental artifacts, mainly filter response.

Correction (c) was made by utilizing the response of the receiver system to a pulse short enough to be a sufficient approximation to a delta function.

C. **Relaxation Measurements.**—T_1. The ability to study relaxation of individual lines in a high-resolution spectrum by Fourier methods is now well established.[44] Freeman, Hill, and Kaptein[58] have described an automated method for measuring spin–lattice relaxation times which is designed to improve the time-scale of the experiment. The basis of the scheme is an adaptive programme and is applicable both to the inversion recovery (180°–τ–90°) method and to the progressive saturation method, although the experiments reported involved the latter. The inversion recovery technique requires that the system has returned to thermal equilibrium when the 180° pulse is applied. The combination of weak resonances and long relaxation times, a situation common in ^{13}C spectroscopy, can make measurements prohibitively long. The time-scale of the experiment is determined by the longest T_1 value, which is of course usually unknown at the beginning of an experiment. In the progressive saturation technique intensities of resonances are measured as a function of the repetition rate of 90° pulses. A dynamic steady state is achieved after only three pulses. Subsequently the only pre-90° pulse condition is that the transverse magnetization must be zero. This is no problem in ^{13}C spectra in the presence of proton noise decoupling when the pulse interval is longer than one second. However, the insertion of a random delay of approximately 30 ms before each pulse effectively destroys transverse magnetization when averaged over many pulses.

Markley, Horsley, and Klein[59] also described an experiment which could improve the time-scale over inversion recovery by a factor of up to 1.9. In this method the spin system is saturated by a burst of non-selective pulses and the 'saturation recovery' monitored by a 90° pulse at suitable time intervals. Transverse magnetization is destroyed by a field gradient pulse. Freeman *et al.* have pointed out the close analogy between the progressive saturation technique

[58] R. Freeman, H. D. W. Hill, and R. Kaptein, *J. Magn. Resonance*, 1972, **7**, 82.
[59] J. L. Markley, W. J. Horsley, and M. P. Klein, *J. Chem. Phys.*, 1972, **55**, 3604.

and the above method and also that the use of identical field gradient pulses does not inhibit the refocusing effect inherent in the steady-state regime.

The repetitive-pulse experiment requires an accurate value for the equilibrium intensities obtained by setting the pulse intervals long enough for return to thermal equilibrium between pulses. This is time consuming; the method of Freeman et al., termed the 'intensity ratio' method, does not require such a measurement. The 'intensity ratio' method is applicable to inversion recovery, saturation recovery, and progressive saturation experiments. The spin–lattice relaxation time is estimated from two measurements taken at pulse intervals $t = a$ and $t = b$. The intensity ratio S_a/S_b for the two measurements (progressive saturation or saturation recovery method) is:

$$\frac{S_a}{S_b} = \frac{1 - \exp(-a/T_1)}{1 - \exp(-b/T_1)}$$

Clearly the choice of the values for a and b influences the accuracy of the value derived for T_1. Freeman et al. plotted T_1/a against S_a/S_b and found that the choice of the ratio b/a was not critical (four was the value adopted). Setting the time too early or too late on the decay curve led to observed ratios S_a/S_b that were insensitive to T_1. This gave T_1 values unduly sensitive to noise. An analysis of the effect of various experimental inadequacies was given (see also ref. 60). The computer programme of Freeman et al. started by setting a at the minimum pulse interval (equal to the acquisition time of 0.2 s). Spectra were compared in pairs with $b/a = 4$. In practice the value of a was doubled between successive experiments. The programme made baseline corrections and estimated r.m.s. noise values from a signal-free region of the 'b' spectrum. The programme continues until the region is found which gives a T_1 value for a given line that is within acceptable error limits. Measurements which took 15 h were made on the ^{13}C spectrum of cortisone acetate, and T_1 values showed good agreement with the normal inversion recovery experiment which took 60 h to perform without allowance for the initial estimation of the longest relaxation time. The experiments were greatly facilitated by the use of a small cassette tape recorder. Thus all the spectra could be stored before read-out, when phase and amplitude adjustments could be made.

Allerhand et al.[61] have demonstrated the application of partially relaxed Fourier-transform spectra to the ^{11}B spectrum of n-B_9H_{15} at 70.6 MHz. Differing ^{11}B spin–lattice relaxation times within the molecule caused intensity differences in the spectra. This enabled the resolution of features which were not distinguished in the ^{11}B spectrum even at this high frequency.

Levy et al.[62] measured T_1 values in phenylacetylenes by the usual 180°–τ–90° method. Since relaxation via the chemical shift anisotropy mechanism depends on the square of the magnetic field it is very useful to measure T_1

[60] D. E. Jones, J. Magn. Resonance, 1972, 6, 191.
[61] A. Allerhand, A. O. Clouse, R. R. Rietz, T. Roseberry, and R. Schaeffer, J. Amer. Chem. Soc., 1972, 94, 2445.
[62] G. C. Levy, D. M. White, and F. A. L. Anet, J. Magn. Resonance, 1972, 6, 453.

Experimental Techniques 179

at different field values (see also Section 7A). One carbon had such a long relaxation time (125 s at 2.3 T) that it was possible to evaluate T_1 in the earth's field by physically removing the sample from the magnet for a time t and then returning it to measure the residual magnetization! The estimated weak-field T_1 value was 340 ± 70 s.

Spin Echoes. Vold and Shoup[63] have studied spin-echo spectra of 1,1-difluoro-2,2-dichloroethane, an ABX_2 spin system, using Fourier-transform techniques, thus extending the earlier work of Freeman and Hill[64] on the homonuclear AMX system of 3-bromothiophen-2-aldehyde. The echo peaks generated by a Carr–Purcell–Meiboom–Gill (CPMG) sequence were sampled by a Nicolet (Fabritek) 1083 Fourier-transform system and subsequently transformed. The frequency spectra obtained displayed components which depended in magnitude and frequency on the pulse repetition rate. The spectral lines are not in general characteristic of a single spectral magnetization, in contrast to normal spectra or a completely selective spin-echo experiment (see also ref. 22). A subspectral analysis in terms of three ab sub-spectra was made; its success implied that cross-relaxation effects were small.

The Reporter feels that the analytical and instrumental problems associated with this type of work still effectively preclude its general use. However, some extremely accurate parameters may be obtainable: *e.g.* J_{AB} in the system was estimated as 3.071 ± 0.011 Hz. Extraction of transverse relaxation rates from the study of spectra obtained by transformation of the free induction decay after the nth echo as a function of the echo number n was investigated but found to be impractical. In a later theoretical paper[65] the difficulty of measuring transverse relaxation times in complex systems was again underlined, particularly in the presence of strong homonuclear coupling and/or transverse cross-relaxation processes.

D. Gated Double Resonance.—It is convenient under this heading to consider the use of a time-shared field–frequency lock channel as a gated double-resonance experiment. It is now common on commercial Fourier spectrometers and leads to a better lock condition. An analysis of time-shared modulation has been given.[66] The distinction between a time-shared lock and a pulse lock system is academic. Redfield and Gupta[67] use a pulse length of a few microseconds at a repetition rate of 4 kHz and lock to a sideband response. The receiver blanking pulse is adjusted until there is no evidence of receiver recovery transients. A simple time-share circuit used in conjunction with a ^{7}Li lock system was used by Ellett *et al.*[68]

[63] R. L. Vold and R. R. Shoup, *J. Chem. Phys.*, 1972, **56**, 4787.
[64] R. Freeman and H. D. W. Hill, *J. Chem. Phys.*, 1971, **54**, 301.
[65] R. L. Vold, *J. Chem. Phys.*, 1972, **56**, 3210
[66] K. Arnold, G. Klose, and P. Herrmann, *J. Magn. Resonance*, 1972, **6**, 136.
[67] Ref. 17, p. 104.
[68] Ref. 18, p. 167.

Gansov and Schittenhelm[69] used a sequence where the proton noise decoupler, after being switched on for approximately one second, was gated off a short time before acquisition of the ^{13}C free induction decay. This technique was also reported at about the same time by Freeman and Hill,[70] and allowed the observation of ^{13}C multiplets with normal splittings but with most of the Overhauser enhancement retained. This distinction between instantaneous appearance and disappearance of multiplets and the T_1^C-governed enhancements was first noted in CW ^{13}C experiments.[71] The opposite experiment has been demonstrated,[72] where the decoupler is switched on t seconds before and left on during the ^{13}C excitation pulse so that a decoupled spectrum is obtained in which the intensities reflect the population build-up during the t seconds, a T_1^C process.

A recent experiment by Schaeffer[73] may also be regarded as a gated selective pulse experiment in which a particular carbon resonance was saturated and removed from the spectrum. Immediately preceding the normal ^{13}C excitation pulse a 2 s, 1 mW burst of rf was centred on the line of interest, causing an equalization of population which persisted during the free induction decay. Various uses of the technique were suggested, including (a) resolution of closely spaced lines by selective attenuation, (b) elimination of a strong (e.g. solvent) line, (c) investigation of (^{13}C,^{13}C) interactions in massively enriched samples, and (d) burning a hole of natural linewidth in an inhomogeneously broadened line. Another simple experiment may be conveniently included here. It consists merely of introducing a delay time after the pulse before acquisition of the free induction decay.[74] The experiment was used to remove broad features (short relaxation times) from the proton spectra of some large biological molecules, revealing narrow lines which had previously been almost completely masked even at 220 MHz.

E. **Difference Spectroscopy.**—Fourier difference spectroscopy was introduced by Ernst[75] as a method which combined the sensitivity of the Fourier technique with almost complete insensitivity to magnetic field variations. The instrumentation was simple except for the digital computer. In a recent paper Ernst[76] has described the use of an analogue Fourier analyser in place of the computer. This scheme does not have the multi-channel advantage of Fourier spectroscopy and produces the same sensitivity as CW experiments. The spin system is subjected to repetitive pulses and the response is amplified, bandwidth-limited, and demodulated in an envelope detector. This causes

---

[69] O. A. Gansov and W. Schittenhelm, *J. Amer. Chem. Soc.*, 1971, **93**, 4294.
[70] R. Freeman and H. D. W. Hill, *J. Magn. Resonance*, 1971, **5**, 278.
[71] J. Feeney, D. Shaw, and P. J. S. Pauwells, *Chem. Comm.*, 1970, 554.
[72] R. Freeman, H. D. W. Hill, and R. Kaptein, *J. Magn. Resonance*, 1972, **7**, 327.
[73] J. Schaeffer, *J. Magn. Resonance*, 1972, **6**, 670.
[74] C. H. A. Seiter, G. W. Feigenson, S. I. Chan, and M. Hsu, *J. Amer. Chem. Soc.*, 1972, **94**, 2535.
[75] R. R. Ernst, *J. Magn. Resonance*, 1971, **4**, 280.
[76] R. R. Ernst, *J. Magn. Resonance*, 1971, **5**, 398.

Experimental Techniques

difference frequencies to be generated, from which the dominant ones caused by the presence of a strong reference line are extracted by means of a low-pass filter. These signals are then passed to the analyser, which may be a phase-sensitive detector referenced at a frequency derived from a voltage-controlled oscillator. The frequency of the analyser is conveniently swept by driving the oscillator by a voltage derived from the x-axis of an XY recorder. The output from the analyser must be averaged over one full period. This was achieved by using a gated integrator which was controlled by a sequence generator. Phase coherence is maintained by synchronizing the exciting pulses to the reference frequency. The ability of the phase detector to produce a d.c. response to odd harmonics of the reference frequency is a nuisance, best overcome by the use of a good analogue multiplier in its place.

7 General Pulsed Spectroscopy

A. Introduction.—The concept called 'Nuclear Magnetic Relaxation Spectroscopy' by Noack[1] is becoming increasingly important as increasing interest (including chemists') is being shown in distinguishing the different mechanisms contributing to overall relaxation times. Thus, ideally one would like to know the dependence of relaxation time on magnetic field or Larmor frequency, pressure, and temperature. Relevant experimental approaches to this goal are given both in the next section and in Section 2.

There has been a steady increase in the use of superconducting solenoids in pulsed experiments. For instance, Spiess et al.[11] were able to measure the ^{13}C spin–lattice relaxation time in $^{13}CS_2$ at 14, 30, and 62 MHz.

B. Spectrometers.—A most comprehensive description of two spectrometers that have been used for multi-pulse studies on solids has been given.[18] These studies required very high-power pulses but most of the techniques are more generally applicable. Thus we have already discussed in previous sections aspects of the generation of frequencies and pulse sequences and of the role of the computer. A detailed discussion with circuit diagrams was given of the duplexer, the circuitry employed to isolate the transmitter pulses from the receiver but at the same time allow optimum operation of each. Two relevant papers have appeared, one on duplexers[77] and the other concerning an improved coil damping circuit.[78] The relative merits of single- and crossed-coil probes are discussed;[18] since in the particular case studied the need for a maximum rf field is paramount the single-coil method was adopted. A probe diagram is given showing the use of the concentric teflon dielectric capacitors necessary to withstand the high (kV) rf voltages (see also Section 2C). The pre-amplifier and receiver system is of broad-band design. To enable most efficient use of the available power the rf carrier is set at the centre of the region of interest. Two rf phase detectors with references in quadrature are

[77] B. M. Moores and R. L. Armstrong, Rev. Sci. Instr., 1971, **42**, 1329.
[78] S. B. W. Roeder, N. L. Rhodes, and G. W. Schmidt, Rev. Sci. Instr. 1971, **42**, 1692.

employed in order to acquire both in-phase and out-of-phase components. This is the same technique as that of Redfield and Gupta[17] (see Section 6B), and the two components are displayed on an XY oscilloscope in a like manner. In the present case it was found more convenient to remove the consequences of slight phase misadjustment by subsequent computer processing. The output signals are digitized and Fourier-transformed. Spectrometer A employs a slower digitizing system, and data has to be dumped on to paper tape for subsequent Fourier processing. In Spectrometer B this task is accomplished conveniently within the on-line PDP-12 computer.

Noack[79] has described a variable-frequency pulse spectrometer specifically designed for operation over a wide range of frequencies: 32 discrete frequencies between 10 kHz and 160 MHz in a high-field electromagnet and from 3 to 750 kHz in a pulsable low-field magnet were available. Eight tunable probes covered the frequency range, and operation between -150 and $+200$ °C was possible. The electromagnet had replaceable polecaps and the air gap was adjustable from 0.5 to 12 cm, enabling the use of large samples at low field. The pulsable low-field magnet allowed the polarization of spin systems in a high field followed by relaxation at a low-field value followed by examination of the residual magnetization at the high-field value. In cases where the value of T_1 was greater than a few seconds the whole probe could be physically displaced from the high-field electromagnet into the low-field solenoid and back again.

A more sophisticated spectrometer has been described[80] in which the sample is moved automatically from low to high field and vice versa. This operation, which took approximately 0.25 s, was achieved by a compressed-air engine whose operation was controlled by the main pulse generator.

Special points concerning probes for experiments (including pulsed) at high pressures or high fields are considered in Section 2.

Mehring and Waugh[81] have analysed the source of phase transients in pulsed spectrometers which arise when a square pulse is applied to the tuned probe circuit; they find that there can be cumulative errors in experiments employing long pulse sequences. These errors could be made to vanish by tuning the probe to its free ringing frequency and ensuring that the rf carrier was coherent with the pulse timing.

C. **Relaxation Measurements in Liquids.**—Most measurements have used traditional pulse sequences and no special mention of these will be made. Ailion[82] has given a critical account of the methods for measuring $T_{1\rho}$.

$T_{1\rho}$. Sears[83] has measured the dependence of $T_{1\rho}$ for ^{19}F in liquid $CHFCl_2$ and $CFCl_3$ on the value of the magnetic field in the rotating frame, from which

---

[79] Ref. 1, p. 83.
[80] G. Parry Jones and J. T. Daycook, *J. Phys. (E)*, 1971, **4**, 641.
[81] M. Mehring and J. S. Waugh, *Rev. Sci. Instr.*, 1972, **43**, 549.
[82] D. C. Ailion, *Adv. Magn. Resonance*, 1971, **5**, 200.
[83] R. E. J. Seers, *J. Chem. Phys.*, 1972, **56**, 983.

Experimental Techniques 183

values for (F, Cl) scalar couplings were deduced. He pointed out the greater convenience of $T_{1\rho}$ experiments for studies on the low-field relaxation compared with low-static-field experiments (see Section 7B).

Spin Echoes. A comprehensive spin-echo study has been made[28] using the CPMG technique. The studies embraced the four nuclei ^{13}C, ^2H, ^1H, and ^{19}F in some small molecules. Very reproducible results were obtained and it was concluded that stability was the important factor in CPMG experiments since the results were not markedly affected by deliberate misadjustment of pulse phases and widths. The stability resulted from the use of a superconducting solenoid and from the use of pulses coherent with the rf. A detailed adjustment procedure was described.

Selective spin-echo spectra have been studied[84] using phase detection and filtering. T_2 was estimated by noting the time constant of an electronically generated exponential voltage which when subtracted from the spin-echo decay placed the echo peaks on the base line (a sophisticated example of analogue processing!).

Muir and Turner[85] have performed selective proton spin-echo experiments at moderate resolution using the CPMG sequence. A more detailed description is promised.

Tokuhiro and Fraenkel[86] were able to calculate spin-echo spectra for dichloroacetaldehyde which agreed well with experiment. A purely theoretical treatment of echo modulation has been given.[87]

T_1. Effects of inhomogeneity in the rf field on T_1 experiments have been considered by Kumar and Johnson,[88] but the Reporter feels that they will be negligible for high-resolution experiments in crossed-coil systems.

A triplet pulse sequence (TPS) has been introduced[89,90] to make measurements of high T_1 values more convenient. The proton T_1 in chloroform is 86 s and its measurement by normal methods would be tedious. The suggested sequence allows the measurement to be made in a single decay. The sequence is

$$180^\circ_y - \tau_2 - \underbrace{90^\circ_y - \tau_1 - 180_x - \tau_1 - 90^\circ_{-y}}_{\text{TPS}} - 2\tau_2 - \text{TPS} - 2\tau_2 \quad \ldots \ldots$$

The magnetization is sampled for times $2\tau_1$ at arbitrary intervals $2\tau_2$. Clearly $2\tau_1 \ll T_1$ and for convenience τ_2 should be approximately $0.1T_1$. A method for obtaining the necessary exact 90° pulse is given.

[84] M. F. Augusteijn, W. M. M. J. Bovee, S. Emid, A. F. Mehlkopf, and J. Smidt, *J. Magn. Resonance*, 1972, **7**, 301.
[85] A. R. Muir and D. W. Turner, *Chem. Comm.*, 1971, 286.
[86] T. Tokuhiro and G. Fraenkel, *J. Chem. Phys.*, 1971, **55**, 2797.
[87] W. B. Mims, *Phys. Rev. (B)*, 1972, **5**, 2409.
[88] A. Kumar and C. S. Johnson, *J. Magn. Resonance*, 1972, **7**, 55.
[89] D. Rogers, M. T. Rogers, and G. D. Vickers, *Rev. Sci. Instr.*, 1972, **43**, 555.
[90] D. Rogers and M. T. Rogers, *J. Chem. Phys.*, 1972, **56**, 542.

A novel method for measuring the composition of mixtures of ortho- and para-deuterium has been described.[91] Free induction decays are measured for a mixture by a pulse saturation recovery technique (saturation–τ–90°, free induction decay). The decay consists of two exponential components differing by several orders of magnitude, a short-lived one from the para-molecules and a very long one from the ortho. For instance, at a mole fraction of 0.04 for para, T_1^p was 0.58 s and T_1^o was 4115 s. The signal amplitude obeyed the relation

$$1 - S(\tau)/S(\infty) = f_o\exp(-\tau/T_1^o) + f_p\exp(-\tau/T_1^p)$$

where f_o, f_p are related to x, the mole fraction of para-deuterium by

$$f_o = \frac{S_o(\infty)}{S(\infty)} = \frac{5(1-x)}{5-3x}$$

$$f_p = \frac{S_p(\infty)}{S(\infty)} = \frac{[S(\infty) - S_o(\infty)]}{S(\infty)} = \frac{2x}{5-3x}$$

Determination of one of these parameters gives x and takes two or three minutes.

D. High-resolution Studies in Solids (See also Chapter 9).—The experiment that has the most potential interest for the chemist is that described by Pines, Gibby, and Waugh.[92] The method allows high-resolution spectra to be obtained for magnetically dilute spins in solids. The resolution reported for ^{13}C in a polycrystalline sample of adamantane was 40 Hz, limited by magnet inhomogeneity! The residual linewidth caused by ($^{13}C, ^{13}C$) dipolar interactions was estimated as 5 Hz. During the acquisition of the ^{13}C free induction decay the broadening influence (dipolar) of the protons is removed by the application of a strong proton rf field. Also, the large total magnetization of the abundant spins I with abundance N_I and magnetic moment $\gamma_I I \hbar$ is transferred to the low-abundance spin system S. This technique was originated by Hartmann and Hahn.[93] The magnetization transfer is achieved by equalizing the precession frequencies of the I and S spins about their respective rf fields. The unique feature of the present experiment arises from the combination of the magnetization transfer and decoupling techniques. The general procedure is: (i) polarize I in high field; (ii) cool I to a low-spin temperature in the rotating frame; (iii) establish contact between the I and S spin systems for time τ; (iv) record the S free induction decay while decoupling I. Steps (iii) and (iv) are repeated about (N_I/N_S) times until the I magnetization is used up and the S decay accumulated. The spectrum is obtained by Fourier transformation. The gain in power sensitivity over conventional Fourier spectroscopy is approximately $(N_I/N_S)(\gamma_I/\gamma_S)^2$ giving a value of 1000 for the ^{13}C–proton

---

[91] R. Wang and D. White, *Rev. Sci. Instr.*, 1971, **42**, 887.
[92] A. Pines, M. G. Gibby, and J. S. Waugh, *J. Chem. Phys.*, 1972, **56**, 1776.
[93] S. R. Hartmann and E. L. Hahn, *Phys. Rev.*, 1962, **128**, 2042.

Experimental Techniques

case. A spectrum of a 50 mg sample of adamantane, the result of an experiment lasting only 0.8 s, displayed an impressive sensitivity even though the apparatus was not optimized. Further gains were envisaged by increasing the initial polarization by prepolarization of the sample outside the apparatus at low temperature and high field.

The technique described above represents a big advance on previous studies of dilute spin systems of low abundance.[94—97] These are all indirect methods that involve monitoring spin I while saturating the spin S. This causes a loss of magnetization from the I system which is studied as a fraction of the S frequency. These experiments take much longer and have yet to equal the linewidth achieved by the direct method.

The study[97] of ^{14}N in single crystals of paraelectric triglycine sulphate by an indirect method produced information on the magnitude and anisotropy of the quadrupole coupling which suggested that the technique may have important biological implications.

In non-dilute spin systems in the solid state, multipulse techniques are used to remove most of the homonuclear dipole–dipole interactions. A four-pulse experiment was used to measure ^{19}F shift anisotropies and was accompanied by a theoretical treatment.[98] A further analysis including some other pulse sequences has appeared.[99] Practical aspects are treated elsewhere.[18] Decoupling of the ^{23}Na spins in a sample of $CaF_2.NaF$ improved the linewidth obtained in fluorine multipulse experiments[100] by removal of the heteronuclear dipolar interactions which are not averaged out by the pulse sequence.

In some situations of intrinsically low resolution certain information of high-resolution type may be gained without recourse to multipulse techniques. Thus it was possible to measure ^{31}P chemical shift anisotropies for chemically shifted resonances in P_4S_3.[101] The phosphorus shifts were large enough and the dipolar interactions small enough to make the technique profitable. In the case of fluorine the method is not so favourable since the dipolar interactions are much larger. However, the work of Carolan[102] illustrates the determination of large shift anisotropy in crystals of $Ca_5F(PO_4)_3$, which have just one type of fluorine.

[94] H. E. Bleich and A. G. Redfield, *J. Chem. Phys.*, 1971, **55**, 5405.
[95] C. S. Yannoni and H. E. Bleich, *J. Chem. Phys.*, 1971, **55**, 5406.
[96] P. Mansfield and P. V. Grannall, *J. Phys. (C)*, 1971, **4**, L197.
[97] R. Blinc, M. Mali, R. Osredker, A. Prelesnik, I. Zupancic, and L. Ehrenberg, *Acta Chem. Scand.*, 1971, **25**, 2403.
[98] M. Mehring, R. G. Griffin, and J. S. Waugh, *J. Chem. Phys.*, 1971, **55**, 746.
[99] U. Haeberlen, J. D. Ellett, and J. S. Waugh, *J. Chem. Phys.*, 1971, **55**, 53.
[100] M. Mehring, A. Pines, W.-K. Rhim, and J. S. Waugh, *J. Chem. Phys.*, 1971, **54**, 3239.
[101] M. G. Gibby, A. Pines, W.-K. Rhim, and J. S. Waugh, *J. Chem. Phys.*, 1972, **56**, 991.
[102] J. L. Carolan, *Chem. Phys. Letters*, 1971, **12**, 389.

5
Spectral Analysis

BY R. G. JONES

1 Introduction

This is the second Report in the series of Specialist Reports on spectral analysis and therefore requires less of a preamble than that provided for the first Report. The following comments are mainly phenomenon-oriented with the exception of the small number of theoretically oriented papers discussed. No new systems have been studied in theoretical detail since the first Report.

2 New Methods of Studying Known Spin Systems

A. Proton Spin-echo N.M.R. Spectra of 2,2-Dichloro-1,1-difluoroethane: Sub-spectral Analysis.—Vold and Shoup[1] have analysed the Fourier Transform of the spin-echo n.m.r. spectrum of 2,2-dichloro-1,1-difluoroethane and used sub-spectral analysis to derive the experimental values for the spectral parameters of this ABX_2 system. The spin-echo spectra were obtained using a single-coil pulsed spectrometer operating at 15.08 MHz. It was modified to include a 14.19 MHz external ^{19}F field/frequency lock system. Spin-echo decays from Carr–Purcell sequences (180°–τ–90°) were accumulated in a Nicolet 1083 FT system by sampling the magnetization at the mid-points of the 180° pulse intervals.

Experimental studies of spin-echo spectra are normally made directly on the decaying envelope itself[2-4] without the benefit of Fourier transformation. The authors[1] have shown that the spin-echo spectra can be broken down into sub-spectra analogous to slow-passage n.m.r. spectra if relaxation can be ignored. Subsequent discussion seeks to establish the validity of the sub-spectral analysis for 2,2-dichloro-1,1-difluoroethane (DCDFE).

Explicit formulae have been derived for the ABX_2 system (when relaxation effects are ignored) with a claim for generality where the X approximation is valid. Each sub-spectrum is given by the formulae describing an ab spin-

---
[1] R. L. Vold and R. R. Shoup, *J. Chem. Phys.*, 1972, **56**, 4787.
[2] R. Freeman and H. D. W. Hill, *J. Chem. Phys.*, 1971, **54**, 301.
[3] J. H. G. Powles and J. H. Strange, *Discuss. Faraday Soc.*, 1960, no. 34, 30.
[4] R. L. Vold and H. S. Gutowsky, *J. Chem. Phys.*, 1967, **47**, 2495.

Spectral Analysis

echo sub-spectrum determined by the coupling constant J_{ab} and the effective chemical shift $\delta_m = \delta + m_X(J_{AX} - J_{BX})$ where $\delta = \omega_A - \omega_B$. The two sub-spectra defined by $m_X = 0$ are degenerate, so there can be at most six lines in the spin-echo spectrum with integrated intensities proportional to $A_m^{\pm}$ (defined below) and corresponding frequencies given by the relations,

$$f_m^+ = \omega_m^+/2\pi \quad \text{(Hz throughout)}$$

and
$$f_m^- = \omega_m^-/2\pi - 1/4\tau$$

Different values of m_X refer to different spin-echo sub-spectra,

$$A_m^{\pm} = (1 \pm \sin 2\theta_m)/2$$

$$\sin 2\theta_m = [\delta_m^2 + J_{AB}^2 \cos(\tau \Delta_m)]\Delta_m^{-2}(1 - G_m^2)^{-\frac{1}{2}}$$

$$\Delta_m = (J_{AB}^2 + \delta_m^2)^{\frac{1}{2}}$$

$$G_m = (J_{AB}/\Delta_m)\sin(\tau \Delta_m)$$

$$\omega_m^{\pm} = (J_{AB}/2) \pm \varepsilon_m/2\tau$$

$$\varepsilon_m = -\sin^{-1} G_m$$

where 2τ is the spacing between successive 180° pulses in the Carr–Purcell sequence.

In contrast to the slow-passage spectrum, an observed spin-echo line is characteristic of no single spectral magnetization. For some weakly coupled spin systems such as the homonuclear AMX described by Freeman and Hill,[2] individual spin-echo lines can be assigned to individual spins and selective detection is not required. In cases where cross-relaxation is important and for strongly coupled spin systems this description does not apply and analysis of spin-echo linewidths in terms of Redfield[5] theory or an equivalent formalism is required. The effect of relaxation mechanisms on the validity of spin-echo sub-spectral analysis in the ABX_2 system studied was examined by Vold and Shoup.[1] It was concluded that sub-spectral analysis is strictly applicable to protons of DCDFE at 15.08 MHz only for pulse spacings > 180 ms. These conclusions are supported by experimental evidence. Figure 2 of the authors' paper[1] is reproduced here in the Figure as an illustration of how the number of observable spin-echo lines for DCDFE, their frequencies, and their intensities vary with pulse spacing.

A single line is observed at zero frequency for $2\tau \leq 10$ ms when the decay is unmodulated. As the pulse spacing is increased, two lines appear which move rapidly away from zero, then oscillate in frequency, sometimes overlapping each other. Two more lines appear at $2\tau \approx 70$—90 ms, one increasing, the other decreasing in frequency as 2τ is increased.

For the purposes of analysis, like signs for δ_{AB}, J_{AX}, J_{BX}, and J_{AB} were

[5] A. G. Redfield, *IBM J. Res. Development*, 1957, **1**, 19.

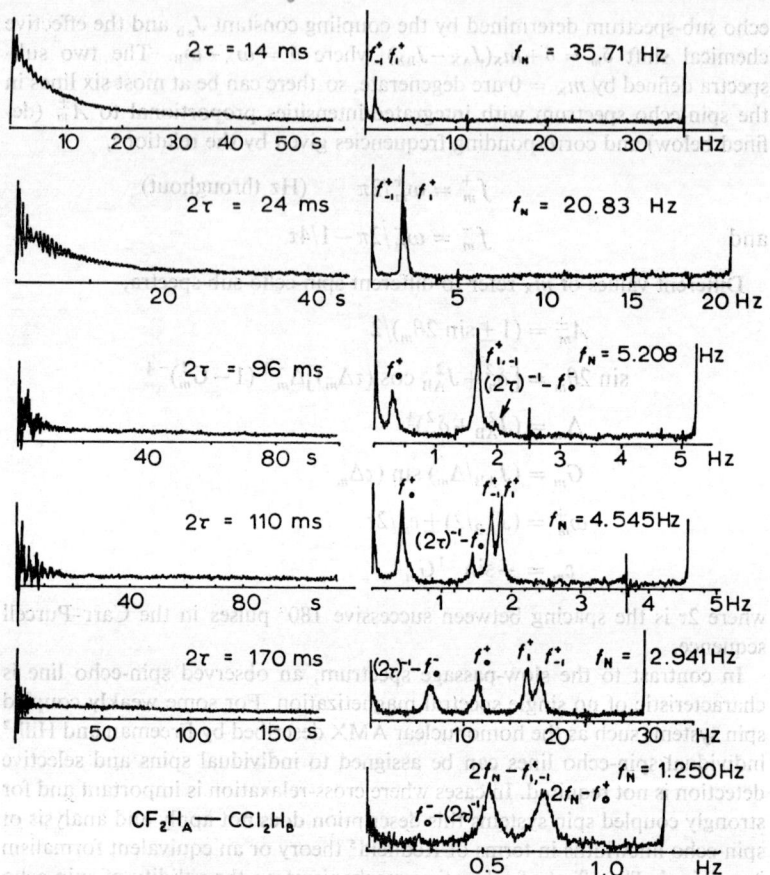

Figure Selected 15.08 MHz proton Carr–Purcell decays and spin-echo spectra of DCDFE with 180° pulse intervals, $2\tau = 14, 24, 96, 110, 180,$ and 400 ms. Individual lines are labelled according to sub-spectrum ($m = 1, 0,$ or -1). Gain settings and abscissae vary from plot to plot, and the sharp lines of random phase observable in most spectra are due to 60 Hz noise, folded back several times. The spike at the end of each spectrum marks the Nyquist frequency, $f_N = 1/4\tau$ Hz, and arises from the beginning of the dispersion mode spectrum
(Reproduced by permission from *J. Chem. Phys.*, 1972, **56**, 4787)

assumed as consistent with the features of the slow-passage spectra. The coupling constant J_{AB} can be determined from spin-echo spectra with great precision:

$$J_{AB} = f_0^+ + f_0^- - 1/4\tau = f_0^+ + 1/4\tau - (1/2\tau - f_0^-)$$

f_0^+ and $1/2\tau - f_0^-$ can be obtained directly from the spin-echo spectra and used to determine a value of J_{AB} for each pulse spacing if both lines are observable,

Spectral Analysis

i.e. if the ab sub-spectrum is sufficiently strongly coupled. The thirteen points obtained at $2\tau \geqslant 125$ ms yielded $J_{AB} = 3.471$ Hz with a standard deviation of 0.011 Hz (perhaps 3.47 ± 0.01 would have been more meaningful). This value for J_{AB} was used with other parameters, $\delta_{AB} = 38.81 \pm 0.10$, $J_{AX} = 55.24 \pm 0.08$, and $J_{BX} = 8.01 \pm 0.08$ Hz, to calculate the spin-echo frequencies as a function of pulse spacings 2τ. Varying δ_{AB} by more than ± 0.05 Hz resulted in significantly poorer visual fits of the $m_X = 0$ lines, and variation of either δ_{AB}, J_{AX}, or J_{BX} by more than 0.3 Hz resulted in a poorer fit for the $m_X = \pm 1$ lines.

The authors point out[1] that valuable applications of spin-echo spectroscopy can be foreseen in studying conformational equilibria, solvent effects, molecular interactions, and other areas where subtle changes of chemical structure can be followed by very accurate measurements of coupling constants and chemical shifts.

B. Spin Hamiltonian for Twofold Symmetry.

Quirt and Martin[6] summarize in their paper the approaches which have been made to the simplification of the analysis of n.m.r. systems using symmetry, distinguishing between point-group symmetry and magnetic equivalence. The work of Woodman,[7] which described a group-theoretical approach giving results entirely equivalent to those obtained by the composite-particle method, is rightly highlighted as an important contribution to the theoretical uses of symmetry. The authors[6] show that by making a suitable choice of basic functions and of *symmetrized spin operators* the use of group quantum numbers to calculate the hamiltonian matrix elements and transition intensities can be extended to the general case of a molecule whose frame has twofold symmetry. The claim is made that the labour involved is substantially reduced since no explicit expansion terms of the original unsymmetrized basis is required. The arguments are logically developed from the hamiltonian for molecules in a mobile, isotropic, fluid phase, in the absence of strong couplings to quadrupolar nuclei or paramagnetic species:

$$h^{-1}\mathcal{H} = \sum_i v_i \hat{I}_{iz} + \sum_{i<j} J_{ij} \hat{I}_i \cdot \hat{I}_j \qquad (1)$$

The $2I+1$ spin states for a nucleus with spin I may be represented by the usual eigenfunctions $|I,m\rangle$ of the square of the nuclear spin angular momentum operator $\hat{I}^2$ and the projection of the total angular momentum on the direction of the applied magnetic field (B_z), $\hat{I}_z$:

$$\hat{I}^2|I,m\rangle = I(I+1)|I,m\rangle \qquad (2)$$

$$\hat{I}_z|I,m\rangle = m|I,m\rangle$$

A simple basis for a collection of nuclei in a molecule is the product basis:

$$\prod_i |I_i,m_i\rangle, \qquad (3)$$

---

[6] A. R. Quirt and J. S. Martin, *J. Magn. Resonance*, 1971, **5**, 318.
[7] C. M. Woodman, *Mol. Phys.*, 1966, **11**, 109.

and transition moments are calculated using the transition operator:

$$\hat{X} = \sum_i \hat{I}_{ix}, \quad (4)$$

in order to predict the intensities of the observed transitions.

In molecules where one or more of the groups meets the conditions for magnetic equivalence, the total spin for the magnetically equivalent group is defined as

$$\hat{\mathbf{F}}_G = \sum_{i \text{ in } G} \hat{\mathbf{I}}_i \quad (5)$$

A new basis set represented by $|F,m\rangle$ can be defined for each group, where

$$\hat{\mathbf{F}}^2 |F,m\rangle = F(F+1)|F,m\rangle \quad (6)$$

and

$$\hat{F}_z |F,m\rangle = m|F,m\rangle$$

The product basis for the molecule is written

$$\prod_i |F_i, m_i\rangle \quad (7)$$

The hamiltonian then becomes

$$h^{-1}\mathcal{H} = \sum_i \nu_i \hat{F}_{iz} + \sum_{i<j} J_{ij} \hat{\mathbf{F}}_i \cdot \hat{\mathbf{F}}_j, \quad (8)$$

and the transition operator becomes

$$\hat{X} = \sum_i \hat{F}_{ix} \quad (9)$$

The numbering is now by magnetically equivalent groups rather than by individual nuclei and any single nucleus is treated as a 'group'. $\hat{\mathbf{F}}^2$ commutes with the hamiltonian and also with $\hat{X}$ so that there can neither be off-diagonal matrix elements nor transitions connecting states between which the F value of any group changes. The matrix elements can be derived using equation (7) directly from equation (8) and each sub-matrix of the simplified hamiltonian matrix becomes a separate problem.

Additional factoring can be achieved when the hamiltonian has a further twofold symmetry, and this is best defined by a rearrangement of equation (8) as follows,

$$\begin{aligned} h^{-1}\mathcal{H} = & \sum_i \nu_i \hat{F}_{iz} + \sum_{i<j} J_{ij} \hat{\mathbf{F}}_i \cdot \hat{\mathbf{F}}_j \\ & + \sum_i \sum_k J_{ik} [\hat{\mathbf{F}}_i \cdot \hat{\mathbf{F}}_k + \hat{\mathbf{F}}_i \cdot \hat{\mathbf{F}}_{k'}] \\ & + \sum_k \nu_k [\hat{F}_{kz} + \hat{F}_{k'z}] + \sum_k J_{kk'} \hat{\mathbf{F}}_k \cdot \hat{\mathbf{F}}_{k'} \\ & + \sum_{k<\ell} \{ J_{k\ell} [\hat{\mathbf{F}}_k \cdot \hat{\mathbf{F}}_\ell + \hat{\mathbf{F}}_{k'} \cdot \hat{\mathbf{F}}_{\ell'}] \\ & + J_{\ell k} [\hat{\mathbf{F}}_k \cdot \hat{\mathbf{F}}_{\ell'} + \hat{\mathbf{F}}_{k'} \cdot \hat{\mathbf{F}}_\ell] \} \quad (10) \end{aligned}$$

In equation (10) magnetically equivalent groups which are transformed into themselves by the twofold symmetry operation ('invariant groups') are labelled i,j.... All other groups must appear in pairs, labelled k and ℓ..., such that the symmetry operation interchanges the primed and unprimed members of each pair. The two different coupling constants, usually *cis* and *trans*, which connect pairs k and ℓ are written $J_{k\ell}$ and $J_{\ell k}$.

Matrix Elements of the Hamiltonian with Symmetrized Spin Operators.

Symmetry has usually been used with advantage in constructing the simplest linear combinations of basic product wavefunctions (3) consistent with the relevant symmetry point group.

The approach here,[6] however, is to define a total spin operator,

$$\hat{A}_k = \hat{F}_k + \hat{F}_{k'}, \qquad (11)$$

for each symmetrically equivalent pair of groups and construct a new set of basis functions which diagonalizes $\hat{A}^2$ and $\hat{A}_z$ for each pair. The new basis functions for each pair may be written

$$|A,m\rangle = \sum_k (F,F',\kappa,M-\kappa|F,F',A,M)|F,\kappa\rangle|F',M-\kappa\rangle \qquad (12)$$

where the $(F,F',\kappa,M-\kappa|F,F',A,M)$ terms are the Wigner or Clebsch–Gordon coefficients which can be read from tables.[8] (M was not defined in the paper[6] but presumably can be written $M = m_k + m_{k'}$). Kappa, κ, can take values $A, A-1, \ldots 0$.

The basis for the entire spin system then consists of a product of magnetic equivalence functions for all invariant groups and pairs total spin functions for all symmetrically equivalent pairs of groups,

$$\prod_i |F_i,m_i\rangle \cdot \prod_k |A_k,M_k\rangle \qquad (13)$$

The hamiltonian is then written in terms of the symmetric spin operators $\hat{A}_k$ and the complementary antisymmetric operators,

$$\hat{B}_k = \hat{F}_k - \hat{F}_{k'} \qquad (14)$$

Such operators have been suggested[9] previously for symmetrical spin systems, but the theory was not developed in detail.

The hamiltonian becomes:

$$h^{-1}\mathcal{H} = \sum_i v_i \hat{F}_{iz} + \sum_{i<j} J_{ij} \hat{F}_i \cdot \hat{F}_j + \sum_i \sum_k J_{ik} \hat{F}_i \cdot \hat{A}_k$$
$$+ \sum_k v_k \hat{A}_{kz} + \tfrac{1}{4}\sum J_{kk'}[\hat{A}_k^2 - \hat{B}_k^2]$$
$$+ \tfrac{1}{2}\sum_{k<\ell}[(J_{k\ell}+J_{\ell k})\hat{A}_k \cdot \hat{A}_\ell + (J_{k\ell}-J_{\ell k})\hat{B}_k \cdot \hat{B}_\ell] \qquad (15)$$

---

[8] E. U. Condon and G. H. Shortley, 'The Theory of Atomic Spectra', Cambridge University Press, London, 1967, pp. 76—9.

[9] R. D. Harris and R. Ditchfield, *Spectrochim. Acta*, 1968, **24A**, 2089.

The transition operator is now written as

$$\hat{X} = \sum_i \hat{F}_{ix} + \sum_k \hat{A}_{kx} \qquad (16)$$

Some details of how the matrix elements can be evaluated by taking advantage of the commutation properties of the operators $\hat{A}, \hat{B}$ and their components are given.[6]

Symmetry Properties of the $|A,M\rangle$ *Basis Functions.* The effect of C_2 operating on $|A,M\rangle$ is to interchange all primed and unprimed groups. Two cases are discussed. The first, for which $F = F'$, involves pairs of functions $|A,M\rangle$ which belong to one of the irreducible representations of C_2. The state of highest spin belongs to the A representation and those of lower spin alternately B and A. The complete product wavefunction (13) will have B symmetry if an odd number of the individual pair functions have B symmetry; otherwise it will have A symmetry. The invariant groups, of course, cannot effect the symmetry of the product function since they always have A symmetry. Every sub-matrix for which $F = F'$ can be separated into two sub-matrices containing A and B basis functions respectively.

The second case involves $F \neq F'$ for at least one pair. Then the product functions (13) are not eigenfunctions of C_2, and the magnetic equivalence sub-matrices cannot be further factorized using symmetry. However, the symmetry of equation (15) suggests that the two sub-matrices related by C_2 are degenerate so that the calculations of energy levels and transition intensities need only be performed once for each related pair.

The form of equation (15) can be simplified and its real structure illustrated more clearly by redefining some of its terms. The definition of $\hat{A}$ is extended to include $\hat{F}$ for invariant groups. Sums over indices $i, j \ldots$ are then extended to include all symmetrically equivalent pairs as well as invariant groups; sums over $k, \ell \ldots$ remain restricted to the symmetrically equivalent pairs. $J_{k\ell}$ is redefined as the average of the two coupling constants between pairs k and ℓ and $J_{\ell k}$ becomes half the difference between the two coupling constants.

$$h^{-1}\mathcal{H} = \sum_i v_i \hat{A}_{iz} + \sum_{i<j} J_{ij} \hat{A}_i \cdot \hat{A}_j + \sum_{k<\ell} J_{\ell k} \hat{B}_k \cdot \hat{B}_\ell$$
$$+ \sum_k J_{kk'} \hat{F}_k \cdot \hat{F}_{k'} \qquad (17)$$

The matrix elements of $\hat{A}$ are identical to those that would be found if all the nuclei in a symmetrically equivalent pair formed a single magnetically equivalent group. However, since $J_{\ell k}$ is not zero these are off-diagonal matrix elements connecting basis functions in which a symmetrically equivalent pair has different A values. This means that the terms in $J_{kk'}$ may not be neglected as they could be if $\hat{A}^2$ commuted with the hamiltonian.

In systems containing symmetrically equivalent pairs of magnetically equivalent groups, not all of the magnetic equivalence sub-matrices can be further factored into symmetric and antisymmetric parts. However, the largest magnetic equivalence sub-matrix, that in which all the magnetically equivalent groups are in their highest spin states, is always factored. This simplifies what would usually be the most difficult calculation.

Some of those interested in the theory of analysis will welcome this paper as a logical extension of the theory of systems such as $[AX_n]_2$ and $[AMX_n]_2$. The final form of the hamiltonian does have an appealing simplicity. However, many spectroscopists will be critical of a paper where there is little emphasis on illustrated examples of the practical application of the method. A number of relevant systems have already been explored and the details of explicit transition energies have been presented. These previous authors[10,11] have also clearly illustrated the ways in which the information can be extracted and it is difficult to see how the new theory can add to, or improve upon, these tested methods. The authors[6] themselves suggest that the use of computers for systems such as $[AX_3]_2$ is a practical necessity and in doing so have failed to give cognisance to the work of previous authors on this specific system. More complicated systems will almost certainly be the subjects of computer analysis.

C. Double-quantum Effects in the Spectrum of 2-Bromothiazole.[12]—It has been shown that for the $^2A^2B^3X$ spin systems ($I_A = \frac{1}{2}$, $I_B = \frac{1}{2}$, $I_X = 1$) measurements of the unsaturated linewidths in the four AB transitions can in principle enable the three parameters associated with the quadrupolar nucleus, J_{AX}, J_{BX}, and T_{1X} (the spin–lattice relaxation time of the X nucleus) to be determined.[13] It was further shown that, in practice, this is not possible in those cases, called white-spectrum cases, in which

$$4\pi^2(J_{AB}^2+\delta_{AB}^2)T_{1X}^2 \ll 1$$

In this limit, as in the case for 2-bromothiazole (1), only two parameters can

(1)

be determined from measurements of the four linewidths, i.e. the $J_{AX}^2 T_{1X}$ product and $|r| = |J_{BX}/J_{AX}|$.

The sign of r has been determined unambiguously[14] from double-resonance experiments. It is claimed by Harris and Pyper that the sign of r can be determined more simply by studying the width of the single-resonance double-

---

[10] B. E. Mann, *J. Chem. Soc. (A)*, 1970, 3050.
[11] R. K. Harris, *Canad. J. Chem.*, 1964, **42**, 2275.
[12] R. K. Harris and N. C. Pyper, *Mol. Phys.*, 1972, **23**, 277.
[13] R. K. Harris and N. C. Pyper, *Mol. Phys.*, 1971, **20**, 467.
[14] A. Kumar, N. R. Krishna, and B. D. N. Rao, *Mol. Phys.*, 1970, **18**, 11.

quantum transition of the perturbed AB system. The claimed advantage is that an explicit expression is available for the half-width of the double-quantum transition with frequency $(v_A + v_B)/2$,

$$\Delta v_{\frac{1}{2}} = (4\pi J_{AX}^2 T_{1X}/3)(1+r)^2$$

provided $(\omega_A - \omega_X)^2 T_{2X}^2 \gg 1$ (as is the case for 2-bromothiazole). The predicted values of $\Delta v_{\frac{1}{2}}$ are:

$$r = +0.3, \quad \Delta v_{\frac{1}{2}} = 0.595 \text{ Hz}$$

$$r = -0.3, \quad \Delta v_{\frac{1}{2}} = 0.17 \text{ Hz}$$

[the value of $(4\pi J_{AX}^2 T_{1X}/3)$ was shown to be 0.35 Hz].[13]

$\Delta v_{\frac{1}{2}}$ was measured ten times at 100 MHz (33.5 °C) at three different values of B_1 to check on saturation of the double-quantum transition. The linewidth of the double-quantum transition was extracted from the experimental spectra using D-MAC pencil follower and program LORENTDECOMP because of overlap between the double- and single-quantum transitions. The experimental value $\langle \Delta v_{\frac{1}{2}} \rangle_{av} = 0.53 \pm 0.03$ was reasonably interpreted as confirming Kumar's result of a positive sign for r.

Conventional double-resonance (tickling) experiments cannot succeed in such cases as this because the resonance lines of the X nucleus are broad.

D. The Direct Method Applied to Nuclear Magnetic Double Resonance.[15]— This paper provides a more 'lucid' approach to the problem originally described by Anderson.[16]

In the 'direct method' the transition frequencies are calculated as eigenvalues of the Hamiltonian derivation superoperator, represented by a matrix in a suitable orthonormal basis of conventional Hilbert-space operators. It is assumed in this paper[15] that the reader is familiar with the work of Banwell and Primas.[17] The improvement claimed is in choosing an orthonormal basis set $\{A_i | A_j\} = \delta_{ij}$. The theory is illustrated for a single-spin case and the AX case.

3 Known Spin Systems

A. AB-Based N.M.R. Systems: Choice of Bounds for θ in Trigonometric Forms of Co-factors in Mixed Wavefunctions.[18]—The problem of bounds for the angle θ in the trigonometric co-factor relationship has been examined in some detail to resolve difficulties found in labelling spectral lines for the purpose of double-resonance experiments carried out in ^{15}N-labelled heterocycles. The

---

[15] B. Gestblom, O. Hartmann, and J. M. Anderson, *J. Magn. Resonance*, 1971, **5**, 174.
[16] J. M. Anderson, *J. Magn. Resonance*, 1969, **1**, 89.
[17] C. N. Banwell and H. Primas, *Mol. Phys.*, 1963, **6**, 225.
[18] M. J. Batterham and C. J. Bigum, *Org. Magn. Resonance*, 1972, **4**, 67.

Spectral Analysis

discussion is limited to the solution of 2×2 determinants associated with ABMX..., systems. The determinant

$$\begin{vmatrix} H_{11}-E, & H_{12} \\ H_{21}, & H_{22}-E \end{vmatrix} = 0$$

with basis wavefunctions ψ_1 and ψ_2 can be written:

$$\begin{vmatrix} x+y, & z \\ z, & x-y \end{vmatrix} = 0,$$

where $x = (H_{11}+H_{22})/2 - E$, $y = (H_{11}-H_{22})/2$, and $z = H_{12} = H_{21}$. The mixing term C then becomes $C = \pm(y^2+z^2)^{\frac{1}{2}}$ and $\tan 2\theta = z/y$, $C\sin 2\theta = z$, and $C\cos 2\theta = y$ define the normal trigonometric forms of the co-factor relationships. The mixed wavefunctions and associated energies become

Wavefunction	Energy level
$\cos\theta\,\psi_1 + \sin\theta\,\psi_2$	$E'_{11} = x' + C$
$-\sin\theta\,\psi_1 + \cos\theta\,\psi_2$	$E'_{22} = x' - C$

where $x' = x + E_{ii}$ for $i = 1$ or 2.

The next step in the procedure, the setting of bounds for the angle θ, has been a contentious problem. Criteria for a satisfactory set of bounds for θ can be formulated: (a) they must provide for a smooth progression from the fully mixed state $[(\alpha\beta - \beta\alpha)/\sqrt{2}]$ towards the appropriate unmixed limits $\alpha\beta$, $\beta\alpha$; (b) as this occurs the system cannot disobey the non-crossing rule which states that, in the process of mixing, energy levels may not become degenerate or cross-over with respect to their associated unmixed levels. The original choice of bounds, $0 \leqslant \theta \leqslant \pi/2$, made by Pople, Schneider, and Bernstein,[19] although quite satisfactory for all positive cases for which it was suggested, is ambiguous when negative parameters are involved, because it implies two situations of maximum mixing and makes a choice between them necessary (for this case $\sin 2\theta$ is positive but $\cos 2\theta$ can have either sign).

An alternative range $-\pi/4 \leqslant \theta \leqslant \pi/4$ has been put forward as necessary to the description of the physical mixing process. Batterham and Bigum,[18] however, state that both sets of bounds adequately describe the system since the bounds of θ can be divided into two completely separate octants. The system is seen as moving from a state of no mixing to a common or equivalent state of maximum mixing in each separate octant. A discontinuity occurs at the point of maximum mixing such that the octants are quite separate and in fact need not be adjacent. For example the octants in the $-\pi/4 \leqslant \theta \leqslant \pi/4$ bounds are separated by $3\pi/2$. Adherence to the non-crossing rule applies only to each octant as a separate entity and has no validity across the discontinuity. The discontinuity at the point of maximum mixing creates labelling problems regardless of the choice of bounds. If the $-\pi/4 \leqslant \theta \leqslant \pi/4$ set is used a change of

[19] J. A. Pople, W. G. Schneider, and H. J. Bernstein, 'High Resolution Nuclear Magnetic Resonance', McGraw-Hill, New York, 1959, p. 120.

sign of $y = \delta_{AB}$ (in ABMX ... systems) causes an inversion of the physical order of the energy levels without any such interchange of level numbers. Thus, the transition numbers of all lines involved with the 2×2 sub-matrix change at the point of maximum mixing. In a hand calculation such a change is disconcerting, but in the iterative use of computer programs variation of line numbers renders effective iteration impossible. The use of bounds $0 \leq \theta \leq \pi/2$ for hand calculation avoids most of these problems. In this case the energy levels are not changed as y passes through zero and no discontinuity occurs in the line patterns.

The behaviour of $\sin\theta$ and $\cos\theta$ co-factors as $y = \delta_{AB}$ changes sign, for all situations between maximum and zero mixing, is described with the help of a diagram. The practical problems associated with the use of the $-\pi/4 \leq \theta \leq \pi/4$ bounds are illustrated with reference to an ABX system. The changes in the line-numbering with changing δ_{AB} from $+10$ Hz to -10 Hz for $J_{AB} = 2$, $J_{AX} = -10$, and $J_{BX} = 1$ Hz are described for the AB and X parts separately with a pair of diagrams which are unfortunately labelled the wrong way round. The four lines of the X multiplet which involve the two second-order determinants show a more dramatic dependence on the sign of the δ_{AB} than the lines in the AB part of the spectrum. Two discontinuities associated with relabelling of transitions are clearly illustrated for the cases when y_1 and y_2 respectively pass through zero (y_1 and y_2 were not defined but are presumably the effective chemical shifts δ_{AB} for $m_X = \pm\frac{1}{2}$). The effects of a change of sign of $z = H_{12}$ and H_{21} upon the signs of C and y are discussed in terms of changes in the line numbers.

It is felt that this paper deals with a fundamental point which may well puzzle many newcomers to the principles of n.m.r. analysis. It is worthy of a closer perusal.

B. Systems involving Protons Only.—*Saturated Acyclic Systems.* (i) *n-Propyl derivatives;* $[AB]_2C_3$.[20] The spectra of eight n-propyl compounds, MeCH$_2$-CH$_2$X, with X = CHO, COMe, CO$_2$Et, CN, or halogen, plus the symmetrical ether (as neat liquids with 8—10% v/v benzene as internal reference), have been measured at 100 MHz and analysed as $[AB]_2C_3$ systems in order to study the mechanism by which, and the extent to which, substituent effects are propagated along a chain of saturated carbon atoms. The approach is claimed to be more rigorous than in a former paper by Cavanaugh and Dailey.[21] The sub-spectral breakdown of $[AB]_2X_3$ systems is described and compared in a qualitative way with the magnetic equivalence factoring in $[AB]_2C_3$ systems. A detailed discussion of the availability of K, M, L, and N values follows but with a very limited bibliography.

The compounds are divided into two groups:

(1) n-propyl halides and di-n-propyl ether where δ_{AB} is large and an $A_2B_2C_3$ analysis(!) reproduced spectra exactly.

[20] G. Schrumpf, *J. Magn. Resonance*, 1972, **6**, 243.
[21] J. R. Cavanaugh and B. P. Dailey, *J. Chem. Phys.*, 1961, **34**, 1094.

(2) X = CHO, COMe, CO₂Et, or CN where θ_{AB} is considerably smaller and the information obtained by the authors was reduced.

Differences between the data obtained are discussed in terms of the electronegativity of the substituent and conformational differences but not in any great depth.

(ii) *2,2′-Dihalogenodiethyl ethers;* $[AB]_2$.[22] The proton spectra of the compounds $(XCH_2CH_2)_2O$, where X = Cl, Br, or I, as neat liquids at ambient temperature (40 °C) and in carbon disulphide at different temperatures, have been measured at 60 MHz and analysed as $[AB]_2$ systems to determine the more populated rotamer in each case. Spectral decomposition methods were employed to follow the variation in the *L* parameter as the conformational equilibrium changed with temperature. An expression

$$3N/2 + L/2 = -3.3\chi + 28.6$$

was derived, where χ is the electronegativity of the halogen.

Saturated Cyclic Systems. (i) *1,3-Dioxans; AB, AMX, ABX,* $[AX]_2$.[23] The 60 MHz and 220 MHz spectra of five substituted 1,3-dioxans which are known to exist in non-chair conformations have been studied to determine the preferred conformation of the two possible structures [(2a) and (2b)] in each case.

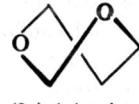   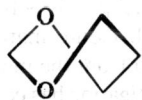

(2a) 1,4-twist   (2b) 2,5-twist

The method of approach to this problem involved the interpretation of vicinal coupling constants ³*J*(H,H), which are well-characterized for the chair conformations but not established in alternative conformations. However, the point was made that generalizations arising from the Karplus relation can be applied, so that for ³*J*(H,H) > 10 Hz a *trans*-periplanar or eclipsed arrangement would be expected and for ³*J*(H,H) < 1 Hz dihedral angles of about ninety degrees would be indicated.

The 2,5-twist form was predicted for *trans*-2,2,4,6-tetramethyl- and 2,2-*trans*, 4-*cis*,5,6,6-pentamethyl-1,3-dioxans and confirmed by $[AMX_3]_2$ and AMX analyses, respectively. An ABX analysis confirmed the predicted 1,4-twist form for 2,4,4,6-tetramethyl-1,3-dioxan. Equilibria between 1,4- and 2,5-twist conformations were postulated for 2,2,4,4,5-pentamethyl-1,3-dioxan and 2,2,4,4,6-pentamethyl-1,3-dioxan following ABX analyses; the 1,4-twist form is favoured. Specific deuteriation was used to simplify some spectra for analysis.

The concentration of the solutions used was ~10% v/v but no solvent was given and no indication was given in the figures of the direction of increasing frequency or magnetic field.

[22] E. Haloni, D. Canet, and P. Granger, *Org. Magn. Resonance*, 1971, **3**, 451.
[23] K. Pihlaja, G. M. Kellie, and R. G. Riddell, *J. C. S. Perkin II*, 1972, 252.

```
        MeCO·O   H
      H  \ / \ / O·COMe
   MeCO·O/   \ /H
        R    H
         \ / \ /
        H  O  H
```

(3) R = OAc, OMe, Cl, Br, or SAc

(*ii*) D-*Aldopentapyranosyl derivatives; ABX.*[24] The conformations of 2,3,4-tri-*O*-acetyl-D-aldopentapyranosyl derivatives (3) have been studied by analysing the proton resonance spectra obtained at 100 MHz in deuterioacetone and deuteriochloroform at 31 °C. Proton–proton coupling constants, measured for a range of tetrasubstituted tetrahydropyran ring systems produced from aldopyranose sugars, have been correlated with values calculated from a generalized version of the Karplus equation which takes into account the electronegativities of the various atoms in the molecule and the configurations of these substituents. Deviations from the main correlation have been explained in terms of a flattening of the ring caused by steric effects. The magnitude of the geminal coupling constant J_{5eq5ax} of the aldopentopyranoso-derivatives provides a rough measure of the conformational population. J_{5eq5ax} is ~ 10.9 Hz when H-4 is axial and is ~ 13.4 Hz when H-4 is equatorial. Conformational populations and equilibrium constants have been derived from the measured coupling constants.

A similar investigation has centred around 1,6-dibromo-1,6-dideoxy-2,3,4,5-tetra-*O*-acetyl-D-mannitol.[25]

(*iii*) N-*Substituted styreneimines; ABX.*[26] The relative importance of reaction-field effects and specific interactions has been studied in *N*-substituted styreneimines (4). The 100 MHz spectra have been obtained for these com-

```
       H     H
    Ph  \ / H
         N
         |
         R
```

(4) R = H or Bu^t

pounds in five solvents (10 mol% and 5 mol%). The samples were degassed and sealed; they included internal tetramethylsilane as reference. The spectra were analysed as ABX systems using LAOCOON 3. The magnitude of the geminal coupling constant increases in dimethyl sulphoxide or deuteriochloroform solutions of styreneimine itself. This is claimed to be the first such observation. Changes in the *cis* and *trans* ^3J(H,H) values appear more or less random and

[24] P. L. Dunette and D. Horton, *Org. Magn. Resonance*, 1971, **3**, 417.
[25] L. A. Mai, O. K. Lukevits, and A. Y. Berzete, *Latv. P.S.R. Zinotnu Akad. Vestis Kim. Ser.*, 1972, 114.
[26] R. H. Cox and L. W. Harrison, *J. Magn. Resonance*, 1972, **6**, 84.

Spectral Analysis

barely exceed experimental error. No obvious correlations between $^2J(H,H)$ and solvent dielectric constant or the corresponding chemical shift were observed. These observations have been interpreted as indicating that some factor other than reaction field is responsible for the solvent-induced variations in $^2J(H,H)$, in particular specific interactions. $^2J(H,H)$ in N-t-butylstyreneimine decreases in magnitude with increasing dielectric constant of the solvent (1.47 Hz→1.01 Hz). $^3J(H,H)$ was again observed to be relatively insensitive. The conclusion reached was that specific interactions (H-bonding in this case?) may over-ride reaction-field effects but the data do not allow a distinction between the two effects. The results further demonstrate the applicability of n.m.r. in investigating weak solute–solvent interactions.

(iv) L-*Azetidine-2-carboxylic acid* (L-*ACA*) *and* N-*acetyl*-L-*azetidine-2-carboxylic acid* (N-*A*-L-*ACA*); *ABCDE*.[27] The conformations of these four-membered ring compounds, (5) and (6), have been investigated by analysing the ABCDE

(5) L-ACA

(6) N-A-L-ACA   trans   cis

spectra, observed at 100 MHz, using the computer program LAME. The spectrum of L-ACA was obtained from a 1.2 mol l^{-1} solution in D$_2$O with t-butyl alcohol as internal reference and locking signal. N-A-L-ACA was dissolved in pyridine to about 1 mol l^{-1} concentration and the first-order spectrum obtained at 90 °C was calibrated. The best iterative fit was obtained for $^3J_{cis} > {}^2J_{trans}$. An overall reduction in the magnitudes of the coupling constants in proceeding from the zwitterion (L-ACA) to the N-acetylamino-acid (N-A-L-ACA) could be explained, it was suggested, either by electron withdrawal from the ring or by conformational effects. A comparison with alanine and N-acetylalanine, where $^3J(H_3CCH)$ remains constant (≈ 7.30 Hz) was interpreted as compatible with a negligible effect due to delocalization of the nitrogen lone pair on acetylation. The changes have therefore been interpreted in terms of a buckled ring, but rapid pseudorotation was not ruled out. It was concluded from the experimental values of the coupling constants around the ring that the *cis*-form of N-A-L-ACA is probably the less stable, especially in polar solvents.

(v) *1,2,3-Substituted prolines; ABCDEF.*[28] The proton n.m.r. spectra of eight substituted prolines (7) have been recorded at 60 MHz and 100 MHz. Three

---

[27] W. A. Thomas and M. K. Williams, *Org. Magn. Resonance*, 1972, **4**, 145.
[28] C. Gallina, M. Paci, and P. Viglino, *Org. Magn. Resonance*, 1972, **4**, 31.

(7) $R^1 = CO_2CH_2Ph$ or H
$R^2 = CO_2Me, CO_2H,$ or H
$R^3 = CO_2Me, CO_2H,$ or H
$R^4 = OH, NH_2,$ or $OSO_2 \cdot C_6H_4 \cdot Me$

amino- and three hydroxy-prolines exhibited *cis–trans* isomerism. A complete spectral analysis was achieved only in the case of the three amino-prolines with the help of LAOCOON 3. Overlapping lines made other spectra rather insensitive to some *J*-value variations. In attempts to establish the conformations of the rings it was concluded that the Karplus relation is not valid in these cases because of (i) the changing nature of the substituents, (ii) conformational distortion, expected to be $\pm 30°$ for the H(C-2)–(C-3)H dihedral angle, and (iii) two quickly interconverting conformations. The results obtained were interpreted as consistent with the proposed synthetic pathway in the substitution reaction employed to prepare the compounds. The impression was of a confused discussion hampered by some translational difficulties which could have been corrected at the referee stage!

Olefinic Compounds: Dichloropentenes; ABCDE.[29] The proton n.m.r. spectra of 2,4-dichloropent-1-ene (8a) and of 2,4-dichloro[1-^{2}H]pent-1-ene [(8b) and (8c)] in 1% deuteriobenzene solution have been recorded at 100 MHz in order to study the short chain branching in poly(vinyl chloride). The analysis of the

(8) a; $R^1 = R^2 = H$
b; $R^1 = H, R^2 = D$
c; $R^1 = D, R^2 = H$
d; $R^1 = R^2 = D$

fully protonated compound was accomplished using LAOCOON 3 and by irradiating the methyl resonance to reduce complexity. An initial ABK treatment (K→C) followed by introduction of coupling constants between H_A, H_B, and the CH_2 protons H_D and H_E as first-order effects, culminated in a full ABCDE iterative analysis. The signs obtained were justified by reference to the literature. In a study of the deuteriated isomers the best 'resolution' of the methylene and olefinic multiplets was obtained using deuteriobenzene as solvent. An analysis in which the methyl protons were considered as only one proton and the deuteron as an infinitely remote nucleus with spin one half was carried out using LAOCOON 3 which in the unmodified form available accepts only nuclei of spin ½. The four components of the mixture (8) arising from the

[29] G. Gurato and A. Rigo, *Org. Magn. Resonance*, 1971, **3** 433

reaction Me·CHCl·CH$_2$·C≡CH + DCl were shown to be in the ratio (HD + DH)2 : (H$_2$)1 : (D$_2$)1, an unexpected result which was not discussed in detail.

Unsaturated Five-membered Ring Compounds. (i) *3,5-Disubstituted 1,2-dihydrofurans and -thiophens; ABX, ABXY$_3$, ABXY$_2$, ABC.*[30] A study of (a) the effect of substituents on the conformations of these unsaturated five-membered ring compounds and (b) the orientations of the substituents themselves, compared with orientations in the saturated compounds, has been pursued by examining the proton spectra at 100 MHz of the neat liquids with 1% TMS added. The double bond eliminates pseudorotation and reduces the possibilities to a choice between the planar and envelope conformations of cyclopentene. LAOCOON 3 was used to refine the spectral parameters. The ratio of the vicinal coupling constants in the CH$_2$CH fragment, J_{cis}/J_{trans}, varied from 0.73 to 0.82 whereas the Karplus relation reportedly predicts a ratio of four. The effect of an electronegative substituent in the 5-position was discounted and the result attributed to the existence of some degree of ring puckering by analogy with the unsubstituted molecules. The relatively high value of J_{AX} was interpreted as favouring isomer (b) of the two alternatives, (9) and (10), shown.

(9a) (9b)

(10a) (10b)

The angle of pucker was deduced to be about the same as that in dihydrofuran (25°).

(ii) *Dimers of cyclopentadienyl compounds.*[31] The proton resonance spectra of the cyclopentadiene *endo*-dimer (11a), its 1$_y$-acetoxy-derivative, and the dimer of spiroheptadiene (11b) have been analysed. The assignment was proved by using spin-decoupling and INDOR. The results are discussed together with literature data on related compounds (norbornane, cyclopen-

---

[30] E. Dradi and G. Gatti, *Org. Magn. Resonance*, 1971, **3**, 479.
[31] N. M. Sergeyev, G. I. Avramenko, V. A. Korenevsky, and Yu. A. Ustynyuk, *Org. Magn. Resonance*, 1972, **4**, 39.

tene, *etc.*) in order to formulate suitable rules to assist in the analysis of the spectra of similar compounds. Better resolution is claimed in this work than that achieved by Foster and McIvor.[32]

(11a)   (11b)

Sesquiterpenoids; ABX{H}.[33] The 100 MHz proton spectra of the title compounds have been analysed with particular interest in long-range coupling constants. The names epoallylic and epohomoallylic interactions have been

(12)

proposed for interactions between olefinic (AB) and aliphatic protons separated by up to six σ-bonds in epoxy-containing sesquiterpenoids (12). These interactions have been shown to exist by using double irradiation and the authors draw a careful distinction between decoupled peak height enhancement and the nuclear Overhauser effect. The discussion is marred by the introduction of the terms 'real coupling mechanism' (*via* σ-electrons) and 'virtual coupling mechanism' (*via* π-electrons). Previous authors[34] referred to in this paper,[33] in discussing the mechanism of long-range couplings in unsaturated compounds, made no mention of 'real' or 'virtual' contributions. The use of 'virtual' is doubly unfortunate since it has undesirable connotations in n.m.r. analysis.

Aromatic Systems. Aromatic substitution patterns.[35] A review of the n.m.r. data for several thousand aromatic compounds 'revealed' that 'between 45% and 55%' of all aromatic compounds can be completely analysed as first order. The point was made that it makes little difference to the practical organic chemist whether the spectrum is truly first order or 'pseudo first order' as long as a valid interpretation and structure assignment can be made. The salient spectral features, empirical correlations, and experimental techniques used in identifying substitution patterns in aromatic systems are dis-

[32] R. G. Foster and M. C. McIvor, *Org. Magn. Resonance*, 1969, **1**, 203.
[33] P. Joseph-Nathan and E. Diaz, *Org. Magn. Resonance*, 1971, **3**, 193.
[34] P. Albriktsen, A. V. Cunliffe, and R. K. Harris, *J. Magn. Resonance*, 1970, **2**, 150.
[35] M. Zanger, *Org. Magn. Resonance*, 1972, **4**, 1.

cussed. An inaccurate limitation is inferred in the use of 'certain complexing agents which can cause enormous changes in chemical shifts of compounds containing amine or hydroxy-groups'.

Biologically important aromatic acids.[36] Precise ring-proton chemical shifts and coupling constants for phenyl acetate, phenylacetic acid, seventeen phenylacetic acid derivatives, and four acetoxy-substituted benzoic acid derivatives have been obtained from computer simulation of the proton spectra recorded at 100 MHz for 1.0 mol l^{-1} solutions in [^{2}H$_6$]acetone ($+5\%$ TMS, v/v). The stated object was to provide greater efficiency and a higher confidence level in the identification of unknown compounds, biological or synthetic, by compiling data obtained under precisely the same experimental conditions. The samples used were obtained commercially and not purified further but checked using mass spectrometry. Samples and solvents were claimed to be dry but no carboxy- or hydroxy-protons were observed. The use of empirical relationships was advocated to predict starting parameters for analysis problems. Diehl's work[37] in this respect was mentioned in a rambling discussion.

Benzocycloheptenes.[38] Computer analysis of the 100 MHz proton spectra of several partially deuteriated benzocycloheptene derivatives (13) at -120 °C (in CHClF$_2$) provided values for all the coupling constants involving protons bonded to C-3 and C-4. An interpretation using the Karplus equation favoured a chair conformation for the seven-membered ring. The validity of the Karplus equation in this situation was determined from relationships between coupling constants. This investigation was stimulated by the controversy over the most stable form of cycloheptene. The deuteriated derivatives are shown (13a—d).

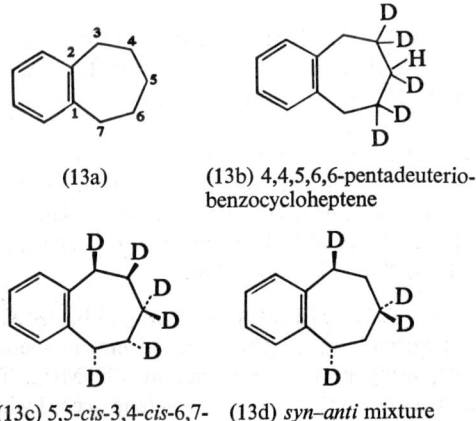

(13a)

(13b) 4,4,5,6,6-pentadeuteriobenzocycloheptene

(13c) 5,5-*cis*-3,4-*cis*-6,7-hexadeuteriobenzocycloheptene

(13d) *syn–anti* mixture

[36] K. N. Scott, *J. Magn. Resonance*, 1972, **6**, 55.
[37] P. Diehl, *Helv. Chim. Acta*, 1961, **44**, 829.
[38] M. St. Jacques and C. Vaziri, *Org. Magn. Resonance*, 1972, **4**, 77.

(i) 4,4,5,6,6-Pentadeuteriobenzocycloheptene analysis; H{D}. Two peaks, other than aromatic, were observed at normal temperatures when the deuteron region was irradiated. At -120 °C an $[AB]_2$ system was observed and analysed using LAOCOON 3 and CALCOMP to give $J_{AA'} \approx 0$, $J_{AB'} = J_{A'B} < 0.5$, and $J_{BB'} < 1.5$ Hz. A doublet attributed to the C-5 proton was observed at -120 °C and coalesced at -58 °C.

(ii) 5,5-cis-3,4-cis-6,7-Hexadeuteriobenzocycloheptene, H{D}. The multiplicity of the spectrum at normal temperatures indicated incomplete stereoselectivity of reduction in the preparative stages. At -120 °C the spectrum consisted of first-order AX spectra, which allowed all parameters for the possible stereochemical arrangements to be obtained. Double-resonance experiments distinguished between the possible doublet combinations.

(iii) 3,5,5,7-Tetradeuteriobenzocycloheptene. The spectrum at -120 °C was analysed as the sum of two ABX spectra where the effects of small cross-ring couplings were taken into account as a broadening of the lines in the experimental spectrum. The linewidth was varied in an iterative fit.

The distinction between the chair and boat forms (14) was made by

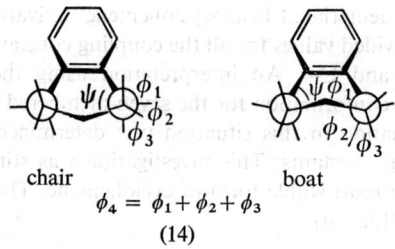

chair boat

$$\phi_4 = \phi_1 + \phi_2 + \phi_3$$
(14)

invoking the Karplus relation: for the boat form $\phi_1 = \phi_3 \approx 20$—$30°$ and $\phi_2 \approx 100°$, so that $\phi_4 = 140$—$160°$; for the chair form $\phi_1 = \phi_3 = 73°$ and $\phi_2 = 44°$, so that $\phi_4 = 189°$. The value obtained from the experimental coupling constants was $\phi_4 = 190°$, confirming that the chair form is more likely. The distortion from the cyclohexane chair form was estimated by using the R parameter, J_{trans}/J_{cis}, which is 2.16 in cyclohexane and 8.7 in benzocycloheptene. The angle ψ was calculated using the relationship $\cos \psi = [3/(2+4R)]^{\frac{1}{2}}$ with the result $\psi = 74°$. This agrees well with the value derived from $\psi = 120° - \phi_2 = 76°$.

Carbanions and Ion Pairing; $[AM]_2X$, $AMRX$, $[AX]_2$, $AA'X$.[39] A study of the environment of carbanions, the effects of solvent, temperature, and cation has been made, using proton resonance at 100 MHz. The spectra were analysed using LAOCOON 3 and the results were in accord with an equilibrium between contact and solvent-separated ion pairs. Some thermodynamic parameters have been derived in favourable cases from chemical shift data.

[39] J. B. Grutzner, J. M. Lawler, and L. M. Jackman, *J. Amer. Chem. Soc.*, 1972, **94**, 2306.

Aryldi-t-butylmethanol.[40] This work represents the extension of similar work where the n.m.r. spectra were interpreted in terms of rotation about an sp^2–sp^3 carbon–carbon bond which is slow on the n.m.r. time scale.[41] Eight further systems have been studied at 60 and 100 MHz in various solvents and a novel type of *syn–anti* isomerism has been found in some members of the series. The proton spectrum of (*p*-methoxyphenyl)di-t-butylmethanol (15) in dimethyl sulphoxide exhibited reversible variations over the temperature range studied (40—174 °C). The spectrum was complex at the lower temperature limit but symmetrical at 174 °C, where it was analysed as [AB]$_2$ using LAOCOON 3.

The (3,4,5-trimethoxyphenyl)di-t-butylmethanol showed three methoxy-resonance signals at ambient temperatures, which simplified to two lines with relative intensities 1:2 at higher temperatures. Two overlapping ABC spectra were observed for (3,4-dimethoxyphenyl)di-t-butylmethanol at ambient temperatures which coalesced into a single ABC spectrum at 185 °C. The two ABC systems were assigned to the different isomers using the nuclear Overhauser effect. Careful integration revealed that the two rotamers were unequally populated. Other molecules showed similar interesting temperature dependence.

Acridine and Five Monomethyl Derivatives; '*ABCDX*'.[42] The opinion was expressed that information about relative signs of coupling constants in condensed aromatic systems are of considerable interest in view of the mechanism of spin interactions and the need to test current theories. The proton spectrum of acridine in deuteriochloroform recorded at 100 MHz was treated as two identical ABCD systems and long-range coupling constants J(H-1,H-9), J(H-4,H-9) were introduced as first-order perturbations. Iterative calculations were made. The near approximation to a first-order spectrum allowed double- and triple-resonance experiments to yield relative signs. The acridine molecule (16) exhibits three different long-range interactions across five bonds. These long-range coupling constants are realized in the inter-ring interactions J_{49}, J_{59}, J_{29}, and J_{79} and in the *para* couplings J_{14} and J_{58}. All ^5J's were shown to have positive sign and were discussed in terms of the

---

[40] R. E. Gall, D. Landman, G. P. Newsoroff, and S. Sternhell, *Austral. J. Chem.*, 1972, **25**, 1.
[41] G. P. Newsoroff and S. Sternhell, *Tetrahedron Letters*, 1967, 2539.
[42] O. Sciacovelli and W. V. Philipsborn, *Org. Magn. Resonance*, 1971, **3**, 339.

σ- and π-mechanisms involved. A consistent value of -0.15 Hz was obtained for J_{39} and J_{69} in acridine and 2-methylacridine. Although small it was attributed to contributions from two independent pathways, *cis,trans,cis*, *trans* and *trans,trans,cis,trans*. The long-range coupling constants involving the methyl groups were used to illustrate the π-bond-order alternation in this system.

Sulphur Heterocycles. Ewing and Scrowston[43] have investigated the effects of varying solvents and concentration on the 100 MHz spectra of thiophen and related systems. These include naphtho-[2,1-*b*]- and -[1,2-*b*]-thiophens, thieno-[2,3-*b*]- and thieno[3,2-*b*]-benzothiophens. The multispin systems constituting ABCD and AB patterns involving long-range coupling constants were analysed using LAOCOON 3. Chemical shifts were extrapolated to infinite dilution and $\Delta^{CS_2}_{CDCl_3}$ solvent shifts reported. The results were discussed in terms of specific solvent–solute interactions but no evidence was presented in support of the otherwise convincing diagrams [(17a) and (17b)].

(17a) Benzo[*b*]thiophen   (17b) Naphtho[2,1-*b*]thiophen

The ABCD protons were rather inaccurately described as 'mutually *ortho*'.

Balkan and Heffernan[44] have analysed the spectrum of benzothiophen recorded at 100 MHz in a number of solvents using LAOCOON 3 and the CALCOMP plotter. The protons of the six-membered ring constitute a decep-

[43] D. F. Ewing and R. M. Scrowston, *Org. Magn. Resonance*, 1971, **3**, 405.
[44] F. Balkan and M. L. Heffernan, *Austral. J. Chem.*, 1972, **25**, 2.

tively simple ABMX system in non-aromatic solvents. The spectra of the neat liquid and of a deuteriobenzene solution were reported as 'more normal'. Inter-ring coupling was evident in the overall complexity manifested in an ABCDEF spectrum. Very small coupling constants which could not be resolved directly were at first estimated from the relative widths of spectral lines, and more accurate values, together with relative signs, were derived from the details of the differential effects observed between different lines assigned to particular protons. Values of $J_{35} = +0.05$ Hz, $J_{36} = -0.12$ Hz, $J_{25} = +0.03$ Hz and $J_{27} = +0.02$ Hz were quoted with particular reservations recorded for the latter two. The obvious effects on the iterative fit of these small values was commented upon. Comparisons were made between these results and the parameters of benzofuran. Attention was drawn to some unexpectedly large differences in J_{24} and J_{27} and also to the fact that all the larger inter-ring couplings have the predicted relative signs if one assumes that the coupling paths are not through the heteroatom.

Long-range coupling constants have also been studied by Blackburn et al.[45] in benzothiophen and benzofuran analogues [(18a) and (18b)] where they are manifest as differentially increased linewidths.

(18a) 7-Methyl(7-chloro)-benzo[1,2-b][4,3-b']-dithiophens
X = Me or Cl

(18b) Thieno[3,2-e]benzofuran

An additional long-range coupling of 0.6 Hz associated with the thiophen part of the molecule was recorded to complement the previously reported coupling attributed to the furan entity. The argument that the inability to decipher a long-range coupling in the symmetrical benzodifuran and benzodithiophen molecules appeared to be associated with the equivalence of the benzenoid protons was not very convincing. The long-range coupling constants were apparently read off directly as first-order perturbations, although this was not explicitly stated. Virtual coupling was invoked to explain what is described as 'masking' of some long-range coupling constants.

The conformations of the five- and six-membered heterocyclic rings of thiochromanones (19a,b), dihydrobenzothiophen dioxide (19c), indanone ethylenedithiol ketal (19d), and related compounds have been investigated, using the proton resonance spectra recorded at 60 and 90 MHz.[46] The

[45] E. V. Blackburn, T. J. Cholerton, and C. J. Timmons, J. C. S. Perkin II, 1972, 101.
[46] J. B. Lambert and F. R. Koeng, Org. Magn. Resonance, 1971, 3, 389.

principle n.m.r. system involved was $[AB]_2$, and this was analysed in each case using LAOCOON 3, NMRPLOT, and ENIT programs.

(19a) 4-Thiochromanone

(19b) 4-Thiochromanone dioxide

(19c) 2,3-Dihydrobenzothiophen dioxide

(19d) Indanone ethylenedithiol ketal

The value of R ($= J_{trans}/J_{cis}$) given for cyclohexanes was 1.9—2.2 and the interpretation of deviations from this range of values suggested that $R<1.9$ implied a flattened structure whereas $R>2.2$ implied a puckered structure. The R value was also used to calculate internal dihedral angles using $\cos \psi = [3/(2+4R)]^{\frac{1}{2}}$. The greatest interest arose in the five-membered rings, where the data were interpreted to suggest that the rings were flattened to so great an extent that they gave R values outside the previously recorded range. No discussion of the validity of the Karplus equation was given. The paper was illustrated with a surfeit of diagrams.

syn- *and* anti-2-*Furanaldoximes*.[47] An investigation of conformational preferences in DMSO, C_6D_6, and $CDCl_3$ solutions of the *syn*- (20) and *anti*- (21) isomers of 2-furanaldoximes using the proton spectra recorded at 100 MHz has been reported. The spectra were presented as close to first order and the program LAME was used to refine the spectral data. The assignments of isomers were made on the basis of chemical shifts in agreement with other physical methods. The zig-zag stereospecific coupling over five bonds was used to elucidate the preferred conformations. $^5J_{25} = 0$ in the *anti*-isomer in all three solvents but $^5J_{24} = 0.70 \pm 0.01$ Hz in each solvent, suggesting the *sym-cis*-form for the *anti*-isomer. $^5J_{25}$ in the *sym-cis*-form and $^5J_{24}$ in the *sym-trans*-form were assumed negligible for the *syn*-isomer. $^5J_{24}$ in the *sym-cis*-form and $^5J_{25}$ in the *sym-trans*-form were assumed to be 0.61 Hz (observed as an experimental splitting?). On this basis the *syn*-isomer was shown to exist 41% in the *sym-cis*-form. ASIS were used to support the assignments of the *sym-cis*-form to the *anti*-isomer, and are qualitatively consistent with the proportions of *sym-cis*- and *sym-trans*-forms of the *syn*-isomer inferred from the

---

[47] R. Wasylischen and T. Schaefer, *Canad. J. Chem.*, 1972, **50**, 274.

Spectral Analysis

sym-cis   *syn* m.p. 75 °C   *sym-trans*
(20)

sym-cis   *anti* m.p. 91 °C   *sym-trans*
(21)

five-bond coupling constants. The effects of chemical shift reagents and CNDO calculations of long-range coupling constants were presented in support of the conclusions reached.

C. Compounds containing Fluorine.

—Fluorinated alkanes have been the subjects of n.m.r. studies designed to investigate conformational equilibria. De Marco and Gatti[48] have interpreted proton (90 MHz) and fluorine (84.6 MHz) spectra in favour of a low barrier to rotation in 2,2,4,4,4-pentafluorobutane. The spectra were analysed as first order although it was stated that the spectrum could be more realistically interpreted as a deceptively simple $[AX]_2$ case. Assignments were confirmed by heteronuclear decoupling and the spectrum was simulated using LAOCOON 3. Relative signs were determined and the barrier to rotation about the C-2—C-3 bond was calculated for twelve solvents. Long-range couplings involving the CF_3 and CH_3 groups were measured ($^5J_g = 0.97$ Hz, $^5J_t = 0.16$ Hz) and the result, $^5J_g > ^5J_t$, was rationalized by invoking a through-space interaction. A similar study involved 1,2-disubstituted tetrafluoroethanes.[49]

Schaumburg[50] has reported work involved with the fluorination of organic compounds and an attendant study of the n.m.r. spectra. Data for the (2-chloro-1,1,2-trifluoroethyl)diethylamine and the amide of chlorofluoroacetic acid are given. The preferred configurations are commented upon. In the same paper the n.m.r. spectral data for a series of aromatic heterocyclic carbonyl fluorides are reported, with particular emphasis on long-range

---

[48] A. De Marco and G. Gatti, *J. Magn. Resonance*, 1972, **6**, 200.
[49] L. Cavalli, *J. Magn. Resonance*, 1972, **6**, 298.
[50] K. Schaumburg, *J. Magn. Resonance*, 1972, **7**, 177.

couplings involving the —COF fluorine and ring protons. It was characteristic of the data that the ring coupling constants were virtually unchanged in value compared with the corresponding acids. The long-range couplings, however, varied considerably in magnitude (and some in sign) in the carbonyl fluorides studied. The data were time-averaged over two configurations (22)

(22) X = O, S, NMe, *etc.*

and it was implied that they were not equally weighted. A 'crude' correlation between ^3J(H,H) and ^3J(H,F) in aldehydes and carbonyl fluorides was claimed which was difficult to see! In general, the conclusion, that there was 'increased evidence' for some empirical correlations useful in interpretation of spectra, was not convincing.

4-Fluoromethylated 5(and 6)-methyl-1,3-dioxans. The 100 MHz proton spectra of these compounds have been reported in two separate papers.[51,52] Vicinal coupling constants (between ring protons) were used to evaluate the conformational energy barriers. Long-range coupling constants $^5J(2ax,\text{F})$ and $^5J(2eq,\text{F})$ were reported and discussed. Experimental details were limited.

Ethyl 2,3-Difluoropropionate; ABCXY.[53] The proton (100 MHz) and fluorine (94.075 MHz) spectra of ethyl 2,3-difluoropropionate have been re-investigated to improve on the results of a previous first-order analysis of these spectra. A sub-spectral analysis was carried out based on an approximate ABCMX system where each abc sub-spectrum was iteratively fitted using LAOCOON 4 SPIN. Seventy-two lines of the full five-spin system were then included in an iterative calculation using LAOCOON 1968. Relative signs agreed with previous results. A re-examination of the dependence of ^3J(H,F) on substituent electronegativity led to the equation

$$\langle ^3J(\text{H,F})\rangle = 53.39(\pm 0.24) - 3.23(\pm 0.02)\sum_j \psi_j$$

based on CF_3CHXY and CH_3CFXY compounds.

The results of a temperature-dependence study of three-bond coupling constants at 220 MHz were interpreted as consistent with the dominance of one isomer, namely (23), at ambient temperature, but probably with an admixture of (24). All three $\langle ^3J(\text{H,F})\rangle$ parameters were larger than expected, however, and it was concluded that this could be explained by a distortion from a perfectly staggered conformation owing to repulsion between the electric dipoles of the fluorine atoms.

[51] P. Dirinck and M. Anteunis, *Canad. J. Chem.*, 1972, **50**, 412.
[52] M. Anteunis and P. Dirinck, *Canad. J. Chem.*, 1972, **50**, 423.
[53] G. Hägele, R. K. Harris, and P. Sartori, *Org. Magn. Resonance*, 1971, **3**, 463.

Spectral Analysis

(23) (24)

Fluorinated Benzofurans; ABC, ABCD, ABCDRX$_3$.[54] A proton resonance (60 MHz) and fluorine resonance (56 MHz) study of the products (25), arising

(25) a; R = CO_2Et, X = F, Cl, CN, or H
b; R = CO_2H, X = F or Cl
c; R = H, X = F
d; R = CO_2Me, X = Cl

from reactions between pentafluorophenyl derivatives and ethyl acetoacetate, was initiated to determine unambiguously the position of substituents and to characterize the substituent dependence of fluorine chemical shifts and coupling constants. Previous quantitative predictions of (F,F) coupling constants provided the foundations but the point was made that problems have arisen because of the approximate equivalence of $J^o(F,F)$ and $J^p(F,F)$. The broadening of the lines of one ^{19}F quartet and the splitting of the methyl protons of the ethyl group were interpreted in terms of a through-space (H,F) coupling. No such coupling was observed for the methylene protons in this same molecule or for the methyl protons in the corresponding methyl ester. This allowed a positive assignment of F_4 and the approximate equivalence of J_{45} and J_{47} confirmed the position of the chlorine substituent as position 6.

Substituent chemical shifts (SCS) were discussed in their application to the elucidation of the substitution pattern of the benzenoid ring and a comparison was made between the values of SCS parameters in heteroaromatic systems and those in benzene compounds. The additivity of the substituent contributions to the *meta* and *para* (F,F) couplings in the heteroaromatic series was investigated and it was found that substituent contributions in the benzene

---

[54] R. J. Abraham, D. F. Wileman, G. R. Bedford, and D. Greatbanks, *Org. Magn. Resonance*, 1972, **4**, 343.

compounds, benzofurans, and benzothiophens were essentially unchanged. A long-range coupling between proton or fluorine in position 6 and the methyl protons in position 2 was favoured by the typical zig-zag coupling path.

7,7-*Difluorobenzocyclopropene;* $[AB]_2X_2$.[55] The proton spectrum of 7,7-difluorobenzocyclopropene (26) has been analysed using LAOCOON 3 to obtain chemical shifts and spin–spin coupling constants. Heteronuclear double-resonance experiments were used to show that the sign of $^4J(H,F)$ was positive whereas that of $^5J(H,F)$ was negative. The results were discussed in terms of the structure compared with that of benzocyclopropene.

(26)

Fluorothiophens have been the subject of two papers[56,57] devoted to experimental determination of the spectral parameters and discussions of substituent effects. Similar work has centred around mono- and di-fluorocyclopropanes.[58]

D. Compounds containing Phosphorus.—1,3,2-*Dithiaphospholans;* $[AB]_2X$.[59]

Sub-spectral analysis of the proton (60 MHz) and phosphorus (24.3 MHz) spectra of a series of 1,3,2-dithiaphospholans (27a—c) has yielded the spectral

(27a)   (27b)   (27c)

R = Cl, Ph, Me, NEt_2, OMe, SMe, or C_6F_5

parameters in a study of the structure and stereochemistry of these compounds. The results of analysis were interpreted as indicating an increased dihedral angle (nearly staggered configuration), consistent with the greater degree of flexibility conferred by the presence of sulphur rather than oxygen atoms in the ring. The result $^3J_{AA'} \neq {}^3J_{BB'}$ was claimed not to have been pre-

---

[55] F. W. Van Deursen, *Org. Magn. Resonance*, 1971, **3**, 221.
[56] S. Rodmar, S. Gronowitz, and U. Rosen, *Acta Chem. Scand.*, 1971, **25**, 3841.
[57] S. Rodmar, L. Moraga, S. Gronowitz, and U. Rosen, *Acta Chem. Scand.*, 1971, **25**, 3309.
[58] K. L. Williamson, S. Mosser, and D. E. Stedman, *J. Amer. Chem. Soc.*, 1971, **93**, 7208.
[59] S. C. Peake, M. Field, R. Schmutzler, R. K. Harris, J. M. Nichols, and R. G. Rees, *J. C. S. Perkin II*, 1972, 380.

Spectral Analysis

viously reported and it was suggested that this implies either long-range effects of the lone pair of the phosphorus on $^3J(H,H)$ or an imbalance in populations or geometries of envelope conformations with phosphorus 'up' or 'down'.

1,3,2-Oxazaphospholans; ABXY.[60] The analysis of the proton ABXY {P} spectrum was achieved following the method developed by Diehl and Chuck. It was verified that the energy level diagram permitted the choice between two solutions to the spectral analysis. Identification of the abxy sub-spectra was made and the $^3J(P,H)$ coupling constants were easily obtained. Relative signs were determined by tickling experiments. A half-chair conformation (28) was preferred in a choice between an envelope equilibrium and the half-chair equilibrium. Similar work has been reported for 1,3,2-diazaphospholans,[61]

$$O\!\!\overset{\frown}{}\!\!P\!\!-\!\!N\!\!\overset{\frown}{} \rightleftharpoons O\!\!\overset{\frown}{}\!\!P\!\!-\!\!N\!\!\overset{\frown}{}$$

(28)

1,3,2-dioxaphospholans,[62] and diazadiphosphetidines and phosphadiazetidinones.[63]

E. Miscellaneous Compounds.—*Methyl- and Chlorine-substituted Diboranes.*[64] This study constituted a re-investigation of known methyl- and halogen-substituted diborane derivatives in search of a quantitative relationship between the bridge proton chemical shifts and the degree and position of substitution. It was also hoped to test further the generality of the existence of 'complex' long-range homo- and hetero-nuclear coupling in the proton spectra of the diboranes. The proton spectra were recorded at 100 MHz with and without irradiation at 32.1 MHz where the boron spectrum was measured. The terminal methyl group protons were observed (in nearly all of the compounds investigated) as triplets when they were decoupled from the boron nuclei. The splitting was attributed to coupling with the bridge protons. Details of the spectral parameters were listed.

Tetrafuranyl- and Tetrathienyl-lead Compounds.[65] The 60 MHz proton spectra of tetra-2(and 3)-furanyl- and -thienyl-lead have been analysed using LAOCOON 3. The principal interest was in the ^{207}Pb proton satellite spectra and the authors claim a 'first' in reporting a complete set of long-range lead–proton coupling constants. Two sign combinations, (*a*) all *J*'s same sign, and (*b*) $^5J(Pb,H)$ negative with respect to all other *J*'s, reproduced the experimental

---
[60] J. Devillers, J. Navech, and J.-P. Albrand, *Org. Magn. Resonance*, 1971, **3**, 177.
[61] J.-P. Albrand, A. Cogne, D. Gagnaire, and J. B. Robert, *Tetrahedron*, 1972, **28**, 819.
[62] M. Revel, M. Bon, and J. Navech, *Compt. rend.*, 1972, **274**, C, 430.
[63] R. K. Harris, J. R. Woplin, R. E. Dunmur, M. Murray, and R. Schmutzler, *Ber. Bunsengesellschaft phys. Chem.*, 1972, **76**, 44.
[64] J. B. Leach, C. B. Ungermann, and T. P. Onak, *J. Magn. Resonance*, 1972, **6**, 74.
[65] G. Barbieri and F. Taddei, *J. C. S. Perkin II*, 1972, 262.

spectra, but alternative (*a*) was preferred because the mercury–proton coupling constants in analogous compounds all have the same sign. Proton chemical shift evidence for $(d-p)\pi$ interactions was not thought as convincing as in the corresponding tin compounds. A correlation between long-range (Pb,H) and (H,H) coupling constants in these systems was taken as confirmation that the Fermi contact term is predominant.

1,3,2-Dioxarsolans; $[AX_3]_2$.[66] This paper is the first of a proposed series of 60 MHz proton studies of dioxarsolans. Isomeric forms of 2-chloro-4,5-dimethyl-1,3,2-dioxarsolan (29) have been prepared in a reaction between butane-2,3-diol and arsenic trichloride.

trans from DL-butanediol

cis from *meso*-butanediol
(29)

(*i*) *trans-2-Chloro-4,5-dimethyl-1,3,2-dioxarsolan.* An ABX_3Y_3 spectrum was expected but the observed spectrum was described as $[AX_3]_2$. Inversion at the arsenic atom was discounted as a possible averaging process because the activation energy is *ca.* 105 kJ mol^{-1}. Chemical exchange of the chlorine atoms was invoked in comparison with ethylene chlorophosphite. In an attempt to detect the exchange phenomenon no line-broadening was observed between 30 and -20 °C in dilute toluene solution. It was deduced that the bimolecular rate of chlorine exchange was faster than in the phosphorus analogue, a reasonable conclusion since the arsenic–chlorine bond is more ionic than the phosphorus–chlorine bond.

(*ii*) *cis-2-Chloro-4,5-dimethyl-1,3,2-dioxarsolans.* The spectra observed were both described as $[AX_3]_2$ but in each case chlorine exchange again gave time-averaged parameters. $J_{XX'}$ was put equal to zero and all parameters were derived except that J_{AX} and $J_{AX'}$ were not separated in $N = J_{AX} + J_{AX'}$.

[66] D. W. Aksnes and O. Vikane, *Acta Chem. Scand.*, 1972, **26**, 835.

Spectral Analysis

Integration showed that there was 54% of the *trans*-form and 46% of the *cis*-form present. The fact that $J_{AA'}(cis) \ll J_{AA'}(trans)$ was interpreted as indicating two rapidly inconverting twist envelope conformers in each case.

o-*Phenylenebis(p-ethoxyphenyl telluride)*; $[AB]_2$.[66] The proton spectra recorded at 100 MHz for this compound (30), o-benzenedithiol, and p-hydroxy-

(30)

benzenethiol have been fully analysed with the aid of LAOCOON 3 to investigate the effects of the groups TeR, SR, and OR on chemical shifts and coupling constants in these analogous benzene compounds. The results were discussed in terms of specific substituent contributions.

4 General Comments

The large number of papers appearing which are concerned with the use of n.m.r. spectral parameters in studies of molecular structures and interactions emphasizes the trend by which n.m.r. analysis is becoming a common tool facilitated in its application by the accessibility of computer programs.

It is not possible to claim that the previous Report has had any influence but the incorrect labelling of second-order splitting effects has 'virtually' disappeared in the past twelve months.

Books.—One book has appeared in the 'N.M.R. Basic Principles and Progress' series with the title, 'Computer Assistance in the Analysis of High Resolution N.M.R. Spectra,' written by P. Diehl, H. Kellerhals, and E. Lustig. It provides a broad survey of the basic approaches to the computer-assisted iterative methods for the refinement of n.m.r. parameters. The section which deals with errors, supported by a comprehensive appendix, should help to clarify the anomaly which often appears between computerized and realistic errors. It has a useful bibliography.

6
Bandshape Phenomena for Fluids

BY R. K. HARRIS

1 General Introduction

A. Current Developments.—Research work reported in this area has increased substantially during the current review period compared to that of Volume 1 of these Reports. This is particularly true for so-called total bandshape analysis of exchanging spin systems. As predicted in Volume 1 this has been largely due to the increased use of spectral digitization and the application of sophisticated theory relevant to multi-spin systems.

However, such treatment has not yet been extended very much to the study of relaxation effects, and it is surprising that for most work involving lines of assumed Lorentzian shape the information is still obtained by the 'single parameter' method of measuring the linewidth. Whether improved accuracy would come from a study of the full digitized bandshape (or, at least, whether such studies are worth the effort!) is presumably matter for debate. At all events the instrumental advances mentioned in Volume 1 (p. 211) have not yet had much impact on two obvious areas of application: gas-phase and saturation bandshape studies. However, there are some indications[1-3] that this situation will change. A further area where surprisingly little research has been reported is that of bandshapes from Fourier transform spectra, although many research groups now have equipment for such work. It has been suggested[4,5] that the rather novel area of Fourier transform spin-echo spectroscopy may provide much detailed n.m.r. information, including some from linewidth studies. It is not, perhaps, clear whether such studies fall within the proper scope of this chapter, but some further comments are made in Section 1E.

B. Coverage of the Report.—In general the coverage is similar to that of Volume 1, *i.e.* bandshape studies for fluids, with special reference to exchange of magnetic sites and to relaxation studies. Bandshapes in solids

---
[1] R. E. Carter and T. Drakenberg, *J.C.S. Chem. Comm.* 1972, 582.
[2] T. Drakenberg and J. M. Lehn, *J. C. S. Pekrin II*, 1972, 532.
[3] R. K. Harris and K. M. Worvill, *Chem. Phys. Letters*, 1972, **14**, 598.
[4] R. L. Vold, *J. Chem. Phys.* 1972, **56**, 3210.
[5] R. L. Vold and R. R. Shoup, *J. Chem. Phys.*, 1972, **56**, 4787.

Bandshape Phenomena for Fluids

(including glasses) and in heterogeneous systems are discussed in Chapter 9. This year has shown a marked increase in the number of applications of bandshape studies, particularly those of systems containing paramagnetic species (see, for instance, ref. 6), to molecules of biochemical interest. However, in general such applications will not be discussed in detail in this chapter, nor will liquid-state bandshape studies of synthetic polymers (see Chapter 8).

There have also been several studies involving bandshapes for species in liquid crystal solvents or similar environments. Tiddy[7] has shown that n.m.r. linewidths of water protons in the lamellar phases of the system sodium caprylate–decanol–water contain contributions from both dipolar relaxation and magnetic susceptibility broadening (see also ref. 8). Similarly there have been measurements of linewidths of quadrupolar nuclei in solutions forming micelles (*e.g.* of ^{85}Rb in aqueous soap solutions[9]). All such applications will in general be held to be outside the scope of the present chapter. They will be covered in a separate chapter in Volume 3.

As in Volume 1, there will be little mention of applications involving trivial use of bandshapes; in particular, magnetic site exchange studies giving coalescence temperature data only will in general not be discussed. However, an attempt is made to mention all studies involving the quantitative measurement of linewidths (*e.g.* in work concerning the effects of paramagnetic solutes, as with research using the Swift–Connick[10] treatment).

The division of the chapter into sections is, to some extent, as arbitrary and subjective as is the limitation on the material to be included in the chapter. In particular the sub-division of Sections 2, 3, and 4 into theoretical work and applications is impossible to apply rigidly. Some attempt is made to cross-reference material, but in general a decision is taken to discuss a given paper in detail in one section only. The important case of Swift–Connick studies of exchange[10] is treated in Section 2C rather than Section 4B, but some related work is felt to be better included in Section 4B. It is not always possible to unambiguously classify exchange problems as intramolecular or intermolecular, though this is the basis of a sub-division of Section 2. In this context it should be pointed out that in Volume 1, Chapter 6, the isomerization between tetrahedral and planar forms of certain transition-metal complexes[11] was somewhat anomalously included under intermolecular processes. The basic exchange mechanism is, in fact, thought to be intramolecular, although excess ligand increases the isomerization *via* second-order intermolecular ligand exchange.

The general coverage of the literature for this report is July 1971 to May 1972, inclusive, but any relevant references from the literature of the preceding eighteen months which was not mentioned in Volume 1 are discussed here

---

[6] I. Salmeen and G. Palmer, *Arch. Biochem. Biophys.*, 1972, **150**, 767.
[7] G. J. Tiddy, *J. C. S. Faraday I*, 1972, **68**, 369.
[8] T. Drakenberg, A. Johansson, and S. Forsén, *J. Phys. Chem.*, 1970, **74**, 4528.
[9] B. Lindman and I. Danielsson, *J. Colloid Interface Sci.* 1972, **39**, 349.
[10] T. J. Swift and R. E. Connick, *J. Chem. Phys.*, 1962, **37**, 307: 1964, **41**, 2553.
[11] Refs. 137 and 138 of Vol. 1 Chap. 6, of these Reports.

whenever possible. Similarly, it has not proved possible, for a variety of reasons, to include here some work published during the current review period. Some of this will be covered in Volume 3. References 12—22 (mostly of Russian work) may contain material relevant to this chapter, but the only information available in English to this Reporter at the appropriate time was the titles, which are therefore given in the references below.

Reviews R21, R48, R68 and R92, given in the list commencing on page xi, are relevant to bandshape analysis (and were not quoted in Chapter 6 of Volume 1).

C. **Nomenclature, Notation, and Units.**—The term total lineshape (or, preferably, bandshape) analysis is becoming very popular, but it appears to be used to describe a number of different methods. Its strictest interpretation would be as iterative computer fitting of digitized total bandshapes, but it seems to be used whenever simulation of the full bandshape is employed, even when eventual comparison of experimental and theoretical spectra is essentially using a single parameter (see, for example, ref. 23).

It is a great pity that broad-line n.m.r. spectroscopists often continue to dissent from normal n.m.r. practice and quote linewidths in field units.[24—27] This is, of course, done for reasons of experimental convenience, but uniform

[12] M. Myagi, V. Erasho, S. Shevelev, and A. Fainzil'berg, '^{13}C, ^{14}N, and ^{17}O n.m.r. spectra of aliphatic nitro-compounds', *Eesti N.S.V. Teaduste Akad. Toimetised Keem. Geol.*, 1971, **20**, 297.
[13] M. Myagi, '^{13}C, ^{14}N, and ^{17}O n.m.r. spectra of aromatic nitro compounds', *Eesti N.S.V. Teaduste Akad. Toimetised Keem. Geol.*, 1971, **20**, 364.
[14] B. D. Zil'berman, B. A. Malevanyi, N. K. Moroz, and N. S. Poleshchuk, 'System for the two-co-ordinate digital recording of broad n.m.r. lines', *Prib. Tekh. Eksp.*, 1972, **72**, 126.
[15] L. K. Skrunts, B. A. Geller, and A. I. Brodskii. 'Kinetics of proton exchange between phenol hydroxyls and water studied by n.m.r.' *Teor. i eksp. Khim.*, 1970, **6**, 496.
[16] B. A. Geller and L. F. Kulish, 'Kinetics and mechanism of proton exchange in *m*-hydroxybenzoic acid studied by a p.m.r. method', *Teor. i eksp. Khim.*, 1971, **7**, 112.
[17] B. A. Geller, L. F. Kulish, and B. D. Shanina, 'N.m.r. study of the mechanism of proton exchange and calculation of quadrupole interaction in the urea–t-butanol system', *Teor. i eksp. Khim.*, 1971, **7**, 763.
[18] L. V. Sulima and I. F. Franchuk, 'Kinetics of proton exchange and association in α-hydroxy hydroperoxides studied by the n.m.r. method', *Ukrain. khim. Zhur.*, 1972, **38**, 333.
[19] P. M. Borodin and N. K. Ziao, 'About intermolecular interactions and kinetics of the chemical processes in some acid water solutions of Li_2SiF_6 and $HF-SiO_2$ made by the n.m.r. method', *Vestnik. Leningrad Univ. Fiz. Khim.*, 1971, **71**, 123.
[20] N. P. Grigor'ev and A. I. Maklakov, 'Effect of correlation time distributions on the width of a n.m.r. line', *Vysokomol. Soedinenii. Ser. B*, 1971, **13**, 652.
[21] V. N. Kalinin and I. F. Franchik, 'N.m.r. study of restrained internal rotation in *N*-acetyl phenyl hydroxylamine and its methyl ester', *Zhur. priklad Spektroskopii*, 1972, **16**, 124.
[22] T. Shono, M. Godo and K. Shinra, 'High-resolution n.m.r. studies of ring-chain tautomerism in ω-hydroxy aldehydes', *Bunseki Kagaku*, 1971, **20**, 1183.
[23] K. G. Orrell, W. Carruthers, and M. G. Pellatt, *Spectrochim. Acta*, 1972, **28A**, 753.
[24] H. Wennerstrom, B. Lindman, and S. Forsén, *J. Phys. Chem.*, 1971, **75**, 2936.
[25] R. Ader and A. Loewenstein, *J. Magn. Resonance*, 1971, **5**, 248.
[26] Dinesh and M. T. Rogers, *J. Magn. Resonance*, 1972, **7**, 30.
[27] R. G. Brüssau and H. Sillescu, *Ber. Bunsengesellschaft phys. Chem.*, 1972, **76**. 31.

Bandshape Phenomena for Fluids

use of Hz is highly desirable. A related problem is the tendency for some linewidths reported (see, for instance, ref. 26) in broad-line work to be actually peak-to-peak separations for the derivative absorption mode [if measured in frequency units this is $(\sqrt{3}\pi T_2)^{-1}$, whereas the full width at half height for the normal absorption mode is $(\pi T_2)^{-1}$].
Thermodynamic data in this report are presented in SI units, e.g. $\Delta H^{\neq}$ etc. as kJ mol^{-1}, although most literature reports still give kcal mol^{-1}. The conversion factor J→cal used here is 0.2390.

D. Experimental Procedures.—'Total bandshape' methods are becoming increasingly more popular, and 'single parameter' methods are repeatedly shown to give rise to error[28] (see Vol. 1, p. 215). Automatic methods of treating the data are growing in importance, as are schemes of fitting bandshapes iteratively to obtain the relevant parameters (as has been commonplace for some time in the analysis of n.m.r. spectra for chemical shifts and coupling constants alone). Such procedures use spectra in digitized form suitable for computer handling; the digitized spectra may be given in paper-tape form from the spectrometer for off-line processing, or may be fed directly into an on-line computer. These methods have been used for both exchange systems[28] and relaxation problems.[3]

Recognition that errors in $\Delta H^{\neq}$ and $\Delta S^{\neq}$ from n.m.r. measurements may be considerable, but that $\Delta G^{\neq}$ at T_c ($\Delta G_c^{\neq}$) is usually more accurate, has prompted use of $\Delta G_{298.2}$ for comparison purposes within a given class of compound (see, for example, refs. 29—31). However, the errors in $\Delta G_{298.2}$ clearly depend on the difference in temperature $|T_c - 298.2|$, so the method is at best only a partial improvement over use of $\Delta H^{\neq}$. In principle, of course, $\Delta S^{\neq}$ should give further information anyway, though its use can not always be justified at the present because of the large (and normally unknown) errors. The advantages of dealing with $\Delta G^{\neq}$ rather than E_a (partly because the former is less influenced by systematic errors than the latter, and partly because $\Delta G^{\neq}$ 'give a more direct impression of the rotational barrier') have been emphasized by Walter et al.[32] They show that use of $\Delta G^{\neq}$ rather than E_a reveals a trend in the barriers for C—N internal rotation (specifically that $\Delta G^{\neq}$ is higher for thioamides than for the corresponding amides) that was missed by previous workers.[33]

Confusion between Arrhenius and Eyring plots is still common; for instance Spaargaren et al.,[28] after a very sophisticated bandshape analysis, illustrate an Arrhenius plot, but quote only Eyring parameters. Reeves and

[28] K. Spaargaren, P. K. Korver, P. J. van der Haak, and Th. J. de Boer, Org. Magn. Resonance, 1971, **3**, 605.
[29] S. L. Spassov, V. S. Dimitrov, M. Agova, I. Kantschowska, and R. Todorova, Org. Magn. Resonance, 1971, **3**, 551.
[30] L. W. Reeves, R. C. Shaddick, and K. N. Shaw, Canad. J. Chem. 1971, **49**, 3683.
[31] R. C. Newman, jun., V. Jonas, K. Anderson, and R. Barry. Biochem. Biophys. Res. Comm. 1971, **44**, 1156.
[32] W. Walter, E. Schaumann, and J. Voss, Org. Magn. Resonance, 1971, **3**, 733.
[33] K. C. Ramey, D. J. Louick, P. W. Whitehurst, W. B. Wise, R. Mukherjee, and R. M. Moriarty, Org. Magn. Resonance, 1971, **3**, 201.

co-workers,[30,34] among others, also prefer to use only Arrhenius plots and to obtain Eyring parameters from the relationships (1) and (2), although this begs the question of the theoretical basis for the Eyring and Arrhenius equa-

$$\Delta H^{\neq} = E_a - RT \qquad (1)$$

$$\Delta S^{\neq} = R\left[2.303 \log\left(\frac{Ah}{kT}\right) - 1\right] \qquad (2)$$

tions. Some authors even quote mixed parameters, e.g. E_a and $\Delta S^{\neq}$ (but not A or $\Delta H^{\neq}$);[35] it is probable that in such cases a different equation from (2) was used, but the precise formulation employed is often not quoted, so great care is necessary in comparing results from different publications. Uniformity is extremely desirable in this area. This Reporter is in favour of plotting $\log(k/T)$ vs. $1/T$ to obtain $\Delta H^{\neq}$ and $\Delta S^{\neq}$ (with an *additional* plot of $\log k$ vs. $1/T$ for E_a and A, if desired): this has the advantage of assumptions with clear-cut thermodynamic implications (even if they are of debatable validity!), e.g. that $\Delta H^{\neq}$ and $\Delta S^{\neq}$ are temperature independent.

Time-sharing between transmitter and receiver is currently receiving considerable attention as a possible technical innovation (see Chapter 4). Such a system has repercussions on the observed bandshape, as is shown by Arnold et al.,[36] who simulated bandshapes for the case of a single spin using the Bloch equations and an analog computer. When the pulse repetition period, T, is much less than T_1 and T_2, the bandshape is closely similar to that obtained with CW irradiation. However, for $T \approx T_1$ or T_2 perturbations appear, especially sidebands at frequency intervals $(2n+1)/T$ from the main band. The German workers discuss the effects of pulse period, sweep rate, and saturation parameter on such spectra, with several illustrations for both the absorption and the dispersion mode.

Doskocilova and Schneider[37] have shown that broadening of n.m.r. lines owing to magnetic susceptibility differences in liquids containing suspended solid particles can be eliminated by 'magic angle' rotation at ca. 100 Hz. High-resolution spectra of the liquid phase can then be obtained. The case of styrene–divinylbenzene cation exchangers in equilibrium with water or methanol is given as an example (see also Chapter 8, Section 3A).

Degat et al.[38] found that, for an n.m.r. experiment involving prepolarization of a liquid flowing through a coil whose axis is perpendicular to B_0, the spectral bandshape is modulated by sidebands at a frequency inversely proportional to the time an elemental volume of liquid spends in the measuring area.

Buckmaster and Skirrow[39,40] have presented a detailed analysis of the

[34] L. W. Reeves and K. N. Shaw, *Canad. J. Chem.* 1971, **49**, 3671.
[35] R. R. Shoup, H. T. Miles, and E. D. Becker, *J. Phys. Chem.*, 1972, **76**, 64.
[36] K. Arnold, G. Klose, and P. Herrmann, *J. Magn. Resonance*, 1972, **6**, 136.
[37] D. Doskocilova and B. Schneider, *Macromolecules*, 1972, **5**, 125.
[38] F. Diegat, A. Dussanchov, and A. Erbeia, *Compt. rend.*, 1971, **273**, B, 49.
[39] H. A. Buckmaster and J. D. Skirrow, *J. Appl. Phys.*, 1971, **42**, 1225.
[40] H. A. Buckmaster and J. D. Skirrow, *J. Magn. Resonance*, 1971, **5**, 285.

effect of modulation broadening on the shapes of unsaturated Lorentzian absorption and dispersion lines. Exact formulae relating the location and amplitude of the extreme and other observables to the true linewidth are given. This work is of less relevance for high-resolution n.m.r. than for broad-line work or e.s.r., but the authors' conviction[39,40] that dispersion measurements will become much more common should be noted.

E. **Miscellaneous Applications.**—Bandshapes are often used, in a sense, during normal spectral analysis when calculated lines are assigned a shape (usually Lorentzian), the whole spectral appearance being calculated from overlap of the various lines, in order to facilitate comparison of theoretical and experimental spectra. Such uses are commonplace and will not be listed here. However, it is perhaps worthwhile to point out the importance of using such bandshapes when there are small, unresolvable (usually 'long-range') coupling constants. Comparison of the shape of the experimental spectrum with spectra calculated for a range of values of the small coupling constant may provide a reasonable measure of that parameter, as, for instance, for certain benzo-substituted five- and six-membered rings, the spectra of which have been illustrated by Lambert and Koeng.[41] Coupling constants are often estimated rather more qualitatively from linewidths (*e.g.* values of $^5J_{HH} \approx 0.1$ Hz in ref. 42).

Haloui *et al.*[43] have described a method of 'spectral decomposition' for obtaining the parameters L and M from [AB]$_2$ spectra sufficiently close to [A$_2$B$_2$] that lines overlap. The procedure consists of a computer comparison of the experimental digitized bandshape in the overlap region with theoretical spectra, varying L and M to minimize errors in a least-squares fashion.

The use of total bandshape analysis for exchanging systems to obtain relative signs of coupling constants and to measure unresolved coupling constants has been emphasized by Reeves and co-workers.[30,34]

Giger *et al.*[44] list ^{1}H-$\{^{14}$N$\}$ INDOR linewidths for some substituted ethylpyridinium cations. They claim these widths are a measure of T_1^N, but the procedure is very dubious, as the authors themselves note.

Lado *et al.*[45] have presented a detailed theory of n.m.r. bandshapes involving magnetic-moment autocorrelation functions. Although developed expressly for broad-line spectra of solids the formalism is stated to be quite general.

The review period has seen the publication of several articles concerning Fourier transform spin-echo n.m.r. spectroscopy,[4,5,46—48] a novel technique

---

[41] J. B. Lambert and F. R. Koeng, *Org. Magn. Resonance*, 1971, **3**, 389.
[42] H. C. E. McFarlane and W. McFarlane, *Org. Magn. Resonance*, 1972, **4**, 161.
[43] E. Haloui, D. Canet, and P. Granger, *Org. Magn. Resonance*, 1971, **3**, 451.
[44] W. Giger, P. Schauwecker, and W. Simon, *Helv. Chim. Acta*, 1971, **54**, 2488.
[45] F. Lado, J. D. Memory, and G. W. Parker, *Phys. Rev. (B)*, 1971, **4**, 1406.
[46] R. Freeman and H. D. W. Hill, *J. Chem. Phys.*, 1971, **54**, 301
[47] R. L. Vold and S. O. Chan, *J. Chem. Phys.*, 1972, **56**, 28.
[48] R. L. Vold and S. O. Chan, *J. Magn. Resonance*, 1971, **4**, 208.

which enables very small coupling constants to be obtained with high accuracy[5,46] (see also Chapters 2 and 4). Most attention has focused on the frequencies and intensities of the lines, though these cannot always be assigned to transitions of individual nuclei.[5] The linewidths, however, also provide in principle selective information about transverse relaxation rates, although the theoretical interpretation may be complicated and there is some doubt about the adequacy of current experimentation.[5] General expressions have been given[48] for the linewidths, and the results applied to the case of two coupled spins relaxing by intramolecular dipole–dipole interactions and by random magnetic fields.[48] It has been shown[48] that selective rf pulse experiments are superior to non-selective pulse experiments. The theory of Fourier transform spin-echo spectroscopy seems to be well advanced,[4,47,48] but although the method promises to be very powerful, there are not yet enough experimental reports to judge its likely impact on n.m.r. It appears[47] that the linewidths depend in general on rf pulse spacing and on chemical shifts. Moreover, cross-relaxation couples different magnetizations and so linewidths are not determined solely by a single spectral magnetization.[47]

2 Exchange of Magnetic Sites

A. Theoretical Work.—The density matrix method for describing n.m.r. of exchanging systems has been re-examined, simplified, and generalized by Kaplan and Fraenkel.[49] The unified treatment developed may be applied to all intra- and inter-molecular exchanging systems. The authors show the development of a Bloch-like series of equations. The algebraic evaluation may be simplified by a procedure referred to as the permutation of indices method; this is achieved through use of the product basis functions, and by-passes the previous need to define the matrices of the re-organization operator in detail. The method also allows a simple relation between the algebra and the exchange chemistry to be seen. The theory is developed in terms of a system involving mutual exchange of parts, $AB + CD \rightleftharpoons AC + BD$, but the method is general and examples of other cases are also discussed. Only the situation of low rf power is covered, but developments are promised for later publication. A computer program using these procedures is being written, and it will be interesting to see if this supersedes n.m.r. exchange programs currently in popular use. Kaplan[50] has noted that in the density matrix treatment of exchanging systems inclusion of relaxation may be made in a 'compromise' form intermediate between the exact theory and the commonly used non-specific form. The compromise applies if uncorrelated random field relaxation is assumed. This note[50] appears to be reproduced as an appendix of the later paper,[49] but with no cross-referencing.

Kaiser[51] has extended the linear response theory of n.m.r.[52] (see Section

---

[49] J. I. Kaplan, and G. Fraenkel, *J. Amer. Chem. Soc.*, 1972, **94**, 2907.
[50] J. I. Kaplan, *J. Chem. Phys.*, 1971, **55**, 1489.
[51] R. Kaiser, *J. Magn. Resonance*, 1971, **5**, 390.
[52] R. Kaiser, *J. Magn. Resonance*, 1971, **5**, 220.

4A) to deal with the AB spin system involving intra- or inter-molecular exchange. The theory is developed in terms of the poles and zeros (in the complex frequency plane) of the transfer function. The manner in which the spectrum varies with the exchange rate can be seen from root locus diagrams. The results are related to certain features in ^{19}F spectra of perfluorocyclohexane and ^{1}H spectra of diphenylmethanol (examples of intra- and inter-molecular exchange, respectively).

Reeves and Shaw[34] have extended their matrix formulation of the Bloch–McConnell equations to include spin–spin coupling. Although the theory is general it is developed in particular for the ABX⇌BAX mutual exchange system with $J_{AB} = 0$. (Actually the authors confine their attention to the first-order case, so the system is more reasonably denoted as AMX⇌MAX.)

$$\begin{array}{cc}
\text{Me} \diagdown \quad / R & \text{Me} \diagdown \\
\quad N-C & \quad N-C \\
/ \quad \diagdown & / \quad \diagdown \\
\text{Me} \quad O & \text{Me} \quad O
\end{array}$$

(1) a; R = F   (2)
    b; R = Me
    c; R = CF$_3$
    d; R = CCl$_3$

Exchange effects in both the AB and X regions are described, and it is emphasized that the bandshapes are sensitive to the relative signs of J_{AX} and J_{BX} at all exchange rates except at the zero exchange limit. The theory is applied to the hindered internal rotation process in NN-dimethyl carbamyl fluoride (1a). Strictly speaking, this molecule gives an A$_3$B$_3$X ($J_{AB} \sim 0$) spin system, but the authors state that the ^{1}H region may be treated using their ABX theory. The observed digitized spectra are iteratively fitted by computer, and it is demonstrated from 'best fit' errors in the region of coalescence that the signs of the two ^4J(H, F) coupling constants are the same. Unfortunately the total range of the spectrum under slow-exchange conditions is less than 5 Hz, so the case is not a particularly favourable one. The kinetic data (see Table 3, p. 265) are discussed in relation to information on related systems.

The Canadian workers have also[30] applied their multi-site exchange theory to the hindered internal rotation process for dimethylacetamide (1b), dimethyltrifluoroacetamide (1c), and dimethylbenzamide (2; R = H), using ^{1}H n.m.r. spectra of the Me$_2$N protons. The systems for the first two compounds are of the type A$_3$B$_3$X$_3$⇌B$_3$A$_3$X$_3$, requiring a treatment using eight spin sites. Dimethylbenzamide is a simple two-site equal population case. Digitized bandshapes were fitted iteratively by computer, with careful allowance for the variation of the chemical shift difference with temperature. The procedure allows the determination of the relative signs of coupling constants and the measurement of unresolved coupling constants. The authors believe[30] they have largely eliminated the systematic errors previously common in dynamic n.m.r. work. The results of this study (Table 3, p. 265) are discussed[30] in relation to earlier work.

In a series of papers[28,53,54] Spaargaren et al. report results of total n.m.r. bandshape fitting for n.m.r. spectra of NN-dimethylamides. Paper III of their series[28] introduces an iterative computer procedure for analysing bandshapes for the case of equally populated two-site exchange and compares the results of this method with those of the Rogers and Woodbrey[55] intensity ratio method as applied to NN-dimethyl-benzamides (2) and -cinnamides (3). The intensity ratio approach always gives larger exchange rates than those from the total bandshape calculations, in spite of several attempts to improve the former, though $\Delta G^{\neq}$ values at the coalescence temperature from the two methods are

Me\
N—C\
/   \\\
Me   O

(3)

identical. The scarcely surprising conclusion from the study is that total bandshape work is superior to methods depending on a single measurable parameter, as has already been strongly pointed out (see Vol. 1, p. 224).

The computational procedure of these workers[28] merits some discussion since they made a serious attempt to determine whether the choice of parameters for iteration affected the results. It is, of course, well known that in the fast exchange region, the chemical shift difference for an exchanging system is strongly correlated with the lifetime. Consequently the authors devised two versions of the program. In the first form there are seven adjustable parameters, namely: the two chemical shifts, the mean lifetime, the two linewidth parameters, an amplitude factor, and a base-line correction. (It is noteworthy that even for this relatively simple case of bandshape fitting as many as seven variables are needed – one may consider that this feature throws some doubt on the accuracy of results from purely visual comparison and fitting, though the human hand can readily correct for the vertical and horizontal position variables.) The second version of the program also used seven parameters but took the mean chemical shift and the difference instead of the individual values. Moreover, at high temperatures the starting value of the chemical shift difference was calculated using extrapolation of the low-temperature

---

[53] K. Spaargaren, P. K. Korver, P. J. van der Haak, and Th. J. de Boer, *Org. Magn. Resonance*, 1971, **3**, 615.

[54] K. Spaargaren, P. K. Korver, P. J. van der Haak, and Th. J. de Boer, *Org. Magn. Resonance*, 1971, **3**, 639.

Bandshape Phenomena for Fluids   225

data. The second program was also more automated than the first, in that it produced the activation parameters directly. It should be noted that in three cases of data treated by both programs, the results for $\Delta H^{\ne}$ and $\Delta S^{\ne}$ differed considerably ($\Delta H^{\ne}$ by up to 4 kJ mol^{-1} and $\Delta S^{\ne}$ by up to 13 J K^{-1}mol^{-1}. This underlines the impression that statistical errors of Arrhenius or Eyring plots are substantially less than the true total errors. It may be significant that, after digital smoothing, Spaargaren et al.[28] used only forty points per spectrum to obtain seven variables, so the system is not as overdetermined as might be thought.

The results (see Table 3, p. 265) for the activation parameters of the amide internal rotation in NN-dimethyl-benzamides and -cinnamides produced by this work[28] were correlated effectively with substituent parameters, to give

$$
\begin{array}{cc}
\text{Me} \quad \triangle_R & \text{Me} \quad \underset{}{\bigcirc}-R \\
\diagdown \!\!\!\!\!\diagup & \diagdown \!\!\!\!\!\diagup \\
N-C & N-C \\
\diagup \quad \diagdown\!\!\!\!\! \diagdown \ \text{O} & \diagup \quad \diagdown\!\!\!\!\!\diagdown \ \text{O} \\
\text{Me} & \text{Me} \\
(4) & (5)
\end{array}
$$

resonance percentages for the barrier. The next paper in the series[53] discusses literature results for medium effects on amide rotational barriers in the same range of molecules, and correlates the barriers for the series with C—N π-bond order as obtained from Hückel MO calculations. The values of $\Delta G^{\ne}$ were previously correlated with solvent polarity; Spaargaren et al.[28,53] find that chloroform is anomalous as a solvent, since barriers in CHCl$_3$ and CH$_3$CN solutions are about equal in spite of the difference in relative permittivity – this fact is attributed to hydrogen bonding in the chloroform solutions. Paper V of the series[54] contains barrier data from the total bandshape method for some NN-dimethylcyclopropanecarboxamides (4). The effects of substituents are discussed. Mention must also be made of the first paper in the series,[56] since it was not discussed in Volume 1. A number of p- and m-substituted NN-dimethylbenzamides, (2) and (5) respectively, were studied using the Woodbrey and Rogers[55] intensity ratio method. It was found that the results (see Table 3, p. 265) compared very favourably with those measured earlier by total bandshape analysis. However, the substantial negative values for $\Delta S^{\ne}$ appear suspicious, especially in view of the positive values found later by the same authors[28,53] for the p-substituted compounds by total bandshape analysis. The values of $\Delta G^{\ne}$ at 25 °C (treated as more reliable than $\Delta H^{\ne}$) were analysed to determine the influence of the aromatic substituent; the sensitivity of the internal rotation barrier to resonance effects was determined for the two classes of compounds for CDCl$_3$ solution using the method of Swain and Lupton.[57]

[55] M. T. Rogers and J. C. Woodbrey, *J. Phys. Chem.*, 1962, **66**, 540.
[56] P. K. Korver, K. Spaargaren, P. J. van der Haak, and Th. J. de Boer, *Org. Magn. Resonance*, 1970, **2**, 295.
[57] C. G. Swain and E. C. Lupton, jun., *J. Amer. Chem. Soc.*, 1968, **90**, 4328.

I

Shanan-Atidi and Bar-Eli[58] have illustrated graphs enabling $\Delta G_c^{\neq}$ to be obtained from coalescence temperatures and chemical shift differences for two-site exchange with *unequal* populations, provided the population difference is known. The graphs ignore non-exchange linewidths.

The use of coalescence data for exchange rates, from $\sqrt{2}k_c = \pi|\Delta v|$ and $\sqrt{2}k_c = \pi[\Delta v^2 + 6J^2]^{\frac{1}{2}}$ for the equal population A⇌X and AB⇌BA cases respectively, has been compared to the use of total bandshape analysis by Kost et al.[59] It is concluded that the approximate methods are satisfactory (defined as agreement of rate constants within 25%) for the A⇌X case except at low Δv. The procedure of Shanan-Atidi and Bar-Eli[58] is, however, preferable to the use of $\sqrt{2}k_c = \pi|\Delta v|/K$ (where K is the equilibrium constant) for the unequal-population case. In the AB⇌BA situation the second equation above is useful provided that $|\Delta v| > |J|$.

B. Examples of Intramolecular Exchange.—Much of the thermodynamic activation data for the processes discussed in this section may be found in the Tables of the appendix (in most cases without further textual reference). In these Tables, reported values of E_a, log A, $\Delta H^{\neq}$, and $\Delta S^{\neq}$ (without interconversion on the part of the Reporter) are given in preference to $\Delta G^{\neq}$, in spite of the reputedly greater accuracy of $\Delta G^{\neq}$ (see Section 1D), since in principle there is more information in the complete data and in practice values of $\Delta G^{\neq}$ are reported at a variety of temperatures, making comparison between data for different compounds of dubious value. The data are given in SI, usually involving conversion by the Reporter and thus giving rise to some rounding approximations. In cases where two sites of unequal populations are involved the parameters listed are for the lowest energy site. For the meaning of the errors (which are usually purely statistical reflections of the fit to Arrhenius or Eyring equations) and for information regarding solvents *etc.* the reader is referred to the original literature.

Internal Rotation about the Carbon–Carbon Bond. A preliminary communication by Anderson and Pearson[60] gives barrier information about the important case of internal rotation about the C—C bond in substituted ethanes. An earlier publication by the same authors[61] gives a more complete description of work on two t-butyl compounds, Me₃C·CMe₂Cl (6) and Me₃C·CMe- ClCH₂Ph (7), which represent two of the three possible spectral types. The ¹H signals due to the t-butyl group were studied [two-site exchange with populations in the ratio 2:1 for (6) and three-site exchange for (7); few details are given of the spectral simulation] and barrier parameters obtained. An interesting footnote discusses the meaning of coalescence for three-site

[58] H. Shanan-Atidi and K. H. Bar-Eli, *J. Phys. Chem.*, 1970, **74**, 961.
[59] D. Kost, E. H. Carlson, and M. Raban, *Chem. Comm.*, 1971, 656.
[60] J. E. Anderson and H. Pearson, *Chem. Comm.*, 1971, 871.
[61] J. E. Anderson and H. Pearson, *J. Chem. Soc. (B)*, 1971, 1209.
[62] B. L. Hawkins, W. Bremser, S. Borcic, and J. D. Roberts, *J. Amer. Chem. Soc.*, 1971, **93**, 4472.

exchange; the authors were unable to obtain an explicit mathematical expression for a rate constant at a coalescence temperature.

The group of J.D. Roberts has also studied[62] the problem of internal rotation about the C—C single bond, in a series of halogenated methylbutanes. Many of the compounds contain Bu^t groups and the ^1H resonance of these was examined as a function of temperature; they are examples of uncoupled two- or three-site exchange. The spectra of 2,3,3-tribromo-2-methylbutane were also treated as due to two-site exchange, a third possible situation being ignored as the population involved was probably negligible. Two further compounds gave somewhat more complicated spectra because the relative populations of sites were not fixed by symmetry. The spectra were simulated on the basis of one nucleus exchanging since in all cases coupling between methyl groups is negligible. Some of the fitting of the spectra was done

$$\text{(structures shown)}$$

(8) a; X = OMe
b; X = CMe_3

iteratively by computer using digitized experimental data. The kinetic parameters are discussed in terms of the non-bonded interactions involved.

Internal rotation about the sp^2–sp^3 C—C bond in a number of di-t-butylarylcarbinols has been studied using the n.m.r. method by Gall et al.[63] In three cases quantitative kinetic data were obtained. The intensity-ratio method was used to determine lifetimes for the lightly coupled AB⇌BA aromatic proton system (8a); data from both 60 and 100 MHz spectra, plus data from methoxy resonances, were used to justify ignoring the spin–spin coupling. Complete lineshape calculations were also carried out. In the case of the bromodimethoxyphenyl compound (9) (an AB⇌XY case) the effect of chemical shifts varying with temperature had to be incorporated in the procedure of total bandshape simulation.

Swedish workers have continued to be active in studies of exchange by n.m.r. Thus Nilsson et al.[64] have examined the spectra of some unsymmetrically halogen-substituted 1,3,5-trineopentylbenzenes (10). These give rise to three temperature-dependent AB⇌BA spectra. Digitized spectra and iterative total bandshape analysis were employed, taking into account the three AB-type spectra simultaneously. Careful attention was paid to obtaining a value for non-exchange linewidths and chemical shift differences suitable for use at

---

[63] R. E. Gall, D. Landman, G. P. Newsoroff, and S. Sternhell, *Austral. J. Chem.* 1972, **25**, 109.
[64] B. Nilsson, R. E. Carter, K. I. Dahlqvist, and J. Marton, *Org. Magn. Resonance*, 1972, **4**, 95.

(9)

each temperature. In each case the barrier parameters are essentially the same for each neopentyl group. It is suggested that this might be because the magnetic non-equivalence of two of the groups arises from the orientation of the third (that between the substituents).

Dunlop et al.[65] describe briefly variable-temperature ^{1}H n.m.r. experiments on 9-nitroso- and 9-formyl-julolidine [(11a) and (11b) respectively]. The AB⇌BA spectra of the aromatic region were examined and barrier data are given. The barriers are much higher than in simple C-nitroso and C-formyl compounds, indicating that the quinolizidine unit is an extremely effective electron-releasing group.

(10) a; X = Cl, Y = I
b; X = Br, Y = I
c; X = Cl, Y = Br

(11) a; X = N
b; X = CH

Oki et al.[66] studied hindered internal rotation about the pivot bond of several biphenyl derivatives using ^{1}H n.m.r. They list $\Delta G^{\neq}$ data for T_c and for one other temperature (obtained using linewidths above T_c), and compare values within a narrow range of compounds.

Exchange process in triarylmethyl cations have been studied[67] using CF_2H substituents as diastereotopic ^{19}F n.m.r. probes. The spectra of the simplest compounds (12a) and (12b) are of the ABX⇌BAX type, showing enantiomer interconversion of the propeller forms. Activation parameters for the process were obtained via rate constants using comparison of computer-simulated and experimental spectra. With $m\text{-}CF_2H, Z \neq H$ the situation is more complex since there are now diastereomers, for which distinct resonances are seen

[65] R. Dunlop, R. K. Mackenzie, D. D. MacNicol, H. H. Mills, and D. A. R. Williams, Chem. Comm., 1971, 919.
[66] M. Oki, K. Akashi, G. Yamamoto, and H. Iwamura, Bull. Chem. Soc. Japan, 1971, 44, 1683.
[67] J. W. Rakshys, jun., S. V. Mckinley, and H. H. Freedman, J. Amer. Chem. Soc., 1971, 93, 6522.

(12) a; p-CF$_2$H, Z = H
    b; m-CF$_2$H, Z = H
    c; m-CF$_2$H, Z = Me

below about -30 °C. Rate constants for both enantiomer and diastereomer interconversion were obtained for Z = Me (12c) (the theory for this four-site exchange is very briefly mentioned). The amount of barrier information given is curiously small in view of the stated work carried out, but the nature of the interconversion processes is discussed in great detail.

Ring Inversion and Related Processes. The use of the 'fluorine probe' (CF$_2$ substitution) in dynamic n.m.r. investigations has been the subject of a further study by Roberts and co-workers,[68] who examined the temperature-dependent n.m.r. spectra of 3,3-difluoro-*trans*-cyclodecene. As the temperature was lowered from 14 to -30 °C, the singlet ^{19}F n.m.r. resonance changed into an AB pattern, and total bandshape fitting gave lifetimes and hence activation parameters. Further substantial spectral changes occurred on cooling to -152 °C, but quantitative bandshape work was not feasible. It did not prove possible to unequivocally decide which mechanism was responsible for the higher temperature process. In the fitting procedure for this process, corrections for non-exchange linewidths were adjusted for the effects of (H,F) coupling.

Similar total bandshape methods have been applied by the Roberts group to ^{19}F resonance of the CF$_2$ groups in $\gamma\gamma$-difluoro-ε-caprolactone (13a),[69] $\gamma\gamma$-difluoro-ε-caprolactam (13b),[69] and 1,1-difluorocyclodecane.[70] The barrier parameters for (13a) and for the cyclodecane are given in Table 2 (see p. 264), but for (13b) it only proved possible to obtain $\triangle G^{\neq}$ at one temperature. The n.m.r. spectra of the lactone and the lactam were best interpreted[69] in terms of the chair conformation. For difluorocyclodecane there is considerable discussion[70] about possible conformations and about possible pathways for effecting ring inversion. A boat-chair-boat conformation is suggested to be the most stable. (See also the work on perfluorocyclohexane,[51] mentioned in Section 2A above.)

Thermodynamic exchange parameters have been obtained[23] for the ring inversion of the novel S-heterocyclic ring compound (14). The ^{1}H spectra are of the AB$\rightleftharpoons$BA type and were treated by complete bandshape simulation,

[68] E. A. Noe, R. C. Wheland, E. S. Glazer, and J. D. Roberts, *J. Amer. Chem. Soc.*, 1972, **94**, 3488.
[69] E. A. Noe and J. D. Roberts, *J. Amer. Chem. Soc.*, 1971, **93**, 7261.
[70] E. A. Noe and J. D. Roberts, *J. Amer. Chem. Soc.*, 1972, **94**, 2020.

(13) a; X = O
b; X = NH

(14)

(15)

with visual comparison of computed and experimental spectra aided by graphical interpolation of linewidths. The possible nature of the ring inversion is discussed.

The total bandshape n.m.r. method has been used by Bushweller et al.[71] to study ring reversal in cyclopentamethylenedimethylsilane (15). The C-3,4,5-ring proton resonances were used since the methyl proton signal does not split even at -171 °C. A number of rather drastic assumptions were made in bandshape fitting; thus the process was considered to be A⇌B with all coupling accounted for by an effective linewidth. The rate constants could only be obtained over a 15 °C temperature interval.

As mentioned in Volume 1 (p. 231), an increase in dynamic n.m.r. studies by ^{13}C resonance was expected, and several articles have appeared since which make use of this technique. Dalling et al.[72] have studied the ring inversion of a series of dimethylcyclohexanes and cis-decalins using ^{13}C n.m.r. The advantages of the method are that (a) chemical shift differences are usually appreciable, (b) the n.m.r. exchange situations are of the simple A⇌X type (when proton decoupling is employed), and (c) frequently several pairs of lines in the spectra can be used for different temperature ranges. The American authors used simple equations to obtain inversion rates at extremes of fast and slow exchange (correcting for 'natural' linewidths) and at coalescence temperatures. There is a discussion of the transmission coefficient used in the Eyring equation, of errors in the barrier parameters, and of the values of these parameters, which in most cases are very close to those of cyclohexane itself.

Complex n.m.r. exchange processes, of the types ABC⇌ACB, A₂BC⇌A₂CB, and ABCD⇌ABDC, have been studied[73] by Binsch and co-workers, using the DNMR computer programs.[74] The advantages of spectral complexity are emphasized, as is the importance of a large temperature range for study. The latter is particularly borne out by an error analysis of the Arrhenius plots. Non-exchange linewidths were obtained from lines not noticeably affected by the exchange. Such lines were identified from the computations in all cases over the entire temperature range. As the authors say: 'the spectra carry their own built-in linewidth calibrations'. Variability of so-called 'static parameters' with temperature was taken carefully into account in the computations, but fitting of theoretical spectra to experimental spectra was done visually. The compounds studied were monoaza- and diaza-

[71] C. H. Bushweller, J. W. O'Neil, and H. S. Bilofsky, *Tetrahedron*, 1971, **27**, 3065.
[72] D. K. Dalling, D. M. Grant, and L. F. Johnson, *J. Amer. Chem. Soc.*, 1971, **93**, 3678.
[73] A. Steigal, J. Saure, D. A. Kleier, and G. Binsch, *J. Amer. Chem. Soc.*, 1972, **94**, 2770.
[74] Quantum Chemistry Program Exchange, Indiana University, Programs 140 and 165.

analogues of the cycloheptatriene–norcaradiene system. The former (16) prefer the azepine structure, but the latter (17) exist preferentially as diazanorcaradienes; the net topomerization (17) is considered to proceed *via* valence isomerism to the azepine form followed by ring inversion. The activation parameters are derived (the authors also give[73] values of $\Delta G_c^{\ne}$ for several related compounds). The nature of the energy profile for the exchange is discussed; this discussion includes an interesting, though tentative, use of $\Delta S^{\ne}$.

At this point it is convenient to mention another case of valence tauto-

(16)

(17)

merism involving a conformational equilibrium, that of bicyclo[5,1,0]octa-2,5-diene (3,4-homotropilidene). Günther and co-workers[75] have studied the AB⇌CD exchange system provided by a suitably deuteriated species (18), using the computer program developed by Creswell and Harris.[76] Their results allow them to conclude that the rearrangement proceeds through a *cisoid* transition state, *i.e.* as (19).

C—N *Internal Rotation in Amides and Related Compounds.* The current review year 1971-72 has been a bonanza one for studies of this type. The work of Reeves and co-workers[30,34] and of Spaargaren *et al.*[28,53,54,56] have been discussed in Section 2A above. Other work will be described below.

Use has been made[77] of the shift reagent Yb(dpm)₃ to study C—N hindered internal rotation in amides, since the two methyl ¹H resonances in dimethyl-formamide and -acetamide are strongly differentiated. Coalescence data only are given. The authors state that the presence of Yb(dpm)₃ appears to retard the coalescence; however, the data indicate a slight *lowering* of $\Delta G_c^{\ne}$. It is apparent that caution needs to be exercised in using shift reagents to study exchange.

[75] H. Gunther, J. B. Pawliczek, J. Ulmen, and W. Grimine, *Angew. Chem.*, 1972, **84**, 539 (*Angew. Chem. Internat. Edn.*, 1972, **11**, 517).
[76] C. J. Creswell and R. K. Harris, *J. Magn. Resonance*, 1971, **4**, 99.
[77] C. Beaute, Z. W. Wolkowski and N. Thoai, *Chem. Comm.*, 1971, 700.

Exchange studies of NN-dimethyl-*trans*-cinnamides, (3) and (5), have attracted the attention of Bulgarian workers,[29] who employed the total bandshape method with spectra digitized manually and iteratively fitted to theoretical spectra by computer. The question of the chemical shift differences and non-exchange linewidths varying with temperature was examined. The results for the activation parameters are compared with those obtained by other workers (*e.g.*, Spaargaren *et al.*,[28,53] see Section 2A), and differences are discussed in relation to errors. The barriers are correlated with substituent parameters, but values of $\Delta S^{\neq}$ are scarcely discussed, in spite of their large variations (particularly between the m-substituted and p-substituted compounds), presumably because little reliance was placed on values for this parameter. Some results are also given using the method of Nakagawa.[78] Earlier work by Spassov *et al.*[79] (not discussed in Volume 1) used the Nakagawa method to obtain barrier information for three conjugated NN-dimethylamides. The results for NN-dimethylsorbinamide (20), with a large negative $\Delta S^{\neq}$, look dubious. Some data ($\Delta G^{\neq}_{298.2}$) are also given[79] in this paper for NN-dimethylbenzamide and NN-dimethyl-p-bromobenzamide.

Table 3 (p. 265) lists values of $\Delta H^{\neq}$ and $\Delta S^{\neq}$ for (*inter alia*) compounds (2), (3), and (5), obtained during the review period by different research groups. The lack of agreement is striking, and it is clear that the difficulties of obtaining accurate activation parameters by n.m.r. have not yet been overcome (at least, not by all workers in this area!). It is, however, only fair to point out that (a) values of $\Delta G_c^{\neq}$ or $\Delta G^{\neq}_{298}$ are in much better agreement than $\Delta H^{\neq}$, and (b) there is usually good agreement between different workers using total bandshape fitting.

[78] T. Nakagawa, *Bull. Chem. Soc. Japan*, 1966, **39**, 1006.
[79] S. L. Spassov, Ts. Buzova, and B. Chorbanov, *Z. Naturforsch.*, 1970, **25b**, 347.

Bandshape Phenomena for Fluids   233

```
    Me₂N                    CH₂CO₂Me
        \                  /
         >=O        Me — N
        /                  \
       /                    COMe
     (20)                   (21)
```

Love et al.[80] have reported a variable-temperature ^{1}H n.m.r. study of methyl-N-acetylsarcosinate (21). Signals due to all four types of proton exhibited the properties of simple two-site exchange, each giving rise to two signals at room temperature. The study indicated that a single rate process, probably hindered internal rotation about the C(O)—N bond, was responsible. The spectra were fitted by eye to simulated spectra, using values of non-exchange linewidths, chemical shift differences, and equilibrium constants extrapolated from low temperature data.

Ramey et al.[81] have studied n.m.r. bandshapes of amides and thioamides using the halfwidth method, which they extend to cover exchange between two uncoupled sites with different populations. Families of curves of $\Delta\nu_{\frac{1}{2}}$ for a variety of values of the chemical shift difference, both before and after coalescence, are illustrated. The authors comment on the difficulties of the $\Delta\nu_{\frac{1}{2}}$ method in the region of coalescence, especially when additional coupling complicates the spectra, as in the cases studied. However, they claim errors in E_a are only ca. twice as large as those from total bandshape analysis. They champion the $\Delta\nu_{\frac{1}{2}}$ procedure because it is less demanding on instrumentation. This reporter, at any event, remains somewhat sceptical about any 'single observable' method of bandshape analysis for obtaining thermodynamic barrier information; total bandshape analysis would seem to be preferable whenever instrumentation allows it. Ramey et al.,[81] using the methods discussed above, investigated the temperature-dependent ^{1}H n.m.r. spectra (in the methyl region) of certain NN-di-isopropyl-amides and -thioamides to obtain information about intramolecular motion. For the thioamides two rate processes were observed; that at the lower temperature is attributed to internal rotation about the bond between N and the alkyl carbon atoms. Since only two low-temperature isomers are observed, the authors conclude[81] that vicinal steric interactions are the most important. The comments of Walter et al.[32] (see Section 1D) probably need to be recalled here. The variability of $\Delta S^{\neq}$ in ref. 81 gives some cause for concern.

The C—N rotational barriers of a number of amides, thioamides, and amidinium ions have been obtained by Neuman and Jonas[82] using ^{1}H n.m.r. bandshape analysis. The compounds were of the type $CD_3C(X)NMe_2$, the deuteriation being carried out in order to simplify the NMe_2 ^{1}H n.m.r. signals. The results are discussed in terms of certain MO calculations, and of solvation effects.

[80] A. L. Love, T. D. Alger, and R. K. Olsen, J. Phys. Chem., 1972, 76, 853.
[81] K. C. Ramey, D. J. Lovick, P. W. Whitehurst, W. B. Wise, R. Mukherjee, J. F. Rosen, and R. M. Moriarty. Org. Magn. Resonance, 1971, 3, 767.
[82] R. C. Newman, jun. and V. Jones. J. Phys. Chem., 1971, 75, 3532.

Hindered internal rotation in NN-dimethyltrichloroacetamide (1d) has been studied[83] using variable-temperature ^{13}C n.m.r. Rates for this process were obtained by computer simulation of the simple equal population $A \rightleftharpoons X$ system. The authors claim theirs to be the first such use of ^{13}C n.m.r., but they are in error (we quoted from one such study[84] in Volume 1). At all events the authors calculated thermodynamic parameters for the barrier in their compound, and compare their results with those from earlier spin-echo studies and with their own work using ^{1}H n.m.r. Agreement between the ^{1}H and ^{13}C CW work is considered good. Curiously, the authors give E_a and $\Delta G_c^{\ne}$, but not A, $\Delta H^{\ne}$, or $\Delta S^{\ne}$.

Neuman et al.[31] report data from an n.m.r. study of internal rotation in N-methylformamide which, it is suggested, provides a peptide-bond model system. Two deuteriated species were studied, as well as the parent compound. Kinetic analysis was performed using the unequal-population two-site exchange theory, modified to provide for unequal values of non-exchange linewidths (presumably partly to cater for the long-range (H, H) or (H, D) coupling, though this procedure has been criticised[30]). Base-catalysed NH or ND exchange was used to improve the situation. The resulting kinetic parameters were not discussed in detail. The data bear out the common comment that $\Delta G_{298}^{\ne}$ is more accurate and useful for comparison purposes than E_a. Total bandshape n.m.r. analysis of the NMe ^{1}H resonance has also been used[85] (by Drakenberg and Forsén) to obtain information about the barrier to C—N internal rotation in N-methyl-formamide and -acetamide.

Liler[86] has reported a study of solvent effects on the proton n.m.r. spectra of [^{15}N]acetamide as a function of temperature. Exchange rates for C—N internal rotation were obtained from bandshapes by measurements of the separation of peaks. The author gives 'rotational barriers' (without defining these more fully) which range from 71 kJ mol^{-1} in acetone to 63 kJ mol^{-1} in water. It is suggested that the lower barrier in water may be due to intermolecular exchange of protons.

A ^{1}H n.m.r. investigation by Russian workers[87] was concerned with cis-trans isomerism for six dipeptides containing N-methylalanine residues, of which only one (22) was studied in detail. Lifetimes were obtained from linewidths of the NMe signal in the region of slow exchange only. The widths were corrected for non-exchange effects, but in view of the fact that lifetimes were only obtained at four temperatures over a region of 40 °C the resulting thermodynamic parameters should be treated with some caution.

Van der Werf and Engberts[88] used total n.m.r. bandshape analysis to

[83] O. A. Gansow, J. Killough, and A. R. Burke, *J. Amer. Chem. Soc.*, 1971, **93**, 4297.
[84] D. Doddrell, C. Charrier, B. L. Hawkins, W. O. Crain, L. Harris, and J. D. Roberts, *Proc. Nat. Acad. Sci. U.S.A.*, 1970, **67**, 1588.
[85] T. Drakenberg and S. Forsén, *Chem. Comm.*, 1971, 1404.
[86] M. Liler, *J. Magn. Resonance*, 1971, **5**, 333.
[87] S. L. Portnova, V. F. Bystrov, T. A. Balashova, V. T. Ivanov, and Yu. A. Ovchinnikov, *Bull. Acad. Sci. U.S.S.R.* (*Div. Chem. Sci.*), 1970, 776.
[88] S. van der Werf and J. B. F. N. Engberts, *Rec. Trav. chim.*, 1971, **90**, 663.

obtain kinetic information about C—N hindered internal rotation in N-methylcarbamates (23)—(25). The ester alkyl absorptions were used in each case, together with two-site unequal-population exchange theory. The bandshapes were iteratively fitted by computer. It is shown that the barrier parameters depend markedly on the non-exchange linewidths used [*e.g.* changing the width from 0.7 to 1.1 Hz increased E_a by 9.6 kJ mol^{-1} for (23)].

Barrier parameters have been given[89] for internal rotation about the C—N bond in two N-acyl prolines (26) from ^{1}H n.m.r. studies; the A⇌X system of the t-butyl protons was used. Analysis was by means of a computer program, and the curve-fitting procedure involved use of four parameters: linewidth at

$$\text{MeC-NH-CH-C-N-CHCO}_2\text{Me} \quad \text{Me} \text{---} \underset{}{\bigcirc} \text{---} \text{SO}_2\text{CH}_2\text{NMeCO}_2\text{R}$$

(22)   (23) a; R = Me
             b; R = CH$_2$CCl$_3$

Me—⟨◯⟩—SCH$_2$NMeCO$_2$Me   ⟨S◯⟩—SO$_2$CH$_2$NMeCO$_2$Me

(24)   (25)

half-height (above T_c), linewidth at half-height for the stronger signal (below T_c), peak-to-valley ratio, and chemical shift difference. It is noteworthy that the barrier for the urethane (26a) is substantially less than that of the peptide (26b).

Nitrogen Inversion, C—N Internal Rotation, and Related Processes. Investigations of magnetic site exchange for nitrogen-containing compounds are apt to lead to considerable discussion regarding mechanisms, since several processes are often possible (though not necessarily distinguishable). Thus nitrogen inversion, internal rotation about one or more C—N bonds, and intermolecular proton exchange (for NH groups) may all occur for the same compound (see, for example, references 90—93, discussed below). Ring inversion may also be a complicating factor for certain saturated nitrogen heterocycles. Consequently, a clear-cut separation of topics is not feasible. (See also ref. 81, discussed in the preceding sub-section.)

It has been some years since the first report[94] of an n.m.r. study of chemical exchange in the gas phase. During the current period there have been two publications discussing[1,2] nitrogen inversion in aziridines, studied in the gas

[89] H. L. Maia, K. G. Orrell, and H. N. Rydon, *Chem. Comm.*, 1971, 1209.
[90] C. H. Bushweller and W. G. Anderson, *Tetrahedron Letters*, 1972, 129.
[91] D. E. Leyden and W. R. Morgan, *J. Phys. Chem.*, 1971, **75**, 3190.
[92] W. R. Morgan and D. E. Leyden, *J. Amer. Chem. Soc.*, 1970, **92**, 4527.
[93] G. E. Hall, W. J. Middleton, and J. D. Roberts, *J. Amer. Chem. Soc.*, 1971, **93**, 4778.
[94] R. K. Harris and R. A. Spragg, *Chem. Comm.*, 1967, 362.

phase. For aziridine itself a deuterium isotope effect was observed.[1] For several alkylaziridines the difference between $\Delta G^{\neq}$ for the gas phase and for solutions in C_6D_{12} was found to be 0.8 kJ mol^{-1} or less.[2] For solvents other than C_6D_{12} the barrier for a given aziridine varied rather more widely, being increased by up to ca. 20 kJ mol^{-1} for aqueous solutions. In both studies only coalescence temperature data are quoted,[1,2] so a full n.m.r. bandshape analysis for a case involving exchange in the gas phase has still not been published to date.

Nitrogen inversion in aziridines is also the subject of discussion in a paper[95]

$$\underset{R\diagdown C\diagup \diagdown O}{\overset{\diagup N \diagdown CO_2Bu^t}{}} \rightleftharpoons \underset{O \diagup \diagdown C\diagdown R}{\overset{\diagup N \diagdown CO_2Bu^t}{}}$$

(26) a; R = PhCH$_2$O
  b; R = PhCH$_2$OCONHCH(CH$_2$OBu^t)

which gives a limited amount of ^{1}H linewidth data for trans-1-methyl-2-(p-biphenyl)-3-benzoylaziridine (27) as a function of temperature.

The rates of internal rotation about the C—N bond and of nitrogen inversion in some N-t-butyl-NN-dialkylamines have been reported by Bushweller and co-workers,[90,96] who used ^{1}H resonances of the Bu^t group and of the CH$_2$ groups in the other alkyl substituents, respectively, for the study of the two processes. Although the 'total bandshape method' was used, values of $\Delta G^{\neq}$ are quoted at one temperature only,[90] except for t-butyldimethylamine.[96] It is noteworthy that the barriers for the two processes are very similar (23 to 30 kJ mol^{-1}), probably indicating a common transition state. More details are given[96] for the case of t-butyldimethylamine. At -165 °C, under conditions of slow exchange, the t-butyl protons give rise to two peaks with intensities in the ratio 2:1, so the process is treated as three-site uncoupled exchange. Corrections to the theoretical bandshape are made to take account of non-exchange linewidths varying substantially with temperature (from 1.0 Hz at -45 °C to 8.7 Hz at -165 °C), the NMe$_2$ peak being used as a monitor. This broadening is shown not to be connected to quadrupolar effects. The fitting procedure for the exchange-broadened bandshapes was by visual comparison of calculated and observed spectra, and errors in the rate constants are estimated to be $\leqslant 5\%$. The activation parameters are discussed in terms of

(27) [structure: Ph-phenyl with H, N(Me), COPh, H substituents]

(27)

[95] D. L. Nagel, P. B. Woller, and N. H. Cromwell, J. Org. Chem., 1971, 36, 3911.
[96] C. H. Bushweller, J. W. O'Neil, and H. S. Bilofsky, Tetrahedron, 1971, 27, 5761.

internal rotation about the C—N bond assisted by rehybridization at nitrogen.

Leyden and Morgan,[91,92] following their own earlier work,[97] have simulated and fitted bandshapes in proton n.m.r. for a series of asymmetric tertiary benzylamines at a range of pH values in order to evaluate the roles of NH proton exchange and of nitrogen inversion. Several spectral effects appear to have been used: (a) the collapse of coupling in the NCH_3 region due to an NH proton, (b) the AB⇌BA exchange for methylene protons of benzyl groups, and (c) (for α-phenylethylbenzylmethylamine[91]) exchange between methyl groups or phenylethyl protons of two diastereoisomers. Few details are given of the fitting procedure. Corrections were made for non-exchange linewidths. The mechanisms for exchange are discussed, and some rate constants for the individual steps of the processes are evaluated.

Shoup *et al.*[98] have given a detailed evaluation of the n.m.r. total bandshape method of analysis for uncoupled exchanging two-site systems, with special reference to internal rotation about the $C(sp^2)$—N bond in two cytosine derivatives (28) and (29). Attention is paid to the effects of non-exchange linewidths and chemical shift differences varying with temperature. Neglect of these factors, or faulty treatment of them leads to activation parameters in error by *ca.* 10% (much in excess of statistical errors of Arrhenius and Eyring relationships). The authors carry out numerical analysis of computed spectra (varying the lifetime while fixing shift differences or non-exchange linewidths at incorrect values) to illustrate their comments. A detailed discussion of temperature measurement and its influence is also given. Differences in E_a of up to 11% are found between use of the spectrometer manufacturers' (Varian) calibration method and that of Van Geet.[99] The recommended procedure from this study was employed[35] to obtain barrier information for the two cytosine derivatives (28) and (29) from the NMe_2 ^1H resonance bandshapes. A curious combination of Arrhenius and Eyring parameters (E_a and $\Delta S^{\neq}$) is listed. The effects of solvent on the barrier were studied and discussed. Not surprisingly SO_2 as a solvent has a marked effect on E_a, lowering it considerably.

Filleux-Blanchard *et al.*[100] have given a brief report of a study of hindered internal rotation about the *N*-aryl bond in anilines. They quote a few values

(28)   (29)

---

[97] D. E. Leyden and W. R. Morgan, *J. Phys. Chem.*, 1969, **73**, 2924.
[98] R. R. Shoup, E. D. Becker, and M. L. McNeel, *J. Phys. Chem.*, 1972, **76**, 71.
[99] A. L. Van Geet, *Analyt. Chem.*, 1970, **42**, 679.
[100] M. L. Filleux-Blanchard, J. Fieux, and J. C. Halle, *Chem. Comm.*, 1971, 851

of $\Delta H^{\neq}$, but do not say how they were obtained. The effects of ring substituents on the barrier are discussed.

The mechanism of degenerate isomerization about the C=N bond has been a matter of controversy; the two possibilities are internal rotation (30) and lateral shift (31). Roberts and co-workers[93] have used ^{19}F n.m.r. bandshape analysis to study this problem for hexafluoroacetone-N-phenylimine (32a) and a series of para-substituted compounds (32b). The n.m.r. exchange is of the coupled mutual exchange type $A_3X_3 \rightleftharpoons X_3A_3$, and the authors used the Binsch[101] computer program, but worked on sub-systems of the real problem. Account was taken both of apparent loss of quartet structure in each CF_3 region and of merging of the two regions. The resulting barrier data are discussed in terms of Hammett parameters, and throw some light on the question of the mechanism. The p-nitro-compound behaves differently from the others.

The AB⇌BA exchange system of benzylic methylene protons for a number of 1,1-substituted thiosemicarbazides and related compounds (33) has been studied by Svanholm,[102] using density matrix calculations to simulate the spectra. Corrections were made for the chemical shift difference varying with temperature. The barrier parameters are given, and Svanholm argues that the process involved is hindered internal rotation about the N—N bond.

Jennings[103] has reported a study of barriers to internal rotation about the

$$\begin{matrix} X_1 \\ X_2 \end{matrix} C \overleftrightarrow{=} N^{-Y} \qquad \begin{matrix} X_1 \\ X_2 \end{matrix} C = N^{-Y}$$

(30)   (31)

$$\begin{matrix} F_3C \\ F_3C \end{matrix} C = N - \bigcirc - R$$

(32) a; R = H
     b; R = Cl, F, OMe, Me, or NO_2

P—N bond in (34a) and (34b). Although it appears that digitized spectra were treated by automatic computer fitting to an n-site exchange program, few details are given, since the publication was only in preliminary form.

Metallotropic Rearrangements and Related Processes. Sergeyev et al.[104] have examined the intramolecular metallotropic rearrangement of trimethylstannylindene (35) using ^{13}C spectra (using both the C_8/C_9 and the C_4/C_7

[101] G. Binsch, *J. Amer. Chem. Soc.,* 1969, **91,** 1304.
[102] U. Svanholm, *Acta Chem. Scand,* 1971, **25,** 1166.
[103] W. B. Jennings, *Chem. Comm.,* 1971, 867.
[104] N. M. Sergeyev, U. K. Grishin, U. N. Luzikov, and U. A. Ustynyuk, *J. Organometallic Chem.,* 1972, **38,** 1C.

pairs of peaks). Few details are given, but the high value of E_a compared to that for the corresponding cyclopentadiene system is stated to be evidence in favour of a mechanism involving successive 1,2-shifts.

The Russian workers have also[105] studied the metallotropic rearrangement of trimethylsilylcyclopentadiene (36a), using linewidths in ^{1}H n.m.r. of the ring protons for the slow and fast exchange regions (treating the process as a three-position case in the latter region). Errors in the treatment of the high and low temperature extremes are pointed out. The way in which exchange causes broadening in the various spectral regions shows that the rearrangement proceeds by a 1,2-shift. Some more qualitative kinetic information from n.m.r. measurements is given for other molecules of the $C_5H_5SiMe_{3-n}Cl_n$ series. The same research group has carried out[106] a similar treatment for the correspond-

$$\underset{X}{\overset{RHN}{\diagdown}}CNH-N(CH_2Ph)_2 \qquad \underset{Ph}{\overset{Cl}{\diagdown}}\underset{\underset{X}{\parallel}}{P}-NPr^i_2$$

(33) a; X = O, R = Bu^t (34) a; X = S
      b; X = S   b; X = O
      c; X = Se, R = Bu^t

(35)

ing compounds of germanium (36b) and tin (36c). In the case of the tin compound the way in which the $^{117/119}$Sn satellites of the methine proton signal broaden was used to distinguish between two possible values for the coupling constants between Sn and the other protons (these parameters were not directly observable). It should be noted that the equations used by the Russian workers differ slightly from those of Campbell and Green[107] (see below): the former comment that the different formulae lead to slightly different values of $\Delta S^{\neq}$.

An earlier study[108] of metallotropic rearrangements of Sn and Si, for compounds (37)—(39), unfortunately not discussed in Volume 1, was made by Davison and Rakita. Their computer simulations of bandshapes included those for the Me protons of (38); in this case the calculations took into account the effects of ^{117}Sn and ^{119}Sn isotopes (using three simultaneous two-

[105] N. M. Sergeyev, G. I. Avramenko, A. V. Kisin, V. A. Korenevsky, and Yu. A. Ustynyuk, *J. Organometallic Chem.*, 1971, **32**, 55.
[106] A. V. Kisin, V. A. Koranevsky, N. M. Sergeyev, and Yu. A. Ustynyuk, *J. Organometallic Chem.*, 1972, **34**, 93.
[107] C. H. Campbell and M. L. H. Green, *J. Chem. Soc. (A)*, 1970, 1318.
[108] A. Davison and P. E. Rakita, *J. Organometallic Chem.*, 1970, **23**, 407.

(36) a; X = Si
b; X = Ge
c; X = Sn

(37) a; X = Sn
b; X = Si

(38)

(39)

site exchanges occurring with the same mean pre-exchange lifetime). The mechanism for the rearrangement, which is clearly intramolecular, is discussed in detail.

Calderon et al. have studied[109] the fluxional behaviour of tetra(cyclopentadienyl)titanium, and have concluded that the molecule exists as $(h^1\text{-}C_5H_5)_2(h^5\text{-}C_5H_5)_2\text{Ti}$, i.e. there are two bonding situations for the C_5H_5 rings. At ca. $-25\ °C$ the proton n.m.r. spectrum consists of two lines of equal intensity, which coalesce as the temperature is raised; this process is attributed to exchange of the two types of C_5H_5 ring, the first observation of such a change. The spectra were subject to bandshape analysis, though few details are given. Below $-25\ °C$ the rearrangements within the $h^1\text{-}C_5H_5$ groups are slowed, resulting in further broadening. The rearrangement within the $\sigma\text{-}C_5H_5$ groups of $(\pi\text{-}C_5H_5)M(CO)_2(\sigma\text{-}C_5H_5)$, M = Fe or Ru, had earlier been studied by Campbell and Green[107] using n.m.r. bandshapes. Only the fast exchange region was utilized, and equations are quoted for linewidths in this region and in the slow exchange region for three-site exchange. The authors concluded that the rearrangement proceeded by a sequence of 1,2-shifts.

Harrod and Taylor[110] have reported in preliminary fashion the results of a ^1H n.m.r. study of intramolecular rearrangement in a series of cis-bis(alkoxy)-bis(acetylacetonato)titanium(IV) complexes and some related compounds (40). The coalescence of ^1H signals due to non-equivalent methyl groups of the cis-acac ligands was examined, though no details of the n.m.r. procedure are given. The large negative $\Delta S^{\neq}$ values found earlier would appear to be real. It is suggested that the mechanism for the exchange involves a transition state resembling a tightly bound ion-pair. The coalescence of non-equivalent methyl ^1H resonances of isopropoxy- and 2,6-di-isopropylphenoxy-groups in complexes with R = Pr^i or 2,6-$\text{Pr}^i_2\text{C}_6\text{H}_3$ was also examined.

The variable temperature AB⇌BA exchange case for the ligand ethylenic

---

[109] J. L. Calderon, F. A. Cotton, and J. Takats, J. Amer. Chem. Soc., 1971, **93**, 3587.
[110] J. F. Harrod and K. Taylor, Chem. Comm., 1971, 696.

(40)   (41)

protons of the π-allyl palladium complex (41) has been studied by Alexander et al.,[111] using iterative fitting of digitized spectra. The methyl signals were also examined. The three determined rates were the same, and it is concluded that there is a single rate-determining process leading to site exchange of the CH_2 protons concurrently with facial exchange of the ring. A σ–π intermediate is indicated. Only $\Delta G^{\neq}$ data are given.

Other Migration Processes. There continues to be an increase in the amount of research devoted to exchange between multi-spin systems, using density matrix theory. Thus Derendyaev[112] has studied the [ABCD]⇌[ABCD]′ exchange system (42) (the process is referred to as 'automerization'[113]). Presumably the computed and observed spectra were compared visually, though this is not stated. The data, together with information derived earlier[113] from the methyl resonance bandshapes, are used in an Arrhenius plot but, unfortunately, the revised Arrhenius activation parameters do not seem to be given in the later paper.[112]

The sigmatropic migration processes (43) and (44) have been studied by n.m.r.[114] and barrier parameters listed. No n.m.r. details are given, but the process is described in terms of competition between homo-[1,7] and homo-[1,5] paths.

C. **Examples of Intermolecular Exchange.**—At this point in Volume 1, mention was made of the lanthanide 'shift reagents'. At that stage few studies of the effects of these reagents gave linewidth information, but it was assumed by this Reporter that there would be an exchange contribution to the widths. This has now been shown to be the case.[115] Reuben and Leigh[115a] report values of T_2^{-1} for H-2 and H-8 of quinoline in solutions containing $Eu(dpm)_3$, obtained from linewidth measurements. The incremental broadening due to the $Eu(dpm)_3$ was plotted against ρ, the concentration ratio of $Eu(dpm)_3$ to total quinoline, and found to give a non-linear graph, showing that there is a

---

[111] C. W. Alexander, W. R. Jackson and W. B. Jennings, *J. Chem. Soc.* (*B*), 1971, 2241.
[112] B. G. Derendyaev, *Org. Magn. Resonance*, 1972, **4**, 27.
[113] V. G. Shubin, D. V. Korchagina, B. G. Derendyaev, V. I. Mamatyuk, and V. A. Koptyug, *Russ. J. Org. Chem.*, 1970, **6**, 2074.
[114] J. C. Gilbert, K. R. Smith, G. W. Klumpp, and M. Schakel, *Tetrahedron Letters*, 1972, 125.
[115] D. F. Evans and M. Wyatt, *J.C.S. Chem. Comm.*, 1972, 312.

[Structures labeled (42), (43), (44) showing equilibria]

substantial chemical exchange contribution to T_2. Measurements of chemical shifts and T_1 values allowed the mean lifetime of the adduct to be calculated (2.2×10^{-7} s from the H-2 data and 2.0×10^{-7} s for H-8 – satisfactory agreement). The authors also observed[115] bandshape changes resulting in decoupling for some protons. They attribute this phenomenon to the enhancement of T_1 for the nucleus to which coupling occurs. Such 'chemical exchange spin decoupling' (an unfortunate description,[116] since it may be viewed as due entirely to relaxation and not to exchange) occurs when $T_1^{-1} > \sqrt{2}\pi J$ (see ref. 116, Volume 1, p. 251, and this chapter, Section 4B).

Some broadening effects of lanthanide shift reagents were also reported[117] earlier for the methyl resonance of 0.10M solutions of t-butanol in CCl_4 containing a variety of lanthanides, present as tris(dpm) complexes. The broadening observed was such that protons experiencing larger shifts exhibited greater broadening, in agreement with the comments of Reuben and Leigh.[115]

Other references to linewidths (in which there is no mention of exchange effects) for solutions containing lanthanide shift reagents are discussed in Section 4B.

The reader is also directed to ref. 51 (see Section 2A) and to refs. 86, 91, and 92 (see Section 2B) for some other applications of bandshape studies involving inter-molecular exchange (see also Section 3D).

Studies of Ligand Exchange using the Swift–Connick[10] Approach. Such work has once more proved popular. It requires simple measurements (the widths of Lorentzian lines) in solutions involving exchange between bound (normally in

[115a] G. Reuben and J. S. Leigh, *J. Amer. Chem. Soc.*, 1972, **94**, 2789.
[116] L. S. Frankel, *J. Mol. Spectroscopy*, 1969, **29**, 273; *J. Chem. Phys.*, 1969, **50**, 943.
[117] D. R. Crump, J. K. M. Sanders, and D. H. Williams, *Tetrahedron Letters*, 1970, 4419.

a paramagnetic complex) and free ligands. The widths usually depend in general on three parameters: the lifetime of the bound species (τ_m), the effective nuclear spin-lattice relaxation time ($T_{2\,m}$), and the chemical shift difference between resonances of the bound and free forms ($\Delta\omega_m$). Sometimes the 'chemical shift contribution' to the linewidth may be considered negligible; however, it is this part which relates most directly to other cases of dynamic n.m.r.

Degani and Fiat[118] have developed the Swift–Connick[10] treatment to

(45) a; $X = Fe^{II}Cl, R = H$
     b; $X = Mn^{II}Cl, R = Me$

(46)

(47)

(48) a; $X = H$
     b; $X = Me$

cater for the case where there are three environments and exchange can occur between any pair. They produce simplified equations for a number of limiting cases. The theory is applied to systems where transitions between high- and low-spin states of a metal ion occur, as well as ligand exchange. The authors chose to study the case of haemin [presumably (45a)] in pyridine–water solutions. The kinetic and thermodynamic parameters for the ligand exchange rates (one pyridine and one water molecule co-ordinate to the haemin in certain conditions) were obtained from the transverse relaxation times of the water and pyridine protons and their temperature dependence. Similar information for the transition between the high- and low-spin states of the haemin was calculated from T_2 for the ring haemin protons (giving $\Delta H^{\ddagger} = 19 \pm 2$ kJ mol^{-1} and $\Delta S^{\ddagger} = -126 \pm 13$ J K^{-1} mol^{-1}).

[118] H. A. Degani and D. Fiat, *J. Amer. Chem. Soc.* 1971, **93**, 4281.

Rusnak and Jordan[119] have studied exchange rates of methanol and NN-dimethylformamide (DMF), used as solvents, from Mn^{III} protoporphyrin IX dimethyl ester (45b). The ^1H n.m.r. linewidths are analysed assuming the 'chemical shift contribution' is small. In the case of MeOH both CH_3 and OH resonances were used; for DMF the two overlapping CH_3 resonances were curve-resolved into separate Lorentzian bands. Different methods of fitting the data were used; they give quite widely varying values of $\Delta H^{\neq}$ and $\Delta S^{\neq}$ for a given exchange. Jordan and co-workers[120,121] have also carried out Swift–Connick[10] studies for DMF exchange rates of Ni^{II} Schiff-base complexes of (46)—(48). In the last two cases, exchange rates of water were also studied. Corrections were made for outer-sphere broadening. Chemical shift exchange broadening was important. The effects of the diamagnetic⇌paramagnetic equilibrium are discussed.[121] Thermodynamic parameters for the solvent exchanges are given. The authors state[120] among the conclusions that the rigid terdentate chelate (46) does not greatly affect the solvent-exchange kinetics.

Frankel[122] has also used the Swift–Connick[10] approach to study ligand exchange kinetics of Ni^{II} with DMF in DMF–nitromethane as a mixed solvent. The data were corrected for second-sphere relaxation effects by extrapolation of the low-temperature data. The activation parameters for exchange are within experimental error of those for solutions in pure DMF. The results support the so-called D mechanism (requiring a five-co-ordinate intermediate) for the exchange.

As mentioned in Volume 1, use of nuclei other than ^1H is becoming increasingly popular for Swift–Connick[10] studies; the ^{17}O nucleus is particularly favoured, and during the current review period two papers dealing with ^{17}O studies of solutions containing nickel have appeared. Neely and Connick[123] have observed ^{17}O signals from both bulk and bound H_2O molecules in solutions containing Ni^{2+}. They use measured linewidths (from dispersion mode spectra), together with chemical shift data, to determine both the lifetime of water molecules in the first co-ordination sphere and the ^{17}O spin–spin relaxation times of such molecules caused by the paramagnetic ions (the predominant contribution to T_2 arises from scalar coupling). The T_2 values are then used to obtain information about the electronic relaxation. Values of $\Delta H^{\neq}$ and $\Delta S^{\neq}$ for the exchange are calculated as 58 kJ mol^{-1} and 42 J K^{-1} mol^{-1} respectively. ^{17}O N.m.r. has also been used[124] to study water exchange in nickel(II)–edta solutions. The full Swift–Connick treatment is used, since the linewidths are affected by both the chemical shift and 'T_{2m}' mechanisms. In acidic solution the data are consistent with the occurrence of a single octahedral nickel species, probably $Ni(H_2O)Hedta$, i.e. with one (protonated) acetate arm replaced by a water molecule. The usual parameters are tabulated.

[119] L. L. Rusnak and R. B. Jordan, *Inorg. Chem.*, 1972, **11**, 196.
[120] L. L. Rusnak, J. E. Letter, jun., and R. B. Jordan. *Inorg. Chem.*, 1972, **11**, 199.
[121] L. L. Rusnak and R. B. Jordan, *Inorg. Chem.*, 1971, **10**, 2686.
[122] L. S. Frankel, *Inorg. Chem.*, 1971, **10**, 2360.
[123] J. W. Neely and R. E. Connick, *J. Amer. Chem. Soc.*, 1972, **94**, 3419.
[124] M. W. Grant, H. W. Dodgen, and J. P. Hunt. *J. Amer. Chem. Soc.*, 1971, **93**, 6828.

In neutral solution the situation is more complicated, and a tentative analysis is given in terms of two species.

The exchange rates of dimethyl sulphoxide (DMSO) ligands for Cu(DMSO)$_6^{2+}$, Ni(DMSO)$_6^{2+}$, and Fe(DMSO)$_6^{2+}$ have been obtained[125] from proton n.m.r. line-broadening (and contact-shift) studies. In the Ni^{II} case corrections were made for outer-sphere exchange contributions to the broadening. Values of $\Delta H^{\neq}$ and $\Delta S^{\neq}$ are given for the exchange for the three cases quoted above. Other systems were also studied and discussed, but it was concluded that the exchange rates fell outside the n.m.r. range.

The ^{1}H n.m.r. linewidth of the hydroxyl and acetic acid methyl solvent protons in ethanol–acetic acid solutions of copper(II) acetate have been measured[126] from -100 to $+100$ °C, and analysed using the Swift–Connick[10] approach. The exchange rate between free and bound acetic acid is found to be 2.5×10^4 s^{-1} at 25 °C. At low temperatures a further process occurs, with an exchange rate constant 230 s^{-1} at -67 °C.

Zeltmann and Morgan[127] have observed the ^{17}O and ^{35}Cl n.m.r. spectra for complex oxovanadium(IV) species in hydrochloric acid solutions as a function of concentration and temperature. Linewidths were measured and corrected for any contributions not arising from paramagnetic V^{IV} ions by using widths for HCl solutions of the same viscosity. It is assumed there is no shift broadening. The linewidth data were analysed using the Swift–Connick formulation[10] to obtain specific ligand-exchange rate constants for both H$_2$O and Cl$^-$ in the species VO(H$_2$O)$_4^{2+}$, VO(H$_2$O)$_3$Cl$^+$, and VO(H$_2$O)$_2$Cl$_2$, where appropriate, and ^{17}O and ^{35}Cl spin–spin relaxation times in the various species. The authors discuss the problem of fitting their linewidth data with the large number of variables involved.

Lincoln and West[128] have continued their ^{14}N n.m.r. studies of exchange of acetonitrile ligands (see Vol. 1, p. 239; ref. 129 contains corrections to the earlier work). The new study[128] involves the cobalt(II)–triethylenetetramine (trien) system in MeCN. Linewidths (unfortunately in magnetic field units) are reported for a range of concentrations and temperatures. The 'chemical shift contribution' to the line broadening is appreciable, and various exchange rates for MeCN are derived. The rate for exchange of MeCN in the first co-ordination sphere of Co(trien)(MeCN)$_n^{2+}$ is $> 10^3$ times that for Co(MeCN)$_6^{2+}$ at 258 K. There is a useful discussion regarding the choice of a standard for estimation of the linewidth in the absence of paramagnetic ions.

In conclusion to this sub-section it is appropriate to mention two other relevant studies. Bryant[130] has studied ^{25}Mg n.m.r. in aqueous solutions of molecules containing phosphate groups which bind magnesium ion (*e.g.* ATP). In 1.5M-MgCl$_2$ solutions $\Delta \nu_{\frac{1}{2}} = 3.8 \pm 0.5$ Hz, but addition of phos-

---

[125] G. S. Vigee and P. Ng, *J. Inorg. Nuclear Chem.*, 1971, **33**, 2477.
[126] H. Grasdalen, *Acta Chem. Scand.*, 1971, **25**, 1103.
[127] A. H. Zeltmann and L. O. Morgan, *Inorg. Chem.*, 1971, **10**, 2739.
[128] S. F. Lincoln and R. J. West, *Austral. J. Chem.* 1972, **25**, 469.
[129] R. J. West and S. F. Lincoln, *Austral. J. Chem.*, 1972, **25**, 222.
[130] R. G. Bryant, *J. Magn. Resonance*, 1972, **6**, 159.

phate causes substantial broadening. This is interpreted on the basis of rapid exchange between free and complexed Mg^{2+}. Linewidths of up to 915 Hz are quoted for the phosphate complexes. The effects of pH and temperature variation were studied, and the Swift–Connick[10] treatment was applied to obtain a value for the exchange rate of magnesium ion with ATP.

Proton n.m.r. cation hydration studies of diamagnetic $La(ClO_4)_3$ and $Zn(ClO_4)_2$ and of paramagnetic $Ce(ClO_4)_3$, $Er(NO_3)_3$, $Fe(ClO_4)_2$, and $Ni(ClO_4)_2$ have been carried out[131] in aqueous solution. Separate signals were observed for bound and bulk water molecules at low temperatures, and linewidths are given for the paramagnetic solutions. These widths range up to 1300 Hz for bulk H_2O and up to 4200 Hz for bound water. The widths are not discussed in detail.

The reader is referred to Section 4B for a discussion of other articles concerning bandshapes for systems containing paramagnetic substances. Other ligand exchange processes are treated in refs. 132—136 (see below).

Proton Exchange Processes. Benassi *et al.*[137] quote exchange parameters for the imidazole (49). Few details of the total bandshape procedure employed are given. The magnitude of the entropy of activation appears to be unbelievably high.

The rate of NH proton exchange for neat *N*-methylmethanesulphonamide, CH_3SO_2NHMe, has been measured[138] as a function of temperature using the 'single parameter' methods due to Takeda and Stejskal,[139] applied to the NMe proton signal. The non-exchange linewidth was taken from the SMe resonance. Activation parameters are quoted for the process. The protolysis kinetics of 5-dimethylaminonaphthalene-1-sulphonic acid (50a) and its *N*-methylsulphonamide (50b) have been studied[140] by ^1H n.m.r. bandshape measurements of the dimethylamino signal and, in the case of (50b), of the signal due to the *N*-methylsulphonamide moiety. Simple two-site exchange equations, with correction for non-exchange linewidths, were used. The pre-exchange lifetimes were obtained as a function of pH, and a possible rate law for the 5-dimethylamino protolysis is discussed.

Rabenstein[133] has studied the rate of exchange of the peptide proton in acetylglycine, $CH_3CONHCH_2CO_2H$, both for the free amino-acid and for a cadmium complex, by bandshape analysis of the methylene signal as a func-

---

[131] A. Fratiello, V. Kubo, S. Peak, B. Sanchez, and R. E. Schuster, *Inorg. Chem.*, 1971, **10**, 2552.
[132] J. W. Faller and J. W. Sibert, *J. Organometallic Chem.*, 1971, **31**, 5C.
[133] D. L. Rabenstein, *Canad. J. Chem.*, 1972, **50**, 1036.
[134] G. H. Reed and R. J. Kula, *Inorg. Chem.*, 1971, **10**, 2050.
[135] H. Grasdalen, *J. Magn. Resonance*, 1971, **5**, 84.
[136] H. Grasdalen, *J. Magn. Resonance*, 1972, **6**, 336.
[137] R. Benassi, P. Lazzaretti, L. Schenetti, F. Takkei, and P. Vivarelli, *Tetrahedron Letters*, 1971, 3299.
[138] P. Olavi, I. Virtanen, L. Pajari, and E. Rahkamaa, *Suomen Kemi.*, 1971, **B44**, 146.
[139] M. Takeda and E. O. Stejskal, *J. Amer. Chem. Soc.*, 1960, **82**, 25.
[140] J. F. Whidby, D. E. Leyden, C. M. Himel, and R. T. Mayer, *J. Phys. Chem.*, 1971, **75**, 4056.

(49)

(50) a; X = OH
    b; X = NHMe

tion of pH. In comparing simulated and experimental spectra great use was made of linewidths and valley-to-peak ratios. Coupling to the methyl protons was properly taken into account. The data were analysed in terms of an exchange process involving the OH$^-$ ion. Rate constants 3.75 (± 0.18) × 10^6 and 2.0 (± 0.28) × 10^8 l mol^{-1} were obtained for the free amino-acid and for the complex, respectively, showing that complexing substantially enhances the rate.

Faller and Sibert quote[132] some rate constants for the process (51; por ≡ porphyrin system) derived from proton n.m.r. linewidths, but there is little detailed n.m.r. information. The authors do, however, prove that the exchange occurs *via* free imidazole, *i.e.* the mechanism is intermolecular.

(51)

Other Studies of Intermolecular Exchange. Bishop and co-workers[141] have investigated exchange in the system Me$_3$P–Me$_3$B, with Me$_3$P in excess, using a first-order treatment following the Bloch–McConnell equations (bearing in mind the opposite signs of J_{PH} in the free and complexed phosphine). Methyl peaks of both the Me$_3$P and Me$_3$B moieties were used. Fitting was by computer iteration varying the chemical shift differences and the mean pre-exchange lifetime (variation of the spin–spin relaxation time for the BMe$_3$ protons was found not to be necessary). Experiments were carried out for a variety of concentrations and temperatures. It is concluded that exchange proceeds by a dissociative mechanism with $\Delta H^{\neq} = 63$ kJ mol^{-1} and $\Delta S^{\neq} = 36$ J K^{-1} mol^{-1}.

Fogelman and Miller[142] have studied the exchange of excess acetonitrile with CH$_3$CN,BX$_3$ (X = F, Cl, or Br), using ^{1}H n.m.r. linewidth measurements. By studying the variation of pre-exchange lifetimes as a function of various concentrations, they conclude the exchange proceeds by a dissociative

[141] K. J. Alford, E. O. Bishop, P. R. Carey, and J. D. Smith, *J. Chem. Soc. (A)*, 1971, 2574.
[142] J. Fogelman and J. M. Hiller, *Canad. J. Chem.*, 1972, **50**, 1262.

mechanism. Lifetimes and activation data (a mixture of E_a and $\Delta S^{\neq}$) are listed for a range of solution conditions. Mixed solvents had to be used in order to obtain low-temperature solubility, but the results are extrapolated to inert solvent conditions.

The ^{1}H and ^{19}F n.m.r. spectra of the methanol–BF$_3$ system, in sulphur dioxide as a solvent, have been examined[143] for a variety of concentration conditions. Some lifetimes, from ^{19}F linewidth measurements, are given for the exchange between CH$_3$OH,BF$_3$ and free BF$_3$.

The self-exchange of methyl groups in CdMe$_2$ and the cross-exchange in CdMe$_2$–GaMe$_3$ and ZnMe$_2$–InMe$_3$ have been studied[144] by ^{1}H n.m.r. All details of the method of n.m.r. analysis were given in an earlier paper;[145] however, certain inconsistencies in the exchange theory are corrected. The lifetime of each metal–methyl bond may be related to a rate constant for the exchange, and activation parameters are reported for the processes. Particular attention is paid to the influence of solvents. Strongly co-ordinating solvents enhance the self-exchange rate for Group II metal alkyls but decrease the cross-exchange rate between Group II and Group III compounds. There is a detailed discussion of the exchange mechanism, based on these results.

Self-exchange of ligands has been studied[146] by ^{1}H n.m.r. for a number of mercurials, and detailed results presented for (Me$_3$Si)$_2$Hg and (Me$_3$Ge)$_2$Hg. The situation was treated as three-site exchange, the sites being those with the two spin states of ^{199}Hg and those with other Hg isotopes. Exchange from one satellite site to the other was ignored, as was any contribution from ^{29}Si or ^{13}C satellites. Some data were also obtained using the slow exchange approximation. Arrhenius activation energies (ca. 48 kJ mol^{-1}) are given for the exchange; more qualitative data were obtained for some related molecules.

Quantitative information about exchange rates of ligands between free and complexed forms has been obtained[134] by n.m.r. studies for the edta-chelate of zinc, and for related complexes of both cadmium and zinc. The data were obtained from linewidths in the slow-exchange region in acidic solution only. Widths for neutral solutions, where exchange does not occur, were used to correct the data for other broadening influences. Uncertainties in $\Delta v_{\frac{1}{2}}$ of only 0.05 Hz are claimed for measurements of up to 5 Hz (but more typically ~1.1 Hz). The dependence of the pre-exchange lifetimes on solution conditions is used to discuss the mechanism of the exchange, and rate constants are derived.

Proton n.m.r. studies of solvation of Al^{III} in ethanol[135] and n-propanol[136] have been carried out by Grasdalen. Low-temperature studies of the OH peaks show there are two ligand exchange processes with different rate constants. This is also indicated by the linewidth dependence of the bulk OH peak. Exchange rate constants are derived for the ethanol case from such

---

[143] K. L. Servis and L. Jao, *J. Phys. Chem.*, 1972, **76**, 329.
[144] J. Soulati, K. L. Henold, and J. P. Oliver, *J. Amer. Chem. Soc.*, 1971, **93**, 5694.
[145] K. L. Henold, J. Soulati, and J. P. Oliver, *J. Amer. Chem. Soc.*, 1969, **91**, 3171.
[146] T. F. Schaaf and J. P. Oliver, *J. Organometallic Chem.*, 1971, **32**, 307.

linewidths (and shifts) using the three-site exchange equations of Patterson and Ettinger.[147] These constants are ca. 8×10^3 s^{-1} and ca. 3×10^2 s^{-1} for the two types of exchange.

3 Effects of Quadrupolar Nuclei

A. Theoretical Work.—Detailed theoretical work on relaxation in multi-spin systems has been reported by Pyper[148—151] during the past year. The results have important consequences for high-resolution bandshapes and some of these have been developed explicitly by Pyper, especially for cases involving spin-$\frac{1}{2}$ nuclei coupled to quadrupolar nuclei. Pyper points out[148,149] that Liouville space is more appropriate than Hilbert space for many relaxation calculations, particularly those involving symmetry of the spin system, since factorization of the problem is possible using Liouville space but not using Hilbert space. The resulting simplifications have been discussed[148] in particular for spin systems containing several quadrupolar nuclei, especially in the fast relaxation limit, when the spin-$\frac{1}{2}$ system can be considered separately from the quadrupolar system (the lattice). Some simplified expressions for spectral density terms are given.[148] Some surprising features emerge, for instance cross-symmetry transitions may become allowed, e.g. for the [^{2}A^{3}X]$_2$ spin system, even when the relaxations of the X nuclei are fully correlated. A later paper[149] further develops the theory using a simpler method. Bandshapes for the spin-$\frac{1}{2}$ nuclei of a ^{2}A^{3}X^{3}Y spin system have been considered more explicitly in a separate publication,[150] where a number of computed spectra are illustrated. Calculations using both the stochastic approach (analogous to chemical exchange) and the relaxation matrix method were performed, and it is shown that the former treatment is inadequate. The two methods, however, give identical results at both fast and slow relaxation limits. For intermediate relaxation rates near the fast relaxation region, the bandshape may be considered as the superposition of a sharp line and a broad line (the stochastic approach does not reproduce the sharp line); this feature is correctly reproduced by a simplified perturbation treatment. The theory which treats the spin-$\frac{1}{2}$ and quadrupolar systems independently (i.e. theory of scalar relaxation of the second kind) is rederived and fully developed in a separate paper by Pyper[151] which discusses the ^{2}A^{2}B^{3}X system in detail. Corrections up to fourth order in the strength of coupling between the two groups of spins are presented. The limitations of the treatment are discussed with particular reference to the ^{2}A^{3}X^{3}Y spin system and the simplified perturbation treatment mentioned above is developed for this case. Altogether, these papers by Pyper[148-151] represent a formidable contribution to the literature; there is, perhaps, a need for their content to be summarized and simplified for more general consumption before the ideas are greatly used in practice. Unfortunately, most of the curious bandshape features which are

---

[147] A. Patterson, jun., and R. Ettinger, Z. Elektrochem., 1960, 64, 98.

predicted have not yet been observed, and there is a need for more experimental work in this direction. The $^2A^2B^3X$ system has, however, been studied,[152] and recently the importance of the double quantum transition has been emphasized[153] for the fast X relaxation limit. The width of this line gives the relative signs of J_{AX} and J_{BX}, whereas the unsaturated spectrum is insensitive to this parameter. In the case of 2-bromothiazole (52) such width measurements[153] showed J_{AX} and J_{BX} to have the same sign, in agreement with double-resonance work.[154]

(52)

Bogdanov et al.[155] have derived (B,H) coupling constants by measuring linewidths of various vinyl and alkoxy boron compounds in both ^1H and ^{11}B resonance, suitably correcting for non-quadrupolar contributions, and using the equation:

$$J_{BH}^2 = \Delta v_{\frac{1}{2}}(H) \Delta v_{\frac{1}{2}}(B)/5$$

This expression assumes the rapid relaxation limit and that $T_1^B = T_2^B$. The authors used linewidths for $^1H-\{^{11}B\}$ double resonance as the correction values, and employed $^1H-\{^1H\}$ double resonance or $^1H-\{^1H\}-\{^{11}B\}$ triple resonance when (H,H) coupling caused overlap problems. See also refs 49 and 50, discussed in Section 2A.

B. Resonances of Spin-$\frac{1}{2}$ Nuclei Coupled to Quadrupolar Nuclei.—Detailed bandshape work in this area proceeds at a steady pace, though the review period has only seen a handful of relevant papers published. The group of J. M. Lehn have continued[156–158] their studies of chemical applications of quadrupolar relaxation, frequently using selective deuterium labelling. Some of this research has been carried out by spin-echo measurements,[156] but most of the work[157,158] concerns ^1H n.m.r. bandshape studies, arriving at deuterium T_1 values indirectly. The advantages of the (H,D) system are succinctly presented,[157] and values of J_{HD} were readily obtained at high

---

[148] N. C. Pyper, *Mol. Phys.*, 1971, **21**, 1.
[149] N. C. Pyper, *Mol. Phys.*, 1971, **22**, 433.
[150] N. C. Pyper, *Mol. Phys.*, 1971, **21**, 977.
[151] N. C. Pyper, *Mol. Phys.*, 1971, **21**, 961.
[152] R. K. Harris and N. C. Pyper, *Mol. Phys.*, 1971, **20**, 467.
[153] R. K. Harris and N. C. Pyper, *Mol. Phys.*, 1972, **23**, 277.
[154] A. Kumar, N. R. Krishna, and B. D. N. Rao, *Mol. Phys.*, 1970, **18**, 11.
[155] V. S. Bogdanov, A. V. Kessenikh, and V. V. Negrebetsky, *J. Magn. Resonance*, 1971, **5**, 145.
[156] J. P. Kintzinger and J. M. Lehn, *Mol. Phys.*, 1971, **22**, 273.
[157] C. Brevard, J. P. Kintzinger, and J. M. Lehn, *Tetrahedron*, 1972, **28**, 2429.

temperatures (inefficient quadrupolar relaxation). Data for T_1^D were derived[157] from computer fitting of bandshapes for a range of compounds containing the CHDX group (especially for X = OAc) and for a steroid having the —CD═CH— fragment. The fitting included variation of J_{HD} and the non-quadrupolar linewidth as well as T_1^D. Values of the correlation time τ_q were derived by assuming a quadrupolar coupling constant of 170 kHz. Eyring-type plots yielded values for $\triangle H^{\neq}$ and $\triangle S^{\neq}$ (in the ranges 7.1 to 13.8 kJ mol^{-1} and -7.5 to $+30$ J K^{-1} mol^{-1} respectively). The results for τ_q for the various molecules are discussed in terms of anisotropic re-orientation and of structural effects on *local* motion. Some solvent effects on τ_q were investigated. The seventh paper in the series[158] presents work on a further range of deuterated organic molecules, with particular reference to a separation of motional effects into overall and internal rotational contributions, using transferability of correlation times for similar sites in related molecules. Activation parameters for local, overall, and internal (hindered rotational) motion are presented and discussed. Some intermolecular effects are seen.

Shepperd et al.[159] have iteratively fitted ^{1}H bandshapes for methyl nitrate in a variety of solvents and, in two cases, over a range of temperatures. The bandshapes ranged over the transitional region between a Lorentzian single line and a clear-cut triplet. Values of J_{NH} were obtained from ^{15}N satellite lines, and a non-quadrupolar linewidth was taken from the solvent resonance. A computer program used this data as input and iterated to a 'best' value for T_1^N. The results are discussed in terms of various models for rotational correlation times, but none of the models are satisfactory for all solvents. Activation energies for molecular reorientation are given for the dimethylformamide and cyclohexane solutions.

Mellon et al.[160] have carried out a detailed study of the ^{1}H n.m.r. spectra of borazine (53) and [^{10}B]borazine over a range of temperatures, and have analysed the shapes of both the BH and NH signals. The analysis was carried out both from a curve-resolving aspect (to Lorentzian shapes in the slow quadrupolar relaxation region) and by computer simulation. In the simulation the influence of non-quadrupolar widths (especially to simulate unresolved splitting due to coupling) was investigated. Graphs of single-parameter characteristics of the simulated spectra ($vs.$ T_1^N) were then drawn. In fact three different ways of obtaining T_1^N or T_1^B from the spectra are discussed. These models were 'tested' by obtaining activation parameters for molecular reorientation from both T_1^N and T_1^B and seeing which 'model' gave the best agreement. The paper is an interesting attempt to disentangle the various influences on such bandshapes, but the procedures appear to be to some extent unnecessarily complicated and probably insufficiently computer-automated.

Marks and Shimp[161] report measurements of the temperature-variation of

[158] C. Brevard, J. P. Kintzinger, and J. M. Lehn, *Tetrahedron*, 1972, **28**, 2447.
[159] C. M. Shepperd, T. Schaefer, B. W. Goodwin, and J.t'Raa, *Canad. J. Chem.*, 1971, **49**, 3158
[160] E. K. Mellon, B. M. Coker, and P. B. Dillon, *Inorg. Chem.*, 1972, **11**, 852.
[161] T. J. Marks and L. A. Shimp, *J. Amer. Chem. Soc.*, 1972, **94**, 1542.

$$\begin{array}{c} \text{H} \\ \text{N} \\ \text{HB} \quad \text{BH} \\ | \quad | \\ \text{HN} \quad \text{NH} \\ \text{B} \\ \text{H} \end{array}$$

(53)

^{1}H n.m.r. spectra of $Zr(BH_4)_4$ and $Hf(BM_4)_4$; the two compounds show identical effects, ascribed to ^{10}B and ^{11}B relaxation (intramolecular rearrangement is stated to be still rapid at -80 °C, in disagreement with earlier work). Theoretical spectra were computed on the basis of multisite exchange (with the different ^{10}B and ^{11}B spin states as 'sites') using the treatment of Kubo[162] and Sack.[163] Ignoring the 18.8% abundant ^{10}B spectrum was found to be a poor approximation. Simulations using both ^{11}B and ^{10}B information, with $T_1(^{11}B) = 0.651\ T_1(^{10}B)$, gave excellent agreement with the spectra. The resulting relaxation data are discussed in terms of correlation times, using an Arrhenius plot, and of the 'micro-viscosity' concept. Quadrupole coupling constants are derived as 3.5 ± 0.6 MHz (^{10}B) and 1.7 ± 0.3 MHz (^{11}B).

Jordan et al.[164] have measured ^{31}P and ^{35}Cl linewidths for liquid mixtures of PCl_3 and PBr_3 and of PCl_3, $PBrCl_2$, PBr_2Cl, and PBr_3 as a function of temperature. Data are also given for neat PCl_3. Values of spin–spin relaxation times are derived, and used, together with T_1 data, to discuss relaxation mechanisms, and to derive quadrupole, scalar, and spin-rotation coupling constants. The ^{35}Cl linewidths were measured from peak-to-peak separations in the derivative absorption mode, but ^{31}P widths are from measurements at half-height. Results are mostly given in graphical form and related to standard equations, assuming that T_2^{Cl} is dominated by quadrupolar relaxation and that T_2^P is given by:

$$T_2^{-1} = T_{20}^{-1} + BT\exp(-E_a/RT)$$

The second term is attributed to the (P,Cl) or (P,Br) scalar coupling. The results yield $J_{PCl} = 127$ Hz for PCl_3 and $J_{PBr} = 296$ Hz for PBr_3. Values of E_a for the various species range from 5.5 to 8.9 kJ mol^{-1}.

C. **N.M.R. Spectra of Quadrupolar Nuclei.**—Direct observation of n.m.r. spectra for quadrupolar nuclei is definitely becoming more popular, but measurements of bandshapes tend to be rather qualitative (often because of signal-to-noise problems). Almost invariably the only information presented is from the 'single parameter' measurement of linewidths; total bandshape analysis and spectral digitization have made little impact in this area to date. The problems of non-uniform reporting of linewidth data have been discussed in Section 1C. (See also Section 2C.)

[162] R. Kubo, *Nuovo Cimento, Suppl.*, 1957, **6**, 1063.
[163] R. A. Sack, *Mol. Phys.*, 1958, **1**, 163.
[164] A. D. Jordan, R. G. Cavell, and R. B. Jordan, *J. Chem. Phys.*, 1972, **56**, 483.

Nitrogen-14. Dinesh and Rogers[26] report ^{14}N linewidths (unfortunately in magnetic field units) for a variety of organic liquids and have used these to calculate ^{14}N spin–spin relaxation times assuming a Gaussian distribution rather than a Lorentzian one since the former gave better agreement with the T_1^N values from spin-echo experiments (this procedure is not closely argued). The linewidths themselves are the peak-to-peak separations for derivative absorption spectra. For compounds with known solid-state nuclear quadrupole coupling constants, the re-orientational correlation times, τ_q, are derived and compared to values, τ_c, calculated using Debye–BPP theory. The ratios τ_c/τ_q are transferred to similar molecules in the series and hence τ_q and the nitrogen quadrupole coupling constants obtained separately.

Ader and Loewenstein[25] have measured ^{14}N linewidths in magnetic field units (peak-to-peak, derivative absorption mode, corrected for modulation broadening) for three cobalt complexes, $[Co(CN)_6]^{3-}$, $Co(NH_3)_6Cl_3$, and $Co(en)_3Cl_3$. In the first example the cations used were Li, K, Cs, and NMe$_4$. The data were used, together with other information, to obtain the ^{14}N quadrupole coupling constants for the chloro-compounds, and it is concluded that binding to the cobalt reduces the ^{14}N electric field gradient considerably. For $[Co(CN)_6]^-$ the procedure was reversed, and e^2qQ (solid-state value) used to obtain the re-orientational correlation time and hence information about ^{59}Co parameters; the ^{59}Co linewidths were *ca.* 4 Hz.

Witanowski *et al.*[165] report, without comment, ^{14}N linewidths for a range of six-membered aromatic heterocycles. Similar information is given for five-membered aromatic heterocycles in ref. 166. The effects of tautomerism for some of the compounds (*e.g.* pyrazole) on the ^{14}N spectra are discussed. Variations of linewidths with chemical structure and with solvent are large (the widths range from 40 Hz for pyrrole in methanol to 1600 Hz for benzotriazole in methanol). Linewidths in ^{14}N resonance have also been given for pyrroles and imidazoles (mostly with nitro-substituents).[167]

The paper by Goldammer and Hertz,[168] unfortunately not mentioned in this chapter of Volume 1, gave values of ^{14}N and ^{17}O relaxation times (quoted as T_1, assumed to be equal to T_2) derived from linewidth measurements for several organic liquids. The information, together with a large quantity of data from pulse experiments, was used to discuss the microdynamic structure of aqueous mixtures with non-electrolytes.

Oxygen-17. Ziessow *et al.*[169] have tabulated ^{17}O n.m.r. linewidths for acetic acid at a variety of concentrations in cyclohexane as part of a study (mainly using chemical shifts) of hydrogen bonding. Lapidot and Irving[170] found no

[165] M. Witanowski, L. Stefaniak, and H. Januszewski, *Tetrahedron*, 1971, **27**, 3129.
[166] M. Witanowski, L. Stefaniak, H. Januszewski, Z. Grabowski, and G. A. Webb, *Tetrahedron*, 1972, **28**, 637.
[167] E. Lippmaa, M. Magi, S. S. Novikov, L. I. Khmelnitski, A. S. Pridhodko, O. V. Lebedev, and L. V. Epishina, *Org. Magn. Resonance*, 1972, **4**, 153.
[168] E. v. Goldammer and H. G. Hertz, *J. Phys. Chem.*, 1970, **74**, 3734.
[169] D. Ziessow, U. Jentschura and E. Lippert, *Ber. Bunsengesellschaft Phys. Chem.*, 1971, **75**, 901.
[170] A. Lapidot and C. S. Irving, *J. C. S. Dalton*, 1972, 668.

detectable ^{17}O signal due to bound $^{17}O_2$ in molecular oxygen complexes of Ir and Rh. They conclude this is due to large rotational correlation times (*i.e.* slow tumbling), resulting in very broad, unobservable lines. (See also ref. 168.)

Alkali Metals. The effects on ^{23}Na linewidths of adding polyglycol dimethyl ethers or macrocyclic polyethers to a solution of sodium tetraphenylboron ion-pairs in tetrahydrofuran solution have been the subject of a brief report.[171] The strong broadening effect of the ethers was studied as a function of viscosity. The authors suggest it is evidence that the added ethers replace the tetrahydrofuran in the solvent-separated NaBPh$_4$ ion-pair. The broadening is a result either of a change in the rotational correlation time or of the electric field gradient (probably the latter). (See also ref. 172, discussed in Section 3D.)

Herlem and Popov[173] have measured ^{23}Na resonance linewidths for sodium iodide and tetraphenylborate solutions in strongly basic solvents, including liquid ammonia. Although the work concentrates largely on chemical shifts, it was noted that in liquid ammonia at low concentrations the line is extremely narrow. The ratio of linewidths for NaI in NH$_3$ and for a saturated aqueous solution of NaCl (the reason for using a ratio is not explained) is plotted against NaI concentration, and the increase ascribed to viscosity changes.

The effect of other univalent positive ions on the ^{85}Rb n.m.r. linewidth (given in field units) for solutions containing humic acid has been investigated.[174] The line-narrowing effect is smallest for Li$^+$ and increases with the size of the competing cation. (See also ref. 9.)

Halogen Nuclei. Wennerström *et al.*[24] have measured ^{35}Cl, ^{79}Br, ^{81}Br, and ^{127}I linewidths for aqueous solutions of some substituted ammonium, phosphonium, and sulphonium salts. The widths are, unfortunately, quoted in magnetic field units. They were measured as peak-to-peak distances in the derivative absorption mode for Br and I, and as the widths at half-height of the first side-band signal in the simple absorption mode for Cl. The effect of concentration, alkyl-chain length, and substituents in the cation were investigated. The use of D$_2$O as a solvent instead of H$_2$O increases linewidths by *ca.* 20%. The authors suggest that the observed line broadening arises from anion–solvent interactions. The cations stabilize the water structure and this affects the anion–solvent interaction. Polar groups on the cations reduce the structure stabilization and hence reduce the linewidths. This description of the situation is not compatible with some earlier work in this field. Mention should be made at this point of the previous work by the same Swedish authors[175] on ^{79}Br studies of solutions of alkylammonium bromides, and of some bio-

[171] A. M. Crotens, J. Smid, and E. de Boer, *Chem. Comm.*, 1971, 759.
[172] E. Shchori, J. Jagur-Grodzinski, Z. Luz, and M. Shporer, *J. Amer. Chem. Soc.*, 1971, **93**, 7133.
[173] M. Herlem and A. I. Popov, *J. Amer. Chem. Soc.*, 1972, **94**, 1431.
[174] B. Lindman and I. Lindqvist, *Chemica Scripta.*, 1971, **1**, 195.
[175] B. Lindman, H. Wennerstrom, and S. Forsén, *J. Phys. Chem.*, 1970, **74**, 754.

Bandshape Phenomena for Fluids

chemical applications of ^{81}Br and ^{35}Cl n.m.r. linewidths[176] (using the 'halogen ion probe technique'), since these papers were not discussed in Volume 1. Kreshkov et al. report[177] linewidths for ^{35}Cl in alkylchlorosilanes and discuss the influence of substituent polarity. The effects of substituents closely parallel their effects on n.q.r. frequencies. Unfortunately, the tabulated data in this paper do not seem to be entirely in accord with the trends given in graphical form. (For some other ^{35}Cl n.m.r. linewidths see ref. 164, discussed in the preceding sub-section.)

Germanium-73. The study of ^{73}Ge ($I = \frac{9}{2}$) n.m.r. has not been popular in the past; there existed, until last year, only three high-resolution studies, all on GeCl$_4$. However, Kaufmann et al.[178] have now measured ^{73}Ge spectra for a number of compounds, using the Fourier Transform method, and have quoted values for linewidths ranging from 0.58 Hz to 32 Hz. In the case of Ge(OMe)$_4$ a fitting routine was used to correct for the effects of spin–spin coupling. The width for Ge(OMe)$_4$ depends significantly on temperature, but those of the germanium tetra-alkyls do not. The values of T_2 calculated from linewidths compare well with those obtained from spin-echo measurements.

D. Effects of Magnetic Site Exchange on N.M.R. Spectra for Spin Systems Containing Quadrupolar Nuclei.

—(Note: Some of the work reported in Section 4C implicitly involves exchange.) Larsen has studied ^{1}H n.m.r. bandshapes for solutions of tetra-alkylammonium ions[179] and tetra-alkylarsonium ions.[180] In the former case methyl resonances for Me$_3$NR$^+$ (R = Me, Et, Pr^i, Bu^t, n-hexyl, n-octyl, n-decyl, or n-hexadecyl) were examined[179] as a function of temperature (the absorption varies from a singlet to a triplet). The shapes are simulated using the usual Pople–Sack treatment (neglecting non-quadrupolar linewidths) but found to represent the experimental shapes only poorly. Simulations were therefore performed by including (a) exchange between two chemical sites, free and ion-paired, with different values of T_1^N, and (b) 'natural' linewidths, assumed the same for both sites (accounted for by a diagonal addition of T_2^{-1} to the appropriate matrix). Agreement with experiment is much improved – not surprisingly since there are now six parameters instead of one. However, most of these parameters were arbitrarily fixed and values of the parameter giving the nitrogen spin–lattice relaxation time in the free ions were obtained. These values are discussed in terms of quadrupole coupling constants and re-orientational correlation times. Some activation energies for re-orientation are given.

For the tetra-alkylarsonium case, Larsen has studied[180] the methyl ^{1}H resonance of Me$_4$As$^+$X$^-$, X = Cl, I, and OH, as a function of solvent, concentration, and temperature. The spectra were always broad Lorentzian

[176] H. Csopak, B. Lindman, and H. Lilja, *F. E. B. S. Letters*, 1970, **9**, 189.
[177] A. P. Kreshkov, V. F. Andronov, and V. A. Drozdov, *Russ. J. Phys. Chem.*, 1972, **46**, 183.
[178] J. Kaufmann, W. Sahm, and A. Schwenk. *Z. Naturforsch.*, 1971, **26a**, 1384.
[179] D. W. Larsen, *J. Phys. Chem.*, 1971, **75**, 509.
[180] D. W. Larsen, *J. Phys. Chem.*, 1971, **75**, 3880.

singlets. The linewidths at half-height were recorded and corrected for non-quadrupolar contributions. The resulting values were interpreted in terms of two-site exchange (the sites being free and ion-paired Me_4As^+) according to the theory of Gore and Gutowsky,[181] neglecting exchange contributions to the linewidth:

$$(\Delta v_{\frac{1}{2}} - \Delta v_{\frac{1}{2}}^0)^{-1} = (5\pi)^{-1}\left(\frac{p_A}{T_1^A} + \frac{p_B}{T_1^B}\right)J^{-2}$$

where A and B are two sites, and T_1 indicates an arsenic spin-lattice relaxation time. Simulations of the spectra using this theory agree with experimental spectra to better than a few per cent. The data are discussed in terms of anisotropic rotational diffusion, using the macroscopic viscosity as a measure of the 'effective collision time'. Asymptotic values of the re-orientational parameter are taken to refer to the ion-paired form alone. It is concluded that the 'single kinetic entity' of the aqueous ion pair includes the counter-ions with at most a few H_2O molecules between the ions, but the cation hydration layer is excluded.

Larsen[182] has also studied the effects on the ^1H n.m.r. bandshape for the α protons of pyridine of addition of lithium salts. The experimental bandshapes were compared with theoretical ones produced by assuming all peaks have a Lorentzian shape of common width, and have resonance frequencies and intensities unperturbed by the added salt. Visual comparison then gives the common width and thence T_1^N (assuming the fast nitrogen relaxation formula applies). The addition of the salts to neat pyridine sharpens the lines, whereas addition of the chloride to aqueous pyridine substantially broadens the lines (bromide or iodide addition has little effect). The results are interpreted in terms of rapid exchange between free pyridine and a complexed species, with consideration of anisotropic rotational diffusion (limiting values for the linewidths in the presence of excess salt giving information about the complexes), much as in the earlier paper of the series.[180] The author suggests geometric models for the complexes. He also obtains information about the covalent character in the N—Li bond by comparing linewidths for the α and β protons.

Kasugai et al.[183] also report a study of the pyridine–water system, using both ^1H and ^{14}N n.m.r. The linewidth of the α protons shows a minimum at 0.3 mole fraction of pyridine, where the ^{14}N linewidth is at a maximum and the viscosity is a maximum. The data are interpreted in terms of rapid exchange between free and hydrogen-bonded (to water) pyridine molecules, the observed linewidths being the weighted average for the two species. Other possible exchange effects are ignored.

Proton n.m.r. bandshapes for HN_3, $HNCS$, and $HNCO$ have been used[184]

---

[181] E. S. Core and H. S. Gutowsky, *J. Phys. Chem.*, 1969, **73**, 2515.
[182] D. W. Larsen, *J. Phys. Chem.*, 1972, **76**, 53.
[183] Y. Kasugai, Y. Arata, and S. Fujiwara, *Bull. Chem. Soc. Japan*, 1971, **44**, 2557.
[184] J. Nelson, R. Spratt, and S. M. Nelson, *J. Chem. Soc. (A)*, 1970, 583.

(54)

to discuss briefly the processes of nitrogen relaxation and proton exchange, but no firm conclusions were drawn.

An interesting case of exchange in a system involving quadrupolar nuclei has been reported by Shchori et al.,[172] who studied the complexation of sodium ions with dibenzo-18-crown-6 (DBC) (54) in dimethylformamide solution using ^{23}Na resonance. The exchange occurs between the solvated sodium and its DBC complex. Although the ^{23}Na chemical shift difference between the species is small, the linewidth for the solvated ^{23}Na is ca. 25 times less than in the complex, owing to deviations from an environment of cubic symmetry in the latter. Linewidths are reported in the form of values of T_2 for the two species in the absence of exchange as a function of concentration (with SCN$^-$ as a counter-ion to the Na$^+$) and of temperature. The data are discussed in terms of viscosity. The case of two-site exchange, using the Bloch–McConnell approach, is then developed for the particular situation studied, and this theory is applied to experimental linewidth results to obtain information about the pre-exchange lifetime of the solvated sodium. The apparent activation energy for decomplexation is found to be 53 ± 3 kJ mol^{-1}, and the exchange mechanism is discussed (see also ref. 171, discussed in Section 3C).

Brüssau and Sillescu[27] report ^{35}Cl n.m.r. linewidths (as peak-to-peak separations in the derivative absorption mode, in magnetic field units) as part of a relaxation study of solutions of polyvinylpyrrolidone in CDCl$_3$. Equations applicable to relaxation times of solvent molecules in systems involving solute–solvent interaction are developed – the average times observed are functions of the rotational correlation times and of the solvent lifetimes in the free and 'bound' states. The theory is extended to allow for anisotropic rotational motion. It is applied to the experimental system, and the lifetime of hydrogen bonding in the solvated state is calculated.

Bramley and Johnson[185] report ^{1}H and ^{59}Co n.m.r. studies of [Co(en)(NH$_3$)$_4$]$^{3+}$. They show by ^{1}H–{^{59}Co} double resonance that some of the width of the CH$_2$ proton signal is due to partially collapsed (Co,H) coupling. The ^{1}H widths at 60 MHz were greater than those at 100 MHz in a fairly viscous solution; this is taken as evidence that the extra broadening is a relaxation phenomenon. The ^{59}Co spectrum had a linewidth of ca.1 kHz. The authors concluded that the dynamics of ring inversion are obscured by the

[185] R. Bramley and R. N. Johnson, Chem. Comm., 1971, 1309.

effects discussed above.[186] The acid dissociation equilibria of [Co(NH$_3$)$_5$-(H$_2$O)](ClO$_4$)$_3$ and the *cis*- and *trans*-isomers of [Co(en)$_2$(NH$_3$)(H$_2$O)]Br$_3$ have been investigated using ^{59}Co n.m.r. linewidths (in magnetic field units) to monitor the equilibria as the pH is varied. It is assumed that exchange is rapid, that there is no exchange contribution to the linewidths, and that the observed widths are the weighted average of those for the species in equilibrium. The pK values are obtained from the sigmoid curves in the usual way.

Fratiello *et al.*[187] list linewidths (peak-to-peak, derivative mode) ranging from 140 Hz to ~18 000 Hz for ^{115}In resonances of indium salt solutions in water–acetone mixtures at 25 °C. The linewidths presumably arise from both quadrupolar effects (^{115}In has $I = \frac{9}{2}$) and the effects of exchange between the various species present. The data are discussed qualitatively in terms of the latter contribution. The widths are sensitive to acid and solvent concentrations.

4 Relaxation Effects

This section includes all bandshape studies in which relaxation data have been the most important information obtained, except for the case of quadrupolar relaxation (see Section 3) and for certain studies involving paramagnetic species (see Section 2C).

A. Theoretical Work.—Khazanovich and Zitserman[188] have developed the theory of external dipolar relaxation for an AB spin system. This is, of course, relevant to bandshapes, although the authors do not develop their ideas in this direction except to give equations for spin–spin relaxation parameters for individual lines, leading to linewidth information.

Kaiser[52] has described a method for calculating n.m.r. spectra in the linear response limit which involves a transfer function characterized by the location of its poles and zeros in the complex frequency plane. The method gives an immediate understanding of the spectral changes resulting from variation of some parameter of the spin system. An example is given of the effects of relaxation interaction between two lines upon the bandshapes. Spectra of non-Lorentzian character are shown to be possible.

Roberts and Lynden-Bell[189] have described theory for calculating bandshapes for a triplet spin system which interacts both with the molecular environment and with an externally applied magnetic field. Although the theory is applicable in principle for any rate of molecular motion, it is particularly valuable for cases of slow motion where the usual methods of high-resolution n.m.r. are incorrect. It may be relevant for e.s.r. and n.q.r. as well as for certain situations in n.m.r., but the subject is to a large extent outside the scope of this chapter (see Chapter 9). Lynden-Bell[190] has also published

[186] F. Yajima, A. Yamasaki, and S. Fujiwara, *Inorg. Chem.*, 1971, **10**, 2350.
[187] A Fratiello, D. D. Davis, S. Peak. and R. E. Schuster, *Inorg. Chem.*, 1971, **10**, 1627.
[188] T. N. Khazanovich and V. Yu. Zitserman, *Mol. Phys.*, 1971, **21**, 65.
[189] J. Roberts and R. M. Lynden-Bell, *Mol. Phys.*, 1971, **21**, 689.
[190] R. M. Lynden-Bell, *Mol. Phys.*, 1971, **22**, 837.

related theoretical work for slowly reorienting systems using density matrix theory in Liouville space, and given further details for the bandshape of a particle with spin $S = 1$. Another theoretical paper of some relevance here, though it is written mainly with the e.s.r. alternating linewidth problem in mind, is that of Atkins,[191] who shows that the Kubo–Tomita theory[192] of bandshapes is not inferior to the density matrix approach, contrary to some reports.

Theoretical relaxation work by Pyper,[148,149,151] although of general applicability, has been discussed in Section 3A since it has been developed in detail, as far as bandshapes are concerned, mostly for systems containing quadrupolar nuclei.

B. Paramagnetic Effects.—There have been a number of interesting novel applications in this area during the current review period. For instance, equilibria between several different paramagnetic species have not been commonly studied by n.m.r., but Martini and Tiezzi[193] have investigated a system of iron dinitrosyl complexes using both e.s.r. and n.m.r. The n.m.r. part of the study consisted of measuring the OH proton linewidth for the solvent, ethanol, as a function of pH. At high pH the line is narrow since the species in the co-ordination sphere of the $Fe(NO)_2^+$ group is EtO^-. At low pH the resonance broadens to over 10 Hz, since under these conditions neutral EtOH molecules are co-ordinated to iron. The existence of three species is demonstrated by e.s.r.: $[Fe(NO)_2(EtOH)_2]^+$, $[Fe(NO)_2(EtOH)(EtO)]$, and $[Fe(NO)_2(EtO)_2]^-$.

The broadening effects of organic free radicals on the n.m.r. spectra of other solutes has received little attention to date, but Kopple and Schamper report[194] work on solutions of amides and peptides containing the radical 3-oxyl-2,2,4,4-tetramethyloxazolidine (55). The NH proton resonances of *trans* amides are more affected by the radical than those of *cis* amides. Further measurements give rise to the suggestion that the technique may be of use in determining peptide conformation. The physical origin of the broadening is not discussed.

Engel[195] has described 'chemical decoupling' effects in ^1H spectra of a number of nitrogen-containing compounds when paramagnetic ions (cobalt or iron compounds) are added. This arises from rapid relaxation of one (or both) of the coupled spins due to dipole–dipole or scalar interaction with the paramagnetic centre, assumed to be co-ordinated to nitrogen. In some cases the effect was thought to be due to second co-ordination sphere effects. Similar 'chemical decoupling' has been described[196] for paramagnetic ion co-ordination at oxygen. In neither reference are the bandshapes studied quantitatively.

---

[191] P. W. Atkins, *Mol. Phys.*, 1971, **21**, 97.
[192] R. Kubo and K. Tomita, *J. Phys. Soc. Japan*, 1954, **9**, 888.
[193] G. Marini and E. Tiezzi, *Trans. Faraday Soc.*, 1971, **67**, 2538.
[194] K. D. Kopple and T. J. Schamper, *J. Amer. Chem. Soc.*, 1972, **94**, 3644.
[195] R. Engel, *J. Chem. Soc. (C)*, 1971, 3554.
[196] R. Engel and G. Nathan, *J. Chem. Soc. (C)*, 1971, 3844.

(55)

Both the contact and the dipolar contributions to n.m.r. line-broadening arising from the presence of paramagnetic sites are in theory proportional to the square of the magnetogyric ratio of the nucleus concerned. It follows that linewidths from this origin in deuteron resonance should be a factor 42.4 smaller than those of proton resonance in the corresponding compound. Johnson and Everett[197] have tested this proposition by measuring ^{2}H and ^{1}H n.m.r. linewidths for a number of paramagnetic transition-metal acetylacetonate complexes. Substantial differences are indeed seen, e.g. for Ti(acac)$_3$ the widths (corrected for non-paramagnetic effects) are 140 and 4300 Hz for ^{2}H and ^{1}H, respectively. Only for Cr(acac)$_3$ is the ratio ca. 40. However, all the ratios are greater than 6.5 (the ratio of chemical shift differences in ^{2}H and ^{1}H) so that ^{2}H n.m.r. provides the better resolution. Possible causes for the ratio being below the theoretical value are given, but no definite conclusions made.

As anticipated in Volume 1, attention is now being turned to the effect on linewidths of the so-called 'shift reagents' (see also Section 2C). For instance, it has been pointed out by Barry et al.[198] that the full potential of n.m.r. studies of solutions containing lanthanide ions has not been utilized because the line-broadening effects have usually been minimized or ignored. These workers emphasize that the lanthanide ions may be classified according to electronic spin relaxation times, and that whereas, for example, Eu^{III} causes substantial n.m.r. shifts but negligible broadening effects, Gd^{III} causes larger isotropic broadening, and Ho^{III} gives both anisotropic broadening and chemical shift perturbations. However, all ions have similar chemical interactions with organic molecules. Since the line-broadening produced (of both types) depends on an inverse power of the distance between the nucleus studied and the lanthanide ion, structural information is available, and the combination of shift and broadening studies is particularly powerful. The authors[198] apply their ideas, using a computer treatment to search for acceptable conformations, to some mononucleotides; they suggest that the methods will be generally applicable to quite large molecules. Possible broadening effects of exchange (see ref. 115) do not seem to have been considered.

Witanowski et al.[199] have given a preliminary report on the effects of lanthanide dipivalomethanato complexes in ^{14}N n.m.r., using the pyridine signal. They state that since most of the deshielding elements are about

[197] A. Johnson and G. W. Everett, jun., J. Amer. Chem. Soc., 1972, **94**, 1419.
[198] C. D. Barry, A. C. T. North, J. A. Glasel, R. J. P. Williams, and A. V. Xavier, Nature, 1971, **232**, 236.
[199] M. Witanowski, L. Stefaniak, H. Januszewski, and Z. W. Wolkowski, Chem. Comm., 1971, 1573.

equally active, a choice must be based on broadening effects; Yb(dpm)$_3$ is therefore recommended. For a shielding agent Dy(dpm)$_3$ is by far the best. A table of pyridine ^{14}N linewidths is given for a variety of chelates and a range of chelate:pyridine molar ratios. These widths range up to 700 Hz (in the absence of chelate the width is 175 Hz). Tomić et al.[200] also report linewidths induced by tris(dpm) complexes. They studied the ^{1}H n.m.r. spectrum of 1-methylcyclopropylmethanol, and found that line-broadening induced by Ho(dpm)$_3$, as Hz per Hz of shift induced, was ca. 1.5 times as large as that caused by Pr(dpm)$_3$.

C. Gas-phase Studies.—The ^{31}P n.m.r. of elemental phosphorus has been studied[201] in the gas phase. The linewidth is reported to be 50—60 Hz.

D. Saturation Studies.—N.m.r. measurements of relaxation effects by saturation studies have not been popular to date, largely because of experimental difficulties, and, in particular, they have not been applied to complex spin systems, though there is a report[202] of 'unpublished work' in this area. Recently, however, Harris and Worvill[3] have published a preliminary communication regarding saturation studies of an AB spin system, 2-chloroacrylonitrile (56). The sample was doped with nickel chloride and the authors were able (a) to show that the spectra could be fitted by assuming random field relaxation but not with intramolecular dipolar relaxation, and (b) to derive the

$$\begin{matrix} H \\ H \end{matrix} \!\!\!>\!\! C = C \!\!<\!\! \begin{matrix} Cl \\ CN \end{matrix}$$

(56)

parameters governing random field relaxation. The bandshape analysis was done by computer using an iterative fitting procedure to digitized experimental spectra. The iteration involves eight parameters for the case of random field relaxation: the chemical shift difference, the coupling constant, a phase correction, horizontal and vertical scaling factors, and three relaxation parameters (only one relaxation parameter is necessary for dipolar relaxation). It is emphasized that the spectral complexity and the appearance of the double quantum transition are advantages for obtaining detailed relaxation information.

5 Bandshapes in Multiple Resonance (See also Chapter 7)

Bucci and co-workers[203] have published a further discussion of the theory of

[200] L. Tomić, Z. Majerski, M. Tomić, and D. E. Sunko, *Croat. Chem. Acta*, 1971, **43**, 267.
[201] G. Heckmann and E. Fluck, *Mol. Phys.*, 1972, **23**, 175.
[202] R. A. Hoffman, *Adv. Magn. Resonance*, 1970, **4**, 87.
[203] P. Bucci, M. Martinelli, S. Santucci, and A. M. Serra. *J. Magn. Resonance*, 1972, **6**, 281.

double resonance, as applied to a system of two spin-$\frac{1}{2}$ nuclei, using the quantized field formalism, which is particularly valuable for intense rf fields. Equations for the bandshapes are given, and one theoretical spectrum (for a particularly simple relaxation case) is illustrated. This shows the characteristics of (a) some 'emission' lines, and (b) a dispersion-like line in the central region. It is not clear how generally useful the theoretical methods will prove.

Saturation transfer continues to be used to study exchange problems (see, for example, ref. 132), but normally involves intensity changes rather than bandshape studies. The same comment applies to multiple resonance studies of exchange using the nuclear Overhauser effect (*e.g.* ref. 204).

Since this chapter was written papers 205–208, published during the review period, have come to the attention of the Reporter. Although they will not be discussed in detail here, their titles are given below, together with the section of this chapter for which they are relevant.

6 Appendix: Tables of Activation Parameters for Intramolecular Exchange Processes, Derived from N.M.R. Bandshape Analysis

These five Tables effectively summarize the results quoted in some of the papers that are mentioned in this Chapter.

---

[204] J.-C. Duplan, C. Chapelet-Barbier, and J. Delmau, *Mol. Phys.*, 1972, **23**, 609.
[205] N.m.r studies of hindered internal rotation in higher N,N-dialkyl amides and thionamides. T. H. Siddall, W. E. Stewart, and F. D. Knight, *J. Phys. Chem.*, 1970, **74**, 3580 (Section 2B).
[206] Stereochemically non-rigid six-co-ordinate molecules. I. A detailed mechanistic analysis for the molecule FeH$_2$[P(OEt)$_3$]$_4$. P. Meakin, E. L. Muetterties, F. N. Tebbe, and J. P. Jesson, *J. Amer. Chem. Soc.*, 1971, **93**, 4701 (Section 2B).
[207] Influence du marquage quadrupolaire en r.m.n. de ^{19}F. Influence de solvant sur les mouvements inter- et intra-moleculaires dans des solutions de fluorures de benzyle, C. Beguin and R. Dupeyre. *Compt. rend.*, 1971, **273**, *C*, 1658 (Section 3B).
[208] N.m.r study of hydrophobic bonding in (DL-lysine)$_m$-(L-alanine)$_n$-(DL-lysine)$_m$, J. C. Howard and H. A. Scheraga, *Macromolecules*, 1972, **5**, 328 (Section 4D).

Table 1 Activation parameters for internal rotation about the C—C bond

Molecule	E_a/kJ mol^{-1}	$\log_{10}A$	$\Delta H^\ddagger$/kJ mol^{-1}	$\Delta S^\ddagger$/J K^{-1} mol^{-1}	Ref.
Me$_3$C·CMe$_2$Cl (6)	a	a	38	−33	61
Me$_3$C·CMeClCH$_2$Ph (7)	a	a	35	−37	61
Bu^t·CMeRR^1					
R = Cl, R^1 = Me	49±1	14.8±0.2	48±1	34±4	62
R = Br, R^1 = Me	57±1	15.6±0.3	55±1	50±6	62
R = R^1 = Cl	46±1	12.7±0.2	44±1	−7±3	62
R = R^1 = Br	49±1	12.6±0.2	48±1	−16±3	62
R = Cl, R^1 = Br	51±1	13.0±0.1	49±1	−2±2	62
R = Cl, R^1 = CD$_2$CD$_3$	43±1	12.0±0.3	41±1	−21±5	62
R = Cl, R^1 = Bu^t	59±2	15.7±0.3	57±2	43±7	62
Me$_2$BrC·CBr$_2$Me	51±1	12.5±0.2	49±1	−13±3	62
(CMeCl$_2$)$_2$	61 and 66^b	14.0 and 15.0^b	59 and 64^b	11 and 30^b	62
(CMeBr$_2$)$_2$	59±1	11.7±0.2	56±1	−35±3	62
(8a)	61—83^b	a	58—81^b	−21 to −71^b	63
(8b)	a	a	82±21	a	63
(9)	96±3	a	93±3	5±8	63
(10a)	60—63^c	a	58—60^c	−10 to −5^c	64
(10b)	69—70^c	a	66—67^c	−12 to −8^c	64
(10c)	66	a	63 and 64^c	6 and 8^c	64
(11b)	a	a	79±4	16±17	65
(12a)	a	a	a	56	67
(12b)	a	a	a	−26	67
(12c)	a	a	a	−3	67

^{a} Not given
^{b} Solvent dependent
^{c} Values for the different neopentyl groups differ slightly.

Table 2 *Activation parameters for ring inversion and related processes*

Molecule	E_a/kJ mol^{-1}	$\log_{10}A$	$\Delta H^{\neq}$/kJ mol^{-1}	$\Delta S^{\neq}$/J K^{-1} mol^{-1}	Ref.
3,3-Difluoro-*trans*-cyclodecene (13a)	62 ± 2	14.7 ± 0.3	*a*	29	68
	52 ± 1	15.0 ± 0.3	*a*	36	69
1,1-Difluorocyclodecane (14)	29 ± 1	14.5 ± 0.4	*a*	30	70
(15)	38.9 ± 0.2	*a*	37.0	14.9	23
	26 ± 1	*a*	25 ± 1	15 ± 10	71
1,1-Dimethylcyclohexane	*a*	*a*	47 ± 3	16 ± 13	72
cis-1,2-Dimethylcyclohexane	*a*	*a*	39 ± 3	−15 ± 13	72
trans-1,3-Dimethylcyclohexane	*a*	*a*	47 ± 3	19 ± 13	72
cis-1,4-Dimethylcyclohexane	*a*	*a*	46 ± 3	17 ± 13	72
cis-Decalin	*a*	*a*	57 ± 3	15 ± 13	72
9-Methyl-*cis*-decalin	*a*	*a*	52 ± 3	−3 ± 13	72
(16)	61 ± 1[b]	12.6 ± 0.1[b]	58 ± 1[b]	−13 ± 2[b]	73
(17a)	64 ± 1[b]	13.0 ± 0.1[b]	61 ± 1[b]	−3 ± 1[b]	73
(17b)	61 ± 1[b]	12.8 ± 0.1[b]	59 ± 1[b]	−8 ± 2[b]	73
(18)	52 ± 1	11.5 ± 0.1	49 ± 1	−34 ± 1	75

[a] Not given
[b] The authors claim higher accuracy than quoted here

Table 3 Activation parameters for C—N internal rotation in amides and related compounds

Molecule	E_a/kJ mol^{-1}	$\log_{10} A$	$\Delta H^{\neq}$/kJ mol^{-1}	$\Delta S^{\neq}$/J K^{-1}mol^{-1}	Ref.
(1a)	77±3	12.9	74±3	−6±9	34
(1b)	71±2	a	68	−15±5	30
(1; R = CD$_3$)	85±1	14.1±0.2	83±1	17±3	82
(1c)	73±1	a	70	−18±3	30
(1; R = CCl$_3$)	69—73^b	a	a	a	83
(2; R = H)	59±1	a	56	−20±4	30
(2; R = H)	a	a	51±2	−50±4	56
(2; R = H)	a	a	{70±1^c / 73±1^d}	{15±1^c / 24±2^d}	28, 53
(2; R ≠ H)^c	a	a	71—73	15—33	28, 53
(2; R ≠ H)	a	a	32—57	−109 to −25	56
(3; R = H)	a	a	{72±1^c / 73±1^d}	{12±1^c / 14±3^d}	28, 53
(3; R = H)	a	a	59±3	−29±8	29
(3; R = H)	76±5	14.1±0.8	73	13	79
(3; R ≠ H)	a	a	71—76	9—25	28, 53
(3; R ≠ H)	a	a	71—75	2—23	29
(4; R = H)	a	a	74±1	12±2	54
(4; R ≠ H)	a	a	76—82	16—26	54

Table 3 (cont.)

Molecule	E_a/kJ mol^{-1}	$\log_{10}A$	$\Delta H^{\neq}$/kJmol^{-1}	$\Delta S^{\neq}$/JK^{-1}mol^{-1}	Ref.
(5; R ≠ H)	a	a	34—57	−109 to −33	56
(5; R ≠ H)	a	a	59—69	−32—0	29
trans-MeHC = CHCONMe$_2$ (20)	40±2	3.8±0.4	38	−84	79
(21)	77±5	14.3±1.0	74	21	79
	74^e	a	71^e	−19^e	80
NN-Di-isopropylamides	65—74	a	62—71	−18 to −8	81
NN-Di-isopropylthioamides	72—88	a	70—85	−20 to +12	81
CD$_3$CSNMe$_2$	108±4	14.6±0.5	106±4	26±9	82
CD$_3$C(NH$_2$)NMe$_2$	89—95^f	12.7—13.5^f	87—93^f	−11 to +6^f	82
MeHNCHO^g	99—104	14.0—15.1	a	a	31
MeHNCHO	a	a	94±8	21±21	85
MeHNCOMe	a	a	95—97^h	21±21	85
(22)	a	a	83	18	87
(23a), (24) and (25)	65—70	12.4—12.9	63—67	−15 to −8	88
(23b)	75±1	13.18±0.10	72±1	−1±2	88
(26a)	a	a	67	−20	89
(26b)	a	a	88	26	89

^{a} Not given
^{b} Depending on experimental method
^{c} In 0.25M-CDCl$_3$
^{d} In 1.0M-CDCl$_3$
^{e} Averages from two different sets of signals; see ref. 80 for the errors
^{f} Dependent on anion or concentration
^{g} And certain deuteriated species
^{h} Solvent dependent

Table 4 Activation parameters for nitrogen inversion, C—N internal rotation, and related processes

Molecule	Process	E_a/kJ mol^{-1}	$\log_{10}A$	$\Delta H^\ddagger$/kJ mol^{-1}	$\Delta S^\ddagger$/J K^{-1} mol^{-1}	Ref.
Bu^{t}NMe$_2$	C(sp^3)—N internal rotation^c	27 ± 1	a	26 ± 1	5 ± 8	96
NN-Di-isopropylthioamides	C(sp^3)—N internal rotation	51 and 59	a	49 and 57	17 and 8	81
(28)	C(sp^2)—N internal rotation	48—74^b	a	a	-34 to $+31^b$	35, 98
(29)	C(sp^2)—N internal rotation	73 and 76^b	a	a	2 and 18^b	35, 98
(11a)	C(sp^2)—N internal rotation	a	a	72 ± 6	49 ± 25	65
(32a)	degenerate isomerization^d	61 ± 1	12.2 ± 0.1	59 ± 1	-21 ± 3	93
(32b)	degenerate isomerization^d	59—66	12.3—13.0	57—64	-18 to -4	93
(33a)	N—N internal rotation	54 ± 3	11.4 ± 0.4	52 ± 3	-35 ± 9	102
(33b)	N—N internal rotation	64—81	11.2—13.6	61—79	-39 to $+7$	102
(33c)	N—N internal rotation	79 ± 5	13.1 ± 0.7	77 ± 5	-4 ± 14	102

^{a} Not given
^{b} Solvent dependent
^{c} Incorporating rehybridization at nitrogen
^{d} See text

Table 5 Activation parameters for metallotropic rearrangements and for migration processes[c]

Molecule	E_a/kJ mol^{-1}	$\log_{10} A$	$\Delta H^{\neq}$/kJ mol^{-1}	$\Delta S^{\neq}$/J K^{-1} mol^{-1}	Ref.
(35)	58±3	11.7±0.8	a	a	104
(36a)	54±4	11.3	a	−39±21	105
(36b)	39±4	9.9±1	a	−66±17	106
(36c)	33±4	13.8±1	a	8±17	106
(37a)	59±2	11.9±0.3	a	a	108
(h^1-C$_5$H$_5$)(h^5-C$_5$H$_5$)$_2$Ti	67±1	13.5±0.5	a	a	109
(π-C$_5$H$_5$)Fe(CO)$_2$(σ-C$_5$H$_5$)	36±3	9.8±0.7	a	a	107
(π-C$_5$H$_5$)Ru(CO)$_2$(σ-C$_5$H$_5$)	40±4	10.0±0.7	a	a	107
(RO)$_2$Ti(acac)$_2$[d]	25—42	a	a	−134 to −84	110
(Pr^{i}O)$_2$Ti(acac)$_2$	39—43[b]	a	a	−138 to −84[b]	110
(Pr^{i}O)$_2$Ti(quin)$_2$[e]	39±8	a	a	−155	110
Pr^{i}O—TiY$_2$ Pr^i [f]	34—51	a	a	−184 to −146	110
(38)	59±3	12.5±0.6	a	a	108
(39)	109±5	14.4±0.7	a	a	108
(37b)	96±7	11.5±0.8	a	a	108
(42)	40—44[g]	12.1—13.2[g]	a	a	113
(43)	123±1	11.7±0.1	118±1	−50±4	114
(44)	124±1	11.4±1	120±1	−59±4	114

[a] Not given
[b] Solvent dependent
[c] For further information about the nature of the processes see the text
[d] R = substituted phenyl group
[e] quin = 8-hydroxyquinolinate
[f] Y = quin or derivative
[g] Depending on the non-exchange linewidth used

7
Multiple Resonance

BY D. SHAW

1 Introduction

This chapter covers the work published on multiple resonance over the 12 months ending June, 1972. As in the previous report[1] all the resonances involved are nuclear, *i.e.* ENDOR and ELDOR are not considered. The papers have been classified as to their relevant concept. Special attention has been paid to one area of research, that of the nuclear Overhauser effects produced by proton irradiation while observing ^{13}C which, because of its great practical importance, has aroused considerable interest in the year. All papers just reporting double resonance data have not necessarily been included, both as a matter of policy and on considerations of space.

During the year one general article on multiple resonance has appeared, by Johannesen and Coyle,[2] and a book dedicated to the chemical applications of the nuclear Overhauser effect has been published.[3] The former provides a general survey of the possible double resonance experiments and pays particular attention to heteronuclear double resonance and INDOR, citing some excellent examples. One paper has appeared where virtually the whole of the n.m.r. arsenal (including homo- and hetero-nuclear double and triple resonance) is applied to derivatives of 6-amino-6-deoxy-1,2:3,5-di-*O*-propylidene-α-D-glucofuranose.[4] This paper is almost a complete abstract of n.m.r. techniques and demonstrates the enormous volume of (useful?) data which can be generated by spectroscopists.

2 Theory of Multiple Resonance

Very little work on theoretical double resonance has appeared in the year. Those papers where the interest has been judged to be more concerned with relaxation or Overhauser effects are dealt with in the sections devoted to those topics.

---

[1] 'Nuclear Magnetic Resonance', ed. R. K. Harris, (Specialist Periodical Reports), The Chemical Society, London, 1972, Chap. 7.
[2] R. B. Johannesen and T. D. Coyle, *Endeavour*, 1972, **31**, 10.
[3] J. H. Noggle and R. E. Schirmer, 'The Nuclear Overhauser Effect: Chemical Applications', Academic Press, New York, 1971.
[4] B. Coxon and L. F. Johnson, *Carbohydrate Res.*, 1971, **20**, 105.

Bucci and co-workers have extended their work on double resonance spectra using a second quantization formulation to include the AB spin system.[5] This technique has previously been applied to single-resonance spectra and double-resonance spectra of a single spin.[6] In their approach they generate a total hamiltonian and treat the magnetic field and the spin system similarly:

$$\mathcal{H}_T = \mathcal{H}_S + \mathcal{H}_R + \mathcal{H}_I \qquad (1)$$

where $\mathcal{H}_S$ is the normal hamiltonian of the isolated spin system
$\mathcal{H}_R$ is the radiation hamiltonian
$\mathcal{H}_I$ is the interaction hamiltonian between the spin and the radiation applied.

Using this approach it is possible to study the system when arbitrarily high field values and many different frequencies are applied; both these situations are not easily approached by the conventional rotating frame method. For an AB system subject to two rf fields, B_1 and B_2,

$$\mathcal{H}_S = -B_0(\gamma_A \hat{I}_{Az} + \gamma_B \hat{I}_{Bz}) + J\,\hat{I}_A \cdot \hat{I}_B \qquad (2a)$$

$$\mathcal{H}_R = \omega_1^+ \hat{a}_1^+ \hat{a}_1^- + \omega_2\, \hat{a}_2^+ \hat{a}_2^- \qquad (2b)$$

$$\mathcal{H}_I = -\left(\frac{\omega}{2V}\right)^{\frac{1}{2}}(\hat{a}_2^+ + \hat{a}_2^-)(\gamma_A \hat{I}_{Ax} + \gamma_A \hat{I}_{Bx}) \qquad (2c)$$

where $\hat{a}^+$ and $\hat{a}^-$ are the creation and annihilation operators and V is the volume containing the radiation field, other symbols having their conventional meaning. Relaxation effects are included by treating the total hamiltonian using the usual Redfield approach.[7] Bucci *et al.* have analysed, using the above procedure, the AB system and arrived at numerical solutions for certain conditions of irradiation power, offset, *etc.* Some of these results have been successfully compared with experimental spectra obtained from 5-bromo-2-chlorothiophen. The case of 'spin-tickling' has been evaluated and it is shown that, as expected when large enough B_2 field is used, the doublet splitting produced is asymmetric. It is also predicted that, above a critical value of B_2, spin decoupling can no longer be observed.

3 Homonuclear Double Resonance

A. Spin Decoupling and INDOR.—Spin decoupling continues to be extensively used in the assignment of ^{1}H spectra at all levels, ranging from simply demonstrating whether or not lines A and B are linked, to assisting in the assignment of deceptively simple spectra, *e.g.* the ABMX spin systems of

---

[5] P. Bucci, M. Martinelli, S. Santucci, and A. M. Serra, *J. Magn. Resonance*, 1972, **6**, 281.
[6] P. Bucci, M. Martinelli, and S. Santucci, *J. Chem. Phys.*, 1970, **53**, 4524 and refs. therein.
[7] A. G. Redfield, *Adv. Magn. Resonance*, 1970, **5**, 271.

dibenzothiophen and 9,9'-bicarbazyl.[8] The INDOR technique continues to be applied to complex molecules, e.g. see ref. 9. The signs of various proton–proton coupling constants, particularly of long-range couplings, continue to be found by spin tickling, e.g. for cumulenes it has been established that ^2J(H—C—H) is of the same order of magnitude and sign (negative) as ^2J(H—C—H) in allylic compounds; cis and trans ^5J(H—C=C=C=C—H) couplings are similar in magnitude and are positive, whereas ^6J(H—C=C=C=C=C—H) is negative.[10] The INDOR technique can also be used to determine coupling constant signs and has been used to show that all the (^{15}N, H) couplings are of the same sign in [^{15}N]-pyrrole.[11] For the significance and further details of coupling constant sign information see Chapter 2.

B. Nuclear Overhauser Effect.—The nuclear Overhauser effect continues to be widely used to solve stereochemical problems and indeed it has been the subject of a recent book.[3] Typical problems tackled by NOE, dependent on its sensitivity to internuclear distance, have been the confirmation of the axial nature of the N-methyl in tetrahydro-3,3,4,6-tetramethyl-1,3-oxazine,[12] assigning the spectra of the E and C nigabilactones,[13] and deducing the stereochemistry of 1,3,5,7-tetramethyltricyclo[5,1,0,0]octane.[14] The presence of a nuclear Overhauser effect on the aldehyde proton in this cis form but not the trans form of 4-bromo-2-formyl-1-methylpyrrole has been used to study the equilibrium of these two forms as a function of temperature.[15]

Duplan et al. have used the generalized nuclear Overhauser effect (see Vol. 1, p. 264) in conjunction with triple resonance to determine all the transition possibilities in an AB system (the C-2 protons of 4 methyl-1,3-dioxan).[16] They then extended the work to include the case of an AB system subject to slow conformational interconversion (the C-2 protons of 1,3-dioxan) and, via a study of the temperature dependence of this Overhauser effect during triple resonance conditions, deduced the interconversion barrier as 40.5 kJ mol^{-1}. Lawlor and Warren have studied the slowly exchanging system of 1-(1'-pyrazolyl)ethanol-acetaldehyde in water and they are the first to report the transfer of generalized Overhauser effects[17] from the methine proton of the 1-(1'-pyrazoly)ethanol to the aldehyde proton of the acetaldehyde.

[8] F. Balkau, M. W. Fuller, and M. L. Heffernan, Austral. J. Chem., 1971, 24, 2393.
[9] M. Anteunis, R. de Clyne, and M. Verzele, Org. Magn. Resonance, 1972, 4, 407.
[10] M. L. Martin, F. Lefevre, and R. Mantione, J. Chem. Soc. (B), 1971, 2049.
[11] E. Rahkamaa, Z. Naturforsch., 1971, 26a, 1187.
[12] H. Booth and R. U. Lemieux, Canad. J. Chem., 1971, 49, 777.
[13] T. Murae, T. Ikeda, T. Tsuyuki, T. Nishihama, and T. Takahashi, Tetrahedron Letters, 1971, 3897.
[14] M. Gordon, W. C. Howell, C. H. Jackson, and J. B. Stothers, Canad. J. Chem., 1971, 49, 143.
[15] B. Roques, C. Jaureguiberry, M. C. Fournie-Zaluski, and S. Combrisson, Tetrahedron Letters, 1971, 2693.
[16] J. C. Duplan, C. Chapelet-Babbier, and J. Dielman, Mol. Phys., 1972, 23, 609.
[17] J. M. Lawlor and J. P. Warren, J. Magn. Resonance, 1972, 7, 319.

The equations for the NOE in exchanging systems predict relative increases in the proton signals when the exchange lifetimes are equal to the nuclear spin relaxation times, thus providing a technique to study kinetic effects at rates slower than those accessible from chemical shift coalescence data. Combrisson et al. have evaluated this approach theoretically and applied it to the study of cis–trans interconversion in 4-bromo-2-formylfuran.[18]

4 Heteronuclear Double Resonance

A. Spin Decoupling and INDOR.—During the year, interest in and results from heteronuclear spin decoupling experiments have grown with the increasing use of multinuclear spectrometers and the increasing rate of conversion of older instruments for heteronuclear spin decoupling (see, for example, ref. 19 and also Chapter 4).

The simplest use of heteronuclear spin decoupling is to remove the effect of a second nucleus from a spectrum, hence permitting its analysis. This approach has been used to remove ^{14}N broadening from the proton spectra of N-methylpyrazinium,[20] isoquinolines,[21] and the ^{19}F spectrum of pentafluoropyridine,[22] rendering the spectra amenable to analysis, the lines being previously too broad to yield useful information. The nitrogen decoupling was followed by ^{1}H-$\{^{1}$H$\}$ and ^{19}F-$\{^{19}$F$\}$ decoupling respectively, i.e. triple resonance, in order to complete the analysis of the spin system.

The spectra of compounds dissolved in liquid crystals are typically exceedingly complex and any technique of simplification is welcomed. As mentioned above in the case of spectra from isotropic systems, simplification can sometimes be obtained by double resonance; however, owing to the magnitude of the dipole–dipole couplings observed in anisotropic spectra (much larger than the chemical shift range), homonuclear decoupling is impossible as first-order spectra do not occur. Heteronuclear spin decoupling is of course possible. This has been demonstrated for the first time by Emsley and co-workers[23] who have decoupled the deuterium splittings from the proton spectrum of CH_3CD_2OH in a nematic phase (see Figure 1). The deuterium spectrum consists of two groups of lines about 1 kHz wide 1.5 kHz apart; such a grouping of resonances required a special decoupling technique. The larger splitting originates from deuterium electric quadrupole interactions with the electric-field gradients present at the nucleus whereas the smaller splittings come from (H, D) dipolar interactions etc. The decoupling technique employed consisted of coherently phase-modulating the 15 MHz frequency, the most efficient technique for decoupling widely separated lines,[24]

---

[18] S. Combrisson, B. Roques, P. Rigny, and J. J. Basselier, Canad. J. Chem, 1971, **49**, 904.
[19] R. Burton, L. D. Hall, and P. R. Steener, Canad. J. Chem., 1971, **49**, 588.
[20] D. J. Elias, A. G. Moritz, and D. B. Paul, Austral. J. Chem., 1972, **25**, 427.
[21] F. Balkan and M. L. Heffernan, Austral. J. Chem., 1971, **24**, 2311.
[22] V. Barboiu, J. W. Emsley, and J. C. Lindon, J. C. S. Faraday II, 1972, **68**, 241.
[23] J. W. Emsley, J. C. Lindon, J. M. Tabony, and T. H. Wilmshurst, Chem. Comm., 1971, 1277.
[24] W. A. Anderson and F. A. Nelson, J. Chem. Phys., 1963, **39**, 183.

Multiple Resonance

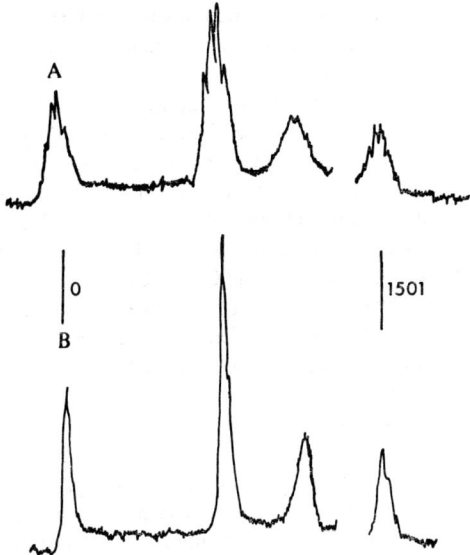

Figure 1 *Spectrum of* CH_3CD_2OH *dissolved in a nematic liquid crystal*, (A) *before, and* (B) *after irradiating at* 15.350 080 MHz, *coherently modulated at* 759 Hz, *and noise modulated at* 434 Hz. *The calibration markers are given in* Hz

and further noise modulating to simultaneously irradiate all the lines within each of the groups.[25]

Heteronuclear spin-tickling continues to be used for coupling constant sign determinations (see also Chapter 2) for which it is ideally suited, even the signs of couplings not apparent in the spectrum being accessible with care. McFarlane has studied the conformational dependence of $^1J(PP)$ in 1,2-dimethyl-1,2-diphenylbiphosphine and found the sign negative by $^1H-\{^{31}P\}$ experiments.[26] Another example of the determination of the sign of a 'hidden coupling' is furnished by Ziessow, who has used transient $^1H-\{^{13}C\}$ techniques (Torrey oscillations) to show that $^2J(COC)$ in $^{13}CH_3O^{13}COCl$ is -2.8 ± 0.2 Hz.[27]

The potential sensitivity of hetero INDOR has been exploited both where sensitivity is naturally limited, *e.g.* in ^{13}C where ^{13}C chemical shifts of methyl and methoxy-compounds have been studied by $^1H-\{^{13}C\}$ INDOR,[19] and when solubility problems appear, *e.g.* the $^1H-\{^{31}P\}$ INDOR study of some trichloro(dimethylphenylphosphine) [ethylenebis (diphenylphosphine)]iridium complexes.[28] This reference reports some interesting sign and magnitude data,

[25] R. R. Ernst, *J. Chem. Phys.*, 1966, **45**, 3845.
[26] H. C. E. McFarlane and W. McFarlane, *Chem. Comm.*, 1971, 1589.
[27] D. Ziessow, *J. Chem. Phys.*, 1971, **55**, 984.
[28] B. E. Mann, C. Masters, and B. L. Shaw, *J. C. S. Dalton*, 1972, 48.

especially $^2J(\text{PIrP})$ *trans* (440 Hz), which has been inferred to be large from the proton spectra of similar compounds for many years.[29]

Certain nuclei, by the nature of their nuclear properties and typical chemistry, lend themselves to chemical shift studies by heteronuclear double resonance; a prime example is ^{119}Sn. The ^{119}Sn chemical shifts in over 70 compounds have been reported, 40 by Zuckerman and co-workers[30,31] and 32 by Ustynyuk and co-workers.[32] The consistency of these slightly overlapping sets of data is interesting but outside the scope of this Report. Tin shifts have been studied as a function of concentration/solvent for Me$_3$SnCl[33] and in relation to the pK_a of RCO$_2$H for Me$_2$PhCCH$_2$SnCO$_2$R.[34]

Other nuclei extensively studied by double resonance techniques are ^{31}P[35-37] and ^{14}N.[38] Mercury and selenium have recently become popular; Ustynyuk and co-workers have reported eleven ^{119}Hg chemical shifts[39] and McFarlane over 80 ^{77}Se shifts.[40]

Paramagnetic shift reagents are now frequently used to expand complex proton spectra to facilitate the analysis of such spectra where lines overlap (see Chapter 10); ^{13}C spectra are frequently assigned by ^{13}C–{^{1}H} decoupling when knowledge of proton assignment is carried over to the ^{13}C spectrum. Feeney and co-workers have combined these techniques to assign the ^{13}C spectrum of ribose 5-phosphate (0.6 mol l^{-1} in D$_2$O).[41] Eu^{3+} ions (0.97 mol l^{-1} at pH 1.1) were added and the expanded proton spectrum assigned by conventional homonuclear decoupling. Selective ^{13}C–{^{1}H} heteronuclear experiments were then possible, having been virtually impossible without the Eu^{3+} owing to overlap of proton signals, and the ^{13}C spectrum thus assigned. Examination of the ^{13}C spectrum as a function of Eu^{3+} concentration corrected for the pseudocontact shifts on the ^{13}C nucleus.

B. ^{13}C–{^{1}H} **Off-resonance Decoupling.**—Heteronuclear ^{13}C–{^{1}H} off-resonance decoupling experiments are widely used in ^{13}C spectral assignment.[42]

---

[29] J. M. Jenkins and B. L. Shaw, *Proc. Chem. Soc.*, 1963, 279.
[30] P. G. Harrison, S. E. Ulrich, and J. J. Zuckerman, *J. Amer. Chem. Soc.*, 1971, **93**, 5398.
[31] P. G. Harrison, S. E. Ulrich, and J. J. Zuckerman, *Inorg. Nuclear Chem. Letters*, 1971, **7**, 865.
[32] A. P. Tupciauskas, N. M. Sergeev, and Yu. A. Ustynyuk, *Org. Magn. Resonance*, 1971, **3**, 655.
[33] V. N. Torocheshnikov, A. Tupciauskas, N. M. Sergeev, and Yu. A. Ustynyuk, *J. Organometallic Chem.*, 1972, **35**, C25.
[34] W. McFarlane and R. J. Wood, *J. Organometallic Chem.*, 1972, **40**, C17.
[35] H. J. Jakobsen and M. Begtrup, *J. Mol. Spectroscopy*, 1971, **40**, 276.
[36] AA. Borisenko, N. M. Sergeyev, and Yu. A. Ustynyuk, *Mol. Phys.*, 1971, **22**, 715.
[37] V. V. Negrebetskii, A. V. Kessenikh, A. F. Vasil'ev, N. P. Ignatova, N. I. Shvetsov-Shilovskii, and N. N. Mel'nikov, *Zhur. strukt. Khim.*, 1971, **12**, 798.
[38] H. C. E. McFarlane and W. McFarlane, *Org. Magn. Resonance*, 1972, **4**, 161.
[39] A. P. Tupciauskas, N. M. Sergeev, Yu. A. Ustynyuk, and A. N. Kashin, *J. Magn. Resonance*, 1972, **7**, 124.
[40] W. McFarlane and R. J. Wood, *J. C. S. Dalton*, 1972, 1397.
[41] B. Birdsall, J. Feeney, J. A. Glasel, R. J. P. Williams, and A. V. Xavier, *Chem. Comm.*, 1971, 1473.
[42] E. W. Randall, *Chem. in Britain*, 1971, 371.

Multiple Resonance

The residual splitting J_R in off-resonance decoupled multiplets was shown by Ernst to be[25]

$$J_R = [(\Delta v - \tfrac{1}{2} J_{CH})^2 + (\gamma B_2)^2]^{\frac{1}{2}} - [(\Delta v + \tfrac{1}{2} J_{CH})^2 + (\tfrac{\gamma}{7} B_2)^2]^{\frac{1}{2}} \quad (3)$$

where Δv is the offset of the decoupler from the frequency of the proton resonance in question, J_{CH} is the true (^1H, ^{13}C) coupling constant and $\tfrac{\gamma}{7} B_2$ is the power of the decoupling field. This function is only linear in Δv for $\Delta v/\tfrac{\gamma}{7} B_2 \approx 0.5$ and requires the strict limiting conditions that $\tfrac{\gamma}{7} B_2 \gg |\Delta v|$. In these circumstances:

$$\Delta v = \gamma B_2 J_R / J_{CH} \quad (4)$$

In this form the equation has to be used to estimate Δv, as an assignment aid, and to measure $\tfrac{\gamma}{7} B_2$. Pachler[43] has pointed out the limitations of the above formulation and suggested that the following form be used, which requires the much less stringent condition $\tfrac{\gamma}{7} B_2 \gg \tfrac{1}{2}|J_{CH} - J_R|$:

$$\Delta v = J_R \frac{[(\gamma B_2)^2 + \tfrac{1}{4}(J_{CH} - J_R)^2]^{\frac{1}{2}}}{(J_{CH}^2 - J_R^2)^{\frac{1}{2}}} \quad (5)$$

A plot of these functions is given in Figure 2.

Reliable estimates of J_{CH} for directly bonded (^{13}C, ^1H) coupling are usually available. In $^{13}C-\{^1H\}$ experiments a plot of the proton chemical shifts versus the corresponding $J_R(J_{CH}^2 - J_R^2)^{\frac{1}{2}}$ values yields a straight line with a slope γB_2, and the exact position of the decoupling field B_2 in the proton spectrum corresponds to $J_R/(J_{CH}^2 - J_R^2)^{\frac{1}{2}} = 0$. Any wrong assignments of carbon resonances will result in deviations from the straight line. Conversely, if the decoupling unit has been calibrated, proton chemical shifts may be calculated from the residual splittings, and assignments of carbon resonances may be made. The equation finally offers a convenient method to calibrate spin decoupling units without the restricting condition $\tfrac{\gamma}{7} B_2 > |\Delta v|$.

Despite the limitations in linearity in equation (4) pointed out above, Birdsall and co-workers[44] have successfully used it as a basis for a graphical assignment of the ^{13}C resonances in nicotinamide adenine dinucleotide. The proton decoupler is incremented at a fixed power through the proton spectrum and the resultant ^{13}C spectra recorded. The data so obtained are plotted with ^{13}C resonance frequencies on one axis and the proton frequency of the decoupler on the other axis. A series of parallel lines, the slope of which is proportional to the decoupler power, are drawn and the intersection of these lines relate a proton with its corresponding directly bonded ^{13}C (see Figure 3). This technique, although tedious, lends itself to the assignment of complex ^{13}C spectra and is ideally suited for computer data handling.

Jakobsen and co-workers have used the proton off-resonance decoupling experiment to measure the signs of $^1J(^{31}P^{13}C)$ in furyl- and triphenyl-

---

[43] K. G. R. Pachler, *J. Magn. Resonance*, 1972, **7**, 442.
[44] B. Birdsall, N. J. M. Birdsall, and J. Feeney, *J. C. S. Chem. Comm.*, 1972, 316.

phosphines.[45] They observed, while searching for the optimum proton decoupling frequency, that offsets from these values resulted in better decoupling for one peak of the phosphorus–carbon doublet than the other; *i.e.* the residual splittings were different. This effect arises as one of the two

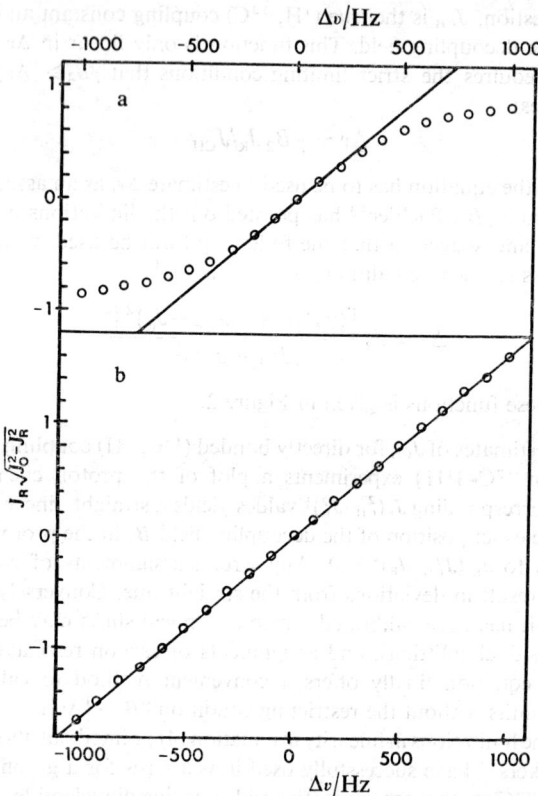

Figure 2 *Experimentally measured residual splittings of the $^{13}CHCl_3$ doublet as a function of the frequency difference Δv. The $^{13}C-\{^1H\}$ experiment was performed in the FT mode on a Varian XL-100 spectrometer equipped with a Gyrocode decoupler. The straight lines in a and b correspond to equations (4) and (5) respectively, with $\gamma B_2 = 615$ Hz*

proton subspectra corresponds to a definite phosphorus spin state more than the other. Thus, with a knowledge of the optimum decoupling frequency, the sign of $^1J(^{13}C^{31}P)$ could be related to that of $J(^1H^{31}P)$. The elegance of this experiment is in its ability to measure more than one sign from a *single* double resonance spectrum.

[45] H. J. Jakobsen, T. Bundgaard, and R. S. Hansen, *Mol. Phys.*, 1972, **23**, 197.

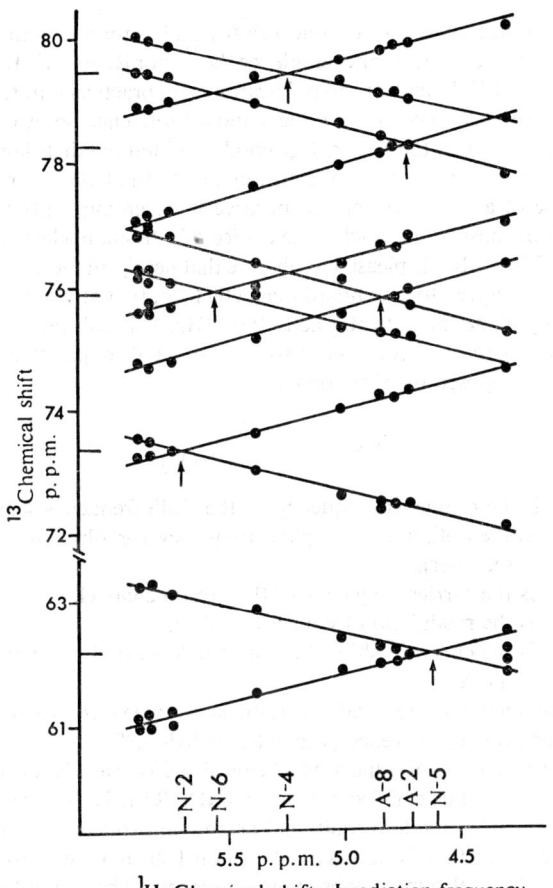

Figure 3 *Plot of peak frequencies in the ^{1}H off-resonance selectively decoupled ^{13}C spectra of NAD$^+$ as a function of position of irradiation in the ^{1}H spectrum, expressed in p.p.m. to high frequency of internal dioxan. The positions of the peaks in the ^{1}H noise-decoupled ^{13}C spectrum are shown by lines on the ordinate and the position of the proton peaks by lines on the abscissa. The arrows ↑ indicate the point of collapse of the ^{13}C doublet and the connection between a given ^{13}C peak and the assigned proton peak. The errors in the position measurements of the ^{13}C peaks are indicated by the size of the points, except near the cross-over positions where the errors are larger (± 0.15 p.p.m.). Small doublet splittings are observed on some of the signals from long-range (C,H) spin-coupling interactions*

5 Chemical Shift Referencing in Multinuclear/Multiresonance Spectra

The subject of referencing for chemical shift data has for a long time been the subject of controversy (just for example see the Senior Reporter's Introduction and Chapter 1 of Volume 1 of this publication!). In practical n.m.r. everything is in effect relative, hence at *least* one standard reference point is necessary. The fewer such points the better. For good or ill tetramethylsilane has been universally accepted as standard for proton n.m.r. (and latterly for carbon-13) with the use of a p.p.m. scale, and positive representing shifts to high frequency.[46] For most 'other nuclei' the choice of a 'homonuclear' reference is not agreed. The author is pleased to observe that nearly all the above chemical shift data are referred to a standard magnetic field, that at which the protons in tetramethylsilane resonate at exactly 100 MHz. This reduces the number of reference points to one. Corrections to the measured frequency are made by the equation proposed by McFarlane.[47]

$$\Delta_{TMS} = \frac{10^8 \Delta_0}{(v_c \pm f - 100\delta)} \qquad (6)$$

where Δ_{TMS} is the resonance frequency on the TMS frequency scale.

Δ_0 is the optimum decoupling frequency (or observing frequency, see later).

v_c is the carrier frequency of the proton channel.

f is the modulation frequency applied.

δ is the chemical shift of the proton locking signal with respect to TMS.

Local 'sub-references' are used for *convenience* when comparing chemical shifts, not as primary references, *e.g.* tetramethyltin.[30]

The above method of reporting chemical shifts was designed for data obtained by heteronuclear decoupling or INDOR; it is, however, ideal for reporting data from modern multinuclear and consequently multiresonance spectrometers. All that is necessary if proton lock is used is to widen the definition of Δ_0 to also mean 'observing frequency'! The method can also be used if a third nucleus, typically deuterium, is used to establish a field-frequency lock. All that is necessary is to determine the frequency at which the field-frequency lock is maintained. This corresponds to $\omega_c \pm f$ in the above formula and can be placed in the above equation when corrected by the magnetogyric ratios of deuterium and the proton. The value of δ can be obtained from a normal proton spectrum of the solution or by the use of known values, depending on the required accuracy. The definition works equally well for Fourier and continuous wave spectrometers, f usually being zero in the pulse mode.

'Sub-references' can of course still be used and a list of the more common

---

[46] *E.g.* see report of Chemical Society N.M.R. Discussion Group, 1971.
[47] W. McFarlane, *Ann. Rev. N.M.R. Spectroscopy*, 1968, **1**, 135.

Table Frequencies of common 'sub-references' in a field defined by the protons in TMS resonating at 100.000 000 MHz

Nucleus	Frequency/Hz	Compound	Solvent	Method[a]	Reference
^{13}C	25 154 002	TMS	neat	DR	19, 49
^{14}N	7 223 881.5 ± 0.3	$Me_4N^+I^-$	[$^2H_6$]DMSO	DR	50
^{15}N	10 133 351.0 ± 0.3	$Me_4N^+I^-$	[$^2H_6$]DMSO	DR	50
^{19}F	94 093 880.4	$CFCl_3$	10% in CCl_4	HL	48
^{29}Si	19 867 220	TMS	neat	DR	47
^{31}P	40 480 740	H_3PO_4	80% in H_2O	DR	47
^{77}Se	19 071 523	Me_2Se	neat	DR	40
^{117}Sn	35 632 290	Me_4Sn	50% CH_2Cl_2	DR	47
^{119}Sn	37 290 662 ± 2	Me_4Sn	20% CCl_4	DR	30
^{127}Te	31 539 860	Me_2Te	neat	DR	47
^{193}Pt	21 401 670	cis-$(Et_3P)_2PtCl_2$	CH_2Cl_2	DR	47
^{119}Hg	17 910 780	Me_2Hg	90% in C_6H_6	DR	39
^{207}Pb	20 920 680	Me_4Pb	neat	DR	47

[a] DR indicates values determined using double resonance techniques; HL indicates values determined using heteronuclear locking spectrometer.

ones are given in the Table. Using this approach, the effects of temperature, concentration, *etc.* on the 'sub-reference' can be eliminated or studied depending on the interest of the spectroscopist, *e.g.* the CFCl$_3$ fluorine resonance moves 0.95 p.p.m., *with respect to* the proton signal due to the 1% of TMS added, on dilution with even CCl$_4$.[48] This raises questions, but always about one compound, TMS, which is reasonably amenable to and subject to intense practical and theoretical study. The Reporter suggests that sanity can be maintained if all n.m.r. data are reported as frequencies at a field defined by the proton resonance for TMS at exactly 100 MHz, while data related to 'sub-references' could be quoted in much the same way as quantities measured in c.g.s. units are reported in an SI journal.

6 Relaxation Effects in Multiple Resonance

Gerace and Kuhlmann have demonstrated the generally assumed fact that the presence of a second rf field as such, *i.e.* as opposed to the decoupling effects it produces on the spin system, has no effect on T_1.[51] To prove this point they measured $T_1^H(CH_3)$ by saturation recovery methods for a 25% solution of methanol in [^{2}H$_6$]DMSO while the spin–spin coupling of the hydroxy-group was removed by (*a*) a second rf field and (*b*) by acid-catalysed exchange. The values of T_1^H, 0.62±0.02 s for the former case and 0.64±0.02 for the latter, agree within experimental error.

Nageswara Rao has continued work on spin–spin relaxation studies by double resonance, by studying weakly coupled three-spin systems[52] and strongly coupled two-spin systems.[53] In the latter case the two-spin system was 2-chloroacrylonitrile; here relaxation was found to be dominated by external random fields with no correlation between sites. The AX$_2$ systems studied are the protons in 1,1,2-trichloroethane and in 2,2-dichloroethanol (excluding the OH proton). The compounds were studied as degassed neat liquids at 300 K and 100 MHz. Spectra were analysed using a theory based on two theorems. Theorem I, for a weakly coupled spin system (AX), states that the double resonance spectrum produced by irradiation of spin X at offset δ (from optimum value) is the mirror image of that produced when an offset of $-\delta$ is used, providing the dominant relaxation mechanisms satisfy the following condition:

$$R_{\phi\alpha,\phi\alpha',\phi\beta,\phi\beta'} = R_{\alpha\alpha'\beta\beta'}, \quad \text{for all } \alpha \text{ and } \beta \tag{7}$$

where $R_{\alpha\alpha'\beta\beta'}$ are Redfield elements of the relaxation matrix and ϕ is a total spin inversion operator.[54] Theorem II, for weakly coupled spin systems,

---

[48] D. Shaw, unpublished results.
[49] E. G. Paul and D. M. Grant, *J. Amer. Chem. Soc.*, 1964, **86**, 2977.
[50] E. D. Becker, *J. Magn. Resonance*, 1971, **4**, 142.
[51] M. J. Gerace and K. F. Kuhlmann, *J. Phys. Chem.*, 1972, **76**, 1152.
[52] N. R. Krishna and B. D. Nageswara Rao, *Mol. Phys.*, 1972, **23**, 1013.
[53] N. R. Krishna and B. D. Nageswara Rao, Proceedings of the 15th Symposium on Nuclear Physics and Solid State Physics, 1971, **3**, 323.
[54] J. M. Anderson and J. D. Baldeschwieler, *J. Chem. Phys.*, 1962, **37**, 39.

states that the relaxation contribution from $\bar{\chi}$ (deviation in the spin density matrix σ_0 in the rotating frame due to B_2) to the intensities of two symmetrical transitions $\alpha \to \lambda \alpha'$; $\alpha \to \lambda \alpha'$ in the double resonance spectrum of the A spins (when the X spins are irradiated) is equal in magnitude but opposite in sign if where λ_μ is a specific spin inversion operator.[54]

$$R_{\lambda_x\alpha,\lambda_x\alpha',\lambda_x\beta,\lambda_x\beta'} = R_{\alpha\alpha'\beta\beta'} \quad \text{for all } \alpha \text{ and } \beta \qquad (8)$$

The spectra obtained from the two compounds studied satisfied the above symmetry conditions, hence only relaxation mechanisms obeying these theorems were considered. The effect of magnet inhomogeneity is considered. The theorems are discussed analytically, and relaxation mechanisms obeying them evaluated. The contributions from such mechanisms were calculated relative to that of the random field mechanism. For 1,1,2-trichloroethane, the spectra were not very sensitive to these parameters. However, use was made of T_1^A and T_1^X, from rapid passage measurements, to obtain a relative contribution of 0.893 for the dipole–dipole mechanism, and with the additional assumption of a value of J(H,Cl) (from similar compounds) contributions of 0.612 and 0.37 for the one- and two-bond scalar (H,Cl) relaxation terms were estimated. For 2,2-dichloroethanol the contribution of chemical exchange was calculated, the linewidths in the spectra being sensitive to this parameter. The relaxation contributions obtained were 0.38 for the dipolar mechanism and 0.093 for scalar (H,Cl) interaction; an exchange correlation time 0.2 ms was found. This latter figure is of interest as it gives a rate of chemical exchange from measurements at *single* temperature, unusual for this type of work.

7 ^{13}C–{^{1}H} Nuclear Overhauser Studies

Carbon-13 spectra are almost invariably run under conditions of proton noise decoupling, thus gaining the simplicity inherent in the removal of spin–spin splittings and the invaluable nuclear Overhauser enhancement. Thus the consequences of the double-resonance experiment are of great importance when considering ^{13}C spectroscopy. This section deals with work on ^{13}C–{^{1}H} experiments or on molecular properties which dictate the consequences of such an experiment, *e.g.* relaxation effects.

Grant and co-workers[55] have shown that the maximum enhancement from proton decoupling of 2.988 can only be obtained if ^{13}C–{^{1}H} dipolar interactions form the dominant relaxation mechanism *and* if molecular rotation is sufficiently rapid—*i.e.*

$$(\omega_C + \omega_H)\tau_r \gg 1 \qquad (9)$$

where ω_C and ω_H are the resonance frequencies of ^{13}C and ^{1}H (in rad s^{-1}) and τ_r is the molecular rotational correlation time (*i.e.* a time in order of which a molecule rotates through 1 radian). If rotation is slow, defined by

[55] K. F. Kuhlmann, D. M. Grant, and R. K. Harris, *J. Chem. Phys.*, 1970, **52**, 3439.

$$\omega_C \tau_r \gg 1 \qquad (10)$$

then the enhancement drops to 1.153. Macromolecules with τ_r about 10^{-8} fall between these limits at normal n.m.r. frequencies. Consequently, Allerhand and co-workers have analysed this situation and derived equations for T_1, T_2, and the ^{13}C NOE, assuming pure dipolar relaxation with one proton in the following cases:[56] (i) isotropic reorientation of a rigid body and (ii) a rotating group (with one degree of internal motion) attached to a body undergoing isotropic rotational reorientation. Their numerical results indicate that the signal-to-noise ratio of a proton-decoupled ^{13}C spectrum, though not the spectrum resolution, *may* decrease when working at higher field, despite the higher Boltzman sensitivity accompanying the increased field. Such conclusions are reached by considering the NOE as a function of field strength and molecular rotation in the region of $\omega_C \tau_r \approx 1$ where the NOE decreases with increasing field. For example if $\tau_r \approx 10^{-9}$ then at 51.7 T (*i.e.* a superconducting magnet) the NOE is only 2, compared with 3 at 14.1 T. In addition, T_2/T_1 will be smaller at the higher field, hence Fourier techniques become less efficient.[57] On the other hand, there will be an increase in resolution at the higher field, not only from greater chemical effect separation but also from a reduction in linewidth (magnet permitting!) owing to changes in T_2, *e.g.* 15 Hz wide methine carbon resonance at 14.1 T compared with 8.5 Hz width at 51.7 T.

The NOE is a very variable and unpredictable quantity, differing even from one carbon to another in the same molecule, and consequently may be undesirable from a quantitative point of view. The Overhauser effect as outlined above only has its maximum value when the dominant ^{13}C relaxation is (C,H) dipolar in nature; it is competition from other mechanisms, *e.g.* spin rotation for methyl groups, which leads to its variability. If the relaxation of all the carbons could be dominated by some other mechanism, then the NOE would be reduced to zero and quantitative spectra would be obtainable. The obvious choice is relaxation *via* a paramagnetic species placed in solution; this approach has been pioneered by La Mar[58] and Natusch.[59] Freeman, Pachler, and La Mar have made a detailed study of this approach using the formic acid- and vinyl acetate–hexa-aquochromium perchlorate systems.[60] They show that beyond about 0.1 mol l^{-1} in paramagnetic substance the Overhauser effect is effectively quenched, only small chemical shifts are induced, and line broadening is only just apparent. The loss in sensitivity from line broadening can be more than overcome by taking advantage of the faster relaxation (see below). The authors confirm that for formic acid the relaxation is predominantly dipolar and that the Cr^{3+} ions act directly on the carbon. The technique is shown to work for vinyl acetate. The authors give a list of eight

---

[56] D. M. Doddrell, V. Glushko, and A. Allerhand, *J. Chem. Phys.*, 1972, **56**, 3683.
[57] D. Gillies and D. Shaw, *Ann. Rev. N.M.R. Spectroscopy*, 1972, **5**, 557.
[58] G. N. La Mar, *Chem. Phys. Letters*, 1971, **10**, 1971.
[59] D. F. S. Natusch, *J. Amer. Chem. Soc.*, 1971, **93**, 2566.
[60] R. Freeman, K. G. R. Pachler, and G. N. La Mar, *J. Chem. Phys.*, 1971, **55**, 4586.

criteria to be met before quantitative data can be obtained from ^{13}C FT spectra.

Gransow, Burke, and La Mar have discussed another consequence of adding Cr^{3+} ions,[61] namely that of reducing the T_1^C of the solute under investigation. When studying ^{13}C spectra by Fourier techniques the presence of carbons with long T_1^C causes sensitivity and relative intensity problems.[57] They show that the addition of tris(acetylacetonato)chromium(III) to organometallic

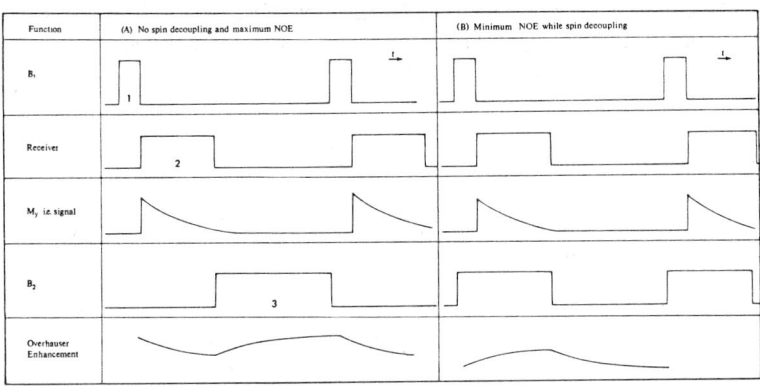

Figure 4 *Duty cycles of observing and decoupling transmitters and the behaviour of the magnetization during gated decoupling experiments, (A) to produce no spin decoupling with maximum NOE, (B) to produce minimum NOE while spin decoupling*

carbonyls minimizes this problem while producing very small (less than 0.1 p.p.m.) contact shifts. Similar results have been obtained by Bareza and Engstrom using the β-diketonates of Cu^{2+} and Fe^{3+}.[62]

The NOE has its origins in relaxation, and hence takes time of the order of T_1^C to either build up or decay, whereas the associated spin decoupling changes are effectively instantaneous, *e.g.* with the presence or absence of a B_2 field. The two effects can thus be separated.[63] Such a separation is ideally suited to the Fourier transform approach where the whole ^{13}C spectrum can be obtained in a fraction of T_1^C. Freeman and Hill have published 'gated decoupling' expriments to yield either enhanced spectra with no decoupling[64] or *vice versa*.[65] The duty cycles for these experiments are shown in Figure 4. Spectra showing enhancement with no decoupling have also been demonstrated by Gansow *et al.*[66,67] The basic approach of these experiments is to have the decoupler

[61] O. A. Gansow, A. R. Burke, and G. N. La Mar, *Chem. Comm.*, 1972, 456.
[62] S. Bariza, and N. Engstrom, *J. Amer. Chem. Soc.*, 1972, **94**, 1762.
[63] J. Feeney, D. Shaw, and P. J. S. Pauwells, *Chem. Comm.*, 1970, 554.
[64] R. Freeman and H. D. W. Hill, *J. Magn. Resonance*, 1971, **5**, 278.
[65] R. Freeman, H. D. W. Hill, and R. Kaptein, *J. Magn. Resonance*, 1972, **7**, 327.
[66] O. A. Gansow and W. Schittenhelm, *J. Amer. Chem. Soc.*, 1971, **93**, 4296.
[67] O. A. Gansow, M. R. Willcott, and R. E. Lenkinski, *J. Amer. Chem. Soc.*, 1971, **93**, 4295.

either on or off depending whether decoupling is required or not during data acquisition, and then off or on during a delay placed between the end of the data acquisition and the start of the next pulse. The efficiency of these experiments depends on the ratio of the delay time to the acquisition time and to T_1^C, but can easily be 90%.[65] These experiments can be easily controlled from the spectrometer's pulse programmer and are already standard on some FT spectrometers.

The ability to turn on or off the Overhauser effect at will, be it at the expense of sensitivity compared with the continuously decoupled spectrum, will add a new dimension to ^{13}C spectroscopy.

8
Macromolecules

BY I. D. ROBB AND G. J. T. TIDDY

1 Introduction

This chapter reviews publications concerned with n.m.r. of macromolecules appearing between June 1971 and May 1972. The Report is largely subdividid according to the nature of the material studied. Papers on synthetic macromolecules, biological macromolecules, and associated small molecules are reviewed in separate sections, which are further subdivided according to the type of nucleus studied and the n.m.r. technique used.

Most of the studies have been concerned with the determination of the structures and conformations of macromolecules and their complexes. The bulk of the publications have reported uses of ^1H n.m.r., although studies of ^{13}C n.m.r. are receiving a rapidly increasing amount of attention. At present ^{13}C n.m.r. is still a relatively new area, and insufficient data have been published for facile spectral assignments of macromolecules to be made. Use of nuclei other than ^1H and ^{13}C is still limited mainly to studies of small molecules or counter-ion binding to macromolecules.

2 Reviews

N.m.r. spectroscopy has an important influence on the field of polymer science, and several reviews dealing with different aspects of the subject have appeared in the past year. Bovey[1] has reviewed the recent literature dealing with the high-resolution spectra of solutions, mainly those of vinyl polymers. A brief outline of the relevant n.m.r. theory and the different stereochemical configurations of substituted vinyl polymers is followed by a discussion of the experimental data relating to di- and mono-substituted vinyl polymers, polyolefins, and copolymers. A review of the application of n.m.r. to biological molecules, covering work mainly in the past five years, has been written by Jardetzky and Wade-Jardetzky.[2] Publications relating to the structure and conformation of amino-acids, peptides, polypeptides, and proteins are critically reviewed together with the theory and results for the interaction of small

---
[1] F. A. Bovey, 'Progress in Polymer Science', ed. A. D. Jenkins, Pergamon Press, Oxford, 1971, Vol. 3.
[2] O. Jardetzky and N. G. Wade-Jardetzky, *Ann. Rev. Biochem.*, 1971, **40**, 605.

molecules, such as metal ions, with macromolecules. A collection of lectures presented at a colloquium on n.m.r. spectroscopy held in April 1970 at Aachen (Western Germany) has been issued[3] as Volume 4 in the series of n.m.r. reviews edited by Diehl, Fluck, and Kosfeld. The book contains short reviews and articles on n.m.r. of polypropylenes, poly(vinyl chloride), methyl methacrylate copolymers, aromatic copolymers, proteins, and water in cellulose. The study of solid polymers by wide-line and spin-echo techniques is also reviewed.

A brief review of the principles and applications of ^{13}C spectroscopy has been published by Breitmaier et al.[4] The article outlines the principles and advantages of the pulsed Fourier-transform technique and lists the ^{13}C chemical shifts associated with chemical groups in organic compounds. Some of the results obtained with natural molecules, such as carbohydrates and polypeptides, mainly of low molecular weight, are discussed.

As part of a general review of research on poly(vinyl alcohol) Fujii[5] has summarized the results of the application of high-resolution n.m.r. to the determination of the configuration of poly(vinyl alcohol) and related polymers such as poly(vinyl acetate), poly(vinyl trifluoroacetate) and poly(vinyl formate).

3 Synthetic Macromolecules

A. ^{1}H High Resolution.—The emphasis of applications of high-resolution n.m.r. to polymer studies continues to be in the areas of polymer characterization, both for simple structure determinations and for the determination of chain configurations. In these studies it is assumed that *all* the polymer present in the sample is in solution. In an important paper on the use of n.m.r. to measure the tacticity of poly(vinyl chloride) (PVC) Wilkes has demonstrated that this is not always true.[6] PVC samples prepared by different techniques were 'dissolved' in *o*-dichlorobenzene together with a standard (for intensity measurements) and the percentage of the polymer giving a high-resolution n.m.r. spectrum was determined from integration of peak areas. The spectra were recorded at temperatures of 413 and 443 K and the range of solubilities was from zero, for a highly crystalline sample of PVC, to 90% or more for some other samples. Samples that gave n.m.r. spectra from greater than about 80% of the polymer were perfectly clear at 413—443 K. Wilkes drew attention to the fact that previous estimates of PVC tacticity using *o*-dichlorobenzene as solvent may be unreliable because of this effect.

Tacticity Determinations. A comparison between i.r. and n.m.r. techniques for determining the stereoregularity of polypropylene has been reported by Kissin

---

[3] 'N.M.R.: Basic Principles and Progress,' ed. P. Diehl, E. Fluck, and R. Kosfeld, Springer-Verlag, Berlin, 1971, Vol. 4.
[4] E. Breitmaier, G. Jung, and W. Voelter, *Angew. Chem. Internat. Edn.*, 1971, **10**, 673.
[5] K. Fujii. *J. Polymer Sci., Part D*, 1971, **5**, 431.
[6] C. E. Wilkes, *Macromolecules*, 1971, **4**, 443.

and co-workers.[7] The i.r. and n.m.r. methods were related by a statistical theory for polymerization, and experimental measurements on polypropylene showed reasonable agreement between n.m.r. and i.r. determinations. The n.m.r. technique involved the determination of the isotactic diad content by the previously reported method,[8] and also from the fractional area of a peak assigned to syndiotactic diads.

A statistical analysis involving a first-order Markovian process has been used to interpret the n.m.r. spectra of chlorinated PVC.[9] The spectra were recorded using $SOCl_2$ as solvent at 348 K and 220 MHz. The tacticity of the unchlorinated polymer was estimated from i.r. spectroscopy, and, with the assumption that the chlorination mechanism was independent of tacticity, the population of the different sequences was determined using both n.m.r. and i.r.

N.m.r. of the polymer prepared from a specifically deuteriated monomer can be of value in determining the conformation of the reactants during the polymerization reaction. Dombroski and Schuerch have reported a detailed study of vinyl methyl ether polymerization using specifically deuteriated *cis* and *trans* monomer isomers.[10,11] In the earlier paper, written with Sarko, the n.m.r. of the methylene region from the deuteriated polymers was interpreted in terms of tetrad sequences.[10] This was possible because of the greatly simplified spectra from the deuteriated polymer compared with that of the undeuteriated material, in particular for the CHD region. The spectrum was also simplified in the region $\delta \sim 3.5$ p.p.m. (methoxy region), because of the elimination of the α-proton resonance. A study of polymers prepared under a range of conditions with different initiators was also reported.[11] From the minor differences in the proportions of *erythro* and *threo-meso* sequence within the isotactic sequences, for polymers prepared by different methods, it was shown that the polymerization process occurs by both isotactic-like and syndiotactic-like presentation of the monomer to the polymer cation. It was suggested that isotactic-like presentation is slightly preferred.

A number of papers have been published on the stereoregularity of poly-(acrylic acid) and its derivatives. Monjol has reported an investigation of the stereoregularity of poly(acrylic acid).[12] The polymer was prepared by low-temperature polymerization which supposedly gave a completely syndiotactic polymer. The stereoregularity of the poly(acrylic acid) was determined by comparison of the spectra of polymethyl esters of the acid with those of polymethyl esters made from hydrolysed poly(isopropyl acrylate) prepared under various conditions, and with the spectra of isopropyl derivatives. The

---

[7] Yu. V. Kissin, V. I. Tsvetkova, and N. M. Chirkov, *European Polymer J.*, 1972, **8**, 529.
[8] E. Lombardi, A. L. Segne, A. Zanbelli, A. Morginangelli, and G. Natta, *J. Polymer Sci., Part C, Polymer Symposia*, 1967, **16**, 2539.
[9] D. Doskočilova, B. Schneider, E. Drakorádová, J. Stoker, and M. Kolínský, *J. Polymer Sci., Part A-1, Polymer Chem.*, 1971, **10**, 2753.
[10] J. R. Dombroski, A. Sarko, and C. Schuerch, *Macromolecules*, 1971, **4**, 93.
[11] J. R. Dombroski and C. Schuerch, *Macromolecules*, 1971, **4**, 449.
[12] P. Monjol, *Bull. Soc. chim. France*, 1972, 1308.

report of Matsuzaki and co-workers is related to this study.[13] They investigated poly(isopropyl [α,β-^{2}H$_2$]-acrylate) and its conversion into poly(methyl [α,β-^{2}H$_2$]acrylate), together with the occurrence of racemization during the hydrolysis of the isopropyl ester. The tacticity of poly(isopropyl acrylate) polymerized at different temperatures was determined and the enthalpy and entropy differences between isotactic and syndiotactic addition were determined. The rate of hydrolysis of poly(isopropyl acrylate) was independent of the tacticity and dependent on the molecular weight whereas n.m.r. indicated that the rate of racemization of the syndiotactic polymer sequences was faster than the rate for isotactic polymer. α-Deuteron–proton exchange was also observed during hydrolysis. In an earlier paper from the same group Uryu and co-workers[14] reported an n.m.r. determination of the stereoregularity of poly(trimethylsilyl acrylates) and poly(alkyl acrylates), where the alkyl group was ethyl or isopropyl. The isopropyl polymer was prepared from the [α,β-^{2}H$_2$]monomer which was polymerized using a radical polymerization mechanism at low temperatures to produce a mainly syndiotactic polymer. Poly(trimethylsilyl acrylate) obtained by the same mechanism also gave a mainly syndiotactic polymer, which was used for the preparation of syndiotactic poly(methyl acrylate).

The tetrad sequences of a series of poly(ethyl α-chloroacrylates) have been reported by Wesslen and co-workers.[15] The report concerns 220 MHz proton spectra of the polymers in chlorobenzene as solvent at 373 K. The relative tetrad intensities were obtained from the backbone methylene resonances, in contrast to the earlier study[16] where the ethyl CH$_3$ resonance was used to obtain triad sequences. The backbone methylene resonance was analysed in terms of six expected tetrad sequences, and use was made of a curve-resolving technique utilizing computer simulation to obtain chemical shifts, coupling constants, and relative intensities. A similar technique was applied by Johnston and Kopf to the determination of the sequence distribution and tacticity in methyl methacrylate–vinyl chloride copolymers.[17] Several polymers were prepared by a free-radical initiation process, with the methyl methacrylate content of the resultant polymers varying from 5.8 to 100%. The α-methyl resonance gave up to five peaks and these werei nterpreted in terms of triad sequence distributions and tacticities, using computer simulation and curve fitting to experimental data.

Information on pentad sequences in poly(acrylonitrile) was obtained[18] from the spectra of samples prepared using [β,β-^{2}H$_2$]acrylonitrile. Spectra were recorded with deuterium decoupling, and the methine proton resonance was

[13] K. Matsuzaki, M. Okada, and K. Hosonuma, *J. Polymer Sci., Part A-1, Polymer Chem.*, 1972, **10**, 1179.
[14] T. Uryu, H. Shiroki, M. Okada, K. Hosonuma, and K. Matsuzaki, *J. Polymer Sci., Part A-1, Polymer Chem.*, 1971, **9**, 2335.
[15] B. Wesslen, R. W. Lenz, and F. A. Bovey, *Macromolecules*, 1971, **4**, 709.
[16] B. Wesslen and R. W. Lenz, *Macromolecules*, 1971, **4**, 20.
[17] N. W. Johnston and P. W. Kopf, *Macromolecules*, 1972, **5**, 87.
[18] K. Matsuzaki, M. Okada, and K. Goto, *J. Polymer Sci., Part A-1, Polymer Chem.*, 1972, **10**, 823.

observed to be a multiplet with the intensities of different peaks differing from sample to sample. The peaks were assigned by comparing the intensities with the probability of the occurrence of different pentad sequences calculated using Bernoullian and Markov statistics. Copolymers of vinylidene chloride (VDC) and methacrylonitrile (MAN), prepared by polymerization with benzoyl peroxide, gave spectra which were analysed in terms of sequence structure.[19] In the note some tetrad assignments were made, and from the diad sequences information was obtained on the copolymerization kinetics. Copolymers of MAN with styrene and α-methylstyrene (MS), together with copolymers prepared from acrylonitrile with MS, were the subject of a study by Patnaik and Gaylord.[20] N.m.r. spectra (100 MHz) were recorded of copolymers in $CDCl_3$ solutions at 348 K, and the spectra were interpreted in terms of co-isotactic, co-syndiotactic, and co-heterotactic units. Both alternating and random copolymers were studied and, as in the previous paper,[19] the results were discussed in terms of the reactivity ratios. In the following paper[21] Patnaik and Gaylord reported a similar study on styrene–α-chloroacrylonitrile copolymers.

The alkaline hydrolysis of methyl methacrylate–isoprene copolymers has been studied by Bevington and co-workers in a series of three papers.[22-24] The first paper[22] outlined the determination of the degree of hydrolysis of a series of copolymers from the relative area of the ester OMe resonance. A limited degree of hydrolysis was observed, and this was correlated with the calculated sequence distributions of the copolymer, assuming that only methacrylate units adjacent to an isoprene unit or in an isotactic methacrylate triad were hydrolysed. The sequence distributions were calculated from reactivity ratios. The second paper[23] described the measurement of the sequence distribution from 220 MHz n.m.r. spectra. In the third paper[24] the formation of γ-lactone rings (by bromination) between methacrylate–isoprene units with specific configuration was used to block certain groups involved in hydrolysis, and further hydrolysis determinations (by n.m.r.) confirmed the previous results.

In an attempt to understand the optical rotatory phenomena of poly-(alkylpropylene oxide) an n.m.r. investigation of poly($trans$-[1-^{2}H]propylene oxide) in a series of solvents has been carried out.[25] The preparation of the monomer was described previously.[26] Polymerization was accomplished using $ZnEt_2$–MeOH in benzene solution. The resultant polymer was separated into two components, one acetone-soluble (amorphous) and the other

[19] R. E. Block and H. G. Spencer, *J. Polymer Sci., Part A-2, Polymer Phys.*, 1971, **9**, 2247.
[20] B. K. Patnaik and N. G. Gaylord, *J. Macromol. Sci., Chem. (A)*, 1971, **5**, 843.
[21] B. K. Patnaik and N. G. Gaylord, *J. Macromol. Sci., Chem. (A)*, 1971, **5**, 859.
[22] J. C. Bevington and J. R. Ebdon, *Makromol. Chem.*, 1972, **153**, 165.
[23] J. C. Bevington and J. R. Ebdon, *Makromol. Chem.*, 1972, **153**, 173.
[24] J. C. Bevington, J. R. Ebdon, and D. J. H. Mallard, *Makromol. Chem.*, 1972, **153**, 181.
[25] T. Hirano, P. H. Khanh, and T. Tsuruta, *Makromol. Chem.*, 1972, **153**, 331.
[26] P. H. Khanh, T. Hirano, and T. Tsuruta, *J. Macromol. Sci., Chem. (A)*, 1971, **5**, 1287.

L

acetone-insoluble (crystalline). The spectra of the two fractions were interpreted in terms of diad tacticities and gave information on the mechanism of polymerization. The coupling constant between the methylene and methine protons (ca. 6 Hz) was interpreted to indicate almost equal populations of *trans* and *gauche* conformations.

The stereochemistry of copolymers prepared from isobutene with maleic anhydride, dimethyl maleate, and dimethyl fumarate has been determined[27] from the multiplicity of the dimethyl resonance. A quartet was observed due to the presence of *threo*-di-isotactic and *threo*-disyndiotactic triads.

A novel method for determining the configuration of poly(vinyl alcohol) has been reported by De Member and co-workers, which involves making use of the OH proton resonance.[28] With [^{2}H$_6$]DMSO as solvent the OH resonance was observed as a multiplet at δ 4.2—4.7 p.p.m. The multiplet consisted of three doublets, and by comparing the spectra of polymers of known tacticity these were assigned to syndiotactic, heterotactic, and isotactic triads in order of increasing chemical shift, δ. The relative populations of the configurations were determined from integration of the peak areas and the authors concluded that this method of tacticity determination was far easier than methods involving the resonances from protons bound to carbon.

Structure Determinations. This section is concerned with studies where n.m.r. has been used as a means of identifying chemical structures of polymers, and the papers are not discussed in any detail unless other features are involved. Chokki and co-workers have determined the structures of polyurethane elastomers by comparison of the polymer spectra with spectra from model compounds.[29] They also measured the number-average molecular weights of the polymers by the relative intensities of *N*-methyl groups attached to the terminal polymer chain groups. In a study of polyethylene branching Lessac and co-workers examined the $CH_3:CH_2$ intensity ratios and obtained good agreement with i.r. measurements for the degree of chain branching.[30] Urea–formaldehyde resins can be analysed by their n.m.r. spectra in [^{2}H$_6$]-DMSO and chemical shifts have been tabulated for a number of different polymer units.[31] Formaldehyde resins containing acetylated t-butylphenol groups are easily studied by n.m.r. because of the t-butyl resonance, and their spectra have been reported together with the peak assignments.[32]

The linewidths of polysulphone derivatives were found to vary with the polarity of the solvent, the lines being sharper in more polar solvents such

[27] R. Bacskai, L. P. Lindeman, and D. L. Rabenstein, *J. Polymer Sci., Part A-1, Polymer Chem.*, 1972, **10**, 1297.
[28] J. R. DeMember, H. C. Haas, and R. L. MacDonald, *J. Polymer Sci., Part B, Polymer Letters*, 1972, **10**, 385.
[29] Y. Chokki, M. Nakabayashi, and M. Sumi, *Makromol. Chem.*, 1972, **153**, 189.
[30] P. Lissac, P. Berticat, and Pham-Quang-Tho, *J. Macromol. Sci., Chem. (A)*, 1971, **5**, 901.
[31] S. M. Kambario and R. C. Vasishth, *J. Appl. Polymer Sci.*, 1971, **15**, 1911.
[32] T. Tanno, Y. Mukouyama, and T. Saito, *J. Polymer Sci., Part B, Polymer Letters*, 1972, **10**, 51.

as $CDCl_3$–[$^2H_6$]DMSO mixtures.[33] A number of polymers, including deuteriated derivatives, were examined to determine the preferential mode of addition. Chlorosulphonated polyethylenes have been studied,[34] to determine the chlorine distribution and hence obtain some knowledge of the mechanism of chlorine addition.

Simplification of poly(propylene–butadiene) copolymer spectra was obtained by investigating [1,1,4,4-$^2H_4$]butadiene at 220 MHz.[35] The spectra of the deuteriated copolymer were simulated using LAOCOON III and the presence of alternating sequences was demonstrated. A naphthalene polymer produced by reaction with anhydrous aluminium chloride had been previously estimated to have a 1,2-dihydro-structure, but Yonemitsu and co-workers have established from the polymer spectrum in $CDCl_3$ that the structure involves a 1,2,3,4-tetrahydro-structure.[36]

The determination of the degree of polymerization of a polymeric amine has been reported by Argentor[37] using statistical analysis involving n.m.r. peak intensity data. The method is applicable to low-molecular-weight polymers only. Mak and Rogers[38] have described a technique for the determination of the degree of chain branching in epoxy-resins using acetylation of hydroxy-groups with trichloroacetyl isocyanate. The trichloroacetyl group shifted the resonances of adjacent CH_2 groups, enabling their intensities to be measured.

Miscellaneous Studies. In this section papers reporting the applications of non-standard techniques are discussed. The choice of papers is very much determined by the personal decisions of the authors and may appear to others to be somewhat arbitrary.

Among the most interesting and novel polymer n.m.r. reports are the papers of Schneider and co-workers on the studies of polymer n.m.r. using magic-angle rotation.[39,40] It has been known for some years[41] that spinning at the magic angle (54.7°) can remove dipolar coupling for solids, if some molecular motion is present and if the spinning speed is great enough. In these reports, sharpening of the spectra of polymer systems by magic-angle rotation has been studied. In the first paper[39] a theory is developed relating the lineshape for magic-angle rotation to the correlation time for molecular motion, the second moment of the resonance line, and the spinning rate. The theory successfully

[33] K. J. Irvin, M. Navratil, and (in part) N. A. Walker, *J. Polymer Sci., Part A-1 Polymer Chem.*, 1972, **10**, 701.
[34] E. G. Brame, jun., *J. Polymer Sci., Part A-1, Polymer Chem.*, 1971, **9**, 3051.
[35] T. Suzuki, Y. Takeyami, J. Furukawa, and R. Hirai, *J. Polymer Sci., Part B, Polymer Letters*, 1971, **9**, 931.
[36] T. Yonemitsu, Y. Matsukama, K. Furukawa, and G. Hazato, *Bull. Chem. Soc. Japan*, 1971, **44**, 3192.
[37] H. Argentar, *J. Polymer Sci., Part B, Polymer Letters*, 1971, **9**, 657.
[38] H. D. Mak and M. D. Rogers, *Analyt. Chem.*, 1972, **44**, 837.
[39] B. Schneider, H. Pivcova, and D. Doskocilova, *Macromolecules*, 1972, **5**, 120.
[40] D. Doskocilova and B. Schneider, *Macromolecules*, 1972, **5**, 125.
[41] E. R. Andrew, 'Progress in Nuclear Magnetic Resonance Spectroscopy,' ed. J. W. Emsley, J. Feeney, and L. H. Sutcliffe, Pergamon Press, Oxford, 1971, Vol. 8, ch. 1.

reproduces the spinning side bands normally observed in these experiments. Spectra of commercial samples of polyethylene were recorded using a Jeolco JNM 3-60 spectrometer with a broad-line measuring attachment and a home-made measuring probe with sample-spinning speeds up to 6800 Hz. Agreement between observed and calculated spectra of the amorphous polymer could only be obtained if two types of amorphous polymer were assumed to be present. The technique can distinguish between isotropic and anisotropic motional narrowing, and even for the most narrow line in the softest polymer sample the motion was found to be anisotropic. The percentage composition of the two types of amorphous polymer and of the crystalline polymer were also determined. In the second paper[40] it was demonstrated that magnetic susceptibility broadening of solvent molecules for samples containing suspended solids could be eliminated by magic-angle rotation as long as the rotation frequency was greater than the linewidth in the static spectrum. Normally this implies rotation at frequencies only of the order of several hundred hertz or less. The technique was used to eliminate broadening of acetonitrile and benzene resonances in the presence of cross-linked polystyrene spheres. The linewidths were reduced from 120 and 58 Hz for static samples to 4.1 and 3.3 Hz respectively. The system consisting of methanol in styrene–divinylbenzene resin (H^+) was also studied, and the methanol hydroxyl proton gave separate resonances for the internal and external hydroxygroups. Magic-angle rotation sharpened and split the CH_3 and sharpened the external OH resonances (to ca. 3 Hz), but the internal OH resonance remained broad. It was suggested that the residual OH broadening was due to proton exchange with the resin. The replacement of methanol by water gave similar spectra, but the position and width of the internal water resonance depended on the degree of cross-linking in the resin. These two papers demonstrate the potential of this type of n.m.r. for the study of both solid polymers and colloidal systems.

A study of partially crystalline polyethylene (solid) using a modified Varian HA 100 spectrometer is reported by Golz and Zachmann.[42] They investigated the spectrum of the polymer over the temperature range 333—413 K. A narrow lorentzian line was observed at 333 K, and above the polymer melting range two lorentzian components were observed. A dilute solution in tetrachloroethane gave a single line at 393 K, and a linewidth one fifth of that of the narrow component in the solid polymer. Possible models to explain the results were proposed.

Chemically induced dynamic nuclear polarization (CIDNP), the process by which free-radical reactions result in enhanced absorption or emission of n.m.r., can be used to obtain information on reaction mechanisms. Bargan has published a note[43] concerning the reinvestigation of reactions where CIDNP had previously been observed, and has interpreted the results in terms of the reaction mechanisms. The CIDNP observed for the free-radical addi-

[42] W. L. F. Golz and H. G. Zachmann, *Kolloid-Z.*, 1971, **247**, 814.
[43] J. Bargon, *J. Polymer Sci., Part B, Polymer Letters*, 1971, **9**, 681.

tion of CBr_4 to styrene was used as a model for the interpretation of CIDNP observed during the telomerization reaction of a series of polymers, including vinyl esters, vinyl bromides, and α-methylstyrene, with CBr_4 and related transfer agents.

Paramagnetic complexes of transition metals can be used to induce large chemical shifts in diamagnetic molecules, enabling overlapping resonances to be separated. The separated resonances can then be used to obtain structural and conformational information about the diamagnetic molecules. Grotens and co-workers have reported[44] the n.m.r. spectra of polyglycol dimethyl ethers (glymes) in the presence of lanthanide complexes. The polymers were of low molecular weight, and partially deuteriated compounds were used to assist spectral assignments. A europium complex was used in an n.m.r. determination of the molecular weight of poly(propylene glycol).[45] The terminal OH groups complex with the europium derivative and this results in a shift of the resonance of protons adjacent to those groups. Integration of the spectrum enables the number-average molecular weight to be determined, though this is applicable only to low-molecular-weight polymers. Since the resonances were well resolved, spin–spin coupling constants could also be determined, and these could give information on polymer chain conformations.

B. Wide-line N.M.R.—The major application of wide-line n.m.r. to polymers is for investigations of polymer molecular motions and their behaviour as a function of temperature and composition. More sophisticated studies include the calculations of theoretical second moments for various models of side-chain and segmental motion and comparison of these with observed spectra to obtain additional insight into polymer dynamics.

In a theoretical study[46] of drawn Nylon-66 fibres, Olf and Peterlin took the previously reported[47] second moments from the wide-line n.m.r. spectra and compared these with calculated values for various models of molecular motion. The previous measurements were made over the temperature range 77—473 K, with the angle between the fibre axis and the magnetic field taking values of 0, 45, and 90°. Detailed studies of the angular dependence of the second moment were made at 77, 293, and 453 K. The data at 77 K were used to calculate the degree of orientation of the fibre and this was found to be in fairly poor agreement with estimates derived from birefringence data. The discrepancy was thought to be due to the approximations implicit in both methods. The relatively sharp resonance of the amorphous chains obscured measurement of second moments due to crystalline chains at high temperatures, reducing the accuracy of the measurement. The second moments fell with

---

[44] A. M. Grotens, J. Smid, and E. de Boer, *Tetrahedron Letters*, 1971, 4863.
[45] F. F.-L. Ho, *J. Polymer Sci., Part B, Polymer Letters*, 1971, **9**, 491.
[46] H. G. Olf and A. Peterlin, *J. Polymer Sci., Part A-2, Polymer Phys.*, 1971, **9**, 1449.
[47] H. G. Olf and A. Peterlin, Technical Report AFML-TR-67-6 to the Air Force Materials Laboratory, 1966.

increasing temperature in two steps, a low-temperature transition at *ca.* 173 K, and two high-temperature transitions beginning at 313 K. A minor contribution to the high-temperature step arose from a transition in the amorphous polymer chains often associated with the glass transition, and this was not considered in detail. The low-temperature transition (γ-process) was ascribed to changes in the amorphous parts of the polymer, and the major part of the high-temperature transition (α_c-process) to changes in the crystalline parts of the polymer. The γ-process was consistent with a model in which non-crystalline chains rotated about axes nearly fixed in space. It was further proposed that the pronounced dependence of the decrease in second moment due to the γ-process on the alignment angle was due to the orientation of these tie molecules during the fibre-drawing process along the fibre axis. The high-temperature decrease in second moment was consistent with models which involved CH_2 segmental oscillation and rotational jumps of the crystalline polymer chain. In an attempt to distinguish between segmental oscillation or rotational jumps as the mechanism of the α_c-process, Olf has reported a further investigation of doubly oriented Nylon-66 in rolled sheet.[48] Wide-line n.m.r. spectra were recorded for a range of angles between the orientation directions and the magnetic field at temperatures of 77, 293, and 453 K. Theoretical second moments were calculated for polymer motions involving double orientation. For the α_c-process the results at 453 K indicate that crystalline chain motion approached full rotation with large oscillation amplitudes of 100° or more. At low temperatures the γ-process was dependent on both orientation directions, suggesting that the non-crystalline chain segments were doubly oriented when they became immobile.

Second-moment data (77 K) for drawn poly(vinyl chloride) (PVC) have been reported,[49] and the data used to derive information about the polymer molecular orientation. Second moments were measured for three polymer samples with different degrees of orientation at a range of values for the angle between the magnetic field and the polymer orientation axis. Although the angular dependence of the second moment was small and the results show some scatter, there was a good correlation with orientation parameters derived from birefringence data.

Drawn poly(ethylene terephthalate) (PET) polymers and polymers prepared by a series of heat treatments at different temperatures were studied by Ito and co-workers.[50] Again, the second moments were recorded as a function of temperature and the angle between the fibre axis and the magnetic field. The dipolar effect of the protons in CH_2 groups dominated the band shapes, and showed that CH_2 segmental motion began at *ca.* 173 K. At 398 K an abrupt decrease in second moment indicated that rotation of aromatic rings was occurring. In highly stretched samples the doublet structure of the CH_2 group was not observed and it was suggested that this was due to an

[48] H. G. Olf, *J. Polymer Sci., Part A-2, Polymer Phys.*, 1971, **9**, 1851.
[49] M. Kashiwagi and I. M. Ward, *Polymer*, 1972, **13**, 145.
[50] S. E. Ito, S. Okajima, and T. Kase, *Kolloid-Z.*, 1971, **248**, 899.

abnormality in the CH_2 group orientation related to the orientation of the benzene ring. However, it is difficult to understand how this can fully explain the results.

A study of the effect of swelling in deuteriated acetone on the second moments of crystallized PET over the temperature range 223—473 K was reported by Eichhoff and Zachmann.[51] The spectra consisted of overlapping broad and narrow lines due to the crystalline and amorphous polymers respectively. Swelling in acetone caused anomalous broadening of the amorphous-polymer line; the precise reason for this was not clear. The broad line became sharper, indicating penetration of the acetone into the crystalline region. An exceptionally sharp line obtained from a sample crystallized in acetone was due to the large amount of acetone in the sample and an increased mobility of the non-crystalline chains.

The effect of the introduction of solid potassium perchlorate as a filler into a polyester containing 50% styrene has been investigated by Price and co-workers.[52] The authors report linewidths of the polymer with a filler for dry polymer, normal polymer, and polymer left exposed to water vapour for two days at room temperature. The variation of linewidth with temperature over the range 163—303 K has been measured and transitions determined from n.m.r. have been compared with those from thermal mechanical analysis. Little correlation was observed between the two sets of results, and the value of the report is lessened by the fact that the identity of the polyester and polymer characterization details are not given.

Ethylene–vinyl acetate copolymers and the homopolymers were investigated at low temperatures by Sobottka and co-workers.[53] The spectra exhibited narrow components assigned to methyl groups rotating at frequencies greater than 10^6 s^{-1}. This was supported by the dependence of the intensity of the narrow line on the concentration of methyl groups, and enabled chain branching in polyethylene to be detected.

An unusual application of broad-line n.m.r. is the report by Cachaty et al.[54] of a study of polymerization in the solid state. The solid monomers of acrylamide and acrylic acid gave broad lines, and their polymers gave a relatively sharp line over the temperature range 261—276 K. The intensity of the sharp line was related to the percentage conversion, and was used to follow the reaction kinetics. The reaction mechanism was further investigated using e.s.r., and the authors report further studies in progress on methacrylic acid and acrylonitrile.

C. Spin-echo N.M.R.

Spin-echo n.m.r. as applied to polymers is generally used for investigation of molecular motions both in solutions and in the solid

---

[51] U. Eichoff and H. G. Zachmann, *Makromol. Chem.*, 1971, **147**, 41.
[52] E. Price, D. M. French, and A. S. Tompa, *J. Appl. Polymer Sci.*, 1972, **16**, 157.
[53] J. Sobottka, A. Buckheim, and A. Heybey, *Plaste Kautschuk*, 1971, **18**, 269 (*Chem. Abs.*, 1971, **75**, 21 241).
[54] C. Chachaty, A. Forchioni, and Buu Ban, *J. Polymer Sci., Part B, Polymer Letters*, 1971, **9**, 483.

state. Changes in molecular motions are related to changes in the relaxation rates T_1^{-1} (spin–lattice relaxation rate), T_2^{-1} (spin–spin relaxation rate), and $T_{1\rho}^{-1}$ (spin–lattice relaxation rate in the rotating frame). The variation of relaxation rates with a change in some property of the system (usually temperature or composition) is monitored and related to other polymer properties. A less popular application of spin-echo n.m.r. is for the measurement of self-diffusion coefficients in the presence of an external field gradient (pulsed or continuous).

Hoch and co-workers[55] have measured T_1, T_2, and $T_{1\rho}$ for protons in solid poly(vinyl acetate) (PVAc) and poly(vinyl deuteroacetate) (PV[^{2}H$_3$]Ac) at 30 MHz. For PVAc the relaxation times were measured over the temperature range 20—473 K at atmospheric pressure and, in addition, over the temperature range 223—500 K at a pressure of 680 atm. The glass transition region for PVAc is at ca. 370 K, and T_1 and $T_{1\rho}$ both exhibited minima in this temperature region, the relaxation times being consistent with theoretical expectations. Dielectric-loss results indicated a similar glass transition temperature, and measurements on other polymers were in agreement with n.m.r. results. At 680 atm the T_1 against temperature curve had a similar shape to the curve at atmospheric pressure, but it was shifted to higher temperatures by an amount equivalent to 27 K/1000 atm. This was close to the $\Delta T/\Delta p$ found for glass transitions and represented an activation volume of 160 cm^3 mol^{-1} and an activation energy of 21 kJ mol^{-1}. T_1 and $T_{1\rho}$ temperature curves exhibited minima in the region 20—60 K due to methyl-group reorientation. The T_1 measurements on PV[^{2}H$_3$]Ac (90% deuteriated) exhibited a much shallower minimum in this region, which confirmed this, and the results were thought to be consistent with a quantum mechanical tunnelling mechanism for rotation.

Christ[56] has measured T_1^{-1} values for atactic polystyrenes (38 MHz) at low temperatures and observed a maximum for T_1^{-1} with low-molecular-weight samples at 130 K. The maximum was not observed with high-molecular-weight polymers, and this was interpreted in terms of spin diffusion from backbone protons, together with rotation of terminal methyl groups. T_1^{-1} values exhibited a 'background' relaxation which was attributed to paramagnetic impurities. A direct relationship between T_1^{-1} and $\bar{M}_n^{-1}$ (number-average molecular weight) was demonstrated, in conflict with a previous report of a relationship between T_1^{-1} and $\bar{M}_n^{-1/3}$ or $M_n^{-2/3}$ which was partly due to the effects of the background relaxation.

The T_1 values for polyethylene glycol (PEG) have been reported[57] over the frequency range 3 kHz—120 MHz at a number of temperatures in the range 258—423 K for both the polymer (melt) and a 50% solution of polymer in D$_2$O. In contrast with the monomer, the polymer T_1-frequency curves showed

---

[55] M. J. R. Hoch, F. A. Bovey, D. D. Davis, D. C. Douglass, D. R. Falcone, D. W. McCall, and W. P. Slichter, *Macromolecules*, 1971, **4**, 712.
[56] B. Christ, jun., *J. Polymer Sci.*, Part A-2, *Polymer Phys.*, 1971, **9**, 1719.
[57] G. Pressing and F. Noack, *Kolloid-Z.*, 1971, **247**, 811.

two dispersion regions, indicating two relaxation processes. These were best resolved in the D_2O solutions, and at 283 K correspond to correlation times (τ) of $\tau > 10^{-4}$ and $\tau < 10^{-8}$ s. Attention was drawn to the similarity of these results to those obtained for biopolymers with helix–coil conformations. Because of the high molecular weight (30 000), methyl end-group rotation could be eliminated as a relaxation mechanism and no other explanation of the results was offered.

Ashworth, Bamford, and Smith have reported T_1 measurements on network copolymers of poly(vinyl trichloroacetate) and poly(methyl methacrylate) as part of a wider study of these polymers.[58] The T_1 variation with temperature (293—433 K) for the copolymers resembled that of poly(methyl methacrylate), the curves being shifted to lower temperatures depending on the poly(vinyl trichloroacetate) concentration. This indicated that the presence of the poly(vinyl trichloroacetate) induced motion in the poly(methyl methacrylate) chains at a lower temperature than was normally observed.

Self-diffusion coefficients (D) have been reported[59] for (polyethylene oxide) (PEO) and polydimethylsiloxane (PDS) in a series of solvents for a range of molecular weights ($M = 590$—3×10^6 and 746—8.0×10^5 respectively) and concentrations, at 306 K. The results were in good agreement with previous studies by other techniques, and the variation of D with M was in agreement with theory for dilute solutions of PEO in $CDCl_3$ and PDS in hexamethyldisiloxane. The concentration dependence of ln D was proportional to polymer concentration for low-molecular-weight samples, but exhibited marked upward curvature for high-molecular-weight samples. This deviation was attributed to polydispersity. Unlike viscosity, D did not exhibit any marked change in its molecular-weight dependence at high molecular weight, indicating that the effects of chain entanglements on D and on viscosity are not the same. The difference between D in good and bad polymer solvents was less than the difference in viscosities.

D. ^{13}C N.M.R.—In recording ^{13}C n.m.r. spectra, proton decoupling is often employed to simplify spectra by removal of proton scalar coupling. Under these conditions a nuclear Overhauser enhancement (n.o.e.) of the ^{13}C resonance intensities is often observed; the enhancement factors have been calculated to be 2.988 and 1.153 for (^{13}C,H) dipolar relaxation in the cases of very fast and very slow molecular motion respectively. Doddrell and co-workers[60] have developed equations to calculate the effect for a number of cases where the molecular correlation time (τ_c) falls between the above extremes. Expressions are derived for T_1, T_2, and the n.o.e. of a ^{13}C–^1H methine group (i) undergoing isotropic reorientation in a rigid molecule and (ii) attached to a rotating group in a rotating rigid body. It was demonstrated that because T_1/T_2 is frequency-dependent for $\tau_c \simeq 10^{-9}$ s, high magnetic field

[58] J. Ashworth, C. H. Bamford, and E. G. Smith, *Pure Appl. Chem.*, 1972, **30**, 25.
[59] J. E. Tanner, K. J. Liu, and J. E. Anderson, *Macromolecules*, 1971, **4**, 586.
[60] D. Doddrell, V. Glushko, and A. Allerhand, *J. Chem. Phys.*, 1972, **56**, 3683.

instruments may not give the best signal-to-noise ratio for ^{13}C spectra, but may improve resolution.

^{13}C n.m.r. has a large number of potential applications to investigations of polymer properties because of the large range of ^{13}C chemical shifts. As the papers reviewed below demonstrate, this is particularly of value in the determination of polymer structures and configurations. Like ^{1}H n.m.r. spectra of polymers, the ^{13}C spectra can give improved resolution at higher temperatures, as has been reported for PVC.[61] The spectra obtained in o-dichlorobenzene at 403 K were much better resolved than spectra at lower temperatures reported previously. Triad configurations were given, based on the methine carbon resonance, but the authors add a cautionary note concerning the effects of crystallinity on solubility (see above, ref. 6). Natural abundance $^{13}C-\{^{1}H\}$ spectra for polypropylene have been reported by Inoue and coworkers.[62] Diad and triad configurations were obtained from the methylene and methyl resonances respectively, and were in good agreement with each other. Additionally, some tetrad configurations were detected in the methylene resonance, and the methyl carbon multiplet corresponded to pentad configurations. Proposed assignments for the methyl resonances were not consistent with observations from the methine resonance, and possible reasons for this were discussed. These included the presence of irregular structures, the possibility that the propagation reaction mechanism was non-stationary, and the contribution of different n.o.e. for different configurations, but no definite conclusions were reached. Spectra of polypropylene-ethylene copolymers prepared from enriched [1-^{13}C]ethylene have been studied[63] in order to gain information on the polymerization mechanism. Comparison of the spectra of different polymers enabled resonances to be assigned, and some evidence on the source of steric control in the polymerization process was obtained.

From the spectra of poly(1,4-butadienes) polymerized with n-butyl-lithium catalyst, Mochel[64] has concluded that the polymer consists of 'blocks' of cis-1,4-units and trans-1,4-units separated by isolated vinyl units. Polymer spectra were compared with those of model compounds, and whereas cis-cis and trans-trans methylene resonances with chemical shifts similar to those in the polymer spectra were observed, the cis-trans resonance was absent from the polymer spectra. Tacticity determinations for polymers with high 1,2-addition contents were also given.

Two papers have reported the $^{13}C-\{^{1}H\}$ n.m.r. of poly(methyl methacrylate).[65,66] Peat and Reynolds[65] have estimated the triad configurations from

[61] C. J. Carman, A. R. Tarply, jun., and J. H. Goldstein, *Macromolecules*, 1971, **4**, 445.
[62] Y. Inoue, A. Nishioka, and R. Chujo, *Makromol. Chem.*, 1972, **152**, 15.
[63] A. Zambelli, G. Gatti, C. Sacchi, W. O. Crain, jun., and J. D. Roberts, *Macromolecules*, 1971, **4**, 475.
[64] V. D. Mochel, *J. Polymer Sci., Part A-1, Polymer Chem.*, 1972, **10**, 1009.
[65] I. R. Peat and W. F. Reynolds, *Tetrahedron Letters*, 1972, 1359.
[66] Y. Inoue, A. Nishioka, and R. Chujo, *Polymer J.*, 1971, **2**, 535.

the methyl, methylene, and carbonyl resonances and made pentad assignments in the carbonyl multiplet. A similar study was published by Inoue and co-workers.[66] Both groups of workers reported identical peak assignments although the polymers used had quite different compositions of isomers.

White[67] has communicated the results for ^{13}C investigations on poly(2,6-dimethyl-1,4-phenylene ethers) and related low-molecular-weight model compounds. Information on the chain branching in polymers prepared at different temperatures was obtained.

E. ^{19}F N.M.R.—Poly(tetrafluoroethylene) (PTFE) has been studied by ^{19}F n.m.r. previously (see references in ref. 68), and McBrierty and co-workers have now reported[68] pulsed n.m.r. T_1, $T_{1\rho}$, and T_2 measurements for oriented PTFE fibres. T_2 values were measured for values of the angle between the fibre axis and the magnetic field of 0, 54, and 90°, over the temperature range 173—473 K. A two-component, free-induction decay was attributed to the presence of crystalline (short decay) and amorphous (long decay) regions in the polymer. T_2 data indicated that the amorphous regions were randomly oriented from an n.m.r. viewpoint, while changes in T_2 from the crystalline chains with temperature were consistent with the onset of rotational motion (at *ca.* 293 K). T_1 values were little different from the values for the unoriented polymer and $T_{1\rho}$ values were too complex to extract useful information at present. Graft copolymers of PTFE and styrene have been studied by both ^1H and ^{19}F wide-line n.m.r.[69] The CF_2 concentration was estimated to be much less than the CF concentration, and the segment mobility of short chains was unaltered by grafting. The concentration of styrene trapped in the polymer was found to vary with the grafting technique.

Apart from the above studies, ^{19}F n.m.r. has been used to study molecules labelled with fluorine-containing groups, using the ^{19}F resonance as a probe for polymer properties. A polyethyleneimine was labelled by acetylation with 10,10,10-trifluorodecanoic acid and the spectra demonstrated[70] that in neutral aqueous solution two environments for the chains were present with approximately equal concentrations; one was assigned to chains surrounded by water molecules and the other to a micelle-like region. The populations in each environment were susceptible to changes caused by raising the pH or adding urea (6M).

4 Biological and Related Macromolecules

A. New Techniques.—Many studies of biological macromolecules involve the use of Fourier Transform n.m.r. on aqueous solutions. Because the residual HDO peak often obscures other peaks of interest, Patt and Sykes have pro-

---

[67] D. M. White, *Polymer Preprints*, 1972, **13**, 373.
[68] V. J. McBrierty, D. W. McCall, D. C. Douglass, and D. R. Falcone, *Macromolecules*, 1971, **4**, 586.
[69] Z. Veksli, J. N. Herak, P. Hedvig, and J. Dobo, *European Polymer J.*, 1971, **7**, 231.
[70] T. W. Johnson and I. M. Klotz, *J. Phys. Chem.*, 1971, **75**, 4061.

posed[71] a technique to eliminate this resonance from the observed spectrum. The method utilizes the large difference in T_1 values typically measured for polymer (T_1 <1.0 s) and HDO in D_2O ($T_1 \simeq 10$ s). A series of 180°–t–90° pulse sequences was used, with the time between pulses (t) adjusted so that the HDO resonance was at the null point, giving no HDO peak in the transformed spectrum. The technique applies to any situation where the differences between T_1 values are large, and modifications enabling measurement of T_1 and T_2 for the remaining resonances were given. The authors have entitled the technique *W*ater *E*liminated *F*ourier *T*ransform or WEFT.

The spectra of polypeptides are usually complex and are composed of many overlapping resonances. Spin decoupling can be used to simplify spectra to some extent, but suffers the disadvantage of perturbing only a few lines in a complex spectrum. One way to overcome this difficulty is the use of homonuclear INDOR (internuclear double resonance). In this technique the intensity changes of one transition are monitored at one frequency while a second irradiating frequency is swept through the remaining part of the spectrum. Perturbation of a transition having a common energy level with the monitored transition leads to either an observed intensity increase or decrease. Because only transitions related by a common energy level give a signal change, INDOR spectra are extremely simple. Gibbons and co-workers have now reported the application of this technique to both amino-acid[72] and polypeptide[73] spectra. In the first paper[72] INDOR was used to analyse the C_α spectrum of amino-acids, which could be uniquely classified by their C_α and side-chain spectra. The coupling constants obtained for the side-chain protons were used to obtain information on conformations. In the second paper[73] the technique was applied to the spectra of gramicidin S-A, tyrocidin A, and a heptapeptide ring fragment of polymyxin B_1. This study demonstrated the reliability of the technique by deriving peak assignments for gramicidin S-A from 90 MHz spectra in agreement with those obtained by decoupling at 250 MHz. The spectra of the remaining compounds were also simplified by INDOR, and in particular a dramatic simplification of the aromatic-amide proton spectra of tyrocidin A was obtained.

B. Synthetic Polypeptides.—N.m.r. coupling constants can be used to obtain information on molecular conformations and this has been applied to side-chain groups. Ramachandran and co-workers[74] have obtained an equation relating vicinal NH–CH coupling constants to the dihedral angle (θ). Using conformational calculations and n.m.r. spectra of a series of amides they derived the relationship:

---

[71] S. L. Pratt and B. D. Sykes, *J. Chem. Phys.*, 1972, **56**, 3182.
[72] W. A. Gibbons, H. Alms, J. Sogn, and H. R. Wyssbrod, *Proc. Nat. Acad. Sci. U.S.A.*, 1972, **69**, 1261.
[73] W. A. Gibbons, H. Alms, R. S. Bockman, and H. R. Wyssbrod, *Biochemistry*, 1972, **11**, 1721.
[74] G. N. Ramachandran, R. Chandrasekaran, and K. D. Kopple, *Biopolymers*, 1971, **10**, 2113.

$$J = 7.9\cos^2\theta - 1.55\cos\theta + 1.35\sin^2\theta$$

This relationship gave reasonable values for the dihedral angle in a number of cyclic oligopeptide structures where conformations were known from other data. Although there are severe limitations on the use of relationships of this type, the equation should prove of value in determining the conformation of polypeptides in favourable circumstances.

The conformations of a series of poly-L-lysine derivatives with different groups substituted at the side-chain nitrogen of the lysine groups have been investigated using circular dichroism (c.d.) and 220 MHz n.m.r.[75] The chemical shifts of the α-CH resonance, and the line broadenings in the spectra of some derivatives confirmed conclusions concerning helix or coil conformations obtained from c.d. Helix–coil transitions induced by changing the $CDCl_3$: trifluoroacetic acid (TFA) ratio in the solvent have been reported for poly(γ-benzyl D,L-glutamates).[76] Both statistical and alternating polymers underwent a helix–coil transition on addition of TFA to $CDCl_3$, the TFA concentration being less than that required to induce the change in poly(γ-benzyl L-glutamate). This indicated that the D,L-polymers formed α-helical structures in $CDCl_3$ in contradiction to previous conclusions. Hardy and co-workers have published[77] a study of alternating polypeptides prepared from the polymerization of dipeptides composed of one L- and one D-residue. The polymers were composed of γ-t-butyl D- and L-glutamates and γ-benzyl D-glutamate–L-leucine units. Helix–coil transitions were detected by n.m.r. in $CDCl_3$–TFA mixtures, together with optical rotatory dispersion (o.r.d.) and i.r. measurements. For polymers prepared by a method in which some racemization occurred, a distorted α-helix conformation was suggested, but where little or no racemization took place the polymer adopted a new conformation, the structure of which was uncertain.

The conformation of the five-membered ring of the proline residues in poly(L-proline) II has been investigated using computer simulation of the 220 MHz spectrum of the ring protons.[78] A Karplus-type relationship was used to determine the most likely conformations from the ^3J values and these differed from the previously determined conformation of the ring in the solid state.

Conti and De Santis have reported[79] a study by n.m.r. of poly(N-methyl-L-alanine) in chloroform–TFA mixtures and have related the conformation obtained from energy calculations to the observed spectrum. They concluded that the polymer exists in two different helical conformations and that the slow exchange between these conformations explains the number of resonances observed for the C_α-CH_3, N-CH_3, and C_α-H resonances. Mixed solvent systems consisting of DMSO with added water or trimethyl phosphate were

---

[75] N. Anand, N. S. R. K. Marthy, F. Naider, and M. Goodman, *Macromolecules*, 1971, **4**, 564.
[76] F. A. Bovey, J. J. Ryan, G. Spach, and F. Hertz, *Macromolecules*, 1971, **4**, 433.
[77] P. M. Hardy, J. C. Haylock, D. I. Marlborough, H. N. Rydon, H. T. Storey, and R. C. Thompson, *Macromolecules*, 1971, **4**, 435.
[78] D. A. Torchia, *Macromolecules*, 1971, **4**, 440.
[79] F. Comti and P. De Santis, *Biopolymers*, 1971, **10**, 2581.

used in conformational studies of poly(L-tyrosine).[80] From the broadening of certain n.m.r. peaks on adding D_2O or trimethyl phosphate to a solution of polymer dissolved in DMSO, together with i.r., u.v. and o.r.d. measurements, it was concluded that the polymer exists in the random coil conformation in DMSO. The changes caused by altering the solvent composition were attributed to the transition of the polymer conformation to a largely right-handed α-helical form. The side-chain conformation of α-helical poly(β-benzyl L-aspartate) has been studied[81] by comparison of the chemical shifts of the benzyl and methyl derivatives. No chemical shift of the side-chain protons due to the effect of the diamagnetic anisotropy of the aromatic ring was observed, enabling the existence of certain side-chain conformations to be discounted. In a further study on esters of poly(L-aspartic acid) Bradbury and co-workers[82] have demonstrated that chemical shifts can be used to characterize the sense (left-handed or right-handed) of the helix structure. By comparing the spectra of copolymers containing different aspartate esters with those of copolymers of β-benzyl L-aspartate with other amides, the dependence of the chemical shifts of the NH and α-CH protons on helix sense could be determined. A curve analyser was used to resolve overlapping resonances of the β-CH_2 peak, and coupling constants obtained from this were used to obtain details of the side-chain conformation. Molecular motion in solid poly(β-benzyl L-aspartate) has been investigated[83] by the measurement of second moments, using broad-line n.m.r. over the temperature range 123—378 K. At 363 K a conformational change from an α-helix to an ω-helix occurred, and the ω-helix was then stable when the temperature was lowered. The second moments were related to different types of polymer motion such as side-chain rotation and backbone rotation. It was concluded that the ω-helix form incorporated helices of random sense at low temperatures, and that increasing the temperature caused rotation of the benzene ring.

A study of block copolymers of poly(γ-benzyl L-glutamate–β-benzyl L-aspartate) has been carried out[84] using the chemical shift dependence of the aspartate NH and α-CH resonances on the helix sense referred to above. The helix sense of the aspartate blocks was sensitive to the incorporation of glutamate residues, less than 4% of the latter being sufficient to switch the helix of the aspartate blocks from the left-handed to the right-handed form. The results were in agreement with o.r.d. measurements and confirmed the relationship between helix sense and chemical shifts. The n.m.r. of polysarcosine has been examined by Sisdo and co-workers,[85] who have interpreted the CH_2

[80] E. M. Bradbury, C. Crane-Robinson, V. Gioncotti, and R. M. Stephens, *Polymer*, 1972, **13**, 33.
[81] D. N. Silverman, G. T. Taylor, and H. A. Scheraga, *Arch. Biochem. Biophys.*, 1971, **146**, 587.
[82] E. M. Bradbury, B. G. Carpenter, C. Crane-Robinson, and H. Goldman, *Macromolecules*, 1971, **4**, 557.
[83] F. Happey, D. W. Jones, and B. M. Watson, *Biopolymers*, 1971, **10**, 2039.
[84] L. Paolillo, P. Temussi, E. Trivellane, E. M. Bradbury, and C. Crane-Robinson, *Biopolymers*, 1971, **10**, 2555.
[85] M. Sisdo, Y. Imonishe, and T. Higashimura, *Biopolymers*, 1972, **11**, 399.

and CH_3 multiplets in terms of different geometrical isomers formed by hindered internal rotation about amide C—N partial double bonds. The peak assignments were made for the different isomers (four dyad sequences) by comparison with the spectra of N-acetylsarcosine dimethylamide. The CH_2 and CH_3 polysarcosine resonances were shown to coalesce to broad singlets at *ca.* 383 K, and the barrier to rotation (free energy) was estimated as 65 kJ mol^{-1}. The n.m.r. spectra of a series of poly(N-alkylglycine)'s, in which the alkyl group varied from methyl to n-butyl, were also measured. A change in the *cis*:*trans* ratio due to the increasing steric requirements of the alkyl group was demonstrated, in agreement with the above assignments.

Howard and Scheraga[86] have made a novel study of relative methyl group intensities recorded under partial saturation conditions in order to obtain information on changes in relaxation times. The area under a resonance peak is a function of $(1+\gamma^2 B_1^2 T_1 T_2)^{-1}$, and if the rf amplitude (B_1) is made large enough this term becomes significantly different from unity, causing resonances with different values of T_1 and T_2 to have different areas. The polymers studied were of the type $(D,L\text{-lysine})_m$–$(L\text{-alanine})_n$–$(D,L\text{-lysine})_m$, n taking values of 160 and 450. The ε-CH_2 protons of lysine gave a separate resonance but the β-, γ-, and δ-CH_2 resonances of lysine overlapped with the β-CH_3 alanine resonance. In order to obtain information on the relative relaxation time changes for alanine and lysine side-chain protons under different conditions, the areas of the combined β-, γ-, and δ-CH_2 resonances of lysine plus the β-CH_3 resonance of alanine were compared with the area of the ε-CH_2 lysine resonance. Relative areas were measured under conditions of partial saturation at constant B_1 levels for solutions of the two polymers at different concentrations and temperatures. Changes in relative peak areas were related to changes in relaxation times, and the latter were related to the existence of hairpin breaks in the conformation of the alanine block with $n = 450$. This interpretation is supported by previous measurements by other techniques. The authors attribute the changes in $T_1 T_2$ to changes in T_1 from the theory of Bloemburgen, Purcell, and Pound. However, it is not possible to draw such clear conclusions since the motion of polymer side-chain protons is anisotropic and is determined by several correlation times. In this situation T_1 and T_2 may be determined by different relaxation processes and simple relaxation theory does not hold. However, this does not invalidate the conclusion concerning the polymer conformations.

C. Natural Polypeptides.—A number of different macromolecules such as simple proteins, enzymes, and hormones are considered in this section. Generally, high-resolution n.m.r. studies have been made, with chemical shifts or line broadening being used to gain information about the conformation of the macromolecules.

Bradbury and King have published two further papers in a series dealing

---

[86] J. C. Howard and H. A. Scheraga, *Macromolecules*, 1972, **5**, 328.

with the denaturation of proteins. In the first paper,[87] they discuss the process and kinetics of the denaturation of proteins, and report measurements of the 100 MHz spectra of lysozyme and α-lactalbumin at different pH and urea concentrations. Unfolding of the protein was monitored by measuring peak heights. Lysozyme showed a multistage process at pH 2.8 and α-lactalbumin underwent a similar process at pH 6.0. In the second paper,[88] the denaturation of ribonuclease-A in aqueous solution, caused by the addition of small molecules such as urea, was studied by high-resolution n.m.r. The high-frequency spectra were simplified by subtracting the spectrum in D_2O solvent from that in H_2O, leaving the N–H resonance. Variations in the shift and intensity of the resonance from the C2 histidines, located at positions 12 and 119, were interpreted as due to binding of urea and guanidine hydrochloride. Unfolding of the protein in acid solutions produced chemical shifts and splitting of the methionine SCH_3 resonance. These results together with the above N–H histidine data were used to establish the existence of different states in the unfolding process.

The 220 MHz spectra of ribonuclease-A in water were examined by Patel et al.,[89] who concentrated on the region between 10 and 15 p.p.m. from TMS. A resonance between 11.5 and 13 p.p.m. was assigned to the ring-nitrogen proton of histidine. Results for the pH and temperature dependences of the linewidth were analysed in terms of proton exchange between histidine and protonated histidine. The binding of α-N-acetyl-α-D-glucosamine to lysozyme was studied[90] by measuring the chemical shift of the saccharide protons. It was assumed that the saccharide exchanged rapidly between two sites, free in solution and bound to the enzyme. The distribution between these two sites was calculated as a function of pH and saccharide concentration. From the results, some conclusions were drawn about the site of adsorption on the enzyme.

The spectra of lysozyme in the native and thermally denatured states have been measured at 220 MHz.[91] In the denatured form, a single resonance appeared about 10 p.p.m. to high frequency from the standard, DSS, but after thermal renaturation, five resolvable peaks appeared between 10 and 11 p.p.m. from DSS. These resonances were assigned to the indole N–H protons. Results of chemical modification, binding of N-acetylglucosamine inhibitor, and deuterium-exchange experiments were used to identify the resonances with specific tryptophan residues.

Cohen et al.[92] have prepared a staphylococcal nuclease analogue with all seven tyrosine residues deuteriated in the positions *ortho* to the OH group,

---

[87] J. H. Bradbury and N. L. R. King, *Austral. J. Chem.*, 1971, **24**, 1703.
[88] J. H. Bradbury and N. L. R. King, *Austral. J. Chem.*, 1972, **25**, 209.
[89] D. J. Patel, C. K. Woodward, and F. A. Bovey, *Proc. Nat. Acad. Sci. U.S.A.*, 1972, **69**, 599.
[90] J. F. Studebaker, B. D. Sykes, and R. Wien, *J. Amer. Chem. Soc.*, 1971, **93**, 4579.
[91] J. D. Glickson, W. D. Phillips, and J. A. Rupley, *J. Amer. Chem., Soc.*, 1971, **93**, 4031.
[92] J. S. Cohen, M. Fiel, and I. Chaiken, *Biochim. Biophys. Acta*, 1971, **236**, 468.

thus producing a single resonance from each residue in the aromatic region. Selective nitration at tyrosine-115 changed the spectra of the deuteriated analogue, and comparison between the nitrated and non-nitrated forms allowed a tentative assignment of the spectra.

The association of ribonuclease S-peptide (and its analogues) with ribonuclease S-protein was studied[93] by comparing the 250 MHz spectra of the S-peptide with and without the S-protein. Association of the two molecules produced a low-frequency shift of the histidine-12 ring-proton resonances and a broadening and loss of intensity of the signals from the arginine 10δ-protons and the methionine SCH_3 protons. It was proposed that the free S-peptide was in a random coil conformation and that a salt-bridge link between glutamate-2 and arginine-10 was formed on binding.

A method for obtaining information about the conformation of small polypeptides by using a mixed solvent system has been described by Pitner and Urry.[94] 2,2,2-Trifluoroethanol was added to methanol or DMSO solutions of the polypeptide, and the peptide protons exposed to the solvent experienced a low-frequency shift. This method might be generally applicable to small, relatively rigid, cyclic polypeptides. An application of the more conventional deuterium-exchange experiment to obtain information about the conformation of polymyxin B, has been reported by Chapman and Golden.[95] It was shown that for some polymyxin B derivatives, such as the hydrochloride, at least five amide protons had not exchanged after twenty minutes, whereas for other derivatives, such as the sulphate, complete exchange of the protons had occurred before the first scan could be taken. The non-exchangeability was explained by a model in which the peptide chain was folded back on itself, making some of the amide protons inaccessible to the solvent.

Two publications,[96,97] relating to the structure determination of gramidicidin A^1 have appeared. A commercial sample of gramicidin A^1 containing three different gramicidin types, A, B, and C was studied. Deuterium-exchange experiments showed that most of the peptide hydrogens were internally hydrogen bonded, and resonances from chemically equivalent protons were shifted by their local magnetic environment. Together with the NH–α-CH coupling constants, which were larger than expected for a random coil, the above results were taken to indicate an ordered structure for gramicidin A^1. This structure was thought to be a π_{LD}-helix rather than an α-helix. An attempt to make a more informative and correct interpretation of the conformational changes that the complex histone molecules can undergo has been made by Bradbury and Rattle.[98] From a knowledge of the amino-acid composition of the histone, a computer-simulated high-resolution spectrum

---

[93] F. M. Finn, J. Dadok, and A. A. Bothner-By, *Biochemistry*, 1972, **11**, 455.
[94] T. P. Pitner and D. W. Urry, *J. Amer. Chem. Soc.*, 1972, **94**, 1399.
[95] T. M. Chapman and M. R. Golden, *Biochem. Biophys. Res. Comm.*, 1972, **46**, 2040.
[96] J. D. Glickson, D. F. Mayers, J. M. Settine, and D. W. Urry, *Biochemistry*, 1972, **11**, 477.
[97] D. W. Urry, J. D. Glickson, D. F. Mayers, and J. Haider, *Biochemistry*, 1972, **11**, 487.
[98] E. M. Bradbury and H. W. E. Rattle, *European J. Biochem.*, 1972, **27**, 270.

was produced and compared with the experimentally determined spectrum. The conformational change induced by the addition of salt was measured experimentally and the resonances of various groups on the computer-simulated spectra were altered to give the best fit with the experimental data. Though limited by some assumptions, such as that all the protons in a segment are equally broadened, this method gives a more realistic and objective analysis of the conformational change than visual methods.

The sites of the main salt-induced interchain and intrachain interactions of histone F2B were investigated by n.m.r. and o.r.d. techniques.[99] The optical studies showed that the salt-induced secondary conformation was in the carboxyl half of the molecule, and that the amino half remained in the random coil form at all ionic strengths. Chemical shift and linewidth results showed that restricted mobility in the carboxyl half was confined to residues between 66 and 102, and that in the amino half of the molecule some interchain interactions occurred between segments 30 and 50. Another attempt to estimate interchain interactions has been made[100] on adrenocorticotropin (ACTH), a 39 amino-acid linear hormone. Fragments of the hormone, 1—10, 11—24, and 1—24 were obtained, and their proton spectra in acid solution, particularly in the NH and CH regions, measured. From a comparison of the spectra it was concluded that there was no interaction between the 1—10 and 11—24 sequences in ACTH 1—24.

Barry et al.[101] have used a lanthanide probe technique to determine the difference in conformation of adenosine-5'-monophosphate in water and in dimethylsulphoxide. Addition of the lanthanide ion can cause a chemical shift and broadening of the spectral lines of the nucleotide. Computer models of the molecule were used to account for the observed perturbations, and a comparison of the conformation of the molecule in the two solvents was made.

High-resolution n.m.r. has been used,[102] with other techniques, to determine the structure of polyprenols isolated from *Aspergillus niger*. It is proposed that the polyprenol was an unsaturated alcohol with a 79-carbon-atom backbone, and its n.m.r. spectrum was used to identify some of the groups present.

D. Polynucleotides.—Polynucleotides are polymers of three types of molecules, namely, a sugar, a phosphomonoester residue, and a nitrogen-containing heterocyclic base. The denaturation of yeast phenylalanyl t-RNA was investigated[103] using high-resolution n.m.r. at 220 MHz. Addition of [^{2}H$_6$]-DMSO to aqueous solutions of the RNA caused it to unfold and become more

---

[99] E. M. Bradbury, P. D. Cary, C. Crane-Robinson, P. L. Riches, and E. W. Jones, *European J. Biochem.*, 1972, **26**, 482.
[100] D. J. Patel, *Macromolecules*, 1971, **4**, 251.
[101] C. D. Barry, J. A. Glasel, A. C. T. North, R. J. P. Williams, and A. V. Xavier, *Biochem. Biophys. Res. Comm.*, 1972, **47**, 166.
[102] R. M. Barr and O. F. W. Hemming, *Biochem. J.*, 1972, **126**, 1193.
[103] J. E. Crawford, S. I. Chan, and M. P. Schweizer, *Biochem. Biophys. Res. Comm.* 1971, **44**, 1.

flexible. At 83% v/v [^{2}H$_6$]DMSO, well-resolved spectra were obtained, but as the proportion of [^{2}H$_6$]DMSO decreased the spectra became less well resolved. Denaturation, as indicated by linewidth changes, was a continuous process with no sharp transition point, although the authors caution that narrowing of a line on denaturation does not necessarily mean more flexibility but may mean that several groups contributing to a line are moving into a more homogeneous environment. The high-resolution spectra at 220 MHz of four purified t-RNA molecules were measured.[104] The temperature and concentration dependences of the exchangeable proton resonances were investigated, and it was concluded that the resonances were from protons involved in intramolecular hydrogen bonds.

A proton high-resolution study of polydeoxyriboadenylic acid at pH 8.0 as a function of temperature has been made.[105] The chemical shift, peak area, and coupling constants were measured for H-2 and H-8 of the base unit and H-1', H-2', and H-2" of the sugar. The results were interpreted in terms of different conformations of the sugar and base units. Cross and Crothers[106] reported the proton high-resolution study of single-stranded and double-helical deoxyribo-oligonucleotides. The chemical shift data from the different oligomers were used to obtain information on their conformation.

Pulsed Fourier Transform spectra at 100 MHz were used, together with circular dichroism, to study the interaction of N-2-acetylaminofluorene to oligonucleotides.[107] Attachment of N-2-acetylaminofluorene caused large low-frequency shifts for the proton resonances of fluorene and bases adjacent to a modified guanosine residue. These shifts were interpreted in terms of changes in the conformation of the oligomers.

E. Proteins containing Paramagnetic Ions.—In this section, high-resolution studies of haemoglobin-type molecules are described. Frequently these have made use of the dramatic effect that a paramagnetic ion can have on the chemical shifts or relaxation times of protons in its close environment. However, the molecules are generally very complex and care is required in extrapolating the behaviour of small groups to that of whole side-chains of the protein.

Lindstrom et al.[108] measured the spectra of different samples of carbon-monoxyhaemoglobin at 250 MHz, concentrating on the methyl resonance. They made use of the shifts caused by close proximity to aromatic rings, and were able to identify the three lowest-frequency resonances as being from methyl groups. From the spectra of the isolated α- and β-chains, it was con-

---

[104] D. R. Kearns, D. Patel, R. G. Shulman, and T. Yamane, *J. Mol. Biol.*, 1971, **61**, 271.
[105] J. L. Alderfer and S. L. Smith, *J. Amer. Chem. Soc.*, 1971, **93**, 7305.
[106] A. D. Cross and D. M. Crothers, *Biochemistry*, 1971, **10**, 4015.
[107] J. H. Nelson, D. Grunberger, C. Cantor, and I. B. Weinstein, *J. Mol. Biol.*, 1971, **62**, 331.
[108] T. R. Lindstrom, I. B. E. Norén, S. Charache, H. Lehmann, and C. Ho, *Biochemistry*, 1972, **11**, 1677.

cluded that the tertiary structure of the haem pocket is not significantly affected by the tetrameric association of the liganded chains.

The proton spectra of deoxyhaemoglobin and modified deoxyhaemoglobins showed[109] three prominent resonances at about -18, -12, and -7 p.p.m. from HDO. The two highest-frequency resonances were assigned to the haem methyl groups, by comparing the spectra of several different modified deoxyhaemoglobins. It was suggested that the two haem groups were non-equivalent and that the highest-frequency line was due to the β-haem methyl group. In a related publication, Lindstrom et al.[110] showed that when deoxyhaemoglobin is titrated with n-butyl isocyanide in the presence of inositol hexaphosphate, the proton resonance at -18 p.p.m. from HDO decreases in intensity more rapidly than that at -12 p.p.m. These results were interpreted as confirming the previously established hypotheses that (a) the higher-frequency line was due to the β-haem methyl resonance and the other line was due to the α-haem methyl, and that (b) n-butyl isocyanide, in the presence of inositol hexaphosphate, is preferentially bound to the β-chains of the haemoglobin. The 220 MHz spectra have been reported[111] of the paramagnetic cyanoferric forms (Fe^{3+}, $S = \frac{1}{2}$) of human haemoglobin A, isolated α- and β-chains, haemoglobin Zurich, isolated β-chains from haemoglobin Zurich, and a tetrameric haemoglobin A carrying haem group on the α-chains only. A comparison of the spectra from these compounds showed that extensive structural changes on one chain had little influence on the spectrum of the haem groups of the neighbouring chains. However, the structural abnormality of haemoglobin Zurich was reflected in the spectra.

The spectra of native and modified cytochrome c were studied by Wüthrich et al.[112] The modifications included formylation of trytophan, alkylation of the methionine, and polymerization by treatment with ethanol. The methionine residue, bound to the haem iron, gave peaks 1.9—3.7 p.p.m. to low frequency of the standard DSS. The n.m.r. data were correlated with the biological activity of the modified cytochrome c. A study of the electron transfer between the oxidized and reduced form of cytochrome c molecules has been made by pulsed Fourier Transform n.m.r.[113] The methionine methyl group protons have a well resolved resonance in both the oxidized and reduced forms, although the unpaired electron of the former causes these protons to relax much more rapidly than in the reduced state. Electron transfer between the two states was detected by selectively saturating the methyl resonance in one oxidation state and observing the transfer of this saturation to the resonance of the same protons in the other oxidation state. The electron transfer could be observed when the electron exchange rate was comparable

[109] D. G. Davis, T. R. Lindstrom, N. H. Mock, J. J. Baldassare, S. Charache, R. T. Jones, and C. Ho, *J. Mol. Biol.*, 1971, **60**, 101.
[110] T. R. Lindstrom, J. S. Olsen, and N. H. Mock, *Biochem. Biophys. Res. Comm.*, 1971, **45**, 22.
[111] K. H. Winterhalter and K. Wüthrich, *J. Mol. Biol.*, 1972, **63**, 477.
[112] K. Wüthrich, I. Aviram, and A. S. Chejter, *Biochim. Biophys. Acta*, 1971, **253**, 98.
[113] R. K. Gupta, S. H. Koening, and A. G. Redfield, *J. Magn. Resonance*, 1972, **7**, 66.

with the value of T_1^{-1} for the proton. Information on the electron transfer as a function of pH and ionic strength was obtained.

The complex formed between haemin, pyridine, and water was investigated by Degani and Fiat.[114] The temperature dependence of the relaxation times and chemical shifts of the haemin protons showed transitions between the high- and low-spin states of the ferric ion. Ligand exchange with the haemin also occurred, and a theoretical treatment was given, based on McConnell's[115] equations. From the temperature dependence of the water and pyridine proton T_2 values, the kinetic and thermodynamic parameters for ligand exchange were calculated. Parameters describing the transition between high- and low-spin states were calculated from the temperature dependence of the haemin T_2 values. The pseudo-contact contribution to the hyperfine coupling constant of the haem protons was calculated, allowing the isotropic contribution to be determined.

F. Membranes.—Biological membranes are made up of a complex mixture of lipids and proteins. The lipids are in a liquid crystalline or gel state with molecular motion intermediate between those of a solid and a liquid. Sheetz and Chan have studied[116] whole erythrocyte membranes obtained from human blood. The spectra were recorded in D_2O over the temperature range 281—348 K at 220 MHz. Little or no structure was observed in the n.m.r. spectra at 281 K, but at 324 K and above a number of resonances were observed, and these were assigned to membrane protein peaks plus a lecithin choline resonance. A calculated spectrum was in good agreement with the observed spectrum. The effects of protein solubilization and added bivalent cations on the spectrum were also investigated.

Davies and Inesi[117] have reported a similar study on sarcoplasmic reticulum membrane from rabbit leg muscle. The high-resolution n.m.r. spectra (D_2O) were mainly from the membrane lipids, and a reversible increase in intensity and sharpening of the choline resonance was observed on raising the temperature from 293 K. This indicated the presence of a conformation transition, and from an estimation of the fraction of choline groups involved at each temperature a heat of transition was determined. Changes in the intensity of the choline resonance were also induced by altering the protein of the membrane, indicating that lipid–protein interactions were involved in the thermal transition.

G. Polysaccharides.—The degree of substitution of hydroxypropylcellulose has been determined by Ho and co-workers[118] using n.m.r. The polymer consists of cellulose substituted with hydroxypropyl groups, and from the intensity of the observed peaks the average number of substituent groups per

[114] H. Asman Degani and D. Fiat, *J. Amer. Chem. Soc.*, 1971, **93**, 4281.
[115] H. M. McConnell, *J. Chem. Phys.*, 1958, **28**, 430.
[116] M. P. Sheetz and S. I. Chan, *Biochemistry*, 1972, **11**, 548.
[117] D. G. Davies and G. Inesi, *Biochim. Biophys. Acta*, 1971, **241**, 1.
[118] F. F. L. Ho, R. R. Kohler, and G. A. Ward, *Analyt. Chem.*, 1972, **44**, 178.

sugar ring was determined. By treating the polymer OH groups with trichloroacetyl isocyanate it was possible to distinguish between terminal and side-chain methyl groups and thus to determine the number-average molecular weight.

Ghosh and Gilbert have reported[119] a study of intramolecular hydrogen bonding in DMSO solutions of polyglucoses. The existence of the intramolecular hydrogen bonds was inferred from the high-frequency chemical shifts of certain hydroxyl resonances.

H. ^{13}C N.M.R.—More information can be obtained from ^{13}C n.m.r. spectra than from proton spectra because of the greater chemical shift of carbon and the absence of exchange effects due to labile protons, and this can outweigh the disadvantage of the low sensitivity of ^{13}C n.m.r.

The helix–coil transitions of proteins and polypeptides have been investigated by ^{13}C n.m.r., as well as by several other techniques. ^{13}C resonances were measured[120] for the homopolypeptide poly(N-δ-benzoxycarbonyl-L-ornithine) in mixtures of deuterochloroform and trifluoroacetic acid (TFA). In pure CDCl$_3$ the backbone polypeptide carbon resonances were broad, owing to restricted backbone mobility of the helix form. At a volume fraction of TFA of about 0.1 the lines narrowed, corresponding to a transition from the helix to the coil form. The linewidths of the side-chain carbon atoms were less influenced by the solvent change than those of backbone atoms. The maximum chemical shift of about 4 p.p.m. during the transition was for the α-carbon. A study of the helix–coil transition of the polypeptide poly(α-benzyl L-glutamate) has also been made[121] using ^{13}C resonance. The transition was produced by adding TFA to a CHCl$_3$ solution. The results showed that the α and β carbon atoms were most sensitive to the transition, undergoing a shift of about 3 p.p.m. compared with about 0.5 p.p.m. for the α-CH in the proton spectra. As expected, the ^{13}C linewidths were greater in the helix form. Mantsch and Smith[122] measured the ^{13}C spectra of polyuridylic acid, uridine, and uridine phosphates. Appreciable coupling was observed between the ^{13}C and ^{31}P, this coupling being sensitive to variations in the dihedral angle and hence the conformation of the molecule.

Strousse et al.[123] prepared chlorophylls a and b enriched in ^{13}C (by feeding green algae on ^{13}CO$_2$) and measured their spectra in solutions of deuterochloroform–methanol, taking the carbon resonance of the chloroform as the reference. Assignments of many of the peaks were made using (^{13}C,^{13}C) coupling constants and comparison with small molecules, although a complete assignment of these complex spectra has still to be made.

[119] K. K. Ghosh and R. D. Gilbert, *Textile Res. J.*, 1971, **41**, 326.
[120] G. Boccalon, A. S. Verdini, and G. Giacometti, *J. Amer. Chem. Soc.*, 1972, **94**, 3639.
[121] L. Paolillo, T. Tancredi, P. A. Temussi, E. Trivellone, E. M. Bradbury, and C. Crane-Robinson, *J. C. S. Chem. Comm.*, 1972, 335.
[122] H. H. Mantsch and I. C. P. Smith, *Biochem. Biophys. Res. Comm.*, 1972, **46**, 808.
[123] C. E. Strousse, V. H. Kollman, and N. A. Matwiyoff, *Biochem. Biophys. Res. Comm.*, 1972, **46**, 328.

A short note[124] on the study of lecithin vesicles and erythrocyte membranes demonstrates some of the advantages of ^{13}C resonance over ^1H. The high-resolution ^{13}C spectra of unsonicated dipalmitoyl-lecithin and human erythrocyte membranes showed some reasonably narrow lines, whereas the equivalent proton spectra show little or no structure. In a second report the ^{13}C spectra of membranes from *Acholeplasma laidlawii* grown on a medium containing [1-^{13}C]-enriched palmitic acid were studied.[125] Spectra of the membrane and a sonicated lipid extract were obtained. Both the intact membrane and the lipid extract gave carbonyl peaks above a certain temperature, the membrane peak being broader than that of the lipid extract. The changes in the spectra with temperature, together with e.s.r. spin-probe experiments, gave information on the thermal transitions of the membrane.

In a different type of application, the pH dependences of the ^{13}C spectra of oxidized and reduced glutathione have been measured by Jung *et al.*[126] Changes in pH caused ionization of different groups along the chain, and the resonance positions of the carbon nuclei closest to the ionized group were shifted. Since each carbon nucleus can be observed independently, the ionization curve for each ionizable group can be measured. Hence, by this method it is possible to determine experimentally the ionization curves for groups with a similar pK value having overlapping titration curves.

I. ^{19}F **and** ^{31}P **N.M.R.**—The large potential for applications of ^{19}F and ^{31}P n.m.r. to studies of biological macromolecules has yet to be realized, and very few papers have been published using these nuclei attached to the macromolecule. Trifluoromethyl labelling has been used to study the properties of haemoglobin.[127] A trifluoroacetyl group was attached to the side-chain of the cysteine-β93 link which had previously been used to bind e.s.r. spin labels. The labelled haemoglobin had properties similar to unlabelled material, and CF_3 chemical shifts were measured for a range of conditions. These included variation of pH and the binding of CN, CO, and O_2 as ligands to the haem iron. Chemical shift changes of the order of 60 Hz were observed and were interpreted in terms of small changes in the tertiary structure of haemoglobin.

Despite the common occurrence of phosphorus compounds in nature and the sensitivity of ^{31}P n.m.r. few investigations using ^{31}P n.m.r. have been reported. Henderson and co-workers[128] have applied ^{31}P n.m.r. to the problem of analysing the phosphonate and orthophosphate components of a sea anemone. The analysis is usually carried out by chemical techniques involving prolonged acid hydrolysis. The large ^{31}P chemical shifts between phosphonate and orthophosphate enable the analysis of each to be determined from the

---

[124] J. C. Metcalfe, N. J. M. Birdsall, J. Feeney, A. G. Lee, Y. K. Levine, and P. Partington, *Nature*, 1971, **233**, 199.
[125] J. C. Metcalfe, N. J. M. Birdsall, and A. G. Lee, *F.E.B.S. Letters*, 1972, **21**, 335.
[126] G. Jung, E. Breitmaier, and W. Voelter, *European J. Biochem.*, 1972, **24**, 438.
[127] W. H. Huestis and M. A. Raftery, *Biochemistry*, 1972, **11**, 1648.
[128] T. O. Henderson, T. Glonek, R. L. Hilderbrand, and T. C. Myers, *Arch. Biochem. Biophys.*, 1972, **149**, 484.

n.m.r. spectra, and it was demonstrated that the chemical technique resulted in incorrect analysis. Multiplet structure for each type of phosphorus resonance indicated the potential of n.m.r. for determining the substituents around the phosphorus atoms.

5 Small Molecules

A. Counterions and Ligands.—In this section we discuss the effect of macromolecules on the behaviour of associated small molecules, such as solvents and counterions. The effects are observed typically as line broadenings, splittings, or changes in relaxation rates. Biological tissue generally contains such polymers as proteins and polysaccharides together with lipids in a complex structure. Previous broad-line studies on the environment of sodium ions in biological tissue [see refs. in ref. 129] were able to detect resonance from only about 30—40% of the sodium ions present. These results were interpreted as indicating that the ions occupied two sites; one containing 30—40% of the ions, corresponding to bulk solution, and the other containing 60—70% of the ions, bound in such a way that the sodium relaxation rate was too great to be observable in the broad-line spectrum. There was necessarily slow exchange between these two sites. Several important papers relating to this phenomenon have been published.

Shporer and Civan[129] measured the broad-line ^{23}Na resonance in a sodium linoleate–water lamellar liquid crystalline phase and found that the line was split into a triplet. The central line had about 34—39% of the total intensity compared with a reference sodium hydroxide solution, and the splitting was due to a first-order quadrupolar interaction. This splitting was caused by fast exchange between bound sodium ions with anisotropic motion and freely rotating bulk ions. Since the splitting depends on the orientation of the electric field gradient, *i.e.* on the environment of the ion, the outer lines may sometimes be too broad to be observed, and the authors point out that in previous broad-line work on sodium in biological systems only the central line was observed. These results appear effectively to disprove the theory that about 60—70% of the sodium ions in biological tissue exist in a bound state with slow exchange. Lindblom[130] reported results similar to those of Shporer and Civan for the liquid crystalline phase of lecithin–sodium cholate–water. Edzes *et al.*[131] studied the resonance of sodium and lithium ions in hydrated DNA, which was oriented in the n.m.r. tube. The broad-line spectrum of both ions was a triplet in which the separation of the outer lines showed a dependence on the angle between the sample and the magnetic field. The work clearly supports Shporer and Civan's theory and provides further experimental evidence against the possibility of there being slow exchange between two sites, *i.e.* bound and free sodium ions, in biological

---

[129] M. S. Shporer and M. M. Civan, *Biophys. J.*, 1972, **12**, 114.
[130] G. Lindblom, *Acta Chem. Scand.*, 1971, **25**, 2767.
[131] H. T. Edzes, A. Rupprecht, and H. J. C. Berendsen, *Biochem. Biophys. Res. Comm.*, 1972, **46**, 790.

tissue. In a further study of sodium resonance, Martinez and Silvidi[132] measured the resonance of ^{23}Na in the saliva of people with cystic fibrosis of the pancreas, an hereditary disease thought to be caused by the binding of sodium ions to macromolecules. The signal showed the same intensity and linewidth as standard samples, indicating that there was no binding, and that the original theory was incorrect.

Civan and Shporer[133] have reported the ^{17}O resonance of frog striated muscle, prepared by incubation in $H_2^{17}O$. A single line was observed, having an intensity of about 75% of that of the same volume of pure $H_2^{17}O$. This loss in intensity was interpreted as due to some of the water molecules being bound or undergoing some form of anisotropic motion. They also looked at the system of sodium linoleate–$H_2^{17}O$, and noted that the ^{17}O resonance showed a quadrupolar splitting analogous to the sodium case.[129]

The binding of chloride ions to the active site, *i.e.* zinc, of horse liver alcohol dehydrogenase was studied[134] by relating the broadening of the ^{35}Cl signal to the extent of binding. The chloride ions undergo fast exchange between the bound and free states. Addition of a coenzyme, which bound preferentially to the zinc sites, removed the broadening of the chloride resonance.

One of the few investigations of the univalent ion ^{205}Tl has been reported,[135] using it as a model counter-ion with pyruvate kinase. Linewidth and chemical shifts measurements of the ^{205}Tl resonance indicated binding of the ion to the enzyme. Addition of magnesium ions produced only small changes in these parameters, though added manganous ions caused large changes, which were used to calculate the distance between the bound manganese and thallous ions.

The interaction between bovine serum albumin and the denaturation agents urea and guanidine hydrochloride has been investigated[136] by observing the ^{14}N resonance of the small molecules. The linewidth was relatively insensitive to changes in concentration of the small molecule but showed a significant increase with increasing protein concentration. Poly-L-lysine produced no change in linewidth, and it was concluded that there was a definite interaction between the protein and the denaturation agents.

Millett and Raftery have used the chemical shift of ^{19}F resonance to characterize interactions in biopolymer systems. They measured[137] the chemical shift of both protons and fluorine in the formation of a complex between β-methylmonofluoro-*N*-acetylglucosamine with lysozyme. The chemical shift of fluorine has been separated into four contributions: (*a*) solvent anisotropy; (*b*) van der Waals forces; (*c*) electric fields; and (*d*) complex formation. Since fluorine is more susceptible to electric field and van der

[132] D. Martinez, and A. A. Silvidi, *Arch. Biochem. Biophys.*, 1972, **148**, 224.
[133] M. M. Civan and M. S. Shporer, *Biophys. J.*, 1972, **12**, 1404.
[134] R. L. Warde and J. A. Hoppe, *Biochem. Biophys. Res. Comm.*, 1971, **45**, 1444.
[135] J. Reuben and F. J. Kayne, *J. Biol. Chem.*, 1971, **246**, 6227.
[136] B. M. Fung and S. G. Sarney, *Biochim. Biophys. Acta*, 1971, **237**, 135.
[137] F. Millett and M. A. Raftery, *Biochem. Biophys. Res. Comm.*, 1972, **47**, 625.

Waals forces, a comparison of the proton and fluorine shifts of the terminal CH_2F group allowed an estimation of (b), (c), and (d). In another study,[138] the ^{19}F chemical shift of the terminal CF_3 group of trifluoroacetylglucosamine oligomers was used to study the binding of that molecule to lysozyme. Fast exchange between the two sites, bound and free, was assumed, and as each site had a different chemical shift, the distribution between the sites could be calculated. Dwek et al.[139] have also studied the binding of a fluorinated sugar to lysozyme. Addition of N-fluoroacetyl-D-glucosamine (1.5×10^{-3} mol l^{-1}) to lysozyme caused a 50% inhibition of the activity of the enzyme. The high-resolution fluorine resonances shifted to high frequency, and it was concluded that since the α- and β-anomers had different shifts, they must have been in different environments. It was suggested that fluorine-substituted sugars may be useful probes for identifying specific binding sites.

Taylor et al.[140] have investigated the binding of carboxylate anions to the enzyme cobalt(II) human carbonic anhydrase. In addition to the normal acetate ion, the mono-, di-, and trifluoro-derivatives were used. The ^1H and ^{19}F high-resolution spectra were measured for the ligand–enzyme system, and the linewidth and chemical shift variations observed were interpreted in terms of the kinetics of the interaction. Frequency and temperature dependence of the ligand T_2 showed four situations where it was controlled by the chemical exchange rate. For mono- and difluoro-acetate ligands, the paramagnetic contribution (from the cobalt) to the proton T_2 was dependent on the chemical shift differences between the bound and free states. Where the ligand relaxation was exchange-rate dependent, the kinetic constants for the overall association and dissociation processes were estimated. The site of 5'-AMP in glycogen phosphorylase b was studied by high-resolution n.m.r.[141] From the linewidth of the resonances of 5'-AMP, its correlation time of 160×10^{-9} s was calculated.

Proton relaxation enhancement has been used in three investigations to examine the binding of ions to biopolymers. Dwek et al.[142] provided a simple and clear summary of the well-established theory of proton relaxation enhancement and applied it to the results for the binding of gadolinium to lysozyme. Proton T_1 values measured at 20 and 35 MHz gave a correlation time for the interaction between the paramagnetic ion and a water proton of 4.5×10^{-11} s. After considering the possible relaxation mechanisms, it was concluded that rotation of the hydrated gadolinium ion was the main relaxation process. Addition of a complexing agent, ethylene glycol-bis(aminoethyl)tetra-acetate, to aqueous gadolinium solution resulted in a reduction in the proton T_1^{-1}. This reduction was due to the removal of the water of hydration of the ion. Proton T_1 measurements of europium in water also

[138] F. Millett and M. A. Raftery, Biochemistry, 1972, 11, 1639.
[139] R. A. Dwek, P. W. Kent, and A. V. Xavier, European J. Biochem., 1971, 23, 343.
[140] P. W. Taylor, J. Feeney, and A. S. V. Bergen, Biochemistry, 1971, 10, 3866.
[141] A. Danchin and H. Buc. F.E.B.S. Letters, 1972, 22, 289.
[142] R. A. Dwek, R. E. Richards, K. G. Moralle, E. Nieboer, R. J. P. Williams, and A. V. Xavier, European J. Biochem., 1971, 21, 204.

indicated that the dominant relaxation process for the ion–proton interaction was the rotation of the hydrated ion. This work is particularly significant in the biological area since the lanthanide series of metal ions exhibits some physicochemical behaviour similar to that of calcium ions.

The interaction between manganese ions and alkaline phosphatase has also been studied[143] by proton relaxation enhancement. For the range of Mn^{2+} (ion):protein (mol) ratios between 0 and 4, the relaxation rate was lower than that for equivalent concentrations of manganese ions in pure water, indicating that the first four manganese ions per mole of protein are bound with little or no contact with the bulk water. At higher ratios enhancement was observed. Addition of cyanide or orthophosphate ions, which inhibit the activity of the enzyme, had no measurable effect on the proton relaxation rate, supporting the model of bulk water being outside the first co-ordination sphere. E.s.r. was used to measure the concentration of the free manganese ions. However, care should be taken in correlating the e.s.r. and n.m.r. data. At 344 K the e.s.r. results indicated that there was sufficient free manganese ion to produce a proton T_2^{-1} greater by a factor of $ca.$ 2 than that measured by n.m.r. Also, at the higher temperatures, the increasing concentration of free Mn^{2+} as indicated by e.s.r. should have produced an increasing ratio of $T_1:T_2$ for protons, whereas this ratio was observed to decrease. Incorrect notation on Figures 2 and 4 of this paper is initially confusing. The binding of manganese to the enzyme isocitrate dehydrogenase has also been studied[144] by proton relaxation enhancement. The data were used in conjunction with e.s.r. and ultrafiltration to suggest a model of a ternary metal bridge for the complex of enzyme, manganese ion, and substrate.

B. Hydration of Macromolecules.—This section deals with the influence of macromolecules on the n.m.r. behaviour of water or aqueous solutions. The proton resonance may be studied by high-resolution, broad-line, or spin-echo techniques. Pesek and Pecsok[145] recorded the proton high-resolution spectra at 243 K of aqueous solutions of resins containing various salts. In considering the different types of gels, they noted that in pure water those gels which had the greatest adsorption power for ions produced the narrowest water line. When different ions were added to one gel type, the ions which adsorbed most strongly produced the narrowest water line. These narrow lines were due to water closely associated with charged groups and not immobilized at 243 K. From these results, it was suggested that the mechanism by which gels separate ions is selective adsorption of the ions rather than by exclusion of the ion from the pores of the gel.

The state of water adsorbed on to wool modified by (i) hydrolysis, (ii) deamination, (iii) ninhydrin addition, and (iv) Dowfax addition, was examined[146] by the spin-echo technique. The temperature dependences of the water proton

[143] G. L. Cottam and B. C. Thompson, *J. Magn. Resonance*, 1972, **6**, 352.
[144] J. J. Villafranca and R. F. Colman, *J. Biol. Chem.*, 1972, **247**, 209.
[145] J. J. Pesek and R. L. Pecsok, *Analyt. Chem.*, 1972, **44**, 620.
[146] L. J. Lynch and I. C. Watt, *Kolloid-Z.*, 1971, **248**, 922.

relaxation times T_1 and T_2 were measured at various degrees of hydration, and the thermodynamic parameters such as free energy and entropy, obtained using Eyring's thermal activation law, were compared with similar data from unmodified wool. The results were discussed in terms of the different binding sites available on the modified and unmodified wool.

In addition to their previously mentioned papers[46-48] on the broad-line spectra of Nylon-66, Olf and Peterlin[147] have investigated the influence of small quantities of water on the chain mobilities of drawn fibre of Nylon-66. They measured the broad-line spectra of water and polymer as a function of water content and the angle between the fibre and the magnetic field. The dry fibre gave a nearly structureless broad line, but above 1.4% (w/w) water a narrow line, due to mobile water molecules, appeared. The width of this narrow water line decreased with increasing water content, and was a function of the alignment of the sample in the magnetic field, indicating that the water molecules were not reorienting isotropically. At water contents close to saturation a line appeared which was intermediate in width between the broad polymer line and the mobile water line. This new line was produced by a mobile portion of the polymer, formed through the reduction in temperature of the α_a ('glass transition') process by the presence of water.

Water adsorbed on to methyl methacrylate grafted cellulose was studied by Ogiwara and Kubota.[148] Increasing amounts of water adsorbed on to the polymer resulted in a decrease in the water proton linewidth, indicating fast exchange between the free, mobile water and the faster relaxing water closely associated with the polymer. Increasing amounts of methacrylate grafting caused the linewidth first to increase and then to decrease, thus giving a maximum in the linewidth at about 10% grafting. The maximum in the linewidth was tentatively explained by the initial graft polymer occupying different sites from the later grafted polymer, though this explanation is not entirely satisfactory. Other possibilities, such as the change in structure caused by grafting, should be considered.

An estimation of the amount of water in the hydration layer of the ribosomes of *Escherichia coli* has been made.[149] Aqueous solutions of the ribosomes were frozen and the water of hydration was calculated from the area under the narrow line of the spectrum. The original 70s samples (this number is related to the particle size and is obtained by ultracentrifugation) were split by dialysis into two fractions, one of 50s and the other of 30s. As the units became smaller, the hydration per gram of macromolecule increased and this was attributed to the fact that splitting the large particle involved the breaking of ionic bonds. When exposed, these ionic sites interacted with water in such a way as to prevent immobilization of the close water molecules just below freezing point. The experimental errors in the data and the complexity of the system prevent a clear, unequivocal interpretation.

[147] H. G. Olf and A. Peterlin, *J. Polymer Sci., Part A-2, Polymer Phys.*, 1971, **9**, 2033.
[148] Y. Ogiwara and H. Kubota, *J. Appl. Polymer Sci.*, 1971, **15**, 3137.
[149] J. P. White, I. D. Kuntz, and C. R. Cantor, *J. Mol. Biol.*, 1972, **64**, 511.

Aqueous solutions of the gelling agent agarose[150] were studied above and below the gel point. Both T_1 and T_2 of the water protons were measured by the spin-echo method. T_2^{-1}, dominated by exchange between bulk water and the macromolecule, showed a sharp increase as gelation took place, whereas T_1^{-1} changed only slightly, indicating little change in the molecular mobility of the water. Hysteresis loops for T_2 with varying temperature were correlated with ageing effects of the gel. The synergistic increase in gel strength resulting from the addition of locust bean gum to agarose solutions was also investigated. A short note[151] describing some spin-echo results on hydrated gelatin, starch, and cellulose at water contents below 15% has been published. An estimation of the number of mobile water protons was made from the height of the observed signal. However, T_2 was not measured by the Gill–Meiboom sequence method so that the values may be in error owing to the contribution from diffusion.

An estimation of the porosity of macroreticular resins has been made.[152] These resins consist of two phases, the resin gel phase and a phase with large pores or voids, the volume of these large pores determining the resin porosity. The high-resolution spectrum showed a single water line, indicating rapid exchange of water between the two phases. Since the pores are large, the chemical shift in the pores was taken to be the same as for bulk water and the average observed chemical shift was determined by the shift in the gel. Comparisons with a gel of the same cross-linking density allowed an estimation of the relative quantities of water in the gel and pore phases to be made. Similar data were obtained from sodium ion chemical shifts.

Kimmich[153] measured the T_1 of aqueous solutions of bovine ferrihaemoglobin and carboxylated haemoglobin over the large frequency range of 3×10^3 to 1.2×10^8 Hz. No sharp dispersion appeared in the T_1-frequency plot, as was the case for other globular proteins. The relative flatness of the curve was attributed to a broad distribution of molecular motion in the protein–water system. Unfortunately, little conclusive information can be obtained from these unusual results. Water proton T_1 and T_2 values have been measured[154] by the spin-echo technique for both native and denatured calf thymus DNA solutions. It was assumed that water molecules were either irrotationally bound to the DNA, and had its correlation time, or were loosely bound and had a correlation time of the order of bulk water. A distribution of water correlation times, although much more realistic, is more difficult to handle theoretically. Also, no account was taken of anisotropic motion of the water molecule, or of exchange between labile DNA and water protons. Correlation times of the order of 2×10^{-8} s were calculated for the DNA. The effect of added salt, variation in temperature, and denaturation on

[150] T. F. Child and N. G. Pryce, *Biopolymers*, 1972, **11**, 409.
[151] M. P. Volarovich, N. I. Gamayunov, and L. Yu. Vasiléva, *Colloid J. (U.S.S.R.)*, 1971, **33**, 771.
[152] L. S. Frankel, *Analyt. Chem.*, 1971, **43**, 1506.
[153] R. Kimmich, *Z. Naturforsch*, 1971, **26b**, 1168.
[154] B. Lubas and T. Wilczok, *Biopolymers*, 1971, **10**, 1267.

the correlation times were studied. However, the model of the hydration of DNA was so simplified that the small changes in the measured correlation times were difficult to interpret precisely.

The state of water in muscle tissue has been investigated by n.m.r. in a number of publications. Cope[155] pointed out that the T_1 values for protons and deuterium for H_2O and D_2O in cells were less than those of the pure liquids, and that this indicated structuring of water in the tissue. Other workers had previously claimed that water in muscle tissue was not structured. It should be noted that T_1 is a function of a correlation time which may or may not be related to structure. Relaxation data taken in isolation cannot unequivocally decide on the rather loosely defined quantity of more or less structuring of water. It, like beauty, to some extent depends on the eye of the beholder. The chemical shift, linewidth, and relaxation time of H_2O protons in muscle tissue and related proteins were investigated.[156] Chemical shifts were quite small and the linewidths of the unsplit signal were about 5—10 Hz, depending on the field strength. T_1^{-1} was directly dependent on protein concentration and both T_1 and T_2 could be described by a single correlation time, indicating that water was exchanging rapidly between one state tightly bound to the protein and a second state, bulk water. The broad-line spectra of deuterium in frozen samples of D_2O–frog muscle and D_2O–tropomyosin have been used[157] to obtain more information about the state of water in the muscle. Below 273 K the muscle and tropomyosin (a protein component of muscle) produced similar spectra, namely a narrow liquid line which decreased in intensity with decreasing temperature and a broad line, which increased in intensity with decrease in temperature. Below 233 K the broad line showed two components corresponding to different coupling constants. The authors have interpreted the results in terms of different sites for the D_2O, such as the ice I structure and being bound to ionic sites on the protein.

C. Other Studies.—Two publications have shown how the behaviour of spin labels on a macromolecule can be combined with conventional n.m.r. data to provide information on the environment of the macromolecule. Rabbit muscle creatine kinase was labelled[158] at the essential sulphydryl group on the active site. The radical showed a broad asymmetric e.s.r. spectrum, and n.m.r. measurements demonstrated that the presence of the spin label increased the relaxation rate of the water protons by about an order of magnitude. Both the e.s.r. and n.m.r. results indicated a high degree of immobilization of the bound spin label. The interaction between the spin-labelled creatine kinase and manganese ions, in the presence of ADP and ATP has been investigated[159] by both n.m.r. and e.s.r. techniques. This double-probe

[155] F. W. Cope, *Nature New Biol.*, 1972, **237**, 215.
[156] R. Cooke and R. Wien, *Biophys. J.*, 1971, **11**, 1002.
[157] W. Derbyshire and J. L. Parsons, *J. Magn. Resonance*, 1972, **6**, 344.
[158] J. S. Taylor, A. McLaughlin, and M. Cohn. *J. Biol. Chem.*, 1971, **246**, 6029.
[159] M. Cohn, H. Diefenbach, and J. S. Taylor, *J. Biol. Chem.*, 1971, **246**, 6037.

experiment gave information on the ternary enzyme–manganese–nucleotide complex, and n.m.r. proton relaxation rate enhancements were used to obtain further knowledge of the manganese environment.

Randall et al.[160] have used both fluorescence and n.m.r. probes to investigate membrane protein fractions. Benzyl alcohol was used as the n.m.r. probe, the linewidth of the aromatic protons being used as an indication of the extent of binding of the probe to the protein. The separation of all the proteins from the erythrocyte membrane by butanol exposed many new binding sites not accessible in the intact protein. Protein fractions I and II, separated by aqueous washings, showed few new binding sites. Exposure of the different protein fractions to lytic concentrations of benzyl alcohol followed by linewidth measurements produced further information on exposed binding sites on the proteins. Benzyl alcohol has been used[161] in a similar way to investigate native and reaggregated membranes. The concentration dependence of the linewidths of the phenyl protons in the native membrane, separated lipid, separated protein, and reaggregated membrane were measured. A difference between the results with native and reaggregated membrane was interpreted as indicating more binding sites in the reaggregated membranes and protein–lipid interaction in the reaggregated membrane. However, since linewidths of signals can be affected by more than the rotational correlation time, which might not be the same for each colloidal system, care is necessary in obtaining quantitative data from such complex systems.

The state of the lipids in membranes has been investigated[162] by deuteron n.m.r. Perdeuteriated lauric and palmitic acids were incorporated into *Acholeplasma laidlawii* and the deuteron broad-line spectra compared with the deuteron spectra of liquid crystalline mixtures of potassium perdeuterolaurate–H_2O and perdeuterolecithin–H_2O. Since the liquid crystalline spectra were narrower than those from the membrane, it was concluded that the membrane lipids were more immobile than the liquid crystalline forms and were similar in mobility to the gel state of the lecithins.

The interaction between tyrosine or tyramine and nucleic acids was investigated[163] by studying the chemical shift of the aromatic protons. The resonance position of the aromatic protons of the purine and pyrimidine rings moved to low frequency in the presence of tyramine, indicating the formation of complexes, with stacking of the aromatic rings. The aromatic protons of tyramine also shifted to low frequency in the presence of purine derivatives and poly(adenylic acid), indicating that the phenol ring is stacked with the adenine ring of poly(adenylic acid). Similar methods were used[164] to investigate

---

[160] R. F. Randall, R. W. Stoddart, S. M. Metcalfe, and J. C. Metcalfe, *Biochim. Biophys. Acta*, 1972, **255**, 888.
[161] J. C. Metcalfe, S. M. Metcalfe, and D. M. Engelman, *Biochim. Biophys. Acta*, 1971, **241**, 412.
[162] E. Oldfield, D. Chapman, and W. Derbyshire, *Chem. Phys. Lipids*, 1972, **9**, 69.
[163] C. Hélène, T. Montenay-Garestier, and J. Dimicoli, *Biochim. Biophys. Acta*, 1971, **254**, 349.
[164] C. Héléne, J. L. Dimicoli, and F. Brun, *Biochemistry*, 1971, **10**, 3802.

the interaction between indole derivatives and poly(adenylic acid). It was concluded that stacked complexes occurred between the serotonin and the adenine ring. The interaction between ethidium bromide (EB) (a three-ring aromatic molecule) and uracil residues, uridylyl(3'-5')uridine and poly-(uridylic acid) was investigated[165] by measuring the chemical shift of the protons of the nucleotides and EB. Changes in the chemical shift were interpreted as indicating complex formation between the nucleotide and EB. Broadening and chemical shift results for the protons on the 1 and 10 positions of EB in the presence of uridylyl(3'-5')uridine or poly(uridylic acid) suggested that these protons were oriented towards the ribose rings of the nucleotide.

Barratt and Rayner[166] have studied the interaction between lysolecithin and several caseins, using both e.s.r. and n.m.r. The high-resolution spectra taken at both 60 and 220 MHz showed that the addition of the caseins to lysolecithin broadened the peaks from the alkyl chains but did not alter those from the choline head group. Similar results were obtained for the system of sodium dodecyl sulphate–casein. It was concluded that the alkyl chains were bound to the protein, but that the head group was relatively free. Caution should be exercised in interpreting data of this type since the choline head groups have two axes of internal rotation as opposed to one for the alkyl chains, and this may result in linewidth differences.

The preferential solvation of polymers in a mixture of two solvents has been studied[167] by measuring the proton T_1 of the two solvents using the method of adiabatic rapid passage. In the system of polystyrene dissolved in benzene and cyclohexane the polymer was preferentially solvated by the former solvent. Poly(methyl methacrylate) dissolved in chloroform and carbon tetrachloride is preferentially solvated by the chloroform. Preferential solvation by a solvent was indicated by an increase in the proton relaxation rate for that solvent. Unfortunately, the relaxation rates are not given directly but are expressed as the ratio $[(T_1^{-1})_{observed} - (T_1^{-1})_{psm}]/(T_1^{-1})_{psm}$, where psm indicates pure solvent mixture, thus limiting the usefulness of the data.

The ^2H and ^{35}Cl relaxation times of dilute D_2O and $CDCl_3$ solutions of poly(vinyl pyrrolidone) (PVP) have been measured.[168] The variation in the relaxation times was interpreted in terms of a two-phase model, *i.e.* the solvent was rapidly exchanging between the bulk state and one closely associated with the polymer. For solutions of PVP in $CDCl_3$ the deuterium relaxation rate increased faster than that of chlorine and this was interpreted as slower rotation of the C_3-axis of the solvent molecule. However, all changes in the relaxation rates were attributed to correlation-time changes and no account was made of any change which may have occurred in the electric field gradients.

The interaction between hexachlorophene and amides or polypeptides in

[165] G. P. Kreishman and S. I. Chan, *J. Mol. Biol.*, 1971, **61**, 45.
[166] M. D. Barratt and L. Rayner, *Biochim. Biophys. Acta*, 1972, **255**, 974.
[167] K. Sato and A. Nishioka, *J. Polymer Sci. Part A-2, Polymer Phys.*, 1972, **10**, 489.
[168] R. G. Brussau and H. Sillescu, *Ber. Bunsengesellschaft phys. Chem.*, 1972, **76**, 31.

non-polar solvents has been studied[169] by measuring the chemical shift and broadening of the methylene and hydroxyl protons. The addition of simple amides or poly(γ-benzyl L-glutamate) to solutions of hexachlorophene in $CDCl_3$ causes the phenolic OH to broaden and shift to low field, though the methylene protons appear unchanged. The results were interpreted in terms of hydrogen bonding between the hydroxy-group and the oxygen of the amide group of the polypeptide.

The molecular dynamics and energies of activation of methylene chloride containing a range of concentrations of polystyrene were studied[170] by pulsed n.m.r. The contributions of solvent–polymer, solvent–solvent, and intramolecular interactions were estimated and the residence times of the solvent molecule on the polymer were discussed. Uncertainties in the assumptions make the conclusions tentative.

Solutions of poly(γ-benzyl L-glutamate) in dichloromethane change from a cholesteric phase to a nematic phase when placed in a magnetic field, the change being slow on the n.m.r. time scale. Small molecules dissolved in these phases exhibit dipolar splitting similar to that observed when p-azoxyanisole is used as a nematic solvent. Orwoll and Vold[171] measured and explained theoretically the different splittings of the dichloromethane resonance in the two types of phases. In the transition from the cholesteric phase to the nematic phase, signals from the dichloromethane from both phases were observed, showing that the solvent molecules were not rapidly exchanging between phases.

Samulski and Berendsen[172] have made a high-resolution study of the proton, deuteron, and nitrogen resonances of dimethylformamide and deuteriated DMF in nematic liquid crystals of poly(L-glutamic acid). Addition of H_2O to the systems changed the degree of orientation. From the proton spectra, the parameters of the order matrices were determined. Comparisons of the different solvent systems gave the amide HCN angle as 107°. By using the quadrupolar splittings of both deuterium and nitrogen, the quadrupole coupling tensors of these nuclei were obtained. Quadrupole coupling constants and asymmetry factors were also found for some of the deuterium and nitrogen nuclei. Splitting of the water deuterium resonance was also observed.

The diffusion coefficient of solvent molecules in a polymer solution is independent of the polymer molecular weight (M) for most large values of M. However, at some low M this is not the case. Tanner[173] measured the diffusion coefficient of solvent molecules in solutions of polymers of different M, using the spin-echo pulsed field gradient technique. The systems studied were poly(ethylene oxide) in chloroform and poly(dimethylsiloxane) in benzene. The solvent diffusion coefficient decreased with increasing M below an M of about 1000, but was independent of M above 1000. In a solution with two

[169] R. Haque and D. R. Buhler, *J. Amer. Chem. Soc.*, 1972, **94**, 1824.
[170] W. G. Rothschild, *Macromolecules*, 1972, **5**, 37.
[171] R. D. Orwoll and R. L. Vold, *J. Amer. Chem. Soc.*, 1971, **93**, 5335.
[172] E. T. Samulski and H. J. C. Berendsen, *J. Chem. Phys.*, 1972, **56**, 3920.
[173] J. E. Tanner, *Macromolecules*, 1971, **4**, 748.

polymers, their diffusion coefficients were dependent on M to a much higher value of about 100 000. Kosfeld and Goffloo[174] used the same pulsed field gradient technique to measure the solvent self-diffusion coefficient in the systems benzene–polystyrene and cyclohexane–polystyrene. Activation energies were calculated from the temperature coefficient of the diffusion coefficient. The results were discussed in terms of the free volume of the polymer in solution and it was concluded that the side groups of the polymer were a major influence on the diffusion of the solvent.

[174] R. Kosfeld and K. Goffloo, *Kolloid-Z.*, 1971, **247**, 801.

9
The Solid State

BY W. DERBYSHIRE

1 General Introduction

This chapter is concerned with work on the n.m.r. of solids published during the period March 1971 to May 1972, with a somewhat abbreviated discussion of papers published in the previous year. Inevitably, many papers involve several aspects of n.m.r. and some papers have relevance to other chapters. This chapter is in general more concerned with CW measurements, but there are points of common interest with Chapter 3, where relaxation in all three states of matter is discussed. Also relevant is Chapter 8, which deals with macromolecules.

Jonas and Gutowsky,[1] reviewing n.m.r. publications for the year 1967, estimated a publication rate of 550 per month. Clearly, the number of publications even on a restricted aspect of n.m.r. is now considerable, and inevitably there will be some omissions. The sources of this literature survey have been Current Contents[2] and the n.m.r. abstracts of the UKCIS.[3] The selection of papers for discussion has been dictated by the availability of journals or reprints and, in general, dissertations have not been included in the survey. Farach and Poole[4] have compiled a list of earlier reviews on all aspects of magnetic resonance, and Andrew[5] has reviewed almost 300 papers on wideline n.m.r. published in 1970. Discussion is brief but the papers are arranged conveniently in tabular form according to topic. Another recent review discussing material relevant to this chapter is that by Allen.[6] This paper discusses the use of n.m.r. techniques to study molecular motion in solids. References are quoted to illustrate the article, but no attempt is made at completeness. The proceedings of the 16th Colloque Ampère[7] provide a reasonable cross-

[1] J. Jonas and H. S. Gutowsky, *Ann Rev. Phys. Chem.*, 1968, **19**, 447.
[2] *Current Contents/Physical and Chemical Sciences*, ed. E. Garfield, Institute for Scientific Information, Philadelphia, Penna, USA.
[3] NMR – Chemical Aspects, United Kingdom Chemical Information Service, Nottingham.
[4] H. A. Farach and C. P. Poole, jun., *Magn. Resonance Rev.*, 1972, **1**, 3.
[5] E. R. Andrew, *Magn. Resonance Rev.*, 1972, **1**, 33.
[6] P. S. Allen, 'NMR Studies of Molecular Motion in Solids' in 'Magnetic Resonance', ed. C. A. McDowell, MTP International Review of Science-Physical Chemistry, Butterworth, 1972, Vol. 4, Chapter 2.
[7] Proceedings XVIth Colloque Ampère, 1971.

section of subjects currently under investigation. The book of Goldman[8] is an authoritative source of information on factors affecting the n.m.r. of solids.

2 Nuclei of Spin $\frac{1}{2}$

A. Introduction.—In general, the n.m.r. spectrum of a nucleus of spin $\frac{1}{2}$ in a diamagnetic solid sample is determined by dipolar interactions with neighbouring spins. The lineshape is difficult to calculate, and will be the subject of a later section, but Van Vleck[9] showed that the moments of absorption lines are readily calculable. Comparisons between observed values and calculated ones can yield structural information. The occurrence of a sufficiently rapid molecular motion reduces the observed second moment by an amount characteristic of that motion. The activation energy of the motion responsible for the reduction in second moment can be determined from the temperature dependence of the linewidth in the transition region, using an expression developed by Gutowsky and Pake[10] from the pioneering paper of Bloembergen *et al.*[11] Smith,[12] for example, has in a number of papers demonstrated the usefulness of this approach. The value of the second moment in the transition region is uncertain, because the observable value depends upon parameters such as the signal-to-noise ratio. Bartko and Feherova[13] have examined the effect of neglecting the wings of absorption curves, be they Gaussian or Lorentzian, on measured second and fourth moments. Andrew and Lipofsky[14] have demonstrated that if second moment values are determined in a controlled manner, their temperature dependence can be used to derive activation energies. However, if the temperature range where motional narrowing occurs is small, the expression relating linewidth or second moment to activation energy is difficult to use. Wert and Marx[15] and Waugh and Fedin[16] related the activation energy directly to the temperature of the transition. Andrew and Canepa[17] have demonstrated the theoretical justification for such a proportionality relationship.

B. Proton N.M.R.—*Clathrate Hydrates.* Garg and Davidson[18] described the use of enclathration in D_2O cages as a method of reducing intermolecular dipolar broadening such that details of previously unresolved spectra might be resolved, and the values of second moment more accurately determined (as intermolecular contributions are notoriously difficult to calculate). Examples

[8] M. Goldman, 'Spin Temperature and Nuclear Magnetic Resonance in Solids', Oxford University Press, 1970.
[9] J. H. Van Vleck, *Phys. Rev.*, 1948, **74**, 1168.
[10] H. S. Gutowsky and G. E. Pake, *J. Chem. Phys.*, 1950, **18**, 162.
[11] N. Bloembergen, E. M. Purcell, and R. V. Pound, *Phys. Rev.*, 1948, **73**, 679.
[12] G. W. Smith, *J. Chem. Phys.*, 1969, **51**, 3569 and references therein.
[13] O. Bartko and J. Feherova, *Czech J. Phys.*, 1971, **A21**, 385.
[14] E. R. Andrew and J. Lipofsky, *J. Magn. Resonance*, 1972, **8**, 217.
[15] C. Wert and J. Marx, *Acta Met.*, 1953, **1**, 113.
[16] J. S. Waugh and I. Fedin, *Soviet Phys. Solid State*, 1963, **4**, 1633.
[17] E. R. Andrew and P. C. Canepa, *J. Magn. Resonance*, 1972, **7**, 429.
[18] S. K. Garg and D. W. Davidson, *Chem. Phys. Letters*, 1972, **13**, 73.

with SF_6, C_3H_6O, and C_4H_6O were given. A number of other papers concerned with ice clathrates have been published. Chassonneau et al.[19] investigated various double type-II clathrates of the form $A,2H_2S,17H_2O$ (where A represents isobutane, CCl_4, tetrahydrofuran, or thiophen). The linewidth of the water protons was similar to that found in ice; an activation energy was extracted from the line narrowing. McDowell and Raghunathan[20] investigated propane clathrate deuterate using two specimens, one richer in guest than the other, but no significant differences between samples were observed in the temperature region 77—260 K. Below 160 K reorientation about the molecular C_2 axes and methyl-group rotation both occurred. At higher temperatures random tumbling occurred, followed by lattice diffusion just below the decomposition temperature. T_1 values were obtained using adiabatic rapid-passage techniques. Khanzada and McDowell[21] extended the work to cyclopropane and acetone. In acetone, methyl-group rotation occurred at 77 K and self-diffusion at 172 K, the activation energy for the latter process being 16.5 kJ mol^{-1}. For cyclopropane they found a restricted motion up to 240 K followed by a free rotation. This result is inconsistent with earlier work[22] where cyclopropane molecules were reported to undergo isotropic reorientation. Davidson[23] considered the general problem of the reorientation of guest molecules in type I and type II cages, and the influence of electrostatic fields of various geometries. He found that there exist a large number of nearly equivalent sites consistent with the motion being describable by a distribution of correlation times. Confirmation was provided by Garg et al.,[24] who reported n.m.r. and dielectric measurements on ethylene oxide clathrate from 2 to 270 K. Above 230 K the ethylene oxide molecules underwent isotropic reorientation. The large temperature range of line narrowing was consistent with considerable variability in the rotational freedom of the ethylene oxide molecules.

Conformational Motion. Andrew and Brookeman[25] have shown that conformational motion, familiar to investigators using high-resolution n.m.r. techniques, can occur in solids, and that the temperature dependence of the ^{1}H n.m.r. of some solid cyclic hydrocarbons is consistent with such motion. The discussion was subsequently extended,[26] the terminology employed being similar to that used in discussing magnetic systems. A further extension, to the situation where the populations of the two conformer states are no longer equal, has been made.[27]

[19] M. A. Chassonneau, J. Dufourcq, and B. Lemanceau, *Compt. rend.*, 1971, **273**, C, 793.
[20] C. A. McDowell and P. Raghunathan, *J. Mol. Structure*, 1970, **5**, 443.
[21] A. W. K. Khanzada and C. A. McDowell, *J. Mol. Structure*, 1971, **7**, 241.
[22] Y. A. Majid, S. K. Garg, and D. W. Davidson, *Canad. J. Chem.*, 1969, **47**, 4697.
[23] D. W. Davidson, *Canad. J. Chem.*, 1971, **49**, 1224.
[24] S. K. Garg, B. Morris, and D. W. Davidson, *J. C. S. Faraday II*, 1972, **68**, 481.
[25] E. R. Andrew and J. R. Brookeman, *J. Magn. Resonance*, 1970, **2**, 259.
[26] E. R. Andrew, *Phys. Letters*, 1971, **34A**, 30.
[27] E. R. Andrew, *J. Magn. Resonance*, to be published.

Carboranes. Baughman[28] investigated the crystalline carboranes $B_{10}H_{10}C_2H_2$ using proton n.m.r., calorimetric, and X-ray diffraction techniques. Second moments for *ortho-* and *meta*-carboranes were 0.711 and 0.822 G^2 (1G = 10^{-4} T). Calculated values for the rigid lattice situation were 34 G^2 and for rapid molecular isotropic reorientation 0.630 G^2. Results were compared and contrasted with those from adamantane.

Hydrogen Sulphide and Selenide. El Saffar and Schultz[29] examined the ^1H lineshapes and second moments of H_2S and H_2Se at 58 K, where the molecules are essentially rigid and, from specific heat measurements, known to be ordered. Earlier workers had reported inconsistent values of the second moments in these phases, but the proton separations determined by El Saffar and Schultz were consistent with the gas-phase values.

Hydrates. El Saffar and his colleagues have also investigated $AlCl_3,6H_2O$.[30] The (H,H) distance of 1.59 Å determined from the polycrystalline sample was inconsistent with the neutron-diffraction value of 1.70 Å. A Pake-type study[31] on the single crystal gave a value of 1.62 ± 0.04 Å. T_1 and second-moment measurements showed that the sample was rigid below room temperature, but that above 300 K the water molecules engaged in rapid 180° flips. Calcium bromate monohydrate has been studied by Cvikl and McGrath,[32] who performed a classical Pake-type analysis using the centres of gravity of the resolved lines to obtain (H,H) separations and directions to an accuracy of 0.02 Å. The temperature dependence of the Pake splittings as a result of molecular vibrations was used to derive an activation energy, confirmed by relaxation measurements. They also observed deuteron resonance, the D_2O molecules undergoing 180° flips down to 248 K. The (H,H) direction and the principal axes of the electric-field gradient tensor determined from the deuteron resonance served to locate the H atoms in the crystal structure, which had not previously been studied by X-ray diffraction. Schnabel and co-workers,[33] taking advantage of better crystals available, studied the mineral kiserite, $MgSO_4,H_2O$. They found a high value of the proton–proton distance, 1.65 Å, attributed to the hydrogen atoms partaking in bifurcated hydrogen bonds. Rochelle salt provides a complicated example, with many molecules of hydration. Kato *et al.*[34] found four proton–proton pairs per unit cell, but had to use only the largest splittings to avoid the regions of overlapping spectra. Their findings were consistent with earlier measurements based on deuterons.[35] Valic and Pintar[36] have examined borax, $Na_2B_4O_5(OH)_4,8H_2O$,

---

[28] R. H. Baughman, *J. Chem. Phys.*, 1970, **50**, 3781.
[29] Z. M. El Saffar and P. Schultz, *J. Chem. Phys.*, 1972, **56**, 2524.
[30] Z. M. El Saffar and W. Mulcahy, *Acta Cryst.*, 1971, **B27**, 1069.
[31] Z. M. El Saffar, W. Mulcahy, and G. Rochau, *J. Chem. Phys.*, 1971, **55**, 3307.
[32] B. Cvikl and J. W. McGrath, *J. Chem. Phys.*, 1970, **52**, 1560.
[33] B. Schnabel, B. Jungnickel, T. Taplick, and K. Heide, *Kristall und Technik*, 1971, **6**, 193.
[34] T. Kato, O. Mizuno, and R. Abe, *J. Phys. Soc. Japan*, 1970, **29**, 393.
[35] J. L. Bjorkstam and J. H. Willmorth, Proceedings XIVth Colloque Ampère, 1966, p. 728.
[36] M. I. Valic and M. M. Pintar, *Phys. Status Solidi*, 1970, **42**, 661.

a one-dimensional proton conductor above 294 K, and used it as a model system for ice. They recorded T_1 and $T_{1\rho}$ values, and the CW lineshape, and found three spectral components attributed to (a) hydroxy-groups, (b) water molecules that were in one phase, rigid or possibly flipping, and (c) water molecules in a further phase, rotating freely. Sergeev et al.[37] considered the n.m.r. of hydrates where water molecules can exchange sites. (H,H) directions are related by crystal symmetry, but the resulting spectra may be complex. Polycrystalline desmine ($CaAl_2Si_7O_{18},7H_2O$) was used as an example.

Ring Compounds. Fyfe et al.[38] have examined molecular motion in coronene, perylene, and triphenylene (extending recent work on pyrene[39]), recording second moments, linewidths, and T_1 values by the adiabatic rapid-passage method. At low temperatures coronene molecules were rigid; at higher temperatures the moment values were consistent with molecular reorientation. The sharpness of the transition was such that the line-narrowing equation could not be used but the activation energy obtained directly from the transition temperature, 24.7 kJ mol^{-1}, was consistent with the value of 24.3 kJ mol^{-1} obtained from T_1 values. Perylene and triphenylene were rigid at 298 K, the lack of motion being attributed to steric hindrance. Molecular motion was observable in pyrene. Brookeman and Rushworth[40] examined 1,3,5-cycloheptatriene, a plastic crystal. Below 80 K a second-moment value of 12.5 G^2 indicated a rigid structure. Above 80 K this was reduced to 0.81 G^2, a reduction not consistent with a simple molecular rotation. The most probable motion is a rotation about the y-axis accompanied by ring inversion.

Biedermann et al.[41] examined the proton resonances of bis(salicylaldehydatopyridazine)cobalt complexes from 114 to 365 K. Above 274 K two components were observed. Fried[42] investigated p-dioxan; below 260 K he obtained the rigid lattice spectrum, but at 272.9 K, the temperature of a thermodynamic transition, narrowing occurred and the spectrum above that temperature consisted of two overlapping lines. The most probable motion is a reorientation about an axis perpendicular to the plane of the molecule.

Bailey and Pittman[43] reported determinations of linewidths and second moments for a series of homologous long-chain fatty acids, in addition to some unsaturated ones. Polymorphism showed itself in the form of different lineshapes, linewidths, and moment values. Some years ago Anderson and Slichter,[44] investigating methyl reorientation in solid alkanes, found that the lineshape at 150 K of alkanes with less than 15 carbon atoms per molecule consisted of two peaks. They suggested that the narrow peak was due to

[37] N. A. Sergeev, O. V. Falaleev, and S. P. Gabuda, *Soviet Phys. Solid State*, 1970, **11**, 1815.
[38] C. A. Fyfe, B. A. Dunell, and J. Ripmeester, *Canad. J. Chem.*, 1971, **49**, 3332.
[39] C. A. Fyfe, D. F. R. Gilson, and K. H. Thompson, *Chem. Phys. Letters*, 1970, **5**, 215.
[40] J. R. Brookeman and F. A. Rushworth, *J. Phys. (C)*, 1971, **4**, 1903.
[41] H. G. Biedermann, P. K. Burkert, and K. E. Schwarzhans, *Z. Naturforsch.*, 1971, **26a**, 968.
[42] F. Fried, *Mol. Crystals Liquid Crystals*, 1971, **13**, 279.
[43] A. V. Bailey and R. A. Pittmann, *J. Amer. Oil Chemists' Soc.*, 1971, **48**, 775.
[44] J. E. Anderson and W. P. Slichter, *J. Phys. Chem.*, 1965, **69**, 3099.

reorienting methyl groups. Van Putte and Van den Enden,[45] re-investigating the situation, found a single line 15.2 G wide at 93 K, and at 133 K three lines of widths 13.2, 6.7, and 0.57 G. The reproducibility of the narrow peak was poor and it was attributed to lattice defects.

Methyl-group Rotation. Chezeau *et al.*[46] have reported a comprehensive n.m.r. investigation of hexamethylethane from 85 to 374 K, involving determination of linewidths, second moments, and T_1 and $T_{1\rho}$ values. Methyl group rotation and rotation of the molecule about the carbon–carbon bond occurred up to 110 K. From the transition temperature (152 K) to 260 K, molecular motion was isotropic; diffusion occurred above this temperature. The activation energy for the diffusion process was calculated using the linenarrowing method. El Saffar *et al.*[47] measured second moments and T_1 values in trimethylacetonitrile, Me₃CCN, from 58 to 292 K. The low-temperature values corresponded to the rigid lattice, but at higher temperatures second-moment values were consistent with rotation of the t-butyl group in addition to the methyl groups. Only one step in the moment value was observed. The activation energy derived from T_1 values was 16.3 ± 0.8 kJ mol^{-1}, and that from line narrowing 12.1 kJ mol^{-1}. This is a plastic crystal. The observation of a single transition was taken as evidence that the three methyl groups are interlocked. Fyfe and Ripmeester[48] examined molecular motion in charge-transfer complexes of trimethylamine with iodine, bromine, and iodine monochloride. Trimethylamine was studied for comparison. At 77 K the trimethylamine spectrum was characteristic of a broadened three-spin system; the second moment was 30.5 G², narrowing to 2.7 G² at 140 K, although other workers had reported 4.1 G². Similar behaviour was observed in all the complexes. The high-temperature moment is too low to be explained by methyl-group rotation; it was proposed that the molecule additionally rotates about the three-fold symmetry axis. For the chlorine and iodine complexes motional narrowing occurred in two distinct stages. Smith[49] studied the plastic crystal hexamethylethane. In previous work, Koide[50] somewhat surprisingly did not report the presence of self-diffusion, and his values of moments and linewidths were not typical of these materials. Smith found that the second-moment value below 80 K is the rigid-lattice value, decreasing smoothly to 2.2 G² at 120 K and then abruptly to 0.9 G², thereafter decreasing smoothly to zero (values consistently lower than those of Koide). At 363 K the values were consistent with methyl-group rotation together with restricted reorientation of the t-butyl group. The lineshape changed through the transition, so it was not possible to extract the activation energy from the linewidth. Evidence for self-diffusion was found below the melting point.

[45] K. van Putte and J. van den Enden, *J. Phys. Chem.*, 1971, **75**, 3901.
[46] J. M. Chezeau, J. Dufourcq, and J. H. Strange, *Mol. Phys.*, 1971, **20**, 305.
[47] Z. M. El Saffar, P. Schultz, and E. F. Meyer, *J. Chem. Phys.*, 1972, **56**, 1477.
[48] C. A. Fyfe and J. Ripmeester, *Canad. J. Chem.*, 1970, **48**, 2283.
[49] G. W. Smith, *J. Chem. Phys.*, 1971, **54**, 174.
[50] T. Koide, *Bull Chem. Soc. Japan*, 1967, **40**, 2026.

Ulrich and Dunell[51] determined proton second-moment and T_1 values in trimethyltin fluoride. Methyl-group rotation occurred at 77 K. The second moment was progressively reduced between 280 and 800 K, a reduction attributed to C_3' group rotation (that is, motion of the trimethyltin group about the chain axis), followed by isotropic reorientation.

Methylammonium halides have been the subject of two papers.[17,52] Tsau and Gilson[52] obtained the wide-line n.m.r. spectra of methylammonium bromide, chloride, and iodide, finding the presence of several phases. With the help of selective deuteriation, the line-narrowing occurring at 150 K was attributed to CH_3 and NH_3 group rotations. Activation energies were extracted from this line-narrowing. Andrew and Canepa[17] examined mono-, di-, tri-, and tetra-methylammonium chlorides, obtaining the mono-compound in a different phase to that reported by Tsau and Gilson. No evidence for diffusion was obtained. Dufourcq and Lemanceau[53] have also studied tetramethylammonium chloride.

Ammonia and Derivatives. As paper 42 in a sequence entitled 'Hydrogen Bond Studies', Tegenfeldt and Olovsson[54] investigated hydrazinium hydrogen oxalate (N_2H_5,HC_2O_4) from 190 to 340 K. Increasing the temperature from 210 to 280 K resulted in the second moment decreasing to 13 G^2 from the rigid-lattice value of 33 G^2. This was attributed to NH_3 group rotation. It was not established that the NH_2 group rotated but from other evidence this was considered improbable. Ammonia has been studied by two groups of workers.[55,56] Relaxation times and moments were determined over extensive temperature regions, although the rigid-lattice value of 48 G^2 was never obtained. This failure was attributed to the existence of quantum-mechanical tunnelling at low temperatures. At 170 K the moment value was consistent with rotation about the triad axis. At higher temperatures self-diffusion occurred. Carolan and Scott reported a change in linewidth, but not second moment, at 65 K. Ammonium ions have been investigated by Hennel and Lalowicz.[57,58] In the earlier paper they investigated ammonium perchlorate at 4.2 and 1.2 K finding second-moment values of 8.9 and 13.5 G^2. Lineshapes were complex and were attributed to two groups of molecules, the one undergoing isotropic reorientation and the second rotating about the molecular C_3 axes. The proportion of each was temperature-dependent; ΔE corresponded to 8 J mol^{-1}. In the second paper, NH_4I was investigated at 1.4 and 77 K; lineshapes were again complicated, with a central line and four pairs of well-resolved satellites. The spectrum also changed when the sample was kept at liquid-helium temperatures for some time. Sharp *et al.*[59] had suggested that

[51] S. E. Ulrich and B. A. Dunell, *J. C. S. Faraday II*, 1972, **68**, 680.
[52] T. Tsau and D. F. R. Gilson, *Canad. J. Chem.*, 1970, **48**, 717.
[53] J. Dufourcq and B. Lemanceau, *J. Chim. phys.*, 1970, **67**, 9.
[54] J. Tegenfeldt and I. Olovsson, *Acta Chem. Scand.*, 1971, **25**, 101.
[55] D. E. O'Reilly, E. M. Peterson, and S. R. Lammert, *J. Chem. Phys.*, 1970, **52**, 1700.
[56] J. L. Carolan and T. A. Scott, *J. Magn. Resonance*, 1970, **2**, 243.
[57] J. W. Hennel and Z. T. Lalowicz, *Acta Phys. Polon.*, 1970, **A38**, 675.
[58] Z. T. Lalowicz and J. W. Hennel, *Acta Phys. Polon.*, 1971, **A40**, 547.
[59] A. R. Sharp, S. Vrscaj, and M. M. Pintar, *Solid State Comm.*, 1970, **8**, 1317.

the central line is due to the molecular isomer of total proton spin $\Sigma I = 2$. Lalowicz and Hennel suggested that there could be some contribution from the $\Sigma I = 1$ isomer. The oscillatory states are split into a number of sub-levels dependent upon symmetry and modulated by thermal vibrations. At low temperatures the lifetimes of the sub-levels become greater. Ratkovic and Forsén[60] determined linewidths and second moments of taurine and ε-aminocaproic acid. Rotation of the NH_3 group occurred. The activation energies extracted from the temperature dependence of the linewidth were related to a mean hydrogen-bond length. Schutte and Heyns[61] investigated the i.r. and n.m.r. spectra of ammonium sulphate between 17 and 298 K, with particular emphasis on the linewidth transition at 163 K. Ammine groups in carbonatotetrammine have been investigated in the complex $[Co(NH_3)_4CO_3]Br$.[62] The spectrum showed fine structure characteristic of a three-spin system. Two spectral components were observed at 77 K. The second moment underwent a transition in the region 100—200 K, but the linewidth was essentially unchanged. An activation energy was calculated from the temperature dependence of the second moment. Calculations indicated that rotation occurs about the Co—N axes.

Rubailo et al.[63] examined trichloroaniline between 293 and 369 K. Line narrowing attributed to a molecular reorientation about the three-fold axis occurred at 329 K. The inter-proton distance in the NH_2 group was determined to be 1.72 Å. Smith and Shafizadeh[64] studied molecular motion in 1,6-anhydro-β-D-glucopyranose, which behaves as a plastic crystal. Below 658 K, rigid-lattice values of second moment were observed. At 658 K a narrow central component appeared. With increasing temperature this grew in intensity at the expense of the broader component, and became narrower.

Diffusion Studies. In addition to the papers discussed earlier involving plastic crystals and diffusion, there has been a sequence of papers dealing with this specific topic. Using line-narrowing techniques, Blum and Sherwood[65] investigated diffusion in camphene, pure and impurity-doped. They were concerned with the lack of agreement in values of diffusion coefficient and activation energy determined by n.m.r. and other methods. Doping of samples produced pronounced changes in the diffusion coefficient as measured by using radiotracer and creep techniques, but not in the n.m.r. spectrum nor its temperature dependence. N.m.r. values of activation energy were consistently lower. Discrepancies were attributed to the different time-scales of the experiments, and the fact that diffusion occurred by a vacancy mechanism. An alternative explanation offered by Chezeau et al.[46] is that discrepancies

---

[60] S. Ratkovic and S. Forsén, *Croat. Chem. Acta*, 1970, **42**, 439.
[61] C. J. H. Schutte and A. M. Heyns, *J. Chem. Phys.*, 1971, **52**, 864.
[62] M. Okabe, Y. Arata, A. Yamasaki, and S. Fujiwara, *J. Phys. Soc. Japan*, 1970, **28**, 935.
[63] A. I. Rubailo, Yu N. Moskvich, S. P. Gabuda, and V. E. Volkov, *J. Struct. Chem.*, 1970, **11**, 960.
[64] G. W. Smith and F. Shafizadeh, *J. Chem. Soc. (B)*, 1971, 908.
[65] H. Blum and J. N. Sherwood, *Mol. Crystals Liquid Crystals*, 1970, **10**, 381.

occur when the Torrey theory[66] is applied in the situation where the motion is describable by a distribution of correlation frequencies. Bladon et al.[67,68] have published two relevant papers. In the first paper they used high-resolution techniques to try to resolve the discrepancies, and in the second paper they reported studies of succinonitrile and tetramethyl succinonitrile, which is known to have peculiar properties as a result of isomerism. Other related papers are those of Hawthorne and Sherwood[69] and of Roeder and Douglass.[70]

Hydrogen diffusion in $TaH_{0.1}$ has been studied by Tanaka and Hashimoto.[71] Above 150 K line-narrowing occurred, consistent with the anomalous change of electrical resistivity. The Knight shift was also measured, and also changed at 150 K.

Ferroelectricity. There is a continuing effort in the study of ferroelectrics and antiferroelectrics. This involves measurements of quadrupolar parameters in addition to those appropriate to this section, and consequently some discussion will be postponed until a later section of this chapter. Blinc has reviewed earlier work.[72] Adriaenssens and Bjorkstam[73] determined second moments in potassium dihydrogen phosphate type crystals but unfortunately the observed second moments were too insensitive to distinguish between static and dynamic ordering of the low-temperature phases; however, they did serve to eliminate one proposal of Nicholson and Soest[74] for a high-temperature transition and supported their contention that crystal decomposition occurs. Blinc et al.[75] investigated diglycine nitrate and tris-sarcosine calcium chloride. Second moments, T_1, and $T_{1\rho}$ were determined. In both compounds, a change in moment occurred at the transition point and was attributed to a flipping of the glycine molecules.

Polymers. Polymers have been extensively investigated using n.m.r. techniques, their CW spectra and relaxation behaviour both being studied; a book on the subject has been published.[76] Peterlin and Olf are undertaking an extensive n.m.r. investigation of polymers. In paper VII[77] they used wide-line techniques on oriented samples of drawn nylon-66 fibres in the temperature range 77—473 K. Second-moment values were compared with calculated ones in an attempt to identify the mechanisms responsible for the low- and high-temperature segmental motion. At high temperatures there is a 60° flip-flop of the chains and torsional motion in the crystalline region. This paper

[66] H. C. Torrey, *Phys. Rev.*, 1953, **92**, 962.
[67] P. Bladon, N. C. Lockhart, and J. N. Sherwood, *Mol. Phys.*, 1971, **20**, 577.
[68] P. Bladon, N. C. Lockhart, and J. N. Sherwood, *Mol. Phys.*, 1971, **22**, 365.
[69] H. M. Hawthorne and J. N. Sherwood, *Trans. Faraday Soc.*, 1970, **66**, 1783.
[70] S. B. W. Roeder and D. C. Douglass, *J. Chem. Phys.*, 1970, **52**, 5525.
[71] K. Tanaka and T. Hashimoto, *J. Phys. Soc. Japan*, 1971, **31**, 1841.
[72] R. Blinc, *Adv Magn. Resonance*, 1968, **3**, 141.
[73] G. J. Adriaenssens and J. L. Bjorkstam, *J. Chem. Phys.*, 1971, **55**, 1137.
[74] J. Y. Nicholson and J. F. Soest, *Bull. Amer. Phys. Soc.*, 1970, **15**, 606.
[75] R. Blinc, M. Jamsek-Vilfan, E. Lahajnar, and G. Hajdukovic, *J. Chem. Phys.*, 1970, **52**, 6407.
[76] I. Ya Slonim and A. N. Lyubimov, 'The NMR of Polymers', Plenum Press, 1970.
[77] H. G. Olf and A. Peterlin, *J. Polymer Sci.*, Part A-2, *Polymer Phys.*, 1971, **9**, 1449.

extended earlier work on polyethylene.[78,79] In paper VIII,[80] Olf extended the measurements to samples of rolled sheet having a double orientation, in order to distinguish between two mechanisms proposed for the high-temperature phase. In paper IX[81] the n.m.r. spectra of aligned fibres were obtained as a function of orientation, temperature, and water content. Not all the water absorbed contributed to the narrow n.m.r. line; increasing water content resulted in increased polymer motion. Folkes and Ward[82] investigated the second-moment anisotropy in drawn low-density polyethylene films from 77 to 333 K. It was proposed that chain sliding occurred at intermediate temperatures. Eichhoff and Zachmann[83] examined the n.m.r. of crystallized poly(ethylene terephthalate) before and after swelling in deuteriated acetone. The width of the glass transition was broader and lower after swelling. The second-moment value of the broad component decreased but the width of the narrow component increased. Kashiwagi and Ward[84] recorded the second-moment anisotropy of a series of drawn poly(vinyl chloride) samples at 77 K, extending determinations of molecular orientation in an amorphous polymer of poly(methyl methacrylate).[85] McBrierty et al.[86] recorded the anisotropy of the fourth moment at 77 K of a number of polyethylene films at different draw ratios. The observed anisotropy was compared with calculated values; the appropriate theory was developed using rotation procedures characterized by Wigner matrices.

Ellis and Clark-Monks[87,88] investigated the n.m.r. of epoxy resins with and without a silica filler, from 110 to 350 K, through the glass transition temperature T_g at 308 K. Above T_g motion was isotropic. The temperature dependence of the linewidth indicated that two processes occur. Price et al.[89] investigated the effect of fillers and moisture on T_g. As a result of introducing the filler, three transition temperatures were observed. The effect of moisture was to reduce T_g. Another relevant paper is that by Shen and Eisenberg.[90]

C. Spin-$\frac{1}{2}$ Nuclei other than ^1H.—As atomic number increases so does the range of chemical shifts. This may result in the observation of separate signals from chemically shifted nuclei and in an additional source of line broadening in the form of anisotropic chemical shifts. The next lightest nucleus of spin $\frac{1}{2}$ to ^1H is ^3He, which is not typical. It belongs to the class of materials known as

[78] H. G. Olf and A. Peterlin, J. Polymer Sci., Part A-2, Polymer Phys., 1970, 8, 771.
[79] H. G. Olf and A. Peterlin, J. Polymer Sci., Part A-2, Polymer Phys., 1970, 8, 753, 791.
[80] H. G. Olf, J. Polymer Sci., Part A-2, Polymer Phys., 1971, 9, 1851.
[81] H. G. Olf and A. Peterlin, J. Polymer Sci., Part A-2, Polymer Phys., 1971, 9, 2033.
[82] M. J. Folkes and I. M. Ward, J. Materials Sci., 1971, 6, 582.
[83] U. Eichhoff and H. G. Zachmann, Makromol. Chem., 1971, 147, 41.
[84] M. Kashiwagi and I. M. Ward, Polymer, 1972, 13, 145.
[85] M. Kashiwagi, M. J. Folkes, and I. M. Ward, Polymer, 1971, 12, 697.
[86] V. J. McBrierty, I. R. McDonald, and I. M. Ward, J. Phys. (D), 1971, 4, 88.
[87] B. Ellis, 'Amorphous Materials', ed. R. W. Douglas and B. Ellis, Wiley, New York, 1972, p. 375.
[88] C. Clark-Monks and B. Ellis, J. Polymer Sci., Part A-2, Polymer Phys., 1970, 8, 2203.
[89] E. Price, D. M. French, and A. S. Tompa, J. Appl. Polymer Sci., 1972, 16, 157.
[90] M. C. Shen and A. Eisenberg, Rubber Chem. Technol., 1970, 43, 95.

quantum solids. Guyer et al.[91] have published a review of n.m.r. experiments in solid helium. In the period under review no papers have been located investigating ^{13}C or ^{15}N n.m.r. using conventional CW techniques, but ^{19}F n.m.r. is still a popular activity.

^{19}F *N.M.R.* Mahajan and Rao[92] investigated motional narrowing in lead fluoride from 223 to 463 K. Second-moment (^{19}F) values indicated that the lattice was rigid below 263 K. Narrowing was attributed to self-diffusion. Habuda and Gagarinsky[93] recorded the ^{19}F resonance in solid HF. They observed a well-resolved doublet of splitting 26 ± 1 G and a second moment of 184 ± 9 G^2, independent of temperature. The proton–fluorine separation was calculated as 1.02 ± 0.03 Å, but when corrections were made for librational motions this was reduced to 0.95 Å. The increase from the gaseous value was attributed to hydrogen-bond formation. The hydrogen bond was found to be linear to within $10°$. Matthews and Gilson[94] examined pyridinium hexafluorophosphate chloride and nitrate using both proton and fluorine n.m.r. The observed narrowing of the proton line was assigned to reorientational motions of the pyridinium ring. The PF_6 group underwent rapid isotropic reorientation down to 175 K where the ^{19}F signal broadened, but even at 77 K there was considerable molecular motion. C_6F_6 was studied between 77 and 270 K by Albert et al.[95] The ^{19}F second moment showed two regions of motional narrowing attributed to two types of C_6F_6 molecules rotating at different rates. The temperature dependence of T_1 and $T_{1\rho}$ was also investigated.

O'Reilly et al.[96] have determined the chemical shift anisotropy of solid fluorine by five separate methods, giving a value of 1050 p.p.m. These methods involve the observation of first and second moments, peak separations, *etc.* as a function of operating frequency. This was reported to be part of a more complete study. Hindermann and Falconer[97] recorded the ^{19}F second moments of XeF_4 as a function of temperature and magnetic field; the chemical shift anisotropy contribution was field-dependent. Measurements at high field allowed the sign of the chemical shift anisotropy tensor to be determined if the tensor was axially symmetric, and also allowed the assumption of axial symmetry to be tested. Hindermann and Falconer found that the resonance narrowed between 230 and 295 K. The low-temperature second moment was surprisingly larger than calculated values, a feature attributed to sample contamination. The high-temperature lineshape was skewed in a manner characteristic of axial symmetry, but not the low-temperature one. Barr and

[91] R. A. Guyer, R. C. Richardson, and L. I. Zane, *Rev. Mod. Phys.*, 1971, **43**, 532.
[92] M. Mahajan and B. D. N. Rao, *Chem. Phys. Letters*, 1971, **10**, 29.
[93] S. P. Habuda and Yu. V. Gagarinsky, *Acta Cryst*, 1971, **B27**, 1677.
[94] C. H. Matthews and D. F. R. Gilson, *Canad. J. Chem.*, 1970, **48**, 2625.
[95] S. Albert, H. S. Gutowsky, and J. A. Ripmeester, *J. Chem. Phys.*, 1972, **56**, 2844.
[96] D. E. O'Reilly, E. M. Peterson, Z. M. El Saffar, and C. E. Scheie, *Chem. Phys. Letters*, 1971, **8**, 470.
[97] D. K. Hindermann and W. E. Falconer, *J. Chem. Phys.*, 1970, **52**, 6198.

Dunell[98] investigated the ^{19}F n.m.r. of the solid addition compounds SF_4AsF_5 and IF_7AsF_5. In each compound they found two distinguishable fluorine sites in the ratios 1:2 and 1:1, consistent with the structures $SF_3^+AsF_6^-$ and $IF_6^+AsF_6^-$. Second-moment values for the first complex indicated the presence of considerable motion, sufficient to average out chemical shift anisotropy. The chemical shift differences between the two sites, determined from the field dependence of second moments, were consistent with solution values. For the iodine complex the second-moment value of 11.6 G^2 (below 220 K) probably corresponds to the rigid lattice. At these temperatures chemical shift anisotropy becomes significant. Forms of the mineral apatite have been the subjects of two investigations.[99,100] Van der Lugt et al.[99] reviewed 12 earlier papers and studied the ^{19}F n.m.r. spectrum of hydroxyapatite. The spectrum consisted of five components in the ratio 1:1:2:1:1, a superposition of a triplet and a doublet. Splittings were orientation-dependent, showing axial symmetry about the c-axis of the crystal. The signals were attributed to O—H - - - F vacancy species and O—H - - - F - - - H—O species. Carolan,[100] using pulsed n.m.r. to monitor resonance frequency, determined the ^{19}F chemical shift tensor in fluoroapatite. It was axially symmetric, with $\sigma_\| - \sigma_\perp = -84$ p.p.m.

^{31}P N.M.R. The other popular spin-½ nucleus is ^{31}P. Dillon and Waddington[101] reported how they used a high-resolution n.m.r. spectrometer to record ^{31}P chemical shifts in nine solid compounds with an accuracy of 1 p.p.m. Phosphorus chemical shifts may well be larger than dipolar linewidths, particularly at high fields. Gibby et al.[102] have examined a single crystal of P_4S_3 and powders of Zn_3P_2, Mg_3P_2, and P_4S_{10} at 99.4 MHz by recording the free induction decay and performing a Fourier Transform to obtain the CW spectrum. The chemical-shift tensors of the four phosphorus atoms in P_4S_3 were readily obtainable. The coincidence in the principal directions of the basal and apical phosphorus chemical shifts was taken as evidence that there is rapid rotation about that axis. The powder patterns were readily interpretable.

Chemical shift anisotropy can be obtained by other methods. These include determinations based on (a) the partial alignment of a molecule in a liquid crystal; (b) the contribution of chemical shift anisotropy to the spin-lattice relaxation rate in a fluid sample, where it may be extracted from the field dependence; (c) the application of various pulse sequences to solids, with the object of reducing or even removing the dipolar broadening which normally masks chemical shift effects. Determinations based on the first two methods are not discussed in this chapter; experiments based on the third are discussed in a later section.

[98] M. R. Barr and B. A. Dunell, Canad. J. Chem., 1970, **48**, 895.
[99] W. Van der Lugt, D. I. M. Knottnerus, and W. G. Perdok, Acta Cryst., 1971, **B27**, 1509.
[100] J. L. Carolan, Chem. Phys. Letters, 1971, **12**, 389.
[101] K. B. Dillon and T. C. Waddington, Spectrochim. Acta, 1971, **27A**, 1381.
[102] M. G. Gibby, A. Pines, W. K. Rhim, and J. S. Waugh. J. Chem. Phys., 1972, **56**, 991.

3 Nuclei of Spin > $\frac{1}{2}$

A. Introduction.—Nuclei with spin $I > \frac{1}{2}$ possess an electric quadrupole moment in addition to a magnetic dipole. Electric-field gradients at the nuclear site attempt to align the electric quadrupole moment of the nucleus in much the same way as a magnetic field attempts to align the magnetic dipole moment. Studies of transitions between different energy levels produced by the electric field at the nucleus give rise to pure nuclear quadrupole resonance (n.q.r.). This quadrupole interaction can be used to investigate the electric-field distribution in the neighbourhood of the nucleus. To obtain full information it is usually necessary to use single-crystal samples, to apply a magnetic field to produce splittings of the pure n.q.r. transitions, and to study the angular dependence of these splittings. The presence of both a magnetic field externally applied and an electric-field gradient, usually produced by the local charge distribution, involves the Zeeman and quadrupole interactions; both interactions attempt to align the nucleus. If one is larger than the other then standard perturbation theory may be applied. For nuclei such as the halogens the quadrupole interaction is normally significantly larger than the Zeeman interaction, whereas for lighter elements ^{2}H, ^{6}Li, ^{7}Li, ^{11}B, *etc.* in typical laboratory fields the converse is true. For ^{14}N, interactions are comparable. Quadrupole interactions over the complete range have recently been reviewed by Lucken.[103] Many papers have been published on the evaluation of quadrupole interactions each year, too many to be included in a chapter of this length without causing a serious imbalance. Consequently, discussion is restricted to the situation where the quadrupole interactions may be considered as a perturbation of the Zeeman interaction. For nuclei with spin $I = 1$ the result is that the n.m.r. transition is split; the splitting is proportional to the product of the nuclear quadrupole moment and the component of the electric-field gradient parallel to the magnetic field. It is thus angle-dependent. The quadrupole interaction is normally an order of magnitude larger than the dipole–dipole interactions responsible for n.m.r. linewidths in solids of spin $\frac{1}{2}$. Consequently, single-crystal measurements can be used to give the quadrupolar interaction tensor to considerable accuracy. Polycrystalline lineshapes tend to be very distinctive. For a nucleus with $I = \frac{3}{2}$ the n.m.r. spectrum typically consists of three lines; the $m = +\frac{3}{2}$ to $+\frac{1}{2}$ and the $m = -\frac{3}{2}$ to $-\frac{1}{2}$ transitions are separated, as in the example of the 1 to 0 and -1 to 0 transitions for a nucleus of $I = 1$. In a first-order perturbation treatment the $m = \frac{1}{2}$ to $-\frac{1}{2}$ transition is unperturbed, but in a second-order treatment angle-dependent shifts of this transition occur, from which the quadrupolar interaction parameters can be derived. Again, polycrystalline lineshapes are distinctive but the problems of calculations of the convoluted pattern of quadrupolar lines, each dipolar broadened, is a formidable one. Motion has a

[103] E. A. C. Lucken, 'Nuclear Quadrupole Coupling Constants', Academic Press, London, 1969. E. A. C. Lucken, 'Physical methods in Heterocyclic Chemistry', Academic Press, London, 1971, Vol. 4.

pronounced effect on the quadrupole tensor, reducing it in a manner characteristic of that particular motion.

B. The Pseudoquadrupole Effect.—Pyykko and Linderberg[104] made the point that electric-field gradients calculated from observed quadrupole interactions may well be in error. The reason is that higher-order terms in the expansion involving nuclear magnetic moments, for example those expansions used in calculating chemical shifts and spin–spin interactions, can give terms proportional to I^2 and of appropriate symmetry. They demonstrated that such terms are not always negligible compared to the normal electric quadrupole interaction. Pyykko[105] subsequently extended the discussion to include the effect of a finite nuclear volume, showing that the major contribution to a term in I^2 arises from the Fermi contact term. In a third paper,[106] Pyykko calculated the size of this nuclear pseudoquadrupolar effect in a number of metals. He developed an equation relating this quantity to the Knight shift and demonstrated that it is typically from 0.1% to a few per cent of the normal quadrupole interaction.

C. Studies of Specific Nuclei.—*Deuterons.* As part of an extensive investigation of the structure of solid and liquid D_2, Meyer and his colleagues[107] reported the n.m.r. lineshape in orientationally disordered hexagonal-close-packed D_2, supplementing earlier work on the cubic-ordered phase.[108,109] Second moments were determined from CW lineshapes, free induction decays, and solid echoes. Extrapolated to a high-temperature limit, the second moment corresponded to the intermolecular nuclear dipole–dipole interaction. At lower temperatures two components were observable; the signal corresponding to $J = I = 1$ molecules broadened more rapidly than that due to $J = 0, I = 2$ molecules. Adriaenssens and Bjorkstam[110] reported the determination of deuteron quadrupole tensors in formic and acetic acid crystals. The formic acid results were as expected, but the acetic acid results were not consistent with molecular orbital calculations. The e^2Qq values were related to the electronegativity of the groups bonded to the CD fragment. Chiba and Soda[111] continued their extensive investigations using deuteron resonance by re-examining the α and β modifications of the oxalic acid dihydrate crystal, corroborating empirical relationships between the e^2Qq value and the hydrogen-bond length, and between the quadrupole principal

---

[104] P. Pyykko and J. Linderberg, *Chem. Phys. Letters*, 1970, **5**, 34.
[105] P. Pyykko, *Chem. Phys. Letters*, 1970, **6**, 479.
[106] P. Pyykko, *J. Phys. (F)*, 1971, **1**, 102.
[107] F. Weinhaus, S. M. Myers, B. Maraviglia, and H. Meyer, *Phys. Rev.*, 1971, **B3**, 3730.
[108] B. Maraviglia, F. Weinhaus, H. Meyer, and R. L. Mills, *Solid State Comm.*, 1970, **8**, 815.
[109] B. Maraviglia, F. Weinhaus, H. Meyer, and R. L. Mills, *Solid State Comm.*, 1970, **8**, 1683.
[110] G. J. Adriaenssens and J. L. Bjorkstam, *J. Chem. Phys.*, 1972, **56**, 1223.
[111] T. Chiba and G. Soda, *Bull. Chem. Soc. Japan*, 1971, **44**, 1703.

axes and the hydrogen-bond direction. Zaucer et al.[112] calculated deuteron quadrupole coupling tensors in the situation where a hydrogen bond is symmetric. Consistency with observed values was good, but the authors concluded that the remainder of the molecular system makes a significant contribution. White and Drago[113] calculated ^{2}H and ^{14}N e^2Qq values in a number of compounds using an extended Hückel-type calculation. As discussed earlier, Cvikl and McGrath[32] supplemented their proton studies on calcium bromate monohydrate by using deuteron resonance, which indicated that the hydrogen bonds were slightly bent. Royston and Smith[114] reported the deuteron resonance of single crystals of deuteriated cupric acetate hydrate, $Cu(CD_3CO_2)_2,D_2O$, at 273 and 77 K. The CD_3 group is one of the few such groups studied. Motional narrowing occurred, but the deuteron quadrupole coupling tensor did not show axial symmetry. O'Reilly and his colleagues[115,116] examined the H_3O^+ group. In the first paper, hydronium perchlorate was studied using deuteron resonance and proton relaxation. Above a phase transition at 243 K the deuteron spectrum consisted of a single line, indicating a rapid isotropic motion. Below 243 K, $e^2Qq = 72$ kHz, and by 130 K it had reached the rigid-lattice value. Below 130 K the proton second moment was 31 G^2; above 160 K it had reduced to 10 G^2. The chlorine n.m.r. line was observable above but not below 160 K. Using this information, bond distances and angles were determined. In the second paper, O'Reilly et al. used ^{1}H and ^{2}H n.m.r. and ^{35}Cl n.q.r. to study chloroauric acid tetrahydrate, $H_5O_2^+AuCl_4^-,2H_2O$, from 180 to 300 K, i.e. through phase transitions at 218 and 290 K.

Burnett and Muller[117] have determined deuteron e^2Qq values in polycrystalline samples of C_2D_6, C_4D_{10}, and C_6D_{12} between 63 and 90 K, with no observable temperature dependence. The C_6D_{12} value was 170 kHz whereas that for ethane was 38 kHz, but at low temperatures the latter did increase to a comparable value. At liquid-helium temperatures the e^2Qq value was extracted from the free induction decay. O'Reilly and Eraker[118] recorded deuteron quadrupole interactions in three phases of polycrystalline D_2S between 77 and 187 K. Wei and Fung[119] determined e^2Qq and asymmetry factors (η) from the powder spectra of $(C_5H_5)_2MoD_2$ and $(C_5H_5)_2WD_2$. The deuteron e^2Qq value is related to the force constant in covalent deuterides. O'Reilly[120] discussed the shift in deuteron and ^{35}Cl e^2Qq values between gaseous and solid phases in HCl and DCl.

[112] M. Zaucer, E. Zakrajsek, J. Koller, D. Hadzi, and A. Azman, *Mol. Phys.*, 1971, **21**, 461.
[113] W. D. White and R. S. Drago, *J. Chem. Phys.*, 1970, **52**, 4717.
[114] J. Royston and J. A. S. Smith, *Trans. Faraday Soc.*, 1970, **66**, 1039.
[115] D. E. O'Reilly, E. M. Peterson, and J. M. Williams, *J. Chem. Phys.*, 1971, **54**, 96.
[116] D. E. O'Reilly, E. M. Peterson, C. E. Scheie, and J. M. Williams, *J. Chem. Phys.*, 1971, **55**, 5629.
[117] L. J. Burnett and B. H. Muller, *J. Chem. Phys.*, 1971, **55**, 5829.
[118] D. E. O'Reilly and J. H. Eraker, *J. Chem. Phys.*, 1970, **52**, 2407.
[119] I. Y. Wei and B. M. Fung, *J. Chem. Phys.*, 1971, **55**, 1486.
[120] D. E. O'Reilly, *J. Chem. Phys.*, 1970, **52**, 2396.

Deuteron resonance has proved to be a powerful technique for the investigation of ferroelectricity. Blinc et al.[121] reported deuteron resonance and relaxation in ferroelectric KD_2PO_4, KD_2AsO_4, and CsD_2AsO_4 crystals at two temperatures for each crystal and each orientation. At particular orientations smaller temperature intervals were used. Deuteron relaxation times were determined by rapid-passage techniques. O'Reilly et al.[122] recorded the deuteron resonance of a single crystal of thiourea at 300 and 120 K in phases V and I, the latter being ferroelectric. Proton T_1 and $T_{1\rho}$ values were also observed. At 300 K a 180° flip motion about the CS axis occurred but the principal axis of the outer deuterons did not lie in the molecular plane.

Sternheimer Antishielding. The Sternheimer antishielding of quadrupolar interactions is significant for heavier atoms. Sternheimer and Peierls[123] calculated improved antishielding factors for a number of alkali isotopes. Accurate calculations of the field gradient can be scaled by a Sternheimer shielding factor and more reliable values of nuclear quadrupole coupling constants determined.

^7Li N.M.R. Ossman and Silvidi[124] determined the ^7Li quadrupole coupling in polycrystalline lithium phosphide, finding $e^2Qq = 68.5$ kHz with axial symmetry, a value slightly less than that for lithium antimonide, 75.2 kHz. Another site in the phosphide crystals had an upper limit of 16 kHz and in the antimonide a value of 17.6 kHz. The temperature dependence was studied in the range 138—343 K. Narrowing occurring at 243 K was attributed to diffusion. Calculations were not consistent with either a purely covalent or a purely ionic model, but with a model used previously in the study of the equivalent sodium compounds.[125] Halstead[126] recorded the ^7Li n.m.r. of lithium niobate from 297 to 953 K. The observed e^2Qq value, somewhat unusually, increased uniformly with temperature. This was attributed to an anisotropic motion. A combination of point-charge models and covalent bonding was used to estimate electric-field gradient values and their temperature dependence. Irwin and Cotts[127] found an electric-field-induced splitting of the ^7Li spectrum of impurity Li in a KCl lattice, confirming observations of stress-induced quadrupolar splittings by Alderman and Cotts.[128]

Svanson and Johansson[129] examined ^7Li n.m.r. in silica-rich lithium silicate glasses. This is part of a larger study involving dielectric measurement. Lineshapes were consistent with there being two components, one due to large clusters and the other to smaller groups in voids. The lineshape was dependent upon heat treatment and composition. Activation energies determined

---

[121] R. Blinc, J. Stepisnik, M. Jamsek-Vilfan, and S. Zumer, *J. Chem. Phys.*, 1971, **54**, 187.
[122] D. E. O'Reilly, E. M. Peterson, and Z. M. El Saffar, *J. Chem. Phys.*, 1971, **54**, 1304.
[123] R. M. Sternheimer and R. F. Peierls, *Phys. Rev.*, 1971, **A3**, 837.
[124] G. W. Ossman and A. A. Silvidi, *J. Chem. Phys.*, 1971, **54**, 979.
[125] G. W. Ossman, A. A. Silvidi, and J. W. McGrath, *J. Chem. Phys.*, 1970, **52**, 509.
[126] T. K. Halstead, *J. Chem. Phys.*, 1970, **53**, 3427.
[127] D. M. Irwin and R. M. Cotts, *Phys. Rev.*, 1971, **B4**, 235.
[128] D. W. Alderman and R. M. Cotts, *Phys. Rev.*, 1970, **B1**, 2870.
[129] S. E. Svanson and B. Johansson, *Acta Chem. Scand.*, 1970, **24**, 735.

from the temperature dependence of the linewidth were low compared with those derived from d.c. conductivity of dielectric measurements. Three-component glasses were also investigated to some extent.

^9Be *N.M.R.* Blinc et al.,[130] using ^9Be n.m.r., investigated the ferroelectric transition in a single crystal of deuteriated triglycine fluoroberyllate. The paraelectric–ferroelectric phase transition occurring between 340 and 360 K was consistent with an order–disorder model. Complete quadrupole tensors were determined at two temperatures.

^{11}B *N.M.R.* Kriz et al.,[131] continuing their work on boron glasses, have reinterpreted earlier work[132-134] on the properties of three- and four-co-ordinated ^{11}B atoms in boron glasses, on the basis of more accurate powder lineshapes for the $+\frac{1}{2}$ to $-\frac{1}{2}$ transition and of measurements of single-crystal specimens (involving, for example, ^{11}B resonance for trigonal BO_3 units with one and two non-bridging oxygen atoms[135]). Kriz and Taylor[136] computer-fitted the ^{11}B spectra of BF_3 reported by Casabella and Oja[137] and discussed the factors determining the prominent features of the powder lineshape of the $+\frac{1}{2}$ to $-\frac{1}{2}$ transitions broadened by dipolar interactions with three fluorine atoms. Values of the ^{11}B e^2Qq value and BF bond distance were determined from the analysis. Rhee and Bray[138] remeasured the ^{11}B spectra of polycrystalline boron oxide at three frequencies (6, 10, and 16 MHz) in an attempt to distinguish between various proposed structural models. Computer-fitted spectra were consistent with a single-site model with $e^2Qq = 2.69$ MHz and $\eta = 0.06$. Values derived from a two-site model converged on that for the single site. Hynes and Alexander[139] reported the observation of the ^{11}B n.m.r. spectrum of β rhombohedral boron and of boron carbide at room temperature. Powder patterns indicated the presence of two species with two quadrupole interactions.

^{23}Na *N.M.R.* D'Alessio and Scott[140] reported the pressure dependence of the ^{23}Na quadrupole interaction in sodium nitrate from 298 to 568 K through a phase transition. In contradiction to the findings[141,142] of Bernheim and Gutowsky, dv_Q/dp was observed to be negative, but the slope was less negative at higher temperatures. Hikita et al.[143] examined the central transition of the

[130] R. Blinc, J. Slak, and J. Stepisnik, *J. Chem. Phys.*, 1971, **55**, 4848.
[131] H. M. Kriz, M. J. Park, and P. J. Bray, *Phys. and Chem. Glasses*, 1971, **12**, 45.
[132] P. C. Taylor and P. J. Bray, *J. Magn. Resonance*, 1970, **2**, 305.
[133] J. F. Baugher, H. M. Kriz, P. C. Taylor, and P. J. Bray, *J. Magn. Resonance*, 1970, **3**, 415.
[134] H. M. Kriz and P. J. Bray, *J. Magn. Resonance*, 1971, **4**, 69.
[135] H. M. Kriz and P. J. Bray, *J. Magn. Resonance*, 1971, **4**, 76.
[136] H. M. Kriz and P. C. Taylor, *J. Chem. Phys.*, 1971, **55**, 2601.
[137] P. A. Casabella and T. Oja, *J. Chem. Phys.*, 1969, **50**, 4814.
[138] C. Rhee and P. J. Bray, *J. Chem. Phys.*, 1972, **56**, 2476.
[139] T. V. Hynes and M. N. Alexander, *J. Chem. Phys.*, 1971, **54**, 5296.
[140] G. J. D'Alessio and T. A. Scott, *J. Magn. Resonance*, 1971, **5**, 416.
[141] R. A. Bernheim and H. S. Gutowsky, *J. Chem. Phys.*, 1960, **32**, 1072.
[142] R. A. Bernheim and H. S. Gutowsky, *J. Chem. Phys.*, 1970, **54**, 1431.
[143] T. Hikita, M. Kasahara and I. Tatsuzaki, *Phys. Letters*, 1971, **37A**, 141.

^{23}Na n.m.r. of ferroelectric AgNa(NO$_2$)$_2$, supplementing earlier work[144] where the ^{23}Na quadrupole interactions had been determined between 298 and 318 K through the phase transition between ferroelectric and paraelectric states. At certain crystal orientations in the ferroelectric phase the central line had been observed to broaden. The more recent measurements show a splitting. Yagi et al.[145] reported measurements of ^{23}Na n.m.r. and dielectric constant in sodium nitrite over a temperature range of 1.2 K through the ferroelectric phase transition. Ossman et al.[125] reported measurements of the central transition in powdered samples of sodium phosphide, showing the existence of two ^{23}Na components with distinctive quadrupole couplings, as in sodium arsenide and antimonide.

Other Nuclei. Borsa et al.[146] determined the e^2Qq values of ^{27}Al and ^{139}La in addition to the temperature dependence of the ^{27}Al T_1 of lanthanum aluminate through a phase transition at 800 K. Hafner et al.[147] determined ^{27}Al quadrupole couplings at two sites in an andalusite crystal. Site assignments were made on the basis of a point-charge model. A notable feature of this paper is that the authors did not employ perturbation theory in their analysis; they diagonalized the spin matrix directly. Electric-field gradient tensors were compared with the hyperfine interactions of paramagnetic ions substituted on the aluminium sites. In a subsequent paper, Raymond and Hafner[148] determined ^{27}Al quadrupole interactions at two sites in sillimanite, Al$_2$SiO$_5$, to supplement those in andalusite.

By observing the central transition, Allerhand[149] determined ^{51}V e^2Qq and η values for VOF$_3$ at 301 K and VOCl$_3$ at 167 K. The values were compared with seven other reported values of vanadium coupling constants. Adriaenssens and Bjorkstam,[73] in an investigation of ferroelectricity in ammonium dihydrogen phosphate and arsenate, measured the quadrupole coupling constants of ^{75}As and ^{14}N in addition to recording the angular dependence of proton second moments above and below the phase transitions.

Niobates have been the subject of a number of investigations.[126,150-152] Schempp et al.[150] investigated the temperature dependence of the ^{93}Nb e^2Qq value in LiNbO$_3$ powder samples from 21 to 515 K. A number of references to earlier work are included. Peterson et al.[151] investigated the ^{93}Nb n.m.r. of sintered powders of the solid solution of LiNbO$_3$ and LiTaO$_3$. The ^{93}Nb e^2Qq value decreased with increasing tantalate concentration in a manner consistent with calculated field gradients based on a point-charge model. Interest in the e^2Qq value and linewidth arises because they are related

---

[144] M. Kasahara, T. Hikita and I. Tatsuzaki, *J. Phys. Soc. Japan*, 1970, **29**, 240.
[145] T. Yagi, I. Tatsuzaki, and I. Todo, *J. Phys. Soc. Japan*, 1970, **28**, 321.
[146] F. Borsa, M. L. Crippa, and B. Derighetti, *Phys. Letters*, 1971, **34A**, 5.
[147] S. S. Hafner, M. Raymond, and S. Ghose, *J. Chem. Phys.*, 1970, **52**, 6037.
[148] M. Raymond and S. S. Hafner, *J. Chem. Phys.*, 1970, **53**, 4110.
[149] A. Allerhand, *J. Chem. Phys.*, 1970, **52**, 162.
[150] E. Schempp, G. E. Peterson, and J. R. Carruthers, *J. Chem. Phys.*, 1970, **53**, 306.
[151] G. E. Peterson, J. R. Carruthers, and A. Carnevale, *J. Chem. Phys.*, 1970, **53**, 2436.
[152] F. Wolf, D. Kline, and H. S. Story, *J. Chem. Phys.*, 1970, **53**, 3538.

to defect concentration, which is greatest at a 50% mix. Wolf et al.[152] measured ^{23}Na and ^{93}Nb e^2Qq values as a function of temperature. ^{93}Nb e^2Qq values are large, so the authors developed expressions based on a third-order perturbation theory, but in practice these were not required. Electric-field gradient calculations based only on point-charge models gave values an order of magnitude too small, demonstrating the importance of covalent bonding.

Cs_2O–B_2O_3 glasses have been examined as a function of composition by Rhee and Bray,[153,154] using both ^{11}B and ^{133}Cs n.m.r. The boron e^2Qq values derived from the powder pattern of the central transition were used to identify the boron–oxygen groups present. Curve-fitting of measurements made at different operating frequencies allowed the ^{133}Cs quadrupole couplings and chemical shift anisotropies to be determined. Both quantities increased up to a 35% Cs_2O composition and then decreased. The existence of chemical shift anisotropy showed that caesium–oxygen bonds had a partial covalent character. In the second paper the temperature dependences of these quantities were investigated from 77 to 773 K. Chemical shift anisotropy and e^2Qq values were constant to room temperature and then decreased, the decrease being attributed to thermally activated diffusion. The activation energies derived from each quantity were lower than that from d.c. conductivity measurements.

Barrere and Leroy[155] studied the rf saturation of ^{139}La and ^{2}H resonance signals in lanthanum deuterides as a function of stoicheiometry. Assumptions of Gaussian and Lorentzian local fields were used to modify the Bloch equations in order to obtain T_1 values.

Fukushima[156] examined the ^{19}F n.m.r. of $KBiF_6$ powder. Curve-fitting enabled both $^1J(BiF)$ and the quadrupolar interaction of the ^{209}Bi nucleus to be determined, together with their relative signs.

There have been a number of determinations of quadrupole interaction tensors based on the technique of pulsed double-resonance. An example of such a determination is that of Blinc et al.,[157] who obtained ^{14}N interactions in triglycine sulphate. In a related but alternative technique, Edmonds and Speight[158] applied a double-resonance technique to ^{14}N quadrupole determinations of a sequence of amino-acids at 77 K. This method involves spin mixing in the laboratory frame. These techniques are discussed in Chapter 3.

4 Lineshape Calculations

The problem of the calculation of the n.m.r. lineshape of a solid is still a challenging one. It has been discussed and treated over a wide range of

---

[153] C. Rhee and P. J. Bray, *Phys. and Chem. Glasses*, 1971, **12**, 165.
[154] C. Rhee and P. J. Bray, *Phys. and Chem. Glasses*, 1971, **12**, 156.
[155] H. Barrere and J. L. Leroy, *Compt. rend.*, 1971, **272**, C, 1533.
[156] E. Fukushima, *J. Chem. Phys.*, 1971, **55**, 2463.
[157] R. Blinc, M. Mali, R. Osredkar, A. Prelesnik, I. Zupančič, and L. Ehrenberg, *J. Chem. Phys.*, 1971, **55**, 4843.
[158] D. T. Edmonds and P. A. Speight, *Phys. Letters*, 1971, **34A**, 325.

complexity and rigour. Parker[159] has written an introductory paper discussing the problems of lineshape calculations, using a local-field approach to demonstrate that shapes should be nearly Gaussian. By doing this he has neglected the effects of spin-flip terms other than by the introduction of a scaling factor. Parker[160] has expressed the free induction decay (a Fourier transform of the CW lineshape) as a Neumann expansion using moments, and Lado et al.[161] calculated CaF_2 lineshapes by relating the free induction decay to its associated Memory function in a general manner. A survey of available experimental data and a review of earlier papers was given by Fornes et al.[162] They applied various theories to the ^{19}F resonance signal in the alkali halides SrF_2, CsF, and NaCl, and concluded that best results are obtained by a combination of the calculations of Evans and Powles[163] and of Lee et al.[164] Betsuyaku[165] calculated the lineshape of CaF_2 on the basis of Kubo–Tomita theory, thereby expanding the calculations of Evans and Powles to higher terms. Powles and Carazza[166,167] applied information theory. They calculated optimum lineshapes based on knowledge of only a limited number of moments. This has been used by Canters and Johnson[168] to calculate second and fourth moments for dipolar-broadened systems at different spin concentrations. At low concentration, moments were used to obtain a truncated Lorentzian, but at high concentration the moments were used to fit expressions given by Powles and Carazza. Canters and Johnson were interested in the use of magnetically dilute samples to improve resolution. Gill[169] presented a statistical method of calculating lineshapes based on a local-field approximation. It is expected to be most effective in dilute systems. Demco and Ceausescu[170] continued an earlier interest in the lineshape problem by using a generalized stationary two-time Greens function.

Van Steenwinkel,[171] investigating the effects of rf irradiation on n.m.r. lines in solids, considered the dynamics of the process leading to the establishment of equilibrium, extending earlier work[172] on the equilibrium situation. Agreement with the Provotorov[173] theory was good except where τ_c was long, when the Provotorov theory was correct only at small B_1 values. Betsuyaku[174] calculated the effect of saturation in solids, using limiting examples of weak

[159] G. W. Parker, Amer. J. Phys., 1970, **38**, 1432.
[160] G. W. Parker, Phys. Rev., 1970, **B2**, 2453.
[161] F. Lado, J. D. Memory, and G. W. Parker, Phys. Rev., 1971, **B4**, 1406.
[162] R. E. Fornes, G. W. Parker, and J. D. Memory, Phys. Rev., 1970, **B1**, 4228.
[163] W. A. B. Evans and J. G. Powles, Phys. Letters, 1967, **24A**, 218.
[164] M. Lee, D. Tse, W. I. Goldburg, and I. J. Lowe, Phys. Rev., 1967, **158**, 246.
[165] H. Betsuyaku, Phys. Rev. Letters, 1970, **24**, 934.
[166] J. G. Powles and B. Carazza, 'Magnetic Resonance', Plenum Press, New York, 1970, p. 133.
[167] J. G. Powles and B. Carazza, J. Phys. (A), 1970, **3**, 335.
[168] G. W. Canters and C. S. Johnson, jun., J. Magn. Resonance, 1972, **6**, 1.
[169] J. C. Gill, J. Phys. (C), 1971, **4**, 1420.
[170] D. Demco and V. Ceausescu, Rev. Roumaine Phys., 1971, **16**, 1093.
[171] R. Van Steenwinkel, Z. Naturforsch., 1971, **26a**, 1825.
[172] J. Haupt and R. Van Steenwinkel, Z. Naturforsch., 1971, **26a**, 260.
[173] B. N. Provotorov, Soviet Phys. JETP, 1962, **14**, 1226.
[174] H. Betsuyaku, J. Phys. Soc. Japan, 1971, **30**, 641.

and strong B_1 fields. He used a system of projection operators. Shimizu[175,176] extended earlier work to discuss saturational narrowing, and to provide a general treatment of spin resonance and relaxation in a system of interacting spins.

Cobb and Johnson[177] discussed the lineshape of a methyl group in a solid, observing the transition from quantum-mechanical tunnelling to thermally activated rotation. In a later paper[178] they used lineshapes calculated for random motion[177] and for tunnelling motion [179] to identify the type of motion occurring in methyl chloroform and methyltrichlorosilane. They found that tunnelling occurred in the silane but not in chloroform. Lineshapes and T_1 values in these compounds were reported by McIntyre and Johnson.[180] Mottley et al.[181] reported the measurement of a tunnelling frequency obtained directly from the n.m.r. lineshape of CH_3CD_2I diluted in CD_3CD_2I, confirming the calculations of Apaydin and Clough.[179] Clough et al.[182] observed weak tunnelling side-bands in the e.s.r. spectrum of a methyl group tunnelling at 4 K, thereby providing further confirmation. Clough[183] extended the discussion to calculate the second moment of a tunnelling group. He predicted that the second moment will approach a value corresponding to one half that of a stationary methyl group.

Lynden-Bell[184] discussed the calculation of the lineshape of a polycrystalline material when a group reorients at a rate comparable to the linewidth. Calculations were of a density-matrix type involving spin operators and Euler angles. The operators were selected to be of suitable symmetry in the Liouville space of spins and orientations. McDonald and Clough[185] and McDonald[186] have calculated lineshapes in the situation where a sample is rotated about axes either perpendicular to the field or at the magic angle, using either the Lowe–Norberg expansion[187] or those due to Clough and McDonald.[188]

Reeves and Wilson[189] computed lineshapes of modulation-broadened absorption or dispersion Lorentzian lines, prompted by their observations of ^{61}Ni n.m.r. in bulk metal.[190] France and Hooper[191] calculated the lineshapes

[175] T. Shimizu, *J. Phys. Soc. Japan*, 1970, **28**, 811.
[176] T. Shimizu, *J. Phys. Soc. Japan*, 1970, **29**, 74.
[177] T. B. Cobb and C. S. Johnson, jun., *J. Chem. Phys.*, 1970, **52**, 6224.
[178] H. M. McIntyre, T. B. Cobb, and C. S. Johnson, jun., *Chem. Phys. Letters*, 1970, **4**, 585.
[179] F. Apaydin and S. Clough, *J. Phys. (C)*, 1968, **1**, 932.
[180] H. M. McIntyre and C. S. Johnson, jun., *J. Chem. Phys.*, 1971, **55**, 345.
[181] C. Mottley, T. B. Cobb, and C. S. Johnson, jun., *J. Chem. Phys.*, 1971, **55**, 5823.
[182] S. Clough, J. Hill, and F. Poldy, *J. Phys. (C)*, 1972, **5**, 1739.
[183] S. Clough, *J. Phys. (C)*, 1971, **4**, 1075.
[184] R. M. Lynden-Bell, *Mol. Phys.*, 1971, **22**, 837.
[185] I. R. McDonald and S. Clough, *Physica*, 1970, **45**, 546.
[186] I. R. McDonald, *Physica*, 1971, **51**, 273.
[187] I. J. Lowe and R. E. Norberg, *Phys. Rev.*, 1957, **107**, 46.
[188] S. Clough and I. R. McDonald, *Proc. Phys. Soc.*, 1965, **86**, 833.
[189] G. K. Reeves and G. V. H. Wilson, *J. Phys. (D)*, 1970, **3**, 1609.
[190] G. K. Reeves, R. Street, and G. V. H. Wilson, *J. Phys. (C), Metal Phys. Suppl. No. 2*, 1970, S241.
[191] P. W. France and H. O. Hooper, *J. Phys. and Chem. Solids*, 1970, **31**, 2223.

of the central transition in the presence of second-order quadrupolar interactions and anisotropic chemical or Knight shifts. Buckmaster and Skirrow[192] considered the effect of modulation broadening on unsaturated Lorentzian lines. The applications of these papers, like those of the following one,[193] are directed towards e.s.r. but the papers are of some relevance to n.m.r. Maruani derived[194] analytical expressions for the spectral lineshape of multi-spin systems as a function of rf level for various amounts of homogeneous and inhomogeneous broadening. The paper is particularly relevant to the observation of triplet species.

5 Line-narrowing Techniques

A. Introduction.—There are basically two techniques for reducing linewidths in solids. The first involves high-speed macroscopic rotation of the sample about an axis inclined at an angle of 54° 44' to the applied field,[195] and the second the application of various pulse sequences.[196]

B. Sample Spinning.—Andrew and Jasinski[197] considered the application of macroscopic rotation in the situation where considerable motion already exists. If motions are orthogonal, macroscopic rotation can produce further narrowing. In a subsequent calculation[198] the theory was extended to include chemical shift anisotropy and the indirect spin–spin interaction. Kubo–Tomita theory was used to calculate a lineshape. Andrew et al.[199] reported the observation of ^{19}F spin multiplets in KPF_6, $NaPF_6$, and $KSbF_6$. In the latter sample, scalar spin–spin interactions with both ^{121}Sb and ^{123}Sb were observed. Andrew et al. applied the technique to metallic samples. They report that the technique yields an order-of-magnitude improvement in the precision of Knight-shift determinations,[200] and show that the reduction of dipolar broadening and Knight-shift anisotropy allow a determination of the Ruderman–Kittel interaction for copper.[201] In a third paper,[202] chemical shifts of ^{63}Cu and ^{65}Cu in insulating materials were measured relative to metallic copper; no isotope dependence was observed. Doskocilova and Schneider[203] have examined some polymers and liquids sorbed on, or dispersed in, solid lattices. In the latter example, resolution equivalent to that observed in liquids was obtained. Doehler and Schnabel[204] examined the

---

[192] H. A. Buckmaster and J. D. Skirrow, *J. Magn. Resonance*, 1971, **5**, 285.
[193] H. A. Buckmaster and J. D. Skirrow, *J. Appl. Phys.*, 1971, **42**, 1225.
[194] J. Maruani, *J. Magn. Resonance*, 1972, **7**, 207.
[195] E. R. Andrew, *Progr. N.M.R. Spectroscopy*, 1971, **8**, 1.
[196] P. Mansfield, *Progr. N.M.R. Spectroscopy*, 1971, **8**, 41.
[197] E. R. Andrew and A. Jasinski, *J. Phys. (C)*, 1970, **4**, 391.
[198] E. R. Andrew and A. Jasinski, Proceedings XVIth Colloque Ampère, 1971, p. 1019.
[199] E. R. Andrew, M. Firth, A. Jasinski, and P. J. Randall, *Phys. Letters*, 1970, **31A**, 446.
[200] E. R. Andrew, J. L. Carolan, and P. J. Randall, *Phys. Letters*, 1971, **35A**, 435.
[201] E. R. Andrew, J. L. Carolan, and P. J. Randall, *Phys. Letters*, 1971, **37A**, 125.
[202] E. R. Andrew, J. L. Carolan, and P. J. Randall, *Chem. Phys. Letters*, 1971, **11**, 298.
[203] D. Doskocilova and B. Schneider, *Chem. Phys. Letters*, 1970, **6**, 381.
[204] H. Doehler and B. Schnabel, *Ann Phys. (Leipzig)*, 1970, **25**, 383.

narrow component of a polymer resonance to investigate the line-narrowing process as a function of rotation speed.

C. Pulse Sequences.—Mansfield[205] reported a pulse cycle suitable for line-narrowing. The sequence was described as a four-pulse 6τ cycle. Mehring et al.[206] reported that the application of their four-pulse sequence to reduce dipolar broadening permits the observation of ^{19}F chemical-shift powder patterns in eight fluoro-compounds. Observed shifts were compared with those calculated on the basis of the theories of Karplus and Das.[207] Ellett et al.[208] reported the resolution of an AB quartet in the ^{19}F n.m.r. of solid perfluorocyclohexane as a consequence of applying a line-narrowing pulse sequence. No evidence of chemical shift anisotropy was obtained. Andrew[209] noted that the molecules undergo isotropic rotation about their lattice site, reducing to zero the intramolecular contributions to dipolar broadening and chemical shift anisotropy, leaving only intermolecular dipolar broadening. Haeberlen et al.[210] discussed the effect of resonance offsets on 4-pulse, 6-pulse, and phase-alternated tetrahedral-angle pulse sequences. These pulse sequences nominally reduce dipolar interactions between the observed spins to zero, and with other spins by a factor of $\sqrt{3}$. Chemical shift anisotropy is similarly reduced by a factor of $\sqrt{3}$. Mehring et al.[211] irradiated a spin species S that is causing dipolar broadening of the I spin in order to produce further line-narrowing. Hanabusa and Kushida[212] and Hanabusa[213] applied these techniques to improve sensitivity by mapping out broad absorption lines. The extended decays permitted a more efficient averaging process. Rhim et al.[214] reported an experiment that violates the spin-temperature hypothesis. Kessemeier and Rhim[215] reported a line-narrowing experiment involving the use of rotary spin echoes. Glycerine and Teflon were used as test materials. The technique averages inhomogeneities in the static magnetic field and in B_1. Bleich and Redfield[216] described a technique for determining the high-resolution n.m.r. of rare spins in solids, involving a combination of the phase-incoherent double-resonance technique of sensitivity enhancement and spin decoupling. When it was applied to CaF_2, ^{43}Ca linewidths of 130 Hz were readily observed.

6 Surface Phenomena

A. Introduction.—The n.m.r. of molecules in dispersed systems is very much

[205] P. Mansfield, Phys. Letters, 1970, 32A, 485.
[206] M. Mehring, R. G. Griffin, and J. S. Waugh, J. Chem. Phys., 1971, 55, 746.
[207] M. Karplus and T. P. Das, J. Chem. Phys., 1961, 34, 1683.
[208] J. D. Ellett, jun., U. Haeberlen, and J. S. Waugh, J. Amer. Chem. Soc., 1970, 92, 411.
[209] E. R. Andrew, Phys. Letters, 1970, 32A, 520.
[210] U. Haeberlen, J. D. Ellett, jun., and J. S. Waugh, J. Chem. Phys., 1971, 55, 53.
[211] M. Mehring, A. Pines, W. K. Rhim, and J. S. Waugh, J. Chem. Phys., 1971, 54, 3239.
[212] M. Hanabusa and T. Kushida, Bull Amer. Phys. Soc., 1970, 15, 105.
[213] M. Hanabusa, J. Appl. Phys., 1971, 42, 1077.
[214] W. K. Rhim, A. Pines, and J. S. Waugh, Phys. Rev. Letters, 1970, 25, 218.
[215] H. Kessemeier and W. K. Rhim, Phys. Rev., 1972, B5, 761.
[216] H. E. Bleich and A. G. Redfield, J. Chem. Phys., 1970, 55, 5405.

influenced by molecular interactions with surfaces and with macromolecules within the medium. The terminology used in describing the n.m.r. of dispersed systems is similar to that used in discussing the n.m.r. of absorbed phases, so dispersed systems will be discussed alongside systems exhibiting various surface phenomena. There have been two recent n.m.r. reviews of surface phenomena.[217,218]

B. **Adsorption on to Silica, Alumina, and Titania Surfaces.**—Bonardet et al.[219] measured chemical shifts of methane, TMS, and benzene adsorbed on to various gels, silica, alumina, and titania as a function of partial pressure. Susceptibility corrections were made. Pearson[220] reported proton wide-line studies of hydroxy-groups and of water adsorbed on to two types of transition alumina. Boddenberg et al.[221,222] measured the diffusion coefficient of benzene adsorbed on to a silica surface. In a later paper[223] an improved treatment was presented, and corrections were made for mobility within the gas phase of the pore volume. There were discrepancies between activation energy values determined from T_2 and diffusion measurements. Mengenhauser[224] was interested in the coalescence of hydrocarbon–water emulsions. He recorded the linewidths and chemical shifts relative to TMS of water adsorbed on to silica, alumina, and titania surfaces. Pickett et al.[225,226] had reported similar measurements; for example, the chemical shifts of methanol adsorbed on to silica were measured as a function of silanization. Fripiat[227] has written an introductory article on the applications of n.m.r. to surface chemistry, taking a number of examples, mostly from his own work, that are concerned mainly with determinations of diffusion rates and activation energy barriers.

C. **Zeolites and Exchange Resins.**—Mestdagh et al.[228] measured the second moment, T_1, and T_2 values of a decationated Y zeolite from 93 to 713 K. Spin–lattice relaxation was attributed to diffusion, allowing the diffusion coefficient D to be evaluated. If it is assumed that the T_1 and T_2 relaxations are determined by the same process, then the mean jump distance and nearest approach to the paramagnetic centre can be determined. Egerton and Green[229] studied the n.m.r. of ethylene adsorbed on to sodium and calcium zeolites. On sodium X zeolite, adsorption gave rise to a broad line which narrowed at

[217] K. J. Packer, *Progr. N.M.R. Spectroscopy*, 1967, **3**, 86.
[218] H. A. Resing, *Adv. Mol. Relaxation Processes*, 1967, **1**, 109.
[219] J. L. Bonardet, A. Snobbart, and J. Fraissard, *Compt. rend.*, 1971, **272**, C, 1836.
[220] R. M. Pearson, *J. Catalysis*, 1971, **23**, 384.
[221] B. Boddenberg, R. Haul, and G. Oppermann, *Ber. Bunsengesellschaft phys. Chem.*, 1971, **75**, 1054.
[222] B. Boddenberg, R. Haul, and G. Oppermann, *Surface Sci.*, 1970, **22**, 29.
[223] B. Boddenberg, R. Haul, and G. Oppermann, *J. Colloid Interface Sci.*, 1972, **38**, 210.
[224] J. V. Mengenhauser, *Technical Report USAMERDC*, No 2002, 1971, Fort Belvoir.
[225] J. H. Pickett and L. B. Rogers, *Separation Sci.*, 1970, **5**, 11.
[226] J. H. Pickett, C. H. Lochmuller, and L. B. Rogers, *Separation Sci.*, 1970, **5**, 23.
[227] J. J. Fripiat, *Catalysis Rev.*, 1971, **5**, 269.
[228] M. M. Mestdagh, W. E. Stone, and J. J. Fripiat, *J. Phys. Chem.*, 1972, **76**, 1220.
[229] T. A. Egerton and R. D. Green, *Trans. Faraday Soc.*, 1971, **67**, 2699.

higher adsorptions. At still higher adsorptions separate peaks were observable and chemical shifts were measurable. On sodium Y zeolite a narrow line was observed, which widened and then narrowed as adsorption increased. On calcium X and Y zeolites there was no uniform behaviour but the detailed explanations seem to be satisfactory. Muha[230] also examined ethylene adsorbed on to KX-, CdX-, and AgX-exchanged zeolites. With the KX zeolites he found a smooth temperature variation of linewidth. With CdX the line broadened below 90 K, showing some structure, but the linewidth was small compared to the rigid-lattice value. On the silver zeolite the line broadened rapidly to the rigid-lattice value. Vucelic et al.[231] reported measurements of the T_1 and T_2 values of methane sorbed on to a synthetic zeolite as functions of temperature and operating frequency. Freude et al.[232] investigated the hydroxy-groups of type Y decationated zeolites. Lineshapes were studied as functions of degree of decationation, pretreatment temperature, and measurement temperature. When pretreated at 673 K, the linewidth decrease occurred at a higher temperature. Second-moment values were used to give proton separations. In a related paper, Freude et al.[233] examined the n.m.r. lineshapes and moments of hydroxy-groups on silica and Y zeolites. They derived a lineshape for a random arrangement of spins on a surface. The lineshapes and moment values indicated that the hydroxy-groups on silica were isolated, the occupation probability of a site decreasing with treatment temperature. In the zeolites, protons were mostly in pairs with pair separations of 3.7 Å. Karger[234,235] investigated diffusion in the 13X zeolite–water system. He found two values of diffusion coefficients (for different types of H_2O) with separate activation energies which were compared with values derived from relaxation studies. Genser[236] recorded ^7Li, ^{23}Na, and ^{27}Al n.m.r. in Y faujasite at different levels of hydration. Na and Al signals were not detected in the dehydrated material.

Frankel[237] attempted to determine the porosity of a hydrated exchange resin. Water exchanged rapidly between gel and pore phases, the chemical shift providing a measure of the relative populations of each phase. This paper is a useful guide providing ten references to earlier work. In another paper Frankel[238] reported chemical shifts and linewidths in a number of ion-exchange resins, including macromolecular ones, which have been less well investigated than gel resins. Headley et al.[239] measured ^7Li T_1 values of LiCl solutions in porous porcelain media. Results were reasonable, the

[230] G. M. Muha, J. Chem. Phys., 1971, 55, 467.
[231] D. Vucelic, M. Susic, I. Zupancic, and M. Huter, J. Chem. Phys., 1971, 55, 4152.
[232] D. Freude, D. Muller, and H. Schmiedel, J. Colloid Interface Sci., 1971, 36, 320.
[233] D. Freude, D. Muller, and H. Schmiedel, Surface Sci., 1971, 25, 289.
[234] J. Karger, Z. phys. Chem. (Leipzig), 1971, 248, 27.
[235] J. Karger, W. Seyd, H. Pfeifer, and D. Geschke, Proceedings XVIth Colloque Ampère, 1970, p. 635.
[236] E. E. Genser, J. Chem. Phys., 1971, 54, 4612.
[237] L. S. Frankel, Analyt. Chem., 1971, 43, 1506.
[238] L. S. Frankel, J. Phys. Chem., 1971, 75, 1211.
[239] L. C. Headley, W. E. Wallace, jun., and P. Waldstein, J. Magn. Resonance, 1971, 5, 168.

relaxation rate increasing with pore size and decreasing with ion concentration. This paper is another useful literature source.

Woessner et al.[240,241] investigated the gelling process in an agar gel. Both T_1 and T_2 showed a large thermal hysteresis; T_2 was multi-component and a function of concentration and temperature, very much reduced from the free-water value The diffusion coefficient measured by the steady gradient technique was similar to that of bulk water. In the second paper[241] an additional dilution experiment was performed and the behaviour reported in both papers explained using a two-phase model. It was suggested that a comparison of proton and deuteron relaxation allows inferences to be made and the validity of various proposed models to be tested. Pesek[242] and Pesek and Pecsok[243] investigated the absorption of salts in chromatographic gels. In dextran gels, ion-exchange is the proposed mechanism. The authors suggested that the absorption observed is related to the findings of Kuntz,[244,245] who observed unfrozen liquid-like n.m.r. lines in frozen solutions of proteins and peptides. The papers by Pesek and Pecsok contain an extensive list of references. Howery et al.[246] studied the proton resonance of water in dowex 50W in 18 ionic forms to investigate the effects of counter-ions on water chemical shifts. They list 17 references, 12 of which occur within the period of this review. These are mostly concerned with the dependence of chemical shift on the nature of the counter-ion, counter-ion concentration, and temperature.

D. Cellulose Systems.—Forslind has reviewed wide-line n.m.r. studies of water adsorption and water binding in cellulose.[247] Not all the water was observable; the spectrum of one fraction was masked by that due to the macromolecule. Departures from Gaussian lineshapes were observed. Ogiwara and Kubota[248] investigated water adsorption on methyl methacrylated grafted cellulose. Linewidths were compared with water content and the grafting procedure, extending the work reported in an earlier paper[249] concerned with the relationship between water-binding capacity and the graft copolymerization process. Pittman and Tripp[250] obtained linewidths, second moments, and T_1 and T_2 values of eight different celluloses of different supramolecular characteristics, both dry and after the addition of 7% water. The second-moment value was found to correlate with crystalline fraction.

[240] D. E. Woessner and B. S. Snowden, jun., *J. Colloid Interface Sci.*, 1970, **34**, 290.
[241] D. E. Woessner, B. S. Snowden, jun., and Y. C. Chiu, *J. Colloid Interface Sci.*, 1970, **34**, 283.
[242] J. J. Pesek, *Analyt. Letters*, 1972, **5**, 127.
[243] J. J. Pesek and R. L. Pecsok, *Analyt Chem.*, 1972, **44**, 620.
[244] I. D. Kuntz, *J. Amer. Chem. Soc.*, 1971, **93**, 514.
[245] I. D. Kuntz, *J. Amer. Chem. Soc.*, 1971, **93**, 516.
[246] D. G. Howery, L. Shore, and B. H. Kohn, *J. Phys. Chem.*, 1972, **76**, 578.
[247] E. Forslind, *N.M.R. Basic Principles and Progress*, 1971, **4**, 145.
[248] Y. Ogiwara and H. Kubota, *J. Appl. Polymer Sci.*, 1971, **15**, 3137.
[249] Y. Ogiwara, H. Kubota, S. Hayashi, and N. Mitomo, *J. Appl. Polymer Sci.*, 1970, **14**, 303.
[250] R. A. Pittman and V. W. Tripp, *J. Polymer Sci., Part A-2, Polymer Phys.*, 1970, **8**, 969.

Addition of water resulted in a reduction in the width of the wide line, a reduction related to the number of exchangeable cellulose protons. The intensity of the line also increased. The increase is presumably due to the existence of bound water. An elaborate method of extracting the difference spectrum usually gave the result that the bound-water signal was the sum of three Gaussians.

E. **Proteins.**—Lalowicz and Remin[251] reported n.m.r. evidence for the existence of strongly bound water in collagen. The n.m.r. lineshape was resolved into several components. A doublet comparable to that observed in crystalline hydrates corresponded to an 8% (by weight) strongly bound water component. This component was independent of temperature and moisture content, and it was suggested that this corresponds to water involved in intramolecular hydrogen bonds. This paper extended earlier work.[252] Edzes et al.[253] examined the ^{7}Li and ^{23}Na n.m.r. spectra of hydrated samples of oriented DNA fibres, showing a well-resolved triplet due to quadrupolar splittings. The angular dependence was investigated. The magnitude of the splitting depended upon the amount of salt and water present. Derbyshire and Parsons[254] reported the deuteron resonance of frozen deuteriated muscle, demonstrating that D_2O molecules show multi-phase behaviour. Similar results had been obtained from frozen solutions of tropomyosin.[255]

Lynch and his colleagues are currently investigating the sorption of water by keratin, usually wool. Lynch and Marsden[256] have written an introductory article, and have considered[257] the applicability of a five-state theory of the keratin–water system (due to Feughelman and Haly) to T_1 measurements; the theory had already been shown to be consistent with T_2 observations. Lynch and Haly[258] investigated the anisotropy of T_2 values of water sorbed on to keratin fibres as a function of water content and temperature. Lynch, Watt, and Marsden[259] considered sorption by modified woods which had been allowed to react with reagents specific to polar groups, and also the influence on the equivalent water content; T_1 and T_2 were plotted as a function of water content and temperature.

F. **Synthetic Polymers.**—Olf and Peterlin,[260] as part of their investigation of

---

[251] Z. T. Lalowicz and M. Remin, Report No. 790/PS/Institute of Nuclear Physics, Cracow, 1972.
[252] Von K. J. Bienkiewicz, Z. Florkowski, J. W. Hennel, and Z. Lalowicz, *Das Leder*, 1970, **21**, 5.
[253] H. T. Edzes, A. Rupprecht, and H. C. Berendsen, *Biochem. Biophys. Res. Comm.*, 1972, **46**, 790.
[254] W. Derbyshire and J. L. Parsons, *J. Magn. Resonance*, 1972, **6**, 344.
[255] T. J. Cyr, W. Derbyshire, J. L. Parsons, J. M. V. Blanshard, and R. A. Lawrie, *Trans. Faraday Soc.*, 1971, **67**, 1887.
[256] L. J. Lynch and K. H. Marsden, *Search*, 1971, **2**, 95.
[257] L. J. Lynch and K. H. Marsden, *J. Textile Inst.*, 1970, **61**, 349.
[258] L. J. Lynch and A. R. Haly, *Kolloid-Z.*, 1970, **239**, 581.
[259] L. J. Lynch, I. C. Watt, and K. H. Marsden, *Kolloid-Z.*, 1971, **248**, 922.
[260] H. G. Olaf and A. Peterlin, *J. Polymer Sci., Part A-2, Polymer Phys.*, 1971, **9**, 2033.

drawn polymers, examined the wide-line n.m.r. spectrum of water sorbed on nylon-66 as a function of temperature, water content, and fibre alignment. Not all water contributed to the narrow signal. The linewidth reached a minimum consistent with Lynch and Haly's measurements.

Applications of n.m.r. investigations of surface phenomena are widespread. One example was provided by Martinez and Silvidi,[261] who investigated the binding of sodium in the saliva of patients with cystic fibrosis of the pancreas to see if this caused the inhibition of sodium transport in such patients. The conclusion was negative.

7 Systems with Unpaired Electron Spins: Para-, Ferro-, and Antiferro-magnets, Metals, and Semiconductors

A. Introduction.—The electronic spin magnetic moment is so large that the presence of unpaired spins in a sample produces very marked effects. Interaction with nuclei may occur by the conventional dipole–dipole interaction or by the Fermi-contact term that arises from the presence of unpaired electrons in *s*-type orbitals on the atom under consideration. The electron spin may be reasonably well localized or delocalized, in free radicals, transition elements, or rare earths. The electron spins may be ordered, as in ferromagnets and antiferromagnets. Electrons present in a conduction band are in turn paramagnetic, but rapid exchange occurs and only a mean electronic magnetic moment is observed. The compilation of papers by Andrew[5] cites a large number of investigations where these interactions have been studied in some detail. In addition, Petrov and Turov[262] have published a comprehensive review of n.m.r. in ferromagnets and antiferromagnets, listing 119 references. In most samples rapid electronic exchange occurs. An example where this does not occur is provided by King *et al.*,[263] who reported the direct observation of the magnetic resonance of nuclei near a paramagnetic centre. For such observation, v_0, the Larmor frequency, must be much greater than τ_c^{-1}, τ_c being the correlation time of the z component of electron spin. The systems studied were $Y(C_2H_5SO_4)_3, 9H_2O$ doped with Yb^{3+} and Nd^{3+} ions, CaF_2 doped with Eu^{2+}, and NH_4Cl doped with Cu^{2+}. It was claimed that this method has advantages over the ENDOR technique.

B. Transition Elements and Rare Earths.—Barnes *et al.*[264] determined the Knight shifts and e^2Qq values of Sc, Y, and B in the diborides of the transition metals Sc, Y, and V at 77 and 300 K. Sc and Y showed no evidence of Knight-shift anisotropy. In the same sequence of papers Creel and Barnes[265] determined e^2Qq and η values in seven transition-metal monoborides. Two models which involved electron transfer in opposite directions were con-

---

[261] D. Martinez and A. A. Silvidi, *Arch. Biochem. Biophys.*, 1972, **148**, 224.
[262] M. P. Petrov and E. A. Turov, *Appl. Spectroscopy Rev.*, 1971, **5**, 265.
[263] A. R. King, J. P. Wolfe, and R. L. Ballard, *Phys. Rev. Letters*, 1972, **28**, 1099.
[264] R. G. Barnes, R. B. Creel, and D. R. Torgeson, *J. Chem. Phys.*, 1970, **53**, 3762.
[265] R. B. Creel and R. G. Barnes, *J. Chem. Phys.*, 1972, **56**, 1549.

sidered. This paper incorporated an extensive discussion of powder lineshapes involving all transitions including the $\frac{3}{2} \to \frac{1}{2}$. The conclusions were that electron transfer from boron to the metal occurs. Other papers in the series have appeared.[266,267] Kopp and Barnes[268] investigated metal borides, using both n.m.r. and Mössbauer techniques. The ^{19}F n.m.r. of three polycrystalline rare-earth trifluorides, HoF_3, ErF_3, and YbF_3, have been studied by Carr and Moulton,[269] who observed the free induction decays over a temperature range. Two distinct lines were detectable. E.s.r. was also studied. HoF_3 did not obey the Curie–Weiss Law, a failure attributed to a low-lying crystal-field state.

Bug et al.[270] performed a comprehensive n.m.r. investigation on single crystals of $CsMnCl_3,2H_2O$, investigating ^1H, ^2H, ^{35}Cl, and ^{123}Cs n.m.r. They performed a Pake-type analysis, and determined the quadrupole interactions of ^2H, ^{35}Cl, and ^{123}Cs, in addition to measuring the hyperfine field interactions from the observed paramagnetic shifts. The flipping motion of water molecules was investigated by examining the temperature dependence of the linewidth. Sheline and his colleagues determined quadrupole coupling constants and chemical-shift tensors in a number of Co and Mn compounds.[271–273] In the first paper, ^{59}Co n.m.r. was studied in single crystals and solutions of cobalt carbonyl complexes, using a pulse spectrometer to map spectra. The quadrupole asymmetry factor η was small, so chemical shifts could be satisfactorily determined from the central transition. Both e^2Qq and chemical shift were compared with calculated values and the isotropic shift was compared with solution values. In the second paper, Mn quadrupole interactions and chemical shifts were investigated in a number of manganese carbonyls, and in paper 3 the ^{59}Co n.m.r. of dicobalt octacarbonyl was studied at 77 K. Eastman et al.[274] reported the ^{17}O n.m.r. of ^{17}O-enriched uranium oxides, KUO_3, and $BiUO_4$, measuring chemical shifts and the negative electron-spin density on the oxygen atoms. Linewidth studies indicated that spin exchange provided the dominant relaxation mechanism. Fukushima and Hecht[275] recorded the temperature dependence of the ^{23}Na and ^{19}F n.m.r. in polycrystalline Na_3UF_8; the paramagnetic susceptibility was found to obey Curie's Law, and the ^{19}F second moments showed no temperature discontinuities, while the ^{23}Na spectra consisted of a singlet and a doublet, the doublet corresponding to an e^2Qq value of 1·5 MHz.

Niesen and Huiskamp[276] recorded the n.m.r. spectra of ^{56}Co, ^{58}Co, and

[266] R. B. Creel and R. G. Barnes, *Phys. Status Solidi*, 1970, **41**, K27.
[267] R. G. Barnes, R. B. Creel, and D. R. Torgeson, *Solid State Comm.*, 1970, **8**, 1411.
[268] J. P. Kopp and R. G. Barnes, *J. Chem. Phys.*, 1971, **54**, 1840.
[269] S. L. Carr and W. G. Moulton, *J. Magn. Resonance*, 1971, **4**, 400.
[270] H. Bug, H. Haas, M. Fleissner, and H. Hartmann, *J. Chem. Phys.*, 1971, **55**, 280.
[271] H. W. Spiess and R. K. Sheline, *J. Chem. Phys.*, 1970, **53**, 3036.
[272] H. W. Spiess and R. K. Sheline, *J. Chem. Phys.*, 1971, **54**, 1099.
[273] E. S. Mooberry, M. Pupp, J. L. Slater, and R. K. Sheline, *J. Chem. Phys.*, 1971, **55**, 3655.
[274] M. P. Eastman, H. G. Hecht, and W. Burton-Lewis, *J. Chem. Phys.*, 1971, **54**, 4141.
[275] E. Fukushima and H. G. Hecht, *J. Chem. Phys.*, 1971, **54**, 4341.
[276] L. Niesen and W. J. Huiskamp, *Physica*, 1972, **57**, 1.

^{60}Co incorporated in a dilute paramagnetic crystal of lanthanum manganese nitrate by cross-relaxation with Ce ions substituted for La. Resonance was observed as a change in the directional anisotropy of the emitted γ-ray intensity. Goto[277] investigated the antiferromagnet $CoCl_2,2H_2O$. Using pulsed n.m.r., Nishihara et al.[278] examined the ^{53}Cr n.m.r. spectrum inferromagnetic CrO_2 to determine the hyperfine field. Two values, 154 and 111 kG, were determined. Saji et al.[279] reported ^{51}V n.m.r. in $CrVO_4$, which undergoes an antiferromagnetic transition at 50 K. The hyperfine field at the vanadium sites and the quadrupole interactions were determined. Takayanagi and Watanabe,[280] using proton n.m.r., investigated the antiferromagnetic clathrate compound $Ni(NH_3)_2,Ni(CN)_4,2C_6H_6$.

Spence and Rao[281] investigated magnetic ordering in $MnCl_2,2H_2O$, extending earlier work on $CsMnCl_3,2H_2O$.[282] They recorded ^{35}Cl and ^{1}H n.m.r. spectra at the Néel temperature of 6.8 K. $MnCl_2,2H_2O$ is isostructural with $CoCl_2,2H_2O$. Both have linear chains of magnetic ions along the crystalline c-axis. Experiments in externally applied zero field showed that the field at the chlorine sites is 40 kG in the ferromagnetic state. In the antiferromagnetic state three quadrupole-split components were observed. De Jonge et al.[283] studied deuteron resonance in $KMnCl_3,2D_2O$ and $CsMnCl_3,2D_2O$ at 77 K, determining the paramagnetic and quadrupolar interactions. El Saffar[284] examined some hydrated manganese chlorides which become antiferromagnetic at low temperatures. His paper overlaps that of Spence and Rao to some extent, but El Saffar was interested in determining hydrogen atom positions. On comparing his results with calculations based on Baur's electrostatic model, he found considerable departures from linear hydrogen bonds. The magnetic properties of copper formate tetrahydrate have been the subject of two papers.[285,286] Ghosh reported the observation of non-exponential decays in ordered magnetic systems and developed an appropriate theory.[287]

C. Semiconductors.—Engelsberg and Norberg[288] investigated the second moment of ^{31}P in the semiconductor InP from 78 to 300 K by both pulse and CW techniques. Lynch et al.[289] investigated several semiconducting phosphate glasses using e.s.r. and ^{31}P and ^{51}V n.m.r. The ^{31}P linewidths in vanadium and molybdate glasses were 1.4 and 0.4 G. The ^{51}V e^2Qq value determined from the lineshape of the central transition was 1.5 MHz. No paramagnetic

[277] T. Goto, J. Phys. Soc. Japan, 1971, **31**, 1842.
[278] H. Nishihara, T. Tsuda, A. Hirai, and T. Shimjo, J. Phys. Soc. Japan, 1972, **32**, 85.
[279] H. Saji, T. Yamadaya, and M. Asanuma, Phys. Letters, 1971, **34A**, 49.
[280] S. Takayanagi and T. Watanabe, J. Phys. Soc. Japan, 1971, **31**, 109.
[281] R. D. Spence and K. V. S. R. Rao, J. Chem. Phys., 1970, **52**, 2740.
[282] R. D. Spence, W. J. M. de Jonge, and K. V. S. R. Rao, J. Chem. Phys., 1969, **51**, 4694.
[283] W. J. M. de Jonge, J. P. A. M. Hijmans, and E. C. A. Gevers, Physica, 1971, **52**, 129.
[284] Z. M. El Saffar, J. Chem. Phys., 1970, **52**, 4097.
[285] A. Dupas and J. P. Renard, Phys. Letters, 1970, **33A**, 470.
[286] K. Yamagata, M. Hayama, and T. Odaka, J. Phys. Soc. Japan, 1971, **31**, 1279.
[287] S. K. Ghosh, Phys. Rev., 1972, **B5**, 174.
[288] M. Engelsberg and R. E. Norberg, Phys. Letters, 1970, **31A**, 311.
[289] G. F. Lynch, M. Sayer, S. L. Segel, and G. Rosenblatt, J. Appl. Phys., 1971, **42**, 2587.

shift was observed. They found some discrepancies with the results of Landsberger and Bray.[290] Using [125]Te n.m.r., Koma et al.[291,292] investigated the effects of quenching and annealing on solid tellurium. In the single crystal three Te lines were observed with a large and anisotropic chemical shift.

D. Metals (General).—Line-narrowing experiments in metals have already been referred to.[200-202] Kanert[293] discussed the influence of sample geometry on the magnitude of the n.m.r. signal from a bulk metallic sample and Von Meerwall and Rowland[294] considered the effect of stress annealing on [27]Al lineshapes. Line broadening on filed samples was attributed to first-order quadrupolar effects. Maintaining a sample at room temperature surprisingly did not restore the original signal. Carver[295] has determined [133]Cs quadrupolar interactions and Knight shifts in a number of caesium–graphite intercalation compounds between 1.3 and 4.2 K, in addition to determining [133]Cs and [13]C T_1 values. Yasuoka et al.[296] examined the metal–insulator transition in V_2O_3 and used pulsed n.m.r. to determine the hyperfine field in the antiferromagnetic state. Other related papers are by Gossard et al.[297] and Rubinstein.[298]

8 Instrumental and Operational Developments

As this is the subject of Chapter, 4 discussion of instrumental developments specific to broad-line measurements will be brief. Dillon and Waddington[101] and Fahnrich et al.[299] discussed modifications of high-resolution spectrometers to permit the observation of n.m.r. in solids.

There have been a number of modifications to Robinson-type and marginal oscillator type spectrometers. Toth et al.[300] discussed the design of a Robinson oscillator with frequency modulation suitable for the measurement of internal magnetic fields and operating over the frequency range 25—100 MHz. Hughes and Smith[301] reported a general mathematical discussion of the factors influencing the operation of marginal and Robinson oscillators. One of the major conclusions was that detection of either amplitude modulation or frequency modulation will involve detection of a mixture of absorption and dispersion components. Jones and Roberts[302] designed a Robinson-type circuit using rf integrated circuitry. This design is suitable for use in a superconducting

[290] F. R. Landsberger and P. J. Bray, *J. Chem. Phys.*, 1970, **53**, 2768.
[291] A. Koma, O. Mizuno, and S. Tanaka, *Phys. Status Solidi*, 1971, **46B**, 225.
[292] A. Koma, *Phys. Status Solidi*, 1970, **40**, 239.
[293] O. Kanert, *Phys. Status Solidi*, 1970, **42**, K19.
[294] E. Von Meerwall and T. J. Rowland, *Scripta Metallurgica*, 1971, **5**, 619.
[295] G. P. Carver, *Phys. Rev.*, 1970, **B2**, 2284.
[296] H. Yasuoka, H. Nishihara, Y. Nakamura, and J. P. Remeika, *Phys. Letters*, 1971, **37A**, 299.
[297] A. C. Gossard, D. B. McWhan, and J. P. Remeika, *Phys. Rev.* 1970, **B2**, 3762.
[298] M. Rubinstein, *Phys. Rev.*, 1970, **B2**, 4731.
[299] J. Fahnrich, B. Sedlak, and E. Repko, *Czech. J. Phys.*, 1971, **A21**, 377.
[300] F. I. Toth, K. Tompa, and G. Gruner, *J. Phys. (E)*, 1972, **5**, 42.
[301] D. G. Hughes and M. R. Smith, *J. Phys. (E)*, 1971, **4**, 13.
[302] G. P. Jones and R. T. Roberts, *J. Phys. (E)*, 1970, **3**, 1003.

magnet. Lazarus et al.[303] described the use of dual-gate MOSFET's at 77 K in rf amplifiers. Iodine and Brandenberger[304] used FET's and integrated circuits in a marginal oscillator, as did Sullivan,[305] whose unit operates from 2 to 60 MHz. Alderman[306] discussed the design of a twin-T bridge spectrometer with the bridge components at liquid-helium temperature. An improvement of 5 in signal-to-noise ratio was obtained. Janzen[307] described a method of determining second-moment values from adiabatic rapid-passage experiments; CaF_2 and lithium stearate were used as examples. There is the restriction that $T_{1\rho}$ must not be too short.

Wollan[308] discussed the problems of lock-in detection, reiterating Goldman's point that the fact that signal intensity is proportional to B_1 is no guarantee that saturation is not occurring. Dianoux et al.[309] reported a method of extracting second moments from free induction decays or echoes. The method appears to have an advantage over the assumption of an analytic function or a series expansion of polynomials in that divergences occur when values become unreliable; $KAsF_6$ was used as an example. Shmyrev and Fedin[310] investigated the accuracy of second-moment measurements with particular emphasis on the effects of truncation. They developed computer programs to determine the optimum point to truncate.

9 Applications

There are a number of applications of n.m.r. to determinations of oil and water content, based on measurements of the intensity of a narrow spectral component. Andersson[311] described Varian's PAT 20 analyser. Mansfield,[312] Wettstrom,[313] Haighton et al.,[314] and Shanbhag et al.[315] recorded measurements of oil content or liquid/solid ratios. Mansfield[316] determined the plasticizer content of PVC plastics with a precision of 0.5% irrespective of the physical condition of the sample. Karras and Rahkamaa[317] related the water content of pulp to n.m.r. linewidth, claiming an accuracy of 0.1%. Samuelsson and Vikelsoe[318] determined the amount of liquid fat in cream and butter.

[303] M. J. Lazarus, M. P. G. Gibson, and M. Ryall, *J. Phys. (E)*, 1971, **4**, 58.
[304] J. D. Iodine, jun., and J. R. Brandenberger, *Rev. Sci. Instr.*, 1971, **42**, 715.
[305] N. Sullivan, *Rev. Sci. Instr.*, 1971, **42**, 462.
[306] D. W. Alderman, *Rev. Sci. Instr.*, 1970, **41**, 192.
[307] W. R. Janzen, *Chem. Phys. Letters*, 1971, **12**, 35.
[308] D. S. Wollan, *Rev. Sci. Instr.*, 1971, **42**, 682.
[309] A. J. Dianoux, S. Sykora, and H. S. Gutowsky, *J. Chem. Phys.*, 1971, **55**, 4768.
[310] I. K. Shmyrev and E. I. Fedin, *J. Struct. Chem.*, 1970, **11**, 964.
[311] L. O. Andersson, *J. Amer. Oil Chemists' Soc.*, 1971, **48**, 47.
[312] P. B. Mansfield, *J. Amer. Oil Chemists' Soc.*, 1971, **48**, 4.
[313] R. Wettstrom, *J. Amer. Oil Chemists' Soc.*, 1971, **48**, 15.
[314] A. J. Haighton, L. F. Kermaas, and C. den Hollander *J. Amer. Oil Chemists' Soc.*, 1971, **48**, 7.
[315] S. Shanbhag, M. P. Steinberg, and A. I. Nelson, *J. Amer. Oil Chemists' Soc.*, 1971, **48**, 11.
[316] P. B. Mansfield, *Chem. and Ind.*, 1971, 792.
[317] M. Karraas and E. Rahkamaa, *Paperi ja Puu*, 1971, **11**, 1.
[318] E. G. Samuelsson and J. Vikelsoe, *Milchwissenschaft*, 1971, **26**, 621.

10
Medium Effects

BY M. I. FOREMAN

1 Introduction

A vast amount of literature relating in some degree or other to medium effects on n.m.r. phenomena has appeared in the past year, so that the present article inevitably reflects to some extent this Reporter's personal interests. The coverage has been extended slightly over the previous article to include effects of the medium which are of a more 'physical' nature, whilst the previous account was restricted rather to 'chemical' interactions. Effects due to the presence of shift-reagents have again been included, since interest in this aspect of n.m.r. spectroscopy has continued unabated. It has also been the deliberate intention of this article to accentuate certain widely held assumptions inherent in much of the work in this field which still remain to be adequately proven.

2 Coupling Constants

Solvent-dependence of coupling constants is generally of two types, which might loosely be termed 'real' and 'apparent'. The latter case arises in systems where rapid internal rotations average out the couplings appropriate to the several rotamers, the effect of the solvent being simply to modify the rotamer populations (ref. 1, p. 568). This situation will not be discussed further here. Many examples of the former situation, where change of solvent effects a real change in the coupling constants of the molecule, exist in the literature.[2] The factors which influence such variations are, however, by no means well understood. For the case of geminal (H,H) couplings, the presence of a solute dipole is important to the observation of solvent-dependent coupling constants.[3,4] The orientation of the dipole with respect to the molecular coordinates is also a factor.[5] It has further been observed that geminal couplings tend to become more negative as the solvent dielectric constant (relative per-

---

[1] J. W. Emsley, J. Feeney, and L. H. Sutcliffe, 'High Resolution Nuclear Magnetic Resonance Spectroscopy', Pergamon Press, 1965, vol. 1.
[2] P. Laszlo, *Progr. N.M.R. Spectroscopy*, 1967, 3, 231.
[3] S. L. Smith and R. H. Cox, *J. Chem. Phys.*, 1966, 45, 2848.
[4] V. S. Watts and J. H. Goldstein, *J. Chem. Phys.*, 1965, 42, 228.
[5] S. L. Smith and A. M. Ihrig, *J. Chem. Phys.*, 1967, 46, 1181.

mittivity) increases,[3,5-7] an effect which is potentially a useful means of assigning absolute signs to the relevant couplings. Clearly, therefore, the 'reaction field'[8] has an important effect on the coupling constants of dissolved material, although dispersion forces have also been identified as a contributory factor.[9] Generally, however, the approximations of the reaction-field approach are too gross, and attempts have been made to refine the theory by introducing the concept of collision complexes.[10,11] A recent attempt to treat solvent-dependent coupling constants by the reaction-field approach was reasonably successful for (H,H) and (H,F) couplings in solvents of low dielec-

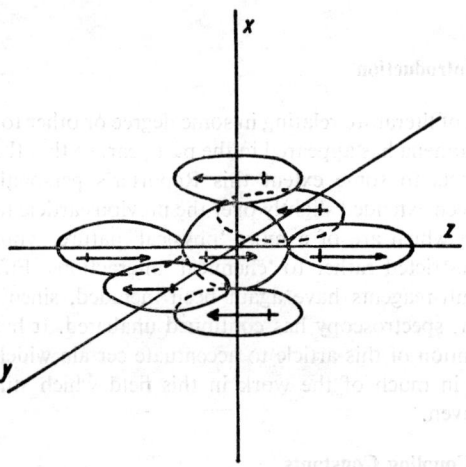

Figure 1 *Cubic close-packing of solvent molecules about a polar solute, the molecular dipoles being arranged as shown*
(Reproduced by permission from *J. Chem. Phys.*, 1971, **55**, 3483)

tric constant.[12] For formaldehyde, however, the method failed to predict both the magnitude and direction of the changes.[12] Subsequently, a more refined approach has been attempted which treats the solute as a dipolar sphere surrounded by solute molecules in a cubic-close-packed arrangement with the molecular dipoles arranged in the manner illustrated in Figure 1.[13] The closest approach between atoms of neighbouring molecules was taken to be the sum of their Van der Waals radii, a distance which was shown to be critical to the success of the calculations. Generally, better agreement between theory and experiment was obtained for this model[13] compared with the earlier results.[12] For formaldehyde in acetonitrile the simple reaction-field

[6] R. H. Cox and S. L. Smith, *J. Magn. Resonance*, 1969, **1**, 432.
[7] C. L. Bell and S. L. Danyluk, *J. Mol. Spectroscopy*, 1970, **35**, 376.
[8] L. Onsager, *J. Amer. Chem. Soc.*, 1936, **58**, 1486.
[9] P. Laszlo and A. Speert, *J. Magn. Resonance*, 1969, **1**, 291.
[10] I. D. Kuntz, jun. and M. D. Johnston, jun., *J. Amer. Chem. Soc.*, 1967, **89**, 6008.
[11] R. L. Schmidt, R. S. Butler, and J. H. Goldstein, *J. Phys. Chem.*, 1969, **73**, 1117.
[12] M. D. Johnston, jun. and M. Barfield, *J. Chem. Phys.*, 1971, **54**, 3083.

treatment[12] yielded a value for $\Delta J(H,H)$ of $+0.11$ Hz, whereas the more refined cluster model[13] gave a value of -2.39 Hz, which is to be compared with the observed $\Delta J(H,H)$ of -2.20 Hz [$\Delta J(H,H)$ is the difference between the coupling for acetonitrile solution and that for cyclohexane solution].

Where, however, more specific interactions are possible between the solute and solvent molecules, such approaches as described above are no longer adequate. For styreneimine (1) and the N-butyl derivative (2) the geminal couplings $J(H_A,H_B)$ show quite different solvent behaviour.[14] For (2) the coupling constant varies with the solvent dielectric constant in the normal way; for styreneimine itself there is, however, no obvious correlation; in fact

$$\begin{array}{c} H_C \quad H_A \\ Ph \diagdown \diagup H_B \\ N \\ | \\ R \end{array}$$

(1) R = H
(2) R = t-butyl

there is a marked increase in $J(H_A,H_B)$ on going from CCl_4 to [$^2H_6$]dimethyl sulphoxide. This behaviour seems to be due to hydrogen-bonding between the NH proton of (1) and the solvent, which overrides the concomitant effects of the reaction field.[14] In the N-substituted case such effects are obviated.

A rather more detailed and comprehensive account of solvent effects on coupling constants, particularly (H,F) couplings, has been reported recently by Smith and Ihrig.[15] (H,F) couplings were selected for particular attention because of their enhanced sensitivity to solvent effects over (H,H) couplings, which tend to be rather less affected by solvent. This is particularly true of the vicinal coupling constants. Difluoroethylenes were chosen as model compounds for study because of the variety of vicinal and geminal (H,F) couplings observable. The behaviour of 1,2-difluoroethylene is of particular interest since it allows simultaneous study of the solvent effect on the vicinal and geminal (H,F) couplings. The authors[15] conclude that the presence of a solute dipole is essential if solvent-dependent couplings are to be observed, and confirm the earlier observation that the orientation of the dipole with respect to the rest of the molecule is important. One interesting point which is brought out in this work is that the factors which affect the vicinal (H,F) couplings, whilst apparently the same for all of the molecules studied, nevertheless seem to be different from those which affect the geminal couplings, even for couplings within the same molecule.

Correlations were also sought[15] between the observed couplings and the reaction field (R) using the Onsager equation:[8]*

[13] M. D. Johnston, jun. and M. Barfield, *J. Chem. Phys.*, 1971, **55**, 3483.
[14] R. H. Cox and L. W. Harrison, *J. Magn. Resonance*, 1972, **6**, 84.
[15] A. M. Ihrig and S. L. Smith, *J. Amer. Chem. Soc.*, 1972, **94**, 34.

* Equations throughout this chapter are in unrationalized three-quantity (c.g.s.) form; modifications may be necessary for conversion to SI usage.

$$R = \frac{\mu}{r^3}\frac{2(\eta^2+2)}{3}\frac{(\varepsilon_r-1)}{(2\varepsilon_r+\eta^2)} \quad (1)$$

and also the dispersion interaction (D) given by[16,17] equation (2):

$$D = (\eta-1)/(2\eta^2+1) \quad (2)$$

where μ is the solute dipole moment in the gas phase, r is the radius of the solute cavity, η the solute refractive index, and ε_r the solvent dielectric constant. It is apparent that both terms contribute to the observed coupling constants (J_{obs}), which are shown to correlate with an expression of the type

$$J_{obs} = J_0 + A.R + B.D \quad (3)$$

where the coefficients A and B are always of opposite sign and J_0 is a constant. Again, where strong solute–solvent interactions are apparent, the above treatment is no longer adequate. Fluorine–fluorine couplings are also shown to be markedly sensitive to solvent and to temperature, and it is suggested[15] that both the $^2J(F,F)$ and $^3J(F,F)$ couplings are dominated by the reaction-field term, although surprisingly $^2J(F,F)$ decreases with a decrease in temperature for solvents of low dielectric constant, whereas the opposite would be expected for a reaction-field effect. For solvents of high dielectric constant, however, $^2J(F,F)$ is virtually temperature-independent. Both $^3J(F,F)$-*cis* and $^3J(F,F)$-*trans* vary in a manner which is inconsistent with a simple treatment in terms of reaction field and dispersion forces. A tentative explanation for this behaviour in terms of temperature-dependent changes in the molecular vibrations, which were thought to have a bearing on (F,F) couplings,[18] and which might act in an opposite sense to the temperature-dependence of the reaction-field term, was proposed.[15] Clearly the development of theories to account for the dependence of coupling constants on the surrounding medium is still at an early stage. Smith and Ihrig,[15] however, have provided a valuable body of experimental data to further this topic.

3 Chemical Shifts

A. Gas-to-solution Shifts.—According to Buckingham, Schaefer, and Schneider[19] the chemical shift of a solute i in a solvent k is $S_{i,k}$ where

$$S_{i,k} = -\delta_g(i) + \sigma_b(k) + \sigma_a(k) + \sigma_w(i,k) + \sigma_e(i,k) \quad (4)$$

In equation (4), $\delta_g(i)$ is the gas-phase shift of i, $\sigma_b(k)$ accounts for the solvent bulk susceptibility, $\sigma_a(k)$ for the solvent anisotropy, $\sigma_w(i,k)$ for the Van der Waals term, and $\sigma_e(i,k)$ for the reaction-field term. In a recent series of papers Malinowski and Weiner have attempted to assess the validity of the above expression and to determine as far as possible which of the terms are of

---

[16] E. G. McRae, *J. Phys. Chem.*, 1957, **61**, 562.
[17] N. S. Bayliss, *J. Chem. Phys.*, 1950, **18**, 292.
[18] K. C. Ramey and W. S. Brey, jun., *J. Chem. Phys.*, 1964, **40**, 2349.
[19] A. D. Buckingham, T. Schaefer, and W. G. Schneider, *J. Chem. Phys.*, 1960, **32**, 1227.

most significance. [Eqation (4) is written above in the standard notation of these Reports. It differs from that of Malinowski and Weiner, whose notation appears to be somewhat confusing, by a change in sign of δ.] The technique of Malinowski and Weiner, which is described in some detail in the first of the papers,[20] is that of 'factor analysis', which might usefully be outlined briefly here. As a first step the chemical shifts, $S_{i,k}$ for a number of substituted-methane solutes were recorded in nine polar and non-polar solvents relative to internal TMS,[20] and the data expressed as a matrix $[S]$ having elements $S_{i,k}$. Each $S_{i,k}$ is expressible as a linear combination of terms [equation (4)], and in order to proceed with the analysis it is further assumed that each individual term is expressible as a product function. Each $S_{i,k}$ is therefore represented by an equation such as (5):

$$S_{i,k} = \sum_{j=1}^{r} U_{ij} \cdot V_{jk} \qquad (5)$$

where each U_{ij} is a solute factor, and each V_{jk} is a solvent factor. The matrix $[S]$ is then given by:

$$[S] = [U][V] \qquad (6)$$

where $[U]$ is the solute factor matrix, and $[V]$ is the solvent factor matrix. It is then possible to determine $[U]$ and $[V]$ in such a way that an estimate can be made for the *minimum value* of r in equation (5) such that the matrix $[S]$ is reproduced to within the error of the experiment. The authors applied this analysis to their data and obtained the result that no more than three factors were required to reproduce the tabulated shifts to within ± 0.5 Hz. At this stage the solute- and solvent-factor matrices have no physical meaning. It is possible, however, to 'rotate' both matrices, thereby generating a second set $[\bar{U}]$ and $[\bar{V}]$ which are related to the original matrices by the 'rotation' matrix $[R]$ where

$$[\bar{U}] = [U][R] \qquad (7)$$

$$[\bar{V}] = [R]^{-1}[V] \qquad (8)$$

The matrices $[\bar{U}]$ and $[\bar{V}]$ may now represent real physical parameters for the system.

In this manner it was found possible to express $[\bar{V}]$ in terms of three solvent vectors corresponding to the solute shifts in acetonitrile, carbon tetrachloride, and methylene bromide; in other words, each shift is expressible in terms of equation (9).

$$S_{i,k} = a_{1,k} \cdot S_{i,\mathrm{CH_3CN}} + a_{2,k} \cdot S_{i,\mathrm{CCl_4}} + a_{3,k} \cdot S_{i,\mathrm{CH_2Br_2}} \qquad (9)$$

It follows, therefore, that the three fundamental terms, which are as yet unidentified, are *all* represented in some degree by the three solvents CH_3CN,

[20] P. H. Weiner, E. R. Malinowski, and A. R. Levinstone, *J. Phys. Chem.*, 1970, **74**, 4537.

CCl_4, and CH_2Br_2 taken together. This latter stipulation is most important, since if solvent anisotropy were a fundamental factor, this is not present in CCl_4, but it is likely to be present in CH_3CN.

An attempt to reproduce [S] in terms of the separated fundamental parameters for the system was next attempted. A new matrix $[\bar{U}]$ was set up which had as one row vector the gas-phase shifts of the solutes, where known, otherwise the vector elements were left blank. It was in fact possible to rotate [U] into $[\bar{U}]$, which identifies the gas-phase shift, $\delta_g(i)$, as indeed being one factor of equation (5). Furthermore, the method provided an estimate of the

Table 1 *Gas-phase shifts,[a] computed from the three-factor analysis, compared with observed values[20]*

Solute	Δv_g(calculated)	Δv_g(observed)
CH_3Cl	168.2	—
$CHCl_3$	427.1	427.3
CH_3Br	147.1	146.9
CH_2Br_2	285.5	285.0
$CHBr_3$	406.8	406.9
CH_3I	118.5	119.0
CH_2I_2	227.6	—
CHI_3	301.5	—
CH_2ClBr	297.7	—

[a] Shifts Δv_g are quoted in Hz at 60 MHz relative to gaseous TMS.

gas-phase shifts for solutes for which these quantities were not previously known, by effectively filling in the gaps of the matrix. Values so obtained are given in Table 1.

An attempt to effect a rotation into calculated[19] reaction-field terms was, however, not successful.[20] This could either be because the reaction field is not one of the three factors, or possibly because the assumptions inherent in the calculation of this term were not valid. Subsequently a study of non-polar solutes in various solvents using an external standard again identified three significant factors [$\sigma_e(i,k)$ can be neglected for non-polar solutes and the authors corrected their data to remove the bulk susceptibility term $\sigma_b(k)$] one of which was shown to be the solvent anisotropy.[21] This was somewhat less satisfactory because of the uncertainty and paucity of reported values of solvent anisotropies. The gas-to-solution shift of methane was also shown to be a fundamental factor.[21] This is to a good approximation equivalent to the Van der Waals term $\sigma_w(i,k)$, of equation (4). The point is pursued further in a later account[22] in which the Van der Waals term $\sigma_w(i,k)$ was expressed as a product function:

$$\sigma_w(i,k) = \sigma_w(i) \cdot \sigma_w(k) \qquad (10)$$

[21] P. H. Weiner and E. R. Malinowski, *J. Phys. Chem.*, 1971, **75**, 1207.
[22] P. H. Weiner and E. R. Malinowski, *J. Phys. Chem.*, 1971, **75**, 3160.

where[23,24]

$$\sigma_w(i) = \frac{8\pi^2 N^2 e^2 \phi}{9V_i} \quad (11)$$

$$\sigma_w(k) = \frac{\eta^2 + 2}{2\eta^2 + 1} \cdot \frac{\sum \langle r_j^2 \rangle}{V_k} \quad (12)$$

where V_i and V_k are the molar volumes of solute and solvent, r_j is the radius of the electron cloud distribution,[25] and ϕ is a constant term. Values of the Van der Waals term so calculated were successfully incorporated into the analysis, thereby identifying the Van der Waals term as a fundamental factor and supporting the validity of the method of calculation.[23,24]

A graphical approach to the factors influencing gas-to-solution shifts has also been described by these same authors.[26] The change in shift on going from the gas phase to solution is $\Delta S_{(i,k)}$, where

$$\Delta S_{(i,k)} = \sigma_w(i,k) + \sigma_e(i,k) + \sigma_a(k) \quad (13)$$

from equation (4), having again corrected for the bulk susceptibility. Since for CCl_4 the term σ_a is effectively zero, it follows from equation (10) that

$$\Delta S_{(i,k)} = \frac{\sigma_w(k)}{\sigma_w(CCl_4)} \cdot \Delta S_{(i,CCl_4)} + \sigma_a(k) \quad (14)$$

for non-polar solutes, and therefore a plot of $\Delta S_{(i,k)}$ versus $\Delta S_{(i,CCl_4)}$ for a given solvent k should be linear, having an intercept which is a measure of the solvent anisotropy. For the case of polar solutes an expression can be obtained which is analogous to equation (14), namely

$$\Delta S_{(i,k)} = \frac{\sigma_w(k)}{\sigma_w(CCl_4)} \cdot \Delta S^*_{(i,CCl_4)} + \sigma_a(k) \quad (15)$$

where now

$$\Delta S^*_{(i,k)} = \Delta S_{(i,k)} - \sigma_e(i,k) \quad (16)$$

By estimating the appropriate $\sigma_e(i,k)$ terms from equation (17),[19]

$$\sigma_e(i,k) = -k \frac{(\varepsilon'_r - 1)}{2\varepsilon'_r + 2.5} \frac{\mu}{\alpha} \cos\theta \quad (17)$$

where α is the solute molecular polarizability, θ the angle between the solute dipole and the appropriate C—H bond, and ε'_r is the effective dielectric constant.[27,28] Again linear plots are to be expected, and are in fact obtained.

[23] B. Linder, *J. Chem. Phys.*, 1960, **33**, 668.
[24] B. B. Howard, B. Linder, and M. T. Emerson, *J. Chem. Phys.*, 1962, **36**, 485.
[25] P. G. Maslov, *J. Phys. Chem.*, 1968, **72**, 1414.
[26] P. H. Weiner and E. R. Malinowski, *J. Phys. Chem.*, 1971, **75**, 3971.
[27] J. K. Becconsall and P. Hampson, *Mol. Phys.*, 1965, **10**, 21.
[28] G. A. Gerhold and E. Miller, *J. Phys. Chem.*, 1968, **72**, 2737.

However, the slopes of the plots for different groups of solutes vary, and it is apparent that an effect, not hitherto allowed for, is operative when the solute and the solvent species both possess permanent dipole moments. The authors believe this effect to be due to a dipole–dipole interaction which must be proportional to the solvent Van der Waals term $\sigma_w(k)$. The authors conclude this series of articles[26] with the suggestion that the chemical shift of a solute i in solvent k is given by equation (18):

$$-\delta_{(i,k)} = -\delta_g(i) + \sigma_a(k) + \sigma_b(k) + \sigma_w(i,k) + \sigma_e(i,k) + \sigma_d(i,k) \quad (18)$$

where the final term accounts for the dipole–dipole interaction discussed above.

B. Hydrogen-bonding Effects.—In the foregoing section, non-specific effects of the solvent on the resonance positions of the nuclei of dissolved material have been discussed. Where, however, rather more specific interactions are possible, either between the solute and the solvent or between solute molecules themselves, several distinct species may co-exist in solution, and such a situation necessitates a quite different approach. If the lifetime of a given nucleus in each of the distinct species present in the solution is sufficiently short (see page 9 of ref. 29 for a more detailed account), a single averaged resonance is observed which is given by

$$\delta_{\text{obs}} = \sum_{i=1}^{n} P_i \delta_i \quad (19)$$

where P_i is the fractional population, and δ_i the chemical shift, of the given nucleus in the ith species. This present section is concerned with studies in which chemical-shift behaviour reflects the formation in solution of rapidly exchanging hydrogen-bonded molecular aggregates.

The behaviour of the hydroxy-proton chemical shift, because of its extreme sensitivity to hydrogen-bonding interactions, has been widely studied. Generally a high-frequency shift is apparent on hydrogen-bond formation, so that the observation of a hydroxy-resonance 1.82 p.p.m. to low frequency of methane for solutions of NaOH in KNO_3–$NaNO_3$ eutectic mixtures at 150 °C, compared with a shift of 14.9 p.p.m. to high frequency of methane for NaOH at infinite aqueous dilution, argues for a considerable lack of hydrogen-bonding in the molten salt system.[30] Qualitative observations of this kind can readily be applied to such systems. Propargyl alcohol, for example, exhibits a solvent- and concentration-dependent hydroxy-resonance, which in the pure liquid state occurs at considerably higher frequencies than expected for the 'free' hydroxy-group.[31] This latter observation suggests extensive hydrogen-bonded self-association in the pure liquid. Addition of the inert solvent

---

[29] R. Foster and C. A. Fyfe, *Progr. N.M.R. Spectroscopy*, 1969, **4**, 1.
[30] A. G. Turnbull, *Austral. J. Chem.*, 1971, **24**, 2213.
[31] L. K. Vasyanina, N. N. Shapet′ko, T. L. Alekseeva, and D. N. Shigorin, *Zhur. strukt. Khim.*, 1971, **12**, 919.

CCl$_4$ disrupts the molecular aggregates, and the hydroxy-resonance shifts to lower frequencies with increasing dilution. The same general trend, although much reduced, occurs with acetone and dioxan. Here the self-association of the alcohol is again reduced, but is offset to some extent by hydrogen-bonding to the solvent. With dimethyl sulphoxide or pyridine, however, the shift is to even higher frequencies. Solute–solvent hydrogen-bonding in this case is even more extensive than self-association in the pure liquid.[31] The CH proton resonance is also reported to be concentration- and solvent-dependent

(3)

in the same sense as that of the hydroxy-group. This proton is therefore also likely to be involved in hydrogen-bond formation.[31] Similar observations have been made for the aqueous methanol system, the concentration- and temperature-dependence of the hydroxy-resonance being reported in some detail,[32] and also for various cyclopropyl alcohols.[33]

The hydroxy-resonance of phenol shows similar behaviour;[34] that of o-aminomethylphenols, on the other hand, is affected by neither solvent nor concentration.[35] Strong intramolecular hydrogen-bonding is evident in these systems, therefore, leading to such structures as (3), which resist both disruption by inert solvents and also competition from other hydrogen-bonding species for the active sites.[35] The general observation has been made that the intramolecularly hydrogen-bonded hydroxy-resonance appears some 9—11 p.p.m. to high frequency of internal TMS in the above systems; for other than o-substitution, where hydrogen-bonds must be intermolecular, the resonance appears about 4 p.p.m. to high frequency of TMS and exhibits the expected solvent dependence.[35]

Easily accessible observations such as those above are proving increasingly valuable in studies of more complex systems, particularly those of biological importance. Hexachlorophene (4) interacts with protein systems, possibly through hydrogen-bond formation. In CCl$_4$ solution the addition of simple amides causes a high-frequency shift of the hydroxy-resonance and some line-broadening is apparent, which is again consistent with a hydrogen-bonding

[32] V. V. Mauk, O. D. Kurilenko, A. A. Barau, and L. M. Men'shova, *Ukrain. khim. Zhur.*, 1971, **37**, 1073.
[33] J.-L. Pierre and R. Perraud, *Compt. rend.*, 1972, **274**, C, 205.
[34] L. K. Vasyanina, Yu. S. Bogachev, N. N. Shapet'ko, and T. L. Alekseeva, *Zhur. obshchei Khim.*, 1971, **41**, 2523.
[35] L. N. Kurkovskaya, A. M. Kuliev, B. Yagshiev, N. N. Shapet'ko, F. N. Mamedov, and A. G. Bairamova, *Doklady Akad. Nauk S.S.S.R.*, 1971, **197**, 842.

(4)

interaction.[36] The detailed behaviour is, however, somewhat unusual. On adding proton-acceptor material the resonance shifts to high frequency initially, the extent of the shift decreasing to a minimum. Ultimately the shift reverses direction and begins to move to low frequency as the proton-acceptor concentration increases. This is not the behaviour expected for simple phenols.[37] Formation of a hydrogen-bonded associated ion-pair complex with subsequent dissociation, as shown in the Scheme, could account for the

Scheme

observations. The latter species would cause the observed shift to lower frequencies, and is consistent with the fact that this reversal of the normal high-frequency shift only occurs in solvents of high dielectric constant.[36] The normal high-frequency shift associated with hydrogen-bonding is observed on adding poly-γ-benzyl-L-glutamate to the system, without the chemical-shift reversal observed for the simple amides. There is some evidence that poly-L-methionine also interacts with hexachlorophene in a similar way.[36]

Much of the work relating to studies of the hydroxy-proton resonance in systems such as those above is of a qualitative nature, as will have been seen. A few rather more quantitative approaches have, however, been reported, and there have been some attempts to relate the observed hydroxy-proton shifts to known thermodynamic quantities of the system.[38] One approach which seems to have been particularly successful is that of Kuppers.[39] For aqueous alcohol solutions where only a single coalesced hydroxy-proton resonance occurs, owing to fast exchange between the various possible species in the medium, it is possible to define an 'ideal' chemical shift for the hydroxy-group, δ_{ideal}, as:

$$\delta_{\text{ideal}} = A_0 \delta_2 + (1 - A_0) \delta_1 \qquad (20)$$

[36] R. Haque and D. R. Buhler, *J. Amer. Chem. Soc.*, 1972, **94**, 1824.
[37] F. Takahashi and N. C. Li, *J. Phys. Chem.*, 1965, **69**, 1622.
[38] W. Storek, R. Radeglia, and W. Köhler, *Z. phys. Chem. (Leipzig)*, 1971, **247**, 151.
[39] J. R. Kuppers, *J. Magn. Resonance*, 1971, **4**, 220.

where δ_1 is the hydroxy-shift in pure water, δ_2 is that of the alcohol, and A_0 is the fraction of hydroxy-protons contributed by the alcohol. An excess chemical shift, δ_{excess}, may then be defined:

$$\delta_{\text{excess}} = \delta_{\text{obs}} - \delta_{\text{ideal}} \qquad (21)$$

which provides an indication of the extent of hydrogen-bonding in the system. Figure 2 shows how the excess chemical shift correlated with other thermodynamic constants for the aqueous t-butyl alcohol system.[39] (For this

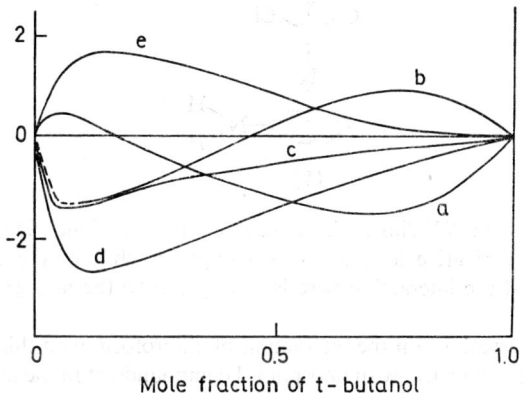

Figure 2 *Correlation of excess chemical shifts with thermodynamic quantities to the aqueous t-butanol system:* (a) δ_{excess} *for the coalesced hydroxy-proton resonances at* 35 °C; (b) *integral heat of mixing per mole of solution at* 26 °C; (c) *relative excess entropy of mixing at* 26 °C; (d) *relative excess volume at* 25 °C; (e) *relative excess viscosity at* 25 °C; *all quantities have been suitably scaled for plotting*
(Reproduced by permission from *J. Magn. Resonance*, 1971, **4**, 220)

correlation *relative excess quantities* were defined, such as the relative excess entropy ΔS^{ER} given by

$$\Delta S^{\text{ER}} = \Delta S^{\text{E}}/\Delta S^{\text{I}} \qquad (22)$$

where ΔS^{I} is the ideal entropy of mixing, and ΔS^{E} is the excess entropy). From Figure 2 it can fairly readily be seen that hydrogen-bonding reaches a maximum throughout the system at about 0.10 mole fraction of t-butyl alcohol. On the other hand, the greatest disruption of hydrogen-bonds occurs in the region of 0.65 mole fraction of alcohol. The same general trends are apparent in a subsequent study of aqueous methanol, ethanol, propan-1-ol, and propan-2-ol systems.[40]

The proton resonance of chloroform has also been widely used recently as a sensitive probe of interactions involving this molecule. It is known from infrared studies that a hydrogen-bond is formed between chloroform and aniline,[41] although it is not clear from this approach whether the bond in-

[40] J. R. Kuppers and N. E. Carriker, *J. Magn. Resonance*, 1971, **5**, 73.
[41] A. G. Moritz, *Spectrochim. Acta*, 1961, **17**, 365.

volves the nitrogen lone-pair electrons or the π-electrons of the benzene ring. Ng has pointed out, however, that such a distinction can in fact be made by observing the direction of the chloroform proton-shift change brought about by the addition of aniline.[42] Co-ordination to the nitrogen atom should result in a high-frequency shift; conversely, the diamagnetic anisotropy of the benzene ring should lead to a low-frequency shift if the interaction is with the ring π-electrons, as in (5). In fact a low-frequency shift is observed on addition

$$\begin{array}{c} Cl \\ Cl-\underset{|}{\overset{|}{C}}-Cl \\ H \\ \end{array} \quad \underset{}{\bigcirc}-N\underset{H}{\overset{H}{\diagdown}}$$

(5)

of either aniline or NN-dimethylaniline. In contrast, cyclohexylamine and its NN-dimethyl derivative lead to a high-frequency shift of the chloroform resonance, and the interaction here is clearly due to the nitrogen lone pair of electrons.

Similar observations on the interaction of chloroform with thiophen have been reported,[43] and also with phosphoryl compounds.[44] In the latter case,[44] association constants and the enthalpy and entropy of the various interactions were determined by the method of Higuchi,[45] and a brief study of the self-association of chloroform in CCl_4 and cyclohexane was reported. In cyclohexane the chloroform resonance is insensitive to temperature; this is not, however, the case for CCl_4, and it is apparent that self-association in the latter solvent is appreciable.[44] Complexation between chloroform and pyrrolid-2-one has also been studied in a quantitative manner; self-association of the lactam complicates the analysis, but a crude estimate of the enthalpy of the interaction gives a value of about -6 kJ mol^{-1}.[46] De Boer and co-workers also note a specific interaction between chloroform and NN-dimethylbenzamides and -cinnamides.[47] From n.m.r. determinations of the barriers to internal rotation in these compounds, the values obtained for chloroform solution appear to be abnormally high, and are explained in terms of chloroform association with the carbonyl group.

Wiley and Miller[48] have reported a very detailed quantitative study of chloroform interactions based on observations of the chloroform proton

---

[42] S. Ng, *Spectrochim. Acta*, 1972, **28A**, 321.
[43] D. F. Ewing and R. M. Scrowston, *Org. Magn. Resonance*, 1971, **3**, 405.
[44] T. Gramstad and O. Mundheim, *Spectrochim. Acta*, 1972, **28A**, 1405.
[45] M. Nakano, N. N. Nakano, and T. Higuchi, *J. Phys. Chem.*, 1967, **71**, 3954.
[46] L. F. Blackwell, P. D. Buckley, K. W. Jolley, and I. D. Watson, *Austral. J. Chem.*, 1972, **25**, 67.
[47] K. Spaargaren, P. K. Korver, P. J. van der Haak, and Th. J. De Boer, *Org. Magn. Resonance*, 1971, **3**, 615.
[48] G. R. Wiley and S. I. Miller, *J. Amer. Chem. Soc.*, 1972, **94**, 3287.

resonance. Equilibrium constants and thermodynamic parameters for a number of systems based on a 1:1 stoicheiometry for the interaction were reported, and the validity of the model was checked by the goodness of fit of the data on the basis of Deranleau's work (this is treated in more detail in the section relating to ASIS effects). Estimates of the shift of the chloroform resonance in the completely complexed state [the δ_t values of equation (19)] are reported, and it is apparent that these shifts show a slight temperature dependence, approximately -3.6 Hz over 60 °C. This behaviour is discussed in terms of the model of Muller and Reiter,[49] which suggests that, as the hydrogen-bond

Table 2 ^{15}N *Chemical shifts and* (^{13}C, ^{15}N) *coupling constants for the pyridine system*[50]

	$\Delta\sigma(^{15}N)^a$	$^1J_{12}{}^b$	$^2J_{13}$	$^3J_{14}$
Pure pyridine	56.8	0.45	2.4	3.6
Pyridine 30% v/v in MeOH	74.4	0.7	2.6	3.8
Pyridinium hydrochloride 30% v/v in MeOH	170.1	12.0	2.1	5.3

^{a} Shifts $\Delta\sigma$ are relative to external $H^{15}NO_3$;
^{b} Coupling constants are quoted in Hz.

length increases with temperature, so the proton tends to approach the environment obtaining for the non-hydrogen-bonded state. This therefore results in a shift to lower frequency with increasing temperature, which is in fact observed in this case.[48]

Hydrogen-bonding in pyridine has been studied recently by Roberts and Lichter[50] by observing the ^{15}N and ^1H spectra of [^{15}N]pyridine. A low-frequency shift of the ^{15}N resonance is observed when the nitrogen is involved in hydrogen-bond formation, and the (^{13}C, ^{15}N) coupling constants are also affected, as can be seen from the figures of Table 2. These authors have also detected a change in the sign of the $^1J_{12}$ coupling of pyridine on dilution with MeOH. The reasoning is as follows: if the pyridinium cation can be regarded as approximating to the fully hydrogen-bonded pyridine situation, then, from a consideration of the chemical shifts, pyridine would appear to be approximately 15% hydrogen-bonded in the methanol solution. From $^3J_{14}$ a figure of about 9% is obtained, giving an average figure of 12% overall. It is apparent, therefore, from the change in the $^1J_{12}$ values, that in order to arrive at the same estimate of the extent of hydrogen-bonding, it is necessary to assume that a change occurs in the sign of $^1J_{12}$. This is thought to be the first reported case of such a change in sign due to an interaction of this type. Liler[51] has reported that the (^{15}N, ^1H) coupling constants of [^{15}N]acetamide are solvent-dependent, being larger in water than in acetone solution. This is again attributed to hydrogen-bonding, which should be more extensive in the latter solvent than in the former.

[49] N. Muller and R. C. Reiter, *J. Chem. Phys.*, 1965, **42**, 3265.
[50] R. L. Lichter and J. D. Roberts, *J. Amer. Chem. Soc.*, 1971, **93**, 5218.
[51] M. Liler, *J. Magn. Resonance*, 1971, **5**, 333.

A method has recently been reported[52] which allows a determination of the absolute configuration of certain chiral species, based on a hypothetical model for a solvent–solute interaction. As an example, N-methyl-N-ethylnaphthylamine oxide interacts with the solvent, (S)-(+)-2,2,2-trifluorophenylethanol, to form solvates (6) which tend to be held in the indicated configuration by the hydrogen-bond, and also by an interaction involving the carbinyl hydrogen and the ring π-electron system. The groups about the nitrogen will therefore be either in the shielding or the deshielding region of the naphthylamine ring system, and their relative chemical shifts allow an assignment of their absolute configuration to be made.

(6)

Theoretical calculations relating to hydrogen-bonded systems are generally very approximate. N.m.r. spectroscopy can, however, provide a useful means of checking such calculations and can also provide valuable information with regard to the general nature of the hydrogen-bond itself. Pople's[53] theoretical approach to the calculation of hydrogen-bonded proton shifts has recently been extended and applied to the formic acid cyclic dimer with some success.[54] A method involving a study of (^{13}C,^{1}H) coupling constants has been described[55] which allows quantitative estimation of the extent of hydrogen-bonding, and Nelson and co-workers[56] have studied the variation of electron distribution in aromatic systems affected by hydrogen-bonding involving ring substituents, using ^{13}C spectroscopy as a probe. Most particularly, however, Morishima, Endo, and Yonezawa have reported a series of n.m.r. studies of hydrogen-bonding involving the stable free radical di-t-butyl nitroxide. The resonance positions of protic molecules in the presence of this radical are generally shifted to lower frequencies.[57] This in itself is an interesting observation, since the shifts must have as their origin the Fermi contact interaction, which requires a degree of covalent character for the hydrogen-bond. Figure 3 illustrates this effect for a series of protic species. The low-shift values observed for acetic acid probably are a result of self-association, which competes effectively with the radical interaction. The unusually high shifts for

[52] W. H. Pirkle, R. H. Muntz, and I. C. Paul, *J. Amer. Chem. Soc.*, 1971, **93**, 2817.
[53] J. A. Pople, *J. Chem. Phys.*, 1962, **37**, 53.
[54] M. Žaucer and A. Ažman, *Rev. Roumaine Chim.*, 1971, **16**, 481.
[55] P. Tuomikoski and K. Blomster, *Suomen Kem.* (B), 1971, **44**, 391.
[56] G. L. Nelson, G. C. Levy, and J. D. Cargioli, *J. Amer. Chem. Soc.*, 1972, **94**, 3089.
[57] I. Morishima, K. Endo, and T. Yonezawa, *J. Amer. Chem. Soc.*, 1971, **93**, 2048.

CHCl$_3$ suggest a greater degree of covalency for hydrogen-bonds involving CH groups over those involving the OH and NH groups.

It has subsequently been shown that aprotic molecules also interact with di-t-butyl nitroxide, leading to marked ^{13}C shifts, although proton shifts are small.[58] It is evident from these observations that the methyl group may act as a weak proton donor in appropriate circumstances. Theoretical models for the interaction have been discussed, based largely on estimates of spin densities throughout the substrate molecules from ^{13}C and ^{1}H n.m.r. results.[59, 60] Of those considered, only the π-model (7) is consistent with the observed behaviour.

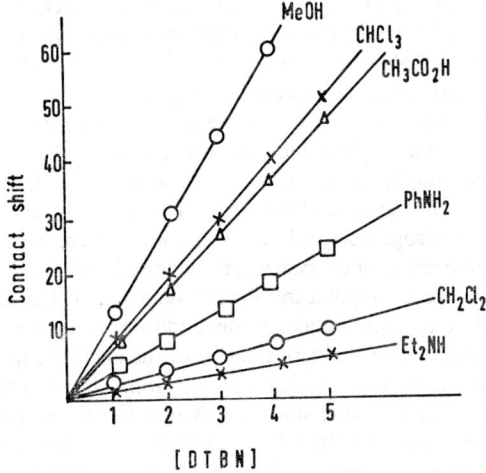

Figure 3 *Plot of the contact shifts observed for* 2.5×10^{-3} mol l^{-1} *solutions of various substrates as a function of the concentration of di-t-butyl nitroxide* (DTBN)
(Reproduced by permission from *J. Amer. Chem. Soc.*, 1971, **93**, 2048)

A correlation has also been noted[60] between the relative values of ^{13}C and proton shifts in a given solvent, and the directly-bonded (^{13}C,H) coupling constants. This is thought to be due to the fact that the (^{13}C,H) coupling constant reflects the spin density in the ^{13}C nucleus resulting from the presence of unpaired spin on the proton.

C. Ionic Solvation and Ion-pairing Effects.—The division between this and the former section is rather arbitrary, since a very common and useful means of studying the behaviour of ionic solutions is to observe by n.m.r. spectroscopy the extent to which the presence of the ions perturbs the structure of the surrounding medium, particularly the extent of hydrogen-bonding. Again, the hydroxy-proton resonance is a powerful probe of suitable systems. In water, for example, a high-frequency shift of the water resonance results from in-

[58] I. Morishima, T. Matsui, T. Yonezawa, and K. Goto, *J.C.S. Perkin II*, 1972, 633.
[59] I. Morishima, K. Endo, and T. Yonezawa, *Chem. Phys. Letters*, 1971, **9**, 143.
[60] I. Morishima, K. Endo, and T. Yonezawa, *Chem. Phys. Letters*, 1971, **9**, 203.

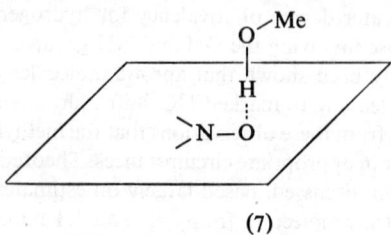

(7)

creased hydrogen-bonding. Ions which effect chemical-shift changes in this direction therefore are classed as 'structure makers'. On the other hand 'structure breakers' have the opposite effect, and reduce the extent of hydrogen-bonding. Results obtained for the alkylammonium salts, however, are not consistent with this simple approach, in that, despite a body of independent evidence which suggests that such ions are structure makers, nevertheless the water proton resonance is shifted to lower frequencies in their presence.[61] The same type of contradiction has also been reported for solutions of sodium alkyl sulphates, and discussed in terms of a change in the extent of covalent character of the hydrogen-bonds in the solution.[62] There has recently been a resurgence of interest in such systems, the anomalous behaviour being attributed to a pronounced temperature effect.[63] Below 25 °C the water resonance does in fact shift to high frequency on addition of tetra-alkylammonium bromides, the shift increasing with the size of the alkyl group, which is consistent with a structure-making effect of the cation. Above 25 °C the converse is true. Subsequently, absolute shifts for these cations were determined by a method described earlier[64] and the effect was confirmed.[65] Another, less detailed, account indicates qualitatively similar observations which were discussed in terms of the formation of multiply charged aggregates which become disrupted with increasing temperature.[66] Other workers have also noted the temperature-dependence of the structuring effect of substituted ammonium ions,[67] the structure-making effect at low temperatures being enhanced as the size of the alkyl group increases. The alkylammonium fluorides seem to be unique in being structure makers at all temperatures.

Ion-pairing in tetra-alkylammonium salt solutions, which must be a factor influencing the effect of the salt on the surrounding medium, has also been studied extensively by n.m.r., particularly where paramagnetic complex anions are involved.[68—70] From such studies it is often possible to gain useful information with regard to the structure of the ion pair when the cation shifts

[61] H. G. Hertz and W. Spalthof, *Z. Elektrochem.*, 1959, **63**, 1096.
[62] J. Clifford and B. A. Pethica, *Trans. Faraday Soc.*, 1964, **60**, 1483.
[63] J. Davies, S. Ormondroyd, and M. C. R. Symons, *Chem. Comm.*, 1971, 1204.
[64] J. Davies, S. Ormondroyd, and M. C. R. Symons, *Trans. Faraday Soc.*, 1971, **67**, 3464.
[65] J. Davies, S. Ormondroyd, and M. C. R. Symons, *J.C.S. Faraday II*, 1972, **68**, 686.
[66] A. LoSurdo and H. E. Wirth, *J. Phys. Chem.*, 1972, **76**, 130.
[67] M. M. Lucas and M.-M. Marciacq-Rousselot, *Compt. rend.*, 1972, **274**, C, 312.
[68] D. W. Larsen, *J. Phys. Chem.*, 1971, **75**, 3880.
[69] I. M. Walker, L. Rosenthal, and M. S. Quereshi, *Inorg. Chem.*, 1971, **10**, 2463.
[70] Y. Y. Lim and R. S. Drago, *J. Amer. Chem. Soc.*, 1972, **94**, 84.

are treated in terms of the dipolar interaction alone (see below). However, it is apparent from detailed studies of substituted ammonium cations that there is also a contribution from the Fermi-contact term,[69-72] which in itself is a most interesting observation, since it necessitates a certain amount of covalent orbital overlap between cation and anion in the ion pair.

One further factor which is of interest in any study of the effect of dissolved ions on the structure of the surrounding medium is the nature and constitution of the solvent co-ordination sphere. A brief review of the n.m.r. methods applicable to this study has appeared recently.[73] The most direct method is to cool the solution so that exchange between the various species becomes sufficiently slow that separate resonances are observed for the free and co-ordinated solvent species. The chemical shifts of the solvent molecules in the co-ordination sphere and also the co-ordination number are then directly observable. In this way it has been shown that Al^{3+} co-ordinates four solvent molecules in ethanol solution.[74] Transition-metal solvation in dimethyl sulphoxide has been studied,[75] and solvation of Al^{3+} in the same solvent, in this case by ^{13}C as well as ^1H n.m.r.[76] Observation of the ^{115}In resonance has been applied to a study of In^{3+} solvation in aqueous acetone solutions.[77] Acetone is frequently used as a diluent in studies of this kind to allow low temperatures to be achieved without freezing the solution. There has so far been no evidence of an interaction between acetone and any of the ions studied.[77] Six-fold co-ordination of In^{3+} by water had been demonstrated earlier by this method for $InClO_4$ solutions[78] over a wide range of concentrations. Subsequent determinations on In^{3+} halide systems yielded co-ordination numbers which decrease as the salt concentration increases,[77] illustrating an alternative method of detecting the extent of ion-pairing in the solution. As the salt concentration is increased, water molecules tend to be displaced from the co-ordination sphere by halide ion, leading to a net reduction in the hydration number. Extensive ion-pairing in the $Er(NO_3)_3$ system has been detected by the same method.[79] Complexation of Group III and Group V metal halides with acetonitrile has been reported,[80] and the hydration numbers of a series of tervalent lanthanide cations in aqueous solution have been determined by this 'direct' approach.[79]

[71] D. G. Brown and R. S. Drago, *J. Amer. Chem. Soc.*, 1970, **92**, 1871.
[72] P. K. Burkett, H. P. Fritz, W. Gretner, H. J. Keller, and K. E. Schwarzhans, *Inorg. Nuclear Chem. Letters*, 1968, **4**, 237.
[73] J. J. Delpuech, A. Péguy, and M. R. Khaddar, *J. Electroanalyt. Chem. Interfacial Electrochem.*, 1971, **29**, 31.
[74] H. Grasdalen, *J. Magn. Resonance*, 1971, **5**, 84.
[75] G. S. Vigee and P. Ng, *J. Inorg. Nuclear Chem.*, 1971, **33**, 2477.
[76] J. C. Boubel, J. J. Delpuech, M. R. Khaddar, and A. Péguy, *Chem. Comm.*, 1971, 1265.
[77] A. Fratiello, D. D. Davis, S. Peak, and R. E. Schuster, *Inorg. Chem.*, 1971, **10**, 1627.
[78] A. Fratiello, R. E. Lee, V. M. Nishida, and R. E. Schuster, *J. Chem. Phys.*, 1968, **48**, 3705.
[79] A. Fratiello, V. Kubo, S. Peak, B. Sanchez, and R. E. Schuster, *Inorg. Chem.*, 1971, **10**, 2552.
[80] I. Y. Ahmed and C. D. Schmulbach, *Inorg. Chem.*, 1972, **11**, 228.
[81] C. Lassigne and P. Baine, *J. Phys. Chem.*, 1971, **75**, 3188.

'Indirect' methods, which avoid the need to observe the solution at low temperatures, are also possible. By plotting the chemical shifts of the methyl resonances of dimethylformamide (DMF) as a function of the relative concentration of $LiClO_4$ in dioxan–DMF–$LiClO_4$ mixtures, a sharp discontinuity is observed when the ratio of DMF to $LiClO_4$ becomes 4:1;[81] the solvated species is therefore likely to be $Li(DMF)_4{}^+$. In the absence of $LiClO_4$, the *cis*-methyl resonance is dependent on the dioxan concentration, although this is not the case for the *trans*-methyl resonance. Co-ordination of dioxan about the *cis*-methyl site is therefore evident.

Beech and Miller[82] have studied the effect of adding Co^{2+} salts on the solvent resonance position for acetone, acetonitrile, acrylonitrile, and dimethyl sulphoxide at various temperatures, and the effect of solvent on the ^{23}Na

(8)

resonance of NaI has been reported.[83] Ehrlich and Popov[84] have remarked on the utility of ^{23}Na resonance studies in investigations of ionic interactions, since the large quadrupole moment of this nucleus renders the linewidth particularly sensitive to the electronic environment, and the resonance position itself is strongly dependent on the solvent.[84] The displacement of DMF from the co-ordination sphere of $^{23}Na^+$ by hexaethylenedimethyl ether or by (8) can be followed by observing the low-frequency shift of the ^{23}Na peak.[85] The concomitant linewidth changes are consistent with an increased electric field gradient at the nucleus; the ion is therefore presumed to be sited at the centre of the species (8), which adopts a planar configuration.[85]

The cation (9) exhibits concentration-dependent proton shifts in dimethyl sulphoxide solution.[85] In this instance it seems to be necessary to consider two effects: non-specific ion pairing between (9) and the gegenanion, and also a hydrogen-bonded ion pair.

Relatively few studies have appeared of the dependence of the spectra of anionic species on solvent and counter-ion. An intensive study of the anions (10)—(13) has, however, appeared,[86] in which it is asserted that the coupling

---

[82] G. Beech and K. Miller, *J.C.S. Dalton*, 1972, 801.
[83] M. Herlem and A. I. Popov, *J. Amer. Chem. Soc.*, 1972, **94**, 1431.
[84] R. H. Ehrlich and A. I. Popov, *J. Amer. Chem. Soc.*, 1971, **93**, 5620.
[85] R. C. Neuman, jun. and V. Jonas, *J. Phys. Chem.*, 1971, **75**, 3550.
[86] J. B. Grutzner, J. M. Lawlor, and L. M. Jackman, *J. Amer. Chem. Soc.*, 1972, **94**, 2306.

constants are dependent on the particular cation present in proportion to the apparent inter-ion distance in the contact ion pair. The coupling constants would therefore seem to depend on the local electric field. The proton shifts depend on both temperature and solvent, and are rationalized in terms of an equilibrium between contact and solvent-separated ion pairs.[86]

Cation fixation by polyphosphate anions in aqueous solution has been studied by determining the extent to which normal hydration of the cation is affected by the presence of polyanion.[87] In Figure 4 the solid line represents the water proton resonance shift as a function of Co^{2+} concentration, and reflects the effect normal cation hydration has on the water structure. The broken line shows the water resonance shift over the same range of Co^{2+} concentrations in the presence of the polyanion as the tetramethylammonium salt. It is clear that, whilst the relative concentration of Co^{2+} to polyanion phosphate groups is less than 50%, the water resonance remains constant. No hydration of the Co^{2+} ions has taken place, therefore, and the cations must be 'fixed' to suitable sites on the polyanion, presumably each Co^{2+} being bound to two neighbouring phosphate groups. Additional Co^{2+} ions introduced into the solution are hydrated in the normal way as no further binding sites remain to them.

D. Aromatic Solvent-induced Shifts (ASIS Effects).—The effect of dissolving polar solute species in an aromatic solvent is now widely known.[2,88] Invariably, the nuclear resonances are shifted from their normal positions as measured in some inert solvent, the shifts being described in terms of an ASIS parameter, $\Delta_{i,AR}$, given by the relation

$$\Delta_{i,AR} = \delta_{i,AR} - \delta_{i,IN} \qquad (23)$$

where $\delta_{i,IN}$ is the ith nuclear resonance position in a suitable inert solvent. Measurements are, of course, made with respect to the same internal reference standard as a general rule, so that the change in chemical shift is regarded as

[87] P. Spegt and G. Weil, *Compt. rend.*, 1972, **274**, C, 587.
[88] J. Ronayne and D. H. Williams, *Ann. Rev. N.M.R. Spectroscopy*, 1969, **2**, 83.

being due to some form of specific interaction between the solute and the solvent which is not experienced by the reference standard. Engler and Laszlo,[89] however, have effectively treated ASIS shifts of certain rigid ketones in terms of a non-specific 'clustering' of solvent molecules about the carbonyl group. In other cases, however, the interaction is clearly more specific and is best discussed in terms of charge-transfer complex formation.[90]

Having determined the ASIS factors for the nuclei of the solute, the next

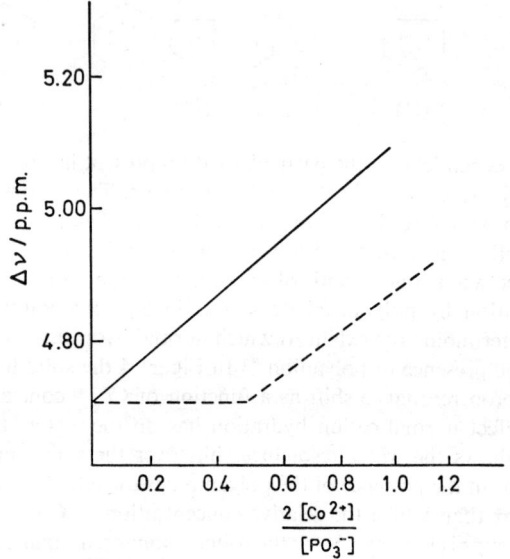

Figure 4 *Chemical shift of the water proton resonance as a function of the concentration of Co^{2+} cation in the absence (full line) and the presence (broken line) of polyphosphate anion*
(Reproduced by permission from *Compt. rend.*, 1972, **274**, C, 587)

step is usually to infer from the direction and magnitude of the shifts certain details relating to the species present in the solution, their stoicheiometry, geometry *etc.*, on the assumption that the ASIS shifts are entirely due to the diamagnetic anisotropy of the aromatic solvent. However, the factors which contribute to a given chemical shift in solution are rather complex [equation (18)], and for the above inferences to be valid it is necessary that the reference solvent be truly inert, and that all other factors which affect the shift be the same for both the reference and aromatic solvent, with the single exception of the interaction which leads to the ASIS effect. Such a requirement can only be approximated in practice, but should always be borne in mind. Ng[91] has

[89] E. M. Engler and P. Laszlo, *J. Amer. Chem. Soc.*, 1971, **93**, 1317.
[90] R. Foster, 'Organic Charge-Transfer Complexes', Academic Press, London and New York, 1969.
[91] S. Ng, *Org. Magn. Resonance*, 1970, **2**, 283.

recently discussed this point with regard to the use of $CHCl_3$ as a reference solvent, pointing out that, since the dielectric constants of benzene and $CHCl_3$ are widely different, the reaction-field term of equation (18) will certainly not be the same for both. Benzene solvent shifts referred to $CHCl_3$ are therefore not a true indication of the diamagnetic anisotropy effect of the aromatic ring. In any case, $CHCl_3$ can hardly be regarded as being inert, as is evident from earlier discussions relating to hydrogen-bonding interactions of this molecule. In this respect CCl_4 is a better reference solvent, since now the dielectric constant is more nearly equal to that of benzene.[91] It has in fact been shown that benzene shifts referred to $CHCl_3$ sometimes differ quite markedly from those referred to CCl_4.[91]

A second major assumption of the method is that a specific solute–solvent complex is formed having a 1:1 stoicheiometry, and which has a very short lifetime on the n.m.r. timescale. It is worth looking a little more closely at the basis of the former assumption, particularly since it has recently been asserted that it has been '... amply verified that there are formed 1:1 collision complexes between the solute molecule and one benzene molecule'.[92] This statement is wholly unfounded. It would in fact be more accurate to state that *no definitive evidence has so far been reported to support the assumption of a 1:1 stoicheiometry in any system so far studied, whilst there is a growing body of evidence to suggest that the 1:1 model is inadequate.* Evidence which is frequently cited to support the 1:1 complex is the observation that plots of the ASIS values, $\Delta_{i,AR}$, are linear with respect to the mole fraction of the solute in a binary mixture of solute and solvent.[93,94] However, such an observation cannot in general be taken as evidence for the proposed model, since, for fast exchange conditions

$$\delta_{i,\text{obs}} = P_{i,S}\delta_{i,S} + P_{i,SB}\delta_{i,SB} \quad (24)$$

where the subscripts S and SB relate to solute and complex respectively, and a 1:1 stoicheiometry is assumed. If the shifts are expressed with reference to the shift of the uncomplexed solute, effectively then $\delta_{i,S} = 0$ and

$$\Delta_{i,\text{obs}} = \frac{[S]}{[S]_o} \Delta_{i,SB} \quad (25)$$

where [S] and [S]$_o$ are the equilibrium and total solute concentrations, respectively. Equation (25) can in no way be expressed as a linear function of solute or solvent mole fraction. Clearly then, even if a 1:1 stoicheiometry were the case, plots of shift *versus* solute mole-fraction should not be linear. In fact, plots reproduced in Ng's account[91] are decidedly non-linear, as indeed is one plot which the authors have described as linear and present as evidence for the 1:1 model.[94] Nor is it satisfactory to argue that the 1:1 model pre-

---

[92] W. W. Paudler and S. A. Humphrey, *Org. Magn. Resonance*, 1971, **3**, 217.
[93] N. E. Alexandrou and A. G. Varvoglis, *Org. Magn. Resonance*, 1971, **3**, 293.
[94] N. E. Alexandrou, P. M. Hadjimihalakis, and E. G. Pavlidou, *Org. Magn. Resonance*, 1971, **3**, 299.

dicts linearity over certain regions of the curve, since this will be true of almost any function.

Deranleau[95,96] has reported a detailed and lucid description of the type of experiment which must be conducted in order to validate any particular theoretical model. Although based on well-known principles,[97-99] the approach has yet to be extensively applied to ASIS systems. Suppose that the behaviour of a given solute S is being investigated in the presence of benzene B as solvent. For a 1:1 association, S + B $\rightleftharpoons$ SB,

$$K = \frac{[SB]}{[S][B]} \qquad (26)$$

Let α be the fraction of substrate molecules not bound to benzene under any given set of conditions, then

$$\alpha = \frac{[S]}{[S]+[SB]} = \frac{1}{1+[SB]/[S]} = \frac{1}{1+K[B]} \qquad (27)$$

and the probability that any substrate molecule chosen at random will be associated with a benzene molecule is therefore s, where

$$s = 1 - \alpha = \frac{K[B]}{1+K[B]} = \frac{[SB]}{[S]_o} \qquad (28)$$

s is the 'saturation factor' referred to by Deranleau.[95] The *total experiment* therefore involves a study of all circumstances ranging from zero probability of the substrate being bound to benzene, to a situation where all the substrate is so bound. By a simple application of the principles of information theory,[100] Deranleau[95] is able to assert that it is necessary to investigate the system over at least 75% of the total range of s values in order to substantiate a particular theoretical model. It is interesting to consider what range of s values has in fact been covered by reported data which purport to establish a 1:1 stoicheiometry for the interaction. For n.m.r. experiments this can be done fairly readily, since it can be shown[95] that

$$s = \frac{\Delta_{obs}}{\Delta_{SB}} \qquad (29)$$

and the range of s values can be assessed from an estimated value of Δ_{SB}. Referring to the lines of ref. 93, simple extrapolation to a benzene mole fraction of 1 provides a value of Δ_{SB} from which it may be seen that only about

[95] D. A. Deranleau, *J. Amer. Chem. Soc.*, 1969, **91**, 4044.
[96] D. A. Deranleau, *J. Amer. Chem. Soc.*, 1969, **91**, 4050.
[97] F. J. C. Rossotti and H. Rossotti, 'The Determination of Stability Constants', McGraw-Hill, New York, 1961.
[98] J. T. Edsall and J. Wyman, 'Biophysical Chemistry', Academic Press, New York, 1958, vol. 1.
[99] G. Weber in 'Molecular Biophysics', ed. B. Pullmann and M. Weissbluth, Academic Press, New York, 1965.
[100] L. Brillouin, 'Science and Information Theory', Academic Press, New York, 1962.

45% of the experiment has been carried out. It should also be noted that the figure of 45% is probably an optimistic estimate, since if the lines in fact curve upwards as the benzene mole-fraction approaches unity, then the true Δ_{SB} values would be greater than those obtained by simple extrapolation, and the s-value range will have been over-estimated. None of the reports to date relating to this topic have covered the range required by Deranleau's treatment.

In quantitative studies of closely similar charge-transfer complex systems, however, it is apparent that the greater the range of s values studied, the less satisfactory is the 1:1 model found to be. In an earlier report,[101] where approximately 60% of the whole experimental range was covered, fairly convincing evidence for the presence of higher complexes was obtained. Other indirect evidence has been cited. Examples are known[90] of equilibrium-constant determinations which yield values that depend on the particular substrate nucleus measured, behaviour to be expected if higher-order complexes were being wrongly neglected in the analysis of the data. Similar observations have been made for systems of trinitroanisole and trinitrotoluene with various indoles.[102]

The method most frequently used in quantitative analyses of systems of this type is that described and used extensively by Foster and co-workers.[29,90] One component is required to be present in large excess and a 1:1 stoicheiometry is assumed. Under such conditions equation (30) holds:

$$\frac{\Delta_{obs}}{[B]} = -K\Delta_{obs} + K\Delta_{SB} \qquad (30)$$

where K is the association constant for the interaction, and B the excess component. A plot of $\Delta_{obs}/[B]$ *versus* Δ_{obs} is therefore linear, and the constant K may thereby be determined. In the past, very good straight lines have in fact been obtained, and quoted in support of the validity of the assumption of the 1:1 stoicheiometry. It is now clear that such good lines were obtained because the range of s-values studied was not representative of the whole experiment. Hanna and Rose[103] have also discussed the validity of a further assumption inherent in equation (30), namely the neglect of the activity coefficients, and have included measured values of the activity coefficients of the excess component, in this case benzene, in their analysis. Homer *et al.*[104] have briefly discussed refinements to alternative n.m.r. methods for the estimation of these association constants, and *state* the approach which must be used. Unfortunately the *statement* requires that a Benesi–Hildebrand type of plot must be used in this method, although compelling reasons why such a plot should not be used have already been presented.[95] Takeuchi[105] has used an

[101] B. Dodson, R. Foster, A. A. S. Bright, M. I. Foreman, and J. Gorton, *J. Chem. Soc.* (B), 1971, 1283.
[102] B. Sabourault and J. Bourdais, *Compt. rend.*, 1972, **274**, C, 813.
[103] M. W. Hanna and D. G. Rose, *J. Amer. Chem. Soc.*, 1972, **94**, 2601.
[104] J. Homer, C. J. Jackson, P. M. Whitney, and M. H. Everdell, *Chem. Comm.*, 1972, 956.

alternative approach[106] to investigate benzene solutions of relatively high solute concentration. Again, only some 35% of the entire experiment was studied, so that the observed fit of the data to the 1:1 complex model cannot be taken as evidence that the model is correct. The association constants reported are, however, of the expected order of magnitude, although some anomalous observations were noted.[105]

Although hardly aromatic, it is perhaps appropriate to consider cyclopropane as an ASIS solvent. Anderson[106] has reported cyclopropane shifts for rigid cyclic ketones which are of the opposite sense to those effected by benzene, and although rather smaller they are nevertheless appreciable. For dimethylformamide, however, the cyclopropane shifts are unusual in that they are of the same sense as the benzene shifts.[106] A solute–solvent complex of some description is thought to be formed, in the same manner as occurs with aromatic solvents.[106] Schenk and Anet[107] have discussed the effect along similar lines, observing that protons located directly above the plane of the cyclopropane ring are generally shielded relative to their normal environment, whilst those lying in the plane of the ring are deshielded. It is thought, therefore, that substrate polar groups possibly form hydrogen-bonds to the 'edge' of the cyclopropane ring which would account for the observed high-frequency shifts of the substrate resonances.[107] For benzene, on the other hand, polar groups appear to interact with the face of the ring, and the substrate resonances are therefore shifted to lower frequencies.

4 Shift Reagents

The term 'shift reagent' is by now well established in the n.m.r. literature. The effect of a shift reagent is to increase the chemical-shift differences between nuclei of suitable polar substrate species, without, if possible, significantly affecting the coupling constants or broadening the lines. In cases where extensive overlapping of the nuclear resonance positions occurs, such reagents can be used greatly to simplify the spectra, in a manner qualitatively similar to that obtained by increasing the applied magnetic field, but at a rather less formidable cost. ASIS effects discussed earlier are of this same general type, and no real distinction is possible. In fact, aromatic solvents are of exactly the same class as the diamagnetic shift reagents referred to below. However, the widespread use of paramagnetic species in this context is sufficient justification for their discussion in a separate section.

A. Lanthanide Complexes.—The utility of paramagnetic metal complexes as shift reagents depends on their ability to bind loosely in certain cases to substrate species (S) having polar groups. An equilibrium is set up:

[105] Y. Takeuchi, *J. Chem. Soc.* (*B*), 1971, 1884.
[106] J. E. Anderson, *J. Chem. Soc.* (*B*), 1971, 2388.
[107] G. E. Schenk and F. A. L. Anet, *Tetrahedron Letters*, 1971, 2779.

Medium Effects

$$nS + SR \rightleftharpoons SR \cdots S_n \quad K = \frac{[SR \cdots S_n]}{[SR][S]^n} \quad (31)$$

Since rapid-exchange conditions generally prevail, the observed resonance positions of the substrate nuclei are the weighted averages of those appropriate to the free and complexed state [equation (19)]. The latter shifts, because of the paramagnetism of the complex, may be very different from the former. The overall effect, therefore, is that the substrate resonances appear to be perturbed from their normal positions by an amount which depends on the relative concentration of shift reagent that is present. Many lanthanide complexes behave in this way, and have become prominent as shift reagents since concomitant effects of line-broadening are often minimal.[108,109]

Another important reason for the widespread interest in lanthanide shift reagents derives from the possibility that the geometry of the substrate might be simply related to the observed shifts. This latter application has, however, given rise to a regrettable number of fragmentary accounts of studies on a few systems, from which conclusions have been drawn purporting to be applicable to lanthanide shift reagent systems as a whole. These, together with often meaningless empirical observations, serve only to increase the confusion which is becoming apparent in studies of such systems.

The Stoicheiometry of the Shift-reagent–Substrate Complex. A case in point relates to the stoicheiometry of the shift-reagent–substrate complex, which is almost universally assumed to be 1:1 in studies of substrate geometries to date. A number of accounts have recently appeared which relate to this most important point. A 1:1 stoicheiometry was inferred initially from the fact that plots of the induced shift *versus* mole ratio of substrate to shift-reagent

(14)   (15)

were linear over large ranges of the curve. Exceptions have been noted,[110] generally in cases where the substrate contains more than one polar group. Non-linear plots are observed[111] for systems of (14) and (15) with $Eu(tmhd)_3$ (tmhd = 2,2,6,6-tetramethylheptane-3,5-dionato), and are almost certainly due to competition between the possible sites of complexation. However, even for the simple case of only one polar group, such plots should not in

---

[108] G. A. Webb, *Ann. Reports N.M.R. Spectroscopy*, 1970, **3**, 211.
[109] H. J. Keller and K. E. Schwarzhans, *Angew. Chem. Internat. Edn.*, 1970, **9**, 196.
[110] H. van Brederode and W. G. B. Huysmans, *Tetrahedron Letters*, 1971, 1695.
[111] I. Fleming, S. W. Hanson, and J. K. M. Sanders, *Tetrahedron Letters*, 1971, 3733.

general be linear, even if the stoicheiometry were 1:1, since the appropriate expression is:[112]

$$\Delta_{i,\text{obs}} = \Delta_{i,\text{SR}\cdots\text{S}}\left(1 - \frac{[\text{SR}]}{[\text{SR}]_o}\right)\frac{[\text{SR}]_o}{[\text{S}]_o} \tag{32}$$

where $\Delta_{i,\text{obs}}$ is the observed lanthanide-induced shift, and $\Delta_{i,\text{SR}\cdots\text{S}}$ is the shift of the ith nucleus in the lanthanide–substrate complex, both measured with respect to the shift in the uncomplexed species. The subscript 'o' signifies total, as opposed to equilibrium concentrations. Armitage et al.[113] have expanded equation (32) to give:

$$\log[\text{S}] = \left(\frac{1}{n}\right)\log\left(\frac{[\text{SR}\cdots\text{S}]}{[\text{SR}]}\right) - \left(\frac{1}{n}\right)\log K \tag{33}$$

so that a plot of log[S] versus log([SR $\cdots$ S]/[SR]) should be linear, of slope $1/n$. A value of n of approximately 1.0 was in fact obtained by this method. However, in order to compute values of [S], [SR], and [SR $\cdots$ S] the authors appear to have used the expression;

$$\Delta_{i,\text{obs}} = \frac{[\text{SR}\cdots\text{S}]}{[\text{S}]_o} \cdot \Delta_{i,\text{SR}\cdots\text{S}} \tag{34}$$

which is only valid if the stoicheiometry is 1:1, so that its use here seems to beg the question somewhat. Studies[114] involving Job's method seem to imply a 1:1 complex; Job plots are, however, insensitive to quite significant amounts of higher complexes, whenever they co-exist with the 1:1 species. Huber and Selig[115] have attempted an analysis of the Eu(tmhd)$_3$–pyridine system, and also conclude that the stoicheiometry is 1:1. An underlying assumption of this approach is that the nuclei of the substrate will experience the same shift in both the 1:1 complex, SR$\cdots$S, and also the 2:1 complex, SR$\cdots$S$_2$.[115] As will be seen later, this is a most unlikely eventuality. In fact, exactly the same problems arise in the determination of shift-reagent stoicheiometries as were discussed in the ASIS section. No conclusive proof exists that the general assumption of 1:1 complexation is valid; on the contrary, in the only case where the stoicheiometry was convincingly determined,[116] by lowering the temperature of the solution until separate resonances were observed for the free and complexed substrate, Eu(fod)$_3$ was clearly shown[116] to complex with *two* molecules of dimethyl sulphoxide (fod = 1,1,1,2,2,3,3-heptafluoro-7,7-dimethyloctane-4,6-dionato).

The Nature of the Lanthanide-induced Shifts. Unpaired electron spins can perturb the nuclear resonance positions of bound substrates either by the Fermi-

[112] D. R. Kelsey, *J. Amer. Chem. Soc.*, 1972, **94**, 1764.
[113] I. Armitage, G. Dunsmore, L. D. Hall, and A. G. Marshall, *Chem. and Ind.*, 1972, 79.
[114] K. Roth, M. Grosse, and D. Rewicki, *Tetrahedron Letters*, 1972, 435.
[115] H. Huber and J. Seelig, *Helv. Chim. Acta*, 1972, **55**, 135.
[116] D. F. Evans and M. Wyatt, *J.C.S. Chem. Comm.*, 1972, 312.

contact interaction or by a dipolar mechanism. The latter interaction, leading to the 'pseudocontact' shifts, is a through-space effect and therefore a function of the substrate geometry, and the anisotropy of the principal molecular magnetic susceptibilities of the complex as a whole. It is often found that the shifts are discussed in terms of the g-tensor anisotropy. This is strictly incorrect, since magnetic susceptibilities are not necessarily related to the g-tensor, and it is the anisotropy of the former which determines the dipolar shifts.[117]

The contact interaction, conversely, is transmitted through the substrate bonds, and requires that the unpaired electron be in an orbital having some degree of s-character. The dipolar and contact effects operate simultaneously, so that:

$$\Delta_{i,\text{obs}} = \Delta_{i,\text{con}} + \Delta_{i,\text{dip}} \tag{35}$$

The interest in the lanthanide shift reagents stems partly from the belief that the contact term can be neglected, and the observed shift therefore treated in terms of the McConnell and Robertson equation for the dipolar term only:[118]

$$\Delta_{i,\text{obs}} = \Delta_{i,\text{dip}} = \frac{K(3\cos^2\theta_i - 1)}{r_i^3} \tag{36}$$

where θ_i is the angle between the principal magnetic axis and the vector r joining the ith nucleus to the paramagnetic center, and K is a constant for the particular complex.

Instances where the contact term may not, however, be neglected are becoming increasingly common. For the N-oxides of pyridine and picoline with Eu(fod)$_3$, transmission of unpaired spin density to the ring *via* the contact interaction is clearly apparent.[119] For these substrates the effect of the contact term increases along the series Pr(fod)$_3$, Yb(fod)$_3$, Eu(tmhd)$_3$, Er(fod)$_3$, Eu(fod)$_3$.[119] The alkoxy-protons of (16) are reported[120] to exhibit

(16)

anomalous shifts with Eu(tmhd)$_3$, which are attributed to a contact interaction, propagated through the conjugated π-electron system. For polyglycodimethyl ethers it appears that shift reagents complex preferentially with the terminal oxygen atoms of the chain, and that the protons of the neighbouring CH$_2$ groups experience appreciable contact shifts.[121] Carbon-13 studies of n-alkylamines provide a further example. With Eu(tmhd)$_3$, shifts are normally

---

[117] W. DeW. Horrocks, jun. and D. DeW. Hall, *Coordination Chem. Rev.*, 1971, **6**, 147.
[118] H. M. McConnell and R. E. Robertson, *J. Chem. Phys.*, 1958, **29**, 1361.
[119] B. F. G. Johnson, J. Lewis, P. McArdle, and J. R. Norton, *J.C.S. Chem. Comm.*, 1972, 535.
[120] S. B. Tjan and F. R. Visser, *Tetrahedron Letters*, 1971, 2833.

to higher frequencies, except where θ_i is such that the angle term $3\cos^2\theta_i - 1$ is negative, in which case the shifts are reversed. Such anomalous low-frequency shifts have been reported[122] for the C-2 carbon of a series of n-alkylamines, despite the fact that the C-2—N—Eu angle must be less than 54°44′ and cannot therefore produce a negative value for the angle-dependent term.[123] This is again proposed as evidence of a significant contact interaction.[122] Alternative explanations are, however, possible, as will become apparent from the following discussion.

Calculation of the Dipolar Interaction. The equation of McConnell and Robertson[118] was first used in the context of lanthanide shift reagents by Hinckley,[124] who believed that the angle-dependent term could be neglected, and that observed shifts were inversely proportional to r_i^3. This was shown subsequently to be invalid, since both low-frequency and high-frequency shifts were sometimes observed in the same substrate. This was then thought to be due to neglect of the angle-dependent term, which can be either positive or negative.[123] Apparent examples of such sign reversal continue to be reported; in studies of 3-deacetylkhivorin,[125] of certain steroidal episulphoxides,[126] and of *cis*-4-t-butyl-1-phenylcyclohexanol,[127] for example.

The McConnell and Robertson equation, as originally derived, is not, however, directly applicable either to the *d*-series transition-metal complexes or to the lanthanide series. Two restrictions were explicitly stated:

(*i*) That only one electronic energy level be available to the complex.
(*ii*) The anisotropy of the magnetic susceptibilities must be axially symmetrical.

The first requirement patently does not hold for the case of Eu^{3+} complexes, for which two electronic energy levels are populated, the $^7F_0$ ground state, and the first excited $^7F_1$ state. Only the latter contributes to the paramagnetism of the complexes. These considerations should, however, affect only the value of K in equation (36), so that the *relative* shifts of the nuclei should still be proportional to the $(3\cos^2\theta_i - 1)/r_i^3$ term. More refined analyses relating to this point have appeared.[128,129] Of possibly greater importance is the assumption of axial symmetry, since in the general case the dipolar shifts are given by

$$\Delta_{i,\text{dip}} = -\frac{1}{3r_i^3}\{[\chi_z - \tfrac{1}{2}(\chi_x - \chi_y)](3\cos^2\theta_i - 1) + \tfrac{3}{2}(\chi_z - \chi_x)(\sin^2\theta_i \cos 2\phi_i)\}$$

(37)

[121] A. M. Grotens, J. Smid and E. de Boer, *Tetrahedron Letters*, 1971, 4863.
[122] R. J. Cushley, D. R. Anderson and S. R. Lipsky, *J.C.S. Chem. Comm.*, 1972, 636.
[123] B. L. Shapiro, J. R. Hlubucek, G. R. Sullivan, and L. F. Johnson, *J. Amer. Chem. Soc.*, 1971, **93**, 3281.
[124] C. C. Hinckley, *J. Amer. Chem. Soc.*, 1969, **91**, 5160.
[125] D. E. U. Ekong, J. I. Okogun, and M. Shok, *J.C.S. Perkin I*, 1972, 653.
[126] M. Kishi, K. Tori, T. Komeno, and T. Shingu, *Tetrahedron Letters*, 1971, 3525.
[127] N. S. Bhacca and J. D. Wander, *Chem. Comm.*, 1971, 1505.
[128] R. J. Kurland and B. R. McGarvey, *J. Magn. Resonance*, 1970, **2**, 286.
[129] B. Bleaney, C. M. Dobson, B. A. Levine, R. B. Martin, R. J. P. Williams, and A. V. Xavier, *J.C.S. Chem. Comm.*, 1972, 791.

where r_i, θ_i, ϕ_i are the polar co-ordinates of the ith nucleus relative to the principal magnetic axes of the complex along which the susceptibilities are χ_x, χ_y, χ_z.

It can now be seen that comparing the relative shifts no longer effectively cancels K, as is the case when axial symmetry obtains. A further assumption which is made in applying equation (36) is not only that the magnetic anisotropy be axially symmetrical, but also that the lanthanide to substrate bond corresponds to the principal magnetic axis, so that values of θ_i may be estimated with respect to this bond. X-Ray structure determinations bear out neither assumption for any system so far studied, although such work is restricted to 2:1 substrate to shift-reagent complexes. The magnetic susceptibility axes have been shown to approximate to the ligand-distribution symmetry axes,[131] which for Eu(tmhd)₃(pyridine)₂ [132] and Ho(tmhd)₃(picoline)₂[133] are not axially symmetrical. The same is likely to be true for 1:1 complexes also. Furthermore, if both 1:1 and 2:1 complexes co-exist in the solution state, quite different dipolar shifts should obtain for both species, if only because of their different symmetries. Honeybourne[134] has outlined an experimental approach to the problem of the correct choice of the magnetic axes.

Temperature-dependence of the Lanthanide-induced Shifts. Lanthanide-induced shifts are in practice temperature-dependent, a fact which can be used to good effect in simplifying n.m.r. spectra by varying the temperature until the required peak separation has been achieved.[135] For the alcohols (17) and (18), shifts induced by Pr(tmhd)₃ are linear with temperature, although only a

(17) R = H

(18) R = Me

limited range was studied.[136] Crossing of the lines shown in Figure 5 was attributed to a temperature-dependent change in the proton positions relative to the metal. Beyond this type of observation there has been no serious study of the variation of the induced shifts with temperature. In principle, tempera-

[130] W. DeW. Horrocks, jun. and J. P. Sipe, tert., *J. Amer. Chem. Soc.*, 1971, **93**, 6800.
[131] B. Bleaney and K. W. H. Stevens, *Reports Progr. Phys.*, 1953, **16**, 108.
[132] R. E. Cramer and K. Seff, *J.C.S. Chem. Comm.*, 1972, 400.
[133] W. De W. Horrocks, jun., J. P. Sipe, tert., and J. R. Luber, *J. Amer. Chem. Soc.*, 1971, **93**, 5258.
[134] C. L. Honeybourne, *Tetrahedron Letters*, 1972, 1095.
[135] R. D. Bennett and R. E. Schuster, *Tetrahedron Letters*, 1972, 673.
[136] L. Tomić, Z. Majerski, M. Tomić, and D. E. Sunko, *Chem. Comm.*, 1971, 719.

ture variations arise from two sources: the temperature-dependent paramagnetism of the complex, and the temperature-dependent equilibrium of equation (31). It should also be stressed that the former term will be very different for Eu^{3+} shift reagents compared with most others of the lanthanide series. This is because the first-order Zeeman effect, which is responsible for the paramagnetism of most of the lanthanides, is not a factor in the Eu^{3+} case because of the non-degeneracy of the $^7F_0$ ground state. Here the para-

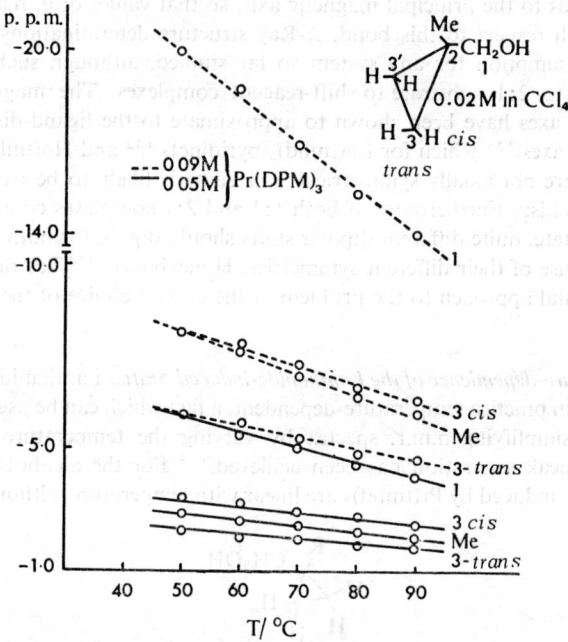

Figure 5 *Temperature- and concentration-dependence of induced proton shifts for different protons in (1-methylcyclopropyl)methanol*
(Reproduced from *Chem. Comm.*, 1971, 720)

magnetism arises from the second-order Zeeman effect which results from occupancy of the first excited state. The two effects have a quite different dependence on temperature.[137,138]

Effects of Isotopic Substitution. From a study of (19), partially deuteriated at the α-carbon, it is apparent that the protons of the deuteriated molecule experience a greater lanthanide-induced shift than do those of the non-deuteriated case.[139] The authors believe this to be due to a stronger shift-reagent–substrate association in the case of the deuteriated species, possibly

[137] S. I. Weissman, *J. Amer. Chem. Soc.*, 1971, **93**, 4928.
[138] J. H. Van Vleck, 'The Theory of Electric and Magnetic Susceptibilities', Oxford University Press, Oxford, 1932.
[139] G. V. Smith, W. A. Boyd, and C. C. Hinckley, *J. Amer. Chem. Soc.*, 1971, **93**, 6319.

Medium Effects

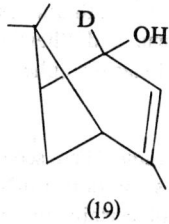

(19)

due to an additional interaction involving the proton or deuteron of the α-carbon and suitably placed oxygen atoms in the ligands of the shift reagent.[140] The possible effect of deuterium substitution on the substrate basicity has also been considered.[140] A further possible explanation is prompted by a report of a pronounced deuterium isotope effect in tris(pentane-2,4-dionato)vanadium(III). The contact shifts of the protons of the diketone ligands are markedly affected by deuterium substitution.[141] If the contact interaction contributes to the shifts of (19), a similar effect might be operative here.

Lanthanide-induced Shifts in Chiral Systems. Whitesides and Lewis[142] first described the use of a chiral Eu^{III} shift reagent to separate the resonances of the two enantiomers of a racemic mixture. The Eu^{III} complex of 3-(trifluoromethylhydroxymethylene)-d-camphor, $Eu(fdc)_3$, has been shown to be a useful reagent in this context.[143] A number of anomalous results have, however, been reported in work of this kind. Goering *et al.*[143] have observed, for example, that the differential shifts obtained for enantiomeric species are not always in the same direction. For 3-methyl-3-phenylpentan-2-one of 27% optical purity, the acyl methyl proton resonance of the major isomer is shifted to lower frequencies relative to that of the minor isomer with $Eu(fdc)_3$. For the 3-methyl proton resonances, however, this order is reversed.[143] The separation of the shifts of the two enantiomers cannot therefore be simply attributed to a different binding constant of the two species with the shift reagent. Other chiral shift reagents have been reported, including Eu^{III} and Pr^{III} complexes of 3-(heptafluoropropylhydroxymethylene)-d-camphor,[143,144] and a series of such reagents has been discussed by Whitesides and Lewis.[145] There is an interesting footnote to the latter article[145] which is credited to McCreary, to the effect that chirality can be induced into a shift reagent. The resonances of the enantiomers of racemic 1-phenylethylamine may be separated by adding a mixture of $Eu(tmhd)_3$ and (R)-N-methyl-1-phenylethylamine. This, apart from suggesting a useful experimental technique, appears to supply further evidence of 2:1 complex formation, since, to effect the observed separation,

---

[140] C. C. Hinckley, W. A. Boyd, and G. V. Smith, *Tetrahedron Letters*, 1972, 879.
[141] R. R. Horn and G. W. Everett, jun., *J. Amer. Chem. Soc.*, 1971, **93**, 7173.
[142] G. M. Whitesides and D. W. Lewis, *J. Amer. Chem. Soc.*, 1970, **92**, 6979.
[143] H. L. Goering, J. N. Eikenberry, and G. S. Koermer, *J. Amer. Chem. Soc.*, 1971, **93**, 5913.
[144] R. R. Fraser, M. A. Petit, and J. K. Saunders, *Chem. Comm.*, 1971, 1450.
[145] G. M. Whitesides and D. W. Lewis, *J. Amer. Chem. Soc.*, 1971, **93**, 5914.

it would seem to be necessary for both amines to become bound to the shift reagent simultaneously. Greene and Shevlin[146] have used Eu(tmhd)$_3$ to separate the CH$_2$ resonances of *meso*-bis(phenylsulphinyl)methane from those of the *dl*-form.

Shift Reagents in Aqueous Solution. There seem to have been few examples of shift reagents used in aqueous solution. Eu(NO$_3$)$_3$,6H$_2$O has been reported to shift the ^{31}P resonances of certain phosphorus-containing anions, and appreciable contact shifts are apparent.[147] Lanthanide perchlorates have also been used to shift the proton resonances of propanoic, 4-aminobutanoic, and 4-hydroxypentanoic acids.[148]

In conclusion, it would seem that the situation with regard to certain applications of the lanthanide shift reagents is by no means as clear-cut as had been hoped. For cases where it can be rigorously demonstrated that the observed shifts are exclusively dipolar in origin, that the magnetic anisotropy is axially symmetric about the lanthanide-to-substrate bond, and that the McConnell and Robertson equation is otherwise valid, various approaches to the refinement of the shift data have been described.[112,149,150] Having thereby obtained the relative lanthanide-induced shifts for a given substrate, computerized methods have been outlined which determine the most likely substrate geometry.[151-153] Bleaney, Williams, and co-workers have described an experimental approach towards assessing the validity of the above assumptions.[129] In many situations such rigour is possibly not necessary, where, for example, the method is being used to distinguish between two grossly different structures. Where such differences are rather more subtle, however, the underlying assumptions should certainly be checked and more sophisticated treatments applied if necessary, particularly where the structure suggested by a shift reagent study is at odds with independent evidence![154] Examples are known where very good correlations have been obtained between a given substrate geometry and the observed lanthanide-induced shifts using equation (36). Such observations do not, however, prove the validity of the approach since, *inter alia*, the position of the lanthanide metal atom is not known, and is possibly acting as a variable parameter, effectively 'smoothing out' the inadequacies of the treatment.

B. *d*-Series Transition-metal Shift Reagents.—The lanthanide shift reagents can in principle provide information relating to the substrate geometry. The

---

[146] J. L. Greene, jun. and P. B. Shevlin, *Chem. Comm.*, 1971, 1092.
[147] J. K. M. Sanders and D. H. Williams, *Tetrahedron Letters*, 1971, 2813.
[148] F. A. Hart, G. P. Moss, and M. L. Staniforth, *Tetrahedron Letters*, 1972, 3389.
[149] I. Armitage, G. Dunsmore, L. D. Hall and A. G. Marshall, *Chem. Comm.*, 1971, 1281.
[150] J. W. ApSimon and H. Bierbeck, *Chem. Comm.*, 1972, 172.
[151] S. Farid, A. Ateya, and M. Maggio, *Chem. Comm.*, 1971, 1285.
[152] M. R. Willcott, tert., R. E. Lenkinski, and R. E. Davis, *J. Amer. Chem. Soc.*, 1972, **94**, 1742.
[153] R. E. Davis, and M. R. Willcott, tert., *J. Amer. Chem. Soc.*, 1972, **94**, 1744.
[154] M. I. Foreman, *J. Organometallic Chem.*, 1972, **39**, 161.

d-series transition metals effect shifts which are in large part due to the Fermi-contact interaction, and are therefore potentially a source of information relating to the substrate molecular orbitals, since the contact shift of nucleus i is a function of the hyperfine interaction constant a_i, where:[108]

$$a_i = \left(\frac{8\pi}{3}\right) g_i \beta_i |\psi(0)_i|^2 \tag{38}$$

$\psi(0)_i$ is the wavefunction which describes the unpaired electron distribution at the ith nucleus. For finite values of a_i, the unpaired electron must necessarily have some degree of s-character.

As before, however, it is essential that the contact and the dipolar shift contributions be clearly separated. Various approximate methods have been employed to this end. Ni^{II} complex shifts, for example, are generally attributed exclusively to the contact interaction. An indirect method for the separation of the two terms has been described by Horrocks,[155] and a similar approach applied to studies of ion-pair interactions.[156] More direct methods have been described in a review by Horrocks and Hall;[117] these allow the principal magnetic susceptibilities of the molecule to be determined. The dipolar term may then be calculated from equation (37). Extrapolating such results to systems in solution must be done with care, taking account of the various equilibria which may be involved. For the pseudotetrahedral dichloro-bis-(triphenylphosphine)-cobalt(II) and -nickel(II) crystalline complexes, the above approach confirms the greater anisotropy of the former species.[157] Magnetic anisotropy, though slight, is clearly apparent for the Ni^{II} complex also, and will therefore result in a finite dipolar term.[157] Large low-frequency dipolar shifts are predicted for the Co^{II} complex and small high-frequency shifts for the Ni^{II} case.[157] In the same manner, large dipolar shifts have been calculated[158] for the $Co(acac)_2(pyridine)_2$ complex (acac = acetylacetonato); this is in contrast to conclusions based on ^{13}C studies alone.[159] The same general conclusions arise from a study of bis-(N-alkylsalicylaldiminato)-cobalt(II) and -nickel(II) species.[160] Again, a significant dipolar contribution is apparent for the Ni^{II} complex.[160] McGarvey[161] has extended an earlier theoretical treatment relating to paramagnetic shifts in complexes of this type.[162] Although, for pure tetrahedral symmetry, no magnetic anisotropy is possible and there is therefore no dipolar shift, it is here shown[161] that quite small distortions from tetrahedral symmetry can lead to appreciable dipolar shifts. In the context of the Ni^{II} shift reagents, bearing in mind the possible distortions from either tetrahedral or octahedral symmetry in the shift-reagent–substrate complex, it would seem to be unwise to neglect entirely the dipolar term.

[155] W. DeW. Horrocks, jun., *Inorg. Chem.*, 1970, **87**, 690.
[156] I. M. Walker, L. Rosenthal, and M. S. Quereshi, *Inorg. Chem.*, 1971, **10**, 2463.
[157] W. DeW. Horrocks, jun. and E. S. Greenberg, *Inorg. Chem.*, 1971, **10**, 2190.
[158] W. DeW. Horrocks, jun. and D. DeW. Hall, *Inorg. Chem.*, 1971, **10**, 2368.
[159] D. Doddrell and J. D. Roberts, *J. Amer. Chem. Soc.*, 1970, **92**, 6839.
[160] C. Benelli, I. Bertini, and D. Gatteschi, *J.C.S. Dalton*, 1972, 661.
[161] B. R. McGarvey, *J. Amer. Chem. Soc.*, 1972, **95**, 1103.
[162] R. J. Kurland and B. R. McGarvey, *J. Magn. Resonance*, 1970, **2**, 286.

When a separation of the paramagnetic shifts into dipolar and contact terms can be satisfactorily carried out, the contact shifts throughout the substrate are frequently used to infer the symmetry, whether σ or π, of the molecular orbital responsible for the distribution of spin density. Horrocks[163] has challenged the basis on which such inferences are drawn, and has demonstrated that characteristics thought previously to be peculiar to a π-delocalization mechanism can in all respects be discussed in terms of the σ-delocalization mechanism alone. It is clear therefore that the assumptions on which many studies involving d-series transition-metal shift reagents are based are as insecure as those of the lanthanide series.

Many reports continue to appear, however, which illustrate the potential utility of such shift reagents in solution. One disadvantage associated with such work is due to broadening of the resonance signals of the substrate species.[108] Johnson and Everett[164] have pointed out that paramagnetic dipolar line-broadening is proportional to the square of the gyromagnetic ratio of the appropriate nucleus, and should therefore be some forty times less for deuterons than for protons, an observation borne out by deuterium resonance studies of suitably deuteriated substrates.[164] Other reports relate to studies of norbornene, nobornadiene, and dimethylindane with Ni^{II} reagents,[165] and Yonezawa and co-workers have extended their work concerning the interaction of Ni^{II} complexes with nitrogen-containing compounds.[166,167] In all of these cases, possible dipolar shifts for the Ni^{II} species were neglected.

Paramagnetic Co^{2+} ions have been used to probe the nature of surfactant systems in aqueous solution, with success. The positions adopted by p-xylene molecules in the interior of suitable micelles have been determined[168] by observing the effect on the p-xylene resonance resulting from the addition of paramagnetic ions to the system, the latter tending to be adsorbed on to the micelle surface. A further application which is frequently encountered involves the location of the potential binding sites on large molecules by observing the effect added paramagnetic ions produce, either in terms of line-broadening or shifting of the resonances. Adenosine 5'-monophosphate has been studied recently in this way.[169]

A potentially very useful application of paramagnetic species in solution has recently been developed by La Mar. For reasons not relevant to this discussion, the use of pulsed n.m.r., followed by Fourier transformation of the data to produce the normal frequency spectrum,[170] is proving extremely

[163] W. DeW. Horrocks, jun. and D. L. Johnston, *Inorg. Chem.*, 1971, **10**, 1835.
[164] A. Johnson and G. W. Everett, jun., *J. Amer. Chem. Soc.*, 1972, **94**, 1419.
[165] K. Tori, Y. Yoshimura, and R. Muneyuki, *J. Amer. Chem. Soc.*, 1971, **93**, 6324.
[166] I. Morishima, K. Okada, T. Yonezawa, and K. Goto, *J. Amer. Chem. Soc.*, 1971, **93**, 3922.
[167] I. Morishima, K. Okada, and T. Yonezawa, *J. Amer. Chem. Soc.*, 1972, **94**, 1425.
[168] K. K. Fox, I. D. Robb, and R. Smith, *J.C.S. Faraday I*, 1972, **68**, 445.
[169] A. W. Missen, D. F. S. Natusch, and L. J. Porter, *Austral. J. Chem.*, 1972, **25**, 129.
[170] T. C. Farrar and E. D. Becker, 'Pulse and Fourier Transform N.M.R.,' Academic Press, New York, 1971.

valuable, particularly in natural-abundance ^{13}C spectroscopy. It has one serious disadvantage, however, in that, as a result of the nuclear Overhauser effect,[171] the normal relationship between the relative peak intensities and the number of nuclei present no longer holds under conditions of proton decoupling. Such conditions are generally desirable in order to reduce the complexity of the spectrum and effect the maximum signal enhancement. For a ^{13}C-{^{1}H} experiment, the nuclear Overhauser enhancement, η, is given by:

$$\eta = \tfrac{1}{2} \frac{T_1^C}{T_{1d}^C} \cdot \frac{\gamma_H}{\gamma_C} S_H \qquad (39)$$

where S_H is the proton saturation factor (~ 1), T_1^C is the ^{13}C spin–lattice relaxation time, and T_{1d}^C is the (C, H) dipolar contribution to T_1^C. It has been shown[172] that T_1^C, and thereby η, can be reduced by introducing an additional mechanism for dipolar relaxation of the ^{13}C nucleus. La Mar[173] has effectively reduced η to zero in this manner by the addition of paramagnetic species such as Co^{2+}, Fe^{3+}, Cr^{3+}, and Mn^{2+}, and also the di-t-butyl nitroxide radical. Mn^{2+} appears to be the most efficient in this respect, effectively reducing η to zero at concentrations below that for which dipolar line-broadening of the resonances becomes objectionable. Under conditions where $\eta = 0$, the relative numbers of the nuclei involved can once again be related directly to the relative intensities of the peaks in the spectrum. A similar application has been reported in ^{13}C studies of metal-bonded carbonyl groups. Normally the spin-lattice relaxation times of such groups are so long that the resonances are not detectable. Addition of Cr(acac)$_3$ dramatically reduces T_1 without apparently affecting the nuclear resonance positions, and therefore allows the spectrum to be obtained by conventional techniques.[174,175]

C. Diamagnetic Shift Reagents.—In addition to shifts effected by paramagnetic complexes, substrate resonances may also be perturbed by complexation to diamagnetic complexes, where such compounds possess ligands having considerable diamagnetic anisotropy. Ru^{II}–porphin complexes form a loose association with pyridine molecules, so that the resonance positions of the latter become subjected to the very large diamagnetic anisotropy of the porphin rings.[176] Chlorophyll acts in a similar manner owing to the presence of the porphyrin ligands.[177] Of a slightly different type are the Fe^{II}–phthalocyanine complexes. Normally paramagnetic, these compounds can bind a further two molecules to form a stable diamagnetic species.[178] The substrate

---

[171] J. H. Noggle and R. E. Schirmer, 'The Nuclear Overhauser Effect', Academic Press, New York and London, 1971.
[172] K. F. Kuhlman, D. M. Grant, and R. K. Harris, *J. Chem. Phys.*, 1970, **52**, 3439.
[173] G. N. LaMar, *Chem. Phys. Letters*, 1971, **10**, 230.
[174] O. A. Gansow, A. R. Burke, and G. N. LaMar, *J.C.S. Chem. Comm.*, 1972, 456.
[175] O. A. Gansow, A. R. Burke, and W. D. Vernon, *J. Amer. Chem. Soc.*, 1972, **94**, 2550.
[176] S. S. Eaton, G. R. Eaton, and R. H. Holm, *J. Organometallic Chem.*, 1971, **32**, C52.
[177] J. J. Katz, T. R. Janson, A. G. Kostka, R. A. Uphaus, and G. L. Closs, *J. Amer. Chem. Soc.*, 1972, **94**, 2883.

resonances are again subjected to the diamagnetic anisotropy of the neighboring ligands, but because the complex is stable, the effect is not dependent on the concentration of the complex.[178] Structural information relating to the substrate can again be inferred, by utilizing iso-shielding contours which have been determined for the analogous silicon complexes.[179]

[178] J. E. Maskasky, J. R. Mooney, and M. E. Kenney, *J. Amer. Chem. Soc.*, 1972, **94**, 2132.
[179] T. R. Janson, A. R. Kane, J. F. Sullivan, K. Knox, and M. E. Kenney, *J. Amer. Chem. Soc.*, 1969, **91**, 5210.

Author Index

Aasen, A. J., 104
Abe, R., 326
Abragam, A., 112
Abraham, R. J., 43, 86, 92, 211
Abushanab, E., 109
Acrivos, J. V., 104
Adams, J. Q., 39
Adcock, W., 43, 109
Ader, R., 130, 218
Adriaenssens, G. J., 331, 336
Agova, M., 219
Agranat, I., 30
Ahmed, I. Y., 371
Ailion, D. C., 113, 160, 182
Ainbinder, N. E., 151
Airey, W., 43, 71
Akashi, K., 228
Aksnes, D. W., 18, 75, 104, 122, 154, 214
Albert, S., 153, 155, 157, 333
Albrand, J. P., 64, 87, 213
Albriktsen, P., 93, 202
Alderfer, J. L., 307
Alderman, D. W., 338, 354
Aleksandsov, A. M., 43
Alekseeva, T. L., 362, 363
Alexander, C. W., 241
Alexander, M. N., 339
Alexandrou, N. E., 375
Alford, K. J., 247
Alger, T. D., 125, 139, 174, 233
Allen, C. W., 81
Allen, F. H., 45
Allen, P. S., 148, 158, 323
Allerhand, A., 40, 111, 128, 137, 140, 141, 172, 175, 178, 282, 297, 340
Allinger, N. L., 20
Almqvist, S. O., 104
Alms, H., 300
Aminova, R. M., 22
Amirkhanov, B. F., 151
Ammon, H. L., 85
Anand, N., 301
Anastassiou, A. G., 30
Ancian, B., 18
Anders, L. R., 33

Anderson, C. H., 116
Anderson, D. R., 171, 382
Anderson, J. E., 28, 126, 153, 226, 297, 327, 378
Anderson, J. M., 194, 280
Anderson, K., 219
Anderson, W. A., 272
Anderson, W. G., 235
Andersson, L. O., 354
Ando, I., 20
Andrew, E. R., 153, 291, 323, 324, 325, 344, 345
Andrews, B. D., 32
Andronov, V. F., 255
Anet, F. A. L., 28, 76, 178, 378
Ang, H. G., 43
Anteunis, M., 76, 87, 96, 210, 271
Anteunis-de Ketelaere, F., 76
Apaydin, F., 343
Appleton, Q., 100
Appleton, T. G., 95
ApSimon, J. W., 20, 21, 386
Arata, Y., 128, 256, 330
Archer, R. A., 41
Argentar, H., 291
Aritomi, M., 64
Armitage, I., 380, 386
Armour, E. A. G. 25
Armstrong, R. L., 115, 151 181
Arnold, K., 179, 220
Asabe, Y., 18
Asanuma, M., 352
Ashworth, J., 297
Asman Degani, H., 309
Assink, R. A., 123
Ateya, A., 386
Atkins, P. W., 138, 259
Augusteijn, M. F., 183
Auramenko, G. I., 76, 107
Avagadro, A., 164
Aviram, I., 308
Avkhutsky, L. M., 43
Avramenko, G. I., 201, 239
Axenrod, T., 33, 46, 64
Aylett, B. J., 43, 71
Azman, A., 67, 337, 368

Babadjamian, A., 40, 64
Bacon, M., 103
Bacskai, R., 290
Bailey, A. V., 327
Bailey, K., 74
Bailey, W. F., 40
Baine, P., 371
Bairamova, A. G., 363
Baird, M. C., 45
Bakker, J., 15
Balashova, T. A., 234
Baldassare, J. J., 308
Baldeschwieler, J. D., 33, 280
Balkan, F., 100, 105, 206, 271, 272
Ballard, R. L., 152, 350
Bamford, C. H., 297
Banwell, C. N., 194
Baranetskaya, N. K., 84
Barau, A. A., 363
Barber, B. H., 104
Barbier, C., 53
Barbieri, G., 18, 90, 213
Barboiu, V., 108, 272
Bar-Eli, K. H., 226
Barfield, M., 58, 60, 62, 104, 356, 357
Bargon, J., 292
Bariza, S., 283
Barnes, R. G., 160, 350, 351
Barnier, S., 106
Baron, D., 88
Barr, M. R., 334
Barr, R. M., 306
Barratt, M. D., 320
Barrere, H., 341
Barry, C. D., 260, 306
Barry, R., 219
Bartko, O., 324
Bartle, K. D., 30, 76, 102, 105
Bartuska, V. J., 40, 56
Barzilay, J., 31
Basch, H., 10
Baskevitch, N., 88
Basselier, J. J., 271
Batterham, M. J., 194
Bauer, L., 86
Baugher, J. F., 339

Baughman, R. H., 161, 326
Bauman, H., 29
Bayer, E., 43
Bayliss, N. S., 358
Beachley, O. T., 47
Beaute, C., 231
Becconsall, J., 361
Beck, B. H., 165
Beck, W., 46
Becker, E. D., 46, 64, 140, 220, 237, 280, 388
Becker, W., 46
Beckmann, P. A., 116
Bedford, G. R., 43, 92, 211
Beeby, P. J., 29
Beech, G., 372
Begtrup, M., 69, 274
Belfort, G., 164
Bell, C. L., 86, 356
Bell, R. A., 35, 39
Benassi, R., 246
Bender, H. J., 119, 120
Benelli, C., 387
Bennett, R. D., 383
Bennett, W. E., 87
Bent, H. A., 63
Berendsen, H. J. C., 173, 312, 321, 349
Berg, A., 103
Bergen, A. S. V., 314
Bergesen, K., 93
Bernander, L., 100
Bernheim, R. A., 54, 339
Bernstein, H. J., 1, 31, 195
Bersohn, R., 53
Berthier, G., 53
Berticat, P., 290
Bertini, I., 387
Bertrand, R. D., 93, 140
Berwick, M. A., 107
Berzete, A. Y., 198
Betsuyaku, H., 342
Beuer, S. W., 78
Beveridge, D. L., 52
Bevington, J. C., 289
Bhacca, N. S., 78, 382
Bichlmeir, B., 88
Biedermann, H. G., 327
Biehl, E. R., 17
Bienkiewicz, Von K. J., 349
Bierbeck, H., 386
Bigum, C. J., 194
Bilofsky, H. S., 230, 236
Binder, H., 72
Binsch, G., 68, 230, 238
Birch, A., 105
Birdsall, N. J. M., 33, 274, 275, 311
Bishop, E. O., 247
Bjorkstam, J. L., 326, 331, 336
Blackburn, E. V., 106, 207
Blackburn, G. M., 45
Blackmer, G. L., 76
Blackwell, L. F., 366
Bladon, P., 162, 331
Blain, M., 106
Blanshard, J. M. V., 349
Bleaney, B., 382, 383

Blears, D. J., 102
Bleich, H. E., 49, 185, 345
Blicharski, B., 144
Blicharski, J. S., 144, 148
Blinc, R., 147, 149, 164, 185, 331, 338, 339, 341
Blizzard, A. C., 54
Block, R. E., 42, 289
Bloembergen, N., 78, 113, 324
Blomster, K., 368
Bloom, M., 113, 114, 116, 159
Bloor, J. E., 21
Blum, H., 330
Boccalon, G., 310
Bock, E., 125, 128
Bockman, R. S., 300
Boddenburg, B., 163, 346
Boden, N., 112, 119, 137, 150
Bogaard, M. P., 21
Bogachev, Yu. S., 363
Bogdanov, V. S., 78, 250
Boicelli, C. A., 105
Boikess, R. S., 27
Bojesen, I. N., 59
Bolesov, I. G., 40
Bon, M., 76, 213
Bonardet, J. L., 346
Bonera, G., 164
Booth, H., 87, 271
Borcic, S., 226
Borisenko, A. A., 34, 40, 65, 95, 274
Borodin, P. M., 218
Borremans, F., 76
Borsa, F., 164, 340
Borstnik, B., 67
Bortone, C., 128
Bothner-By, A. A., 59, 305
Boubel, J. C., 371
Bourdais, J., 377
Bousquet, F. B., 31
Bouznik, V. M., 43, 47
Bovee, W. M. M. J., 183
Bovey, F. A., 27, 285, 288, 296, 301, 304
Boyd, W. A., 384, 385
Bradbury, E. M., 302, 305, 306, 310
Bradbury, J. H., 304
Bradley, C. H., 76
Bradley, R. B., 46, 64
Brame, E. G., jun., 291
Bramley, R., 257
Brandenberger, J. R., 354
Brauman, J. I., 27
Bray, P. J., 339, 341, 353
Breitmaier, E., 41, 175, 286, 311
Bremser, W., 226
Brevard, C., 250, 251
Brey, W. S., jun., 358
Bridges, F. D., 113
Briggs, J. M., 46
Bright, A. A. S., 377
Briguet, A., 173
Brillouin, L., 376
Brini, M., 88

Brink, M., 74, 76, 98
Brodskii, A. I., 218
Brookeman, J. R., 154, 325, 327
Brophy, G. C., 86
Brot, C., 154
Brouwer, H., 40, 76, 100
Brown, D. G., 371
Brown, R. A., 40, 70
Brown, R. D., 10, 131
Brown, R. J. C., 119
Brüssau, R. G., 127, 218, 320
Brun, F., 319
Brune, H. A., 64
Bryant, R. G., 173, 245
Bubb, W. A., 86
Buc, H., 314
Bucci, P., 14, 261, 270
Buchanan, G. W., 87
Buckheim, A., 295
Buckingham, A. D., 3, 21, 31, 358
Buckley, P. D., 366
Buckmaster, H. A., 220, 344
Bug, H., 351
Buhler, D. R., 321, 364
Buishvili, L. L., 152
Bull, T. E., 123
Bullpitt, M. L., 78
Bundgaard, T., 40, 276
Burden, F. R., 10
Burdon, J., 43, 80
Burg, A. B., 66
Burgade, R., 80
Burke, A. R., 110, 234, 283, 389
Burke, T. E., 125
Burkert, P. K., 327
Burkert, P. K., 371
Burnell, E. E., 116
Burnett, L. J., 121, 337
Burns, G., 56
Burton, G. W., 88
Burton, R., 272
Burton-Lewis, W., 351
Bushweller, C. H., 165, 230, 235, 236
Buslaev, Y. A., 47
Butler, R. S., 356
Buu Ban, 295
Buzova, Ts, 232
By, A. W., 74
Bystrov, V. F., 234

Cade, R. F., 5
Cahill, R., 76
Calderon, J. L., 240
Calinaud, P., 97
Callaghan, D., 20, 24
Calvin, M., 103
Campbell, B. S., 76
Campbell, C. H., 78, 239
Canepa, P. C., 324
Canet, D., 197, 221
Canters, G. W., 342
Cantor, C. R., 307, 316
Carazza, B., 134, 342

Author Index

Cardin, A. D., 69
Carey, P. R., 247
Cargioli, J. D., 41, 368
Carlson, E. H., 226
Carman, C. J., 298
Carnevale, A., 340
Carolan, J. L., 49, 159, 185, 329, 334, 344
Carpenter, B. G., 302
Carr, M. D., 88
Carr, S. L., 351
Carriker, N. E., 365
Carroll, M., 41
Carruthers, J. R., 340
Carruthers, W., 218
Carter, R. E., 216, 227
Cartledge, F. K., 64
Carver, G. P., 353
Carver, J. P., 137
Cary, P. D., 306
Casabella, P. A., 339
Castellano, S., 106
Casy, A. F., 40
Caughey, W. S., 41
Cavalieri, E., 103
Cavalli, L., 80, 209
Cavanaugh, J. R., 196
Cavelius, E., 164
Cavell, R. G., 121, 174, 252
Cawley, S., 19
Ceausescu, V., 342
Cernivez, F., 43
Chachaty, C., 295
Chaigneau, M., 33
Chaiken, I., 304
Chakrabarti, B., 59, 104
Chan, C. L., 35
Chan, S. I., 7, 125, 140, 180, 306, 309, 320
Chan, S. O., 135, 136, 171, 221
Chandra, P., 59
Chandrasekaran, R., 88, 300
Chandrasekhar, S., 132
Chapelet-Barbier, C., 262, 271
Chapman, D., 319
Chapman, T. M., 305
Charache, S., 307, 308
Charrier, C., 80, 234
Chassonneau, M. A., 325
Chatterjee, N., 125, 128
Chejter, A. S., 308
Cheng, C. L., 21
Chew, K. F., 46
Chezeau, J. M., 154, 328
Chiba, T., 336
Chien, M., 115
Chihara, H., 151
Child, T. F., 317
Chirkov, N. M., 287
Chisholm, M. H., 40, 70
Chiu, Y. C., 348
Chivers, T., 43
Chokki, Y., 290
Cholerton, T. J., 106, 207
Chorbanov, B., 232
Christ, B., jun., 296
Christl, M., 40

Chujo, R., 298
Chuvylkin, N. D., 61
Civan, M. M., 312, 313
Clar, E., 30
Clark, H. C., 40, 70, 80, 95
Clark-Monks, C., 332
Clifford, J., 370
Clifford, P. R., 41
Closs, G. L., 389
Clough, S., 158, 159, 343
Clouse, A. O., 111, 178
Cobb, T. B., 343
Cochran, D. W., 41
Code, R. F., 62
Cogne, A., 64, 87, 213
Cohen, J., 161
Cohen, J. S., 45, 304
Cohen, M., 131
Cohn, M., 318
Coker, B. M., 251
Collin, P. J., 64
Collins, S. W., 139, 174
Colman, R. F., 315
Combrisson, S., 271
Comti, F., 301
Condon, E. U., 191
Connick, R. E., 217, 244
Cook, D. B., 7
Cooke, D. F., 151
Cooke, R., 318
Cookson, R. C., 74
Cooley, J. W., 176
Cooney, J. D., 21
Cooper, M. A., 85
Cope, F. W., 318
Core, E. S., 256
Corfield, M. G., 21
Cornwell, C. D., 5
Corriccini, L. R., 134
Cottam, G. L., 315
Cotton, F. A., 240
Cotts, R. M., 338
Courtney, J., 115
Coviello, D. A., 76
Cowking, A., 158
Cowley, A. H., 66
Cox, R. H., 75, 76, 198, 355, 356, 357
Coxon, B., 68, 110, 269
Coyle, T. D., 269
Crabb, T. A., 74, 76
Craig, W. G., 20
Crain, W. O., jun. 234, 298
Cramer, E. R., 383
Crane-Robinson, C., 302, 306, 310
Crawford, J. E., 306
Creel, R. B., 350, 351
Cremer, D., 57
Cremer, H., 29
Cremer, S. E., 69, 81
Creswell, C. J., 231
Cribley, K., 38
Crippa, M. L., 164, 340
Crombie, D. A., 105
Cromwell, N. H., 236
Cross, A. D., 307
Crotens, A. M., 254
Crothers, D. M., 307
Crump, D. R., 242

Csizmadia, I. G., 10
Csopak, H., 255
Cunliffe, A. V., 202
Cupas, C. A., 87
Cushley, R. J., 171, 382
Cuvelier, C., 97
Cvikl, B., 326
Cyr, N., 39
Cyr, T. J., 349
Czubryt, J. J., 125, 128

Dabrowski, J., 46, 64
Dadok, J., 305
Dahlquist, K. I., 227
Dailey, B. P., 49, 196
Dale, S. W., 7, 141
D'Alessio, G. J., 339
Dalling, D. K., 60, 230
Damasco, M. C., 66
Danchin, A., 314
Danielsson, I., 217
Danyluk, S. S., 102, 356
Darmon, I., 154
Das, T. P., 6, 7, 53, 345
Davidson, D. W., 324, 325
Davies, A. M., 7
Davies, J., 370
Davies, P. B., 6, 144
Davies, R., 74
Davis, D. D., 258, 296, 371
Davis, D. G., 308, 309
Davis, J., 66
Davis, P. P., 161
Davis, R. E., 386
Davison, A., 239
Dawson, D. A., 86
Day, M. C., 78
Dayan, E., 33
Daycook, J. T., 182
Deavonport, D. L., 35, 105
de Bie, M. J. A., 39
de Boer, E., 254, 293
de Boer, Th. J., 219, 224, 225, 366
De Bruin, K. E., 94
De Clercq, M., 78
de Clyne, R., 271
Degani, H. A., 243
de Graaf, A. M., 4
de Haan, J. W., 39, 41
de Jonge, W. J. M., 352
De-Kok, A. J., 86
de la Mare, P. B. D., 88
Delmau, J., 262
Delpuech, J. J., 83, 371
De Marco, A., 42, 87, 93, 209
Demarco, P. V., 20
Dembech, P., 80, 106
Demco, D., 342
DeMember, J. R., 290
De-Mey, C. A., 86
den Hollander, C., 354
Denis, A., 54
de-Pessemier, F., 76
De Poorter, B., 78
Deranleau, D. A., 376
Derbyshire, W., 46, 318, 319, 349

Author Index

Derendyaev, B. G., 241
Derighetti, B., 340
De Santis, P., 301
Deutch, J. M., 114, 118
Deverell, C., 143
Devillers, J., 213
Dewar, M. J. S., 16
DeZwaan, J., 123
Dianoux, A. J., 354
Diaz, E., 202
Dickson, R. B., 100
Diefenbach, H., 318
Diegat, F. 220
Diehl, P., 102, 203
Dielman, J., 271
Diercksen, G., 7
Dietrich, W., 121
Diez, E., 103
Dillon, K. B., 45, 334
Dillon, P. B., 251
Dimic, V., 147
Dimicoli, J. L., 319
Dimitrov, V. S., 219
Dinesh, 120, 121, 142, 218
Dirinck, P., 76, 210
Ditchfield, R., 42, 54, 61, 191
Dittmer, D. C., 40, 64
Doane, J. W., 144, 145
Dobo, J., 299
Dobosh, P. A., 52
Dobson, C. M., 382
Dobson, G. R., 40, 70
Doddrell, D., 40, 41, 140 172, 175, 234, 297, 387
Doddrell, D. M., 282
Dodgen, H. W., 244
Dodson, B., 377
Doehler, H., 344
Dogadina, A. V., 80
Dombroski, J. R., 287
Domngang, S., 130
Dong, R. Y., 114, 145, 147, 156
Donlan, V. L., 174
Dorie, J. P., 106
Dorman, D. E., 40, 41
Doskocilova, D., 220, 287, 291, 344
Doucet, J. P., 18
Douglass, D. C., 162, 296, 299, 331
Douris, J., 103
Dradi, E., 76, 201
Dräger, M., 80
Drago, R. S., 337, 370, 371
Drake, J. E., 66
Drake, P., 133
Drakenberg, T., 216, 217, 234
Drakorádová, E., 287
Drenth, W., 21, 39, 64
Drinkard, W. C., 10
Drozdov, V. A., 255
Druck, S. J., 140
Dubois. J. E., 18
Duerst, R. W., 43, 77
Düwel, H., 29
Dufourcq, J., 325, 328, 329
Dunell, B. A., 154, 156, 327, 329, 334
Dunette, P. L., 198
Dunlop, R., 228
Dunmur, D. A., 21
Dunmur, R. E., 43, 72, 213
Dunsmore, D., 380, 386
Dupas, A., 352
Duplan, J. C., 262, 271
Durette, P. L., 76
Dussanchov, A., 220
Duval, E., 59
Dwek, R. A., 43, 314
Dybowski, C. R., 115, 145, 146, 147

Eades, R. G., 153, 158
Eary, J. G., 77
Eastman, M. P., 351
Eaton, G. R., 389
Eaton, S. S., 389
Ebdon, A. P., 90
Ebdon, J. R., 289
Ebsworth, E. A. V., 6
Eckert-Maksić, M., 60
Edge, S. N., 21
Edmonds, D. T., 341
Edsall, J. T., 376
Edwards, T. G., 25
Edzes, H. T., 312, 349
Egan, R. S., 76
Egerton, T. A., 346
Ehrenberg, L., 185, 341
Ehrlich, R. H., 372
Eichoff, U., 295, 332
Eikenberry, J. N., 385
Eisenberg, A., 332
Eisenhut, M., 72
Eisenkremer, M., 159
Eisenstadt, M., 159
Ekong, D. E. U., 382
Elguero, J., 106
Elias, D. J., 106, 272
Eliel, E. L., 40
Elleman, D. D., 167
Ellett, J. D., 170, 185, 345
Elliott, R. L., 38
Ellis, B., 332
Ellis, G., 40
Ellis, I. A., 43, 71
Ellis, P. D., 38, 47, 68, 69
El Saffar, Z. M., 326, 328, 333, 338, 352
Emanuel, R. V., 60
Emerson, M. T., 361
Emid, S., 183
Emsley, J. W., 43, 108, 109, 172, 272, 355
Endo, K., 368, 369
Engardt, P. D., 160
Engberts, J. B. F. N., 234
Engel, R., 110, 259
Engelman, D. M., 319
Engelsberg, M., 352
Engler, E. M., 374
Englert, G., 49, 102
Engstrom, N., 283
Enzell, C. R., 104
Epishina, L. V., 39, 64, 173, 253

Eraker, J. H., 337
Erasho, V., 218
Erbeia, A., 173, 220
Ermler, W. C., 9
Ernst, L., 39
Ernst, R. R., 180, 273
Ernstbrunner, E. E., 87
Esumi, N., 106
Ettinger, R., 249
Evans, D. F., 33, 380
Evans, M., 40, 84, 108
Evans, W. A. B., 342
Everdell, M. H., 377
Everett, G. W., jun., 260, 385, 388
Ewing, D. F., 105, 206, 366

Fahnrich, J., 353
Fainzil'berg, A., 218
Fairwell, T., 61
Falaleev, O. V., 327
Falaleeva, L. G., 47
Falcone, D. R., 296, 299
Falconer, W. E., 333
Faller, J. W., 29, 246
Fang, K. N., 87
Farach, H. A., 112, 323
Farid, S., 386
Farnell, L. F., 40, 46
Farrar, T. C., 121, 140, 388
Faure, R., 40, 64
Fedin, E. I., 354
Fedin, I., 324
Fedotov, V. D., 137
Fee, J. A., 131
Feeney, J., 101, 180, 274, 275, 283, 311, 314, 355
Feherova, J., 324
Feicock, F. D., 8
Feigenson, G. W., 180
Feldmann, R., 29
Fiat, D., 243, 309
Fiel, M., 304
Fieux, J., 237
Fild, M., 72, 85, 212
Filleux-Blanchard, M. L., 237
Finch, E. D., 134
Finer, E. G., 79, 82
Finn, F. M., 305
Fiorito, R. B., 132, 167
Firth, M., 344
Fischer, A., 18
Fischer, E. O., 73
Fisher, W. F., 23
Fleischer, E. B., 17
Fleissner, M., 351
Fleming, I., 379
Florkowski, Z., 349
Fluck, E., 45, 82, 261
Flygare, W. H., 6, 143
Fogelman, J., 247
Folkes, M. J., 332
Folland, R., 150, 162
Folsom, T. K., 100
Forbes, W. F., 145, 147, 156
Forchioni, A., 295
Foreman, M. I., 377, 386

Author Index

Fornes, R. E., 342
Forsen, S., 131, 217, 218, 234, 254, 330
Forslind, E., 348
Foster, K. R., 114
Foster, R., 362, 374, 377
Foster, R. G., 202
Fournie-Zaluski, M. C., 271
Fox, K. K., 388
Fraenkel, G., 135, 183, 222
Fraissard, J., 346
France, P. W., 343
Franchuk, I. F., 218
Francis, J., 80
Frankel, L. S., 242, 244, 317, 347
Fransen, J., 76
Fraser, R. R., 74, 75, 385
Fratiello, A., 246, 258, 371
Freed, J. H., 122
Freedman, H. H., 228
Freedman, L. D., 45
Freeman, R., 109, 136, 140, 175, 177, 179, 180, 186, 221, 282, 283
French, D. M., 295, 332
Freude, D., 163, 347
Fried, F., 327
Friedel, R. A., 39
Fripiat, J. J., 163, 346
Fritz, H. P., 371
Frost, D. J., 31
Fruchier, A., 106
Fuganti, G., 41
Fujii, K., 286
Fujimoto, Y., 21
Fujita, S., 30
Fujiwara, F. Y., 67
Fujiwara, S., 128, 256, 258, 330
Fukui, H., 106
Fukushima, E., 71, 341, 351
Fuller, A. M., 146
Fuller, M. W., 105, 271
Fung, B. M., 46, 313, 337
Furukawa, J., 291
Furukawa, K., 291
Fyfe, C. A., 154, 327, 328, 362

Gaasch, J. F., 80
Gaber, B. P., 131
Gabuda, S. P., 327, 330
Gagarinsky, Yu. V., 333
Gagnaire, D., 53, 64, 76, 87, 93, 213
Gall, R. E., 105, 205, 227
Gallina, C., 87, 199
Gamayunov, N. I., 317
Gansow, O. A., 40, 70, 110, 111, 180, 234, 283, 389
Garg, S. K., 324, 325
Garnier, R., 40, 64
Garreau, M., 80
Gates, P. N., 45
Gatow, G., 80

Gatteschi, D., 387
Gatti, G., 41, 42, 76, 86, 87, 93, 201, 209, 298
Gaylord, N. G., 289
Gazzard, V. J., 111
Geens, A., 76
Geissman, T. A., 31
Gelas, J., 97
Geller, B. A., 218
Genser, E. E., 347
Gerace, M. J., 280
Gerhold, G. A., 361
Gerlach, D. H., 45
Gero, S. D., 41
Gerritsen, J., 75, 93
Geschke, D., 347
Gestblom, B., 194
Gevers, E. C. A., 352
Gey, E., 42, 59
Ghose, S., 340
Ghosh, K. K., 310
Ghosh, S. K., 352
Giacometti, G., 310
Gibbons, W. A., 300
Gibby, M. G., 49, 168, 170, 184, 185, 334
Gibson, M. P. G., 354
Gielen, M., 78
Giger, W., 221
Gil, V. M. S., 31, 106
Gilbert, J. C., 241
Gilbert, R. D., 310
Gill, J. C., 342
Gillen, K. T., 7, 124, 144, 166
Gilles, J. M., 29
Gillespie, R. J., 70, 71
Gillies, D. G., 46, 168, 175, 282
Gilson, D. F. R., 327, 329, 333
Gioncotti, V., 302
Giuliani, A. M., 88
Glasel, J. A., 260, 274, 306
Glattli, H., 159
Glazer, E. S., 229
Glickson, J. D., 304, 305
Glonek, T., 311
Glushko, V., 175, 282, 297
Goddard, N., 35, 65
Godo, M., 218
Goering, H. L., 385
Goffin, N., 78
Goffloo, K., 322
Goggin, P. L., 77
Goldammer, E. v., 253
Goldburg, W. I., 342
Golden, M. R., 305
Golden, R., 16, 23
Goldman, H., 302
Goldman, M., 113, 324
Goldstein, J. H., 39, 40, 63, 77, 89, 298, 355, 356
Goldwhite, H., 94
Golz, W. L. F., 292
Gonbeau, B., 40
Goodfellow, R. J., 77
Goodisman, J., 6, 143
Goodman, M., 301
Goodwin, B. W., 126, 251

Gopakumar, G., 61
Gopinathan, M. S., 56, 57, 58
Gordon, M., 92, 271
Gorrichon, J.-P., 76
Gorton, J., 377
Gossard, A. C., 353
Goto, K., 288, 369, 388
Goto, T., 352
Gottlieb, A. M., 168
Govil, G., 56
Grabowski, 2, 253
Graham, K. C., 74
Gramstad, T., 366
Granger, P., 197, 221
Grannall, P. V., 185
Grant, D. M., 40, 70, 82, 119, 120, 125, 139, 140, 172, 174, 230, 280, 281, 389
Grant, M. W., 244
Grasdalen, H., 245, 246, 371
Gray, G. A., 69, 81
Greatbanks, D., 43, 92, 211
Grechkin, N. P., 82
Green, C. H., 21, 97
Green, M. L. H., 78, 239
Green, R. D., 346
Greenberg, E. S., 387
Greene, J. L., jun., 386
Gretner, W., 371
Gribble, G. W., 31
Griffin, R. G., 49, 185, 345
Grigor'ev, N. P., 218
Grimes, R. N., 67
Grimine, W., 231
Grinter, R., 47
Grishin, Yu. K., 34, 40, 107, 238
Grobe, J., 79
Grohman, K. K., 29
Gronowitz, S., 16, 107, 212
Gross, B., 133
Grosse, M., 380
Grotens, A. M., 293, 382
Groves, D., 45
Gründeman, E., 80
Grunberger, D., 307
Gruner, G., 353
Grutzner, J. B., 29, 104, 204, 372
Gubaidullina, R. Z., 22
Günther, H., 29, 57, 76, 85, 231
Guiliani, A. M., 33
Gupta, R. K., 169, 308
Gurato, G., 75, 200
Gutowsky, H. S., 51, 60, 119, 137, 153, 155, 157, 186, 256, 323, 324, 333, 339, 354
Guyer, R. A., 333

Haake, P., 64
Haas, C. K., 63
Haas, H., 351
Haas, H. C., 290

Habuda, S. P., 333
Haddock, S. R., 77
Hadjimihalakis, P. M., 375
Hadzi, D., 337
Haeberlin, U., 49, 136, 140, 153, 167, 168, 170, 172, 185, 345
Hägele, G., 91, 210
Hafner, S. S., 340
Hahn, E. L., 184
Haider, J., 305
Haigh, C. W., 25, 26, 27, 103
Haighton, A. J., 354
Hailstone, R. K., 141
Hajdukovic, G., 331
Hall, D. De W., 381, 387
Hall, G. E., 235
Hall, J. R., 95
Hall, L. D., 272, 380, 386
Halle, J. C., 237
Haloni, E., 197, 221
Halpern, Y., 41
Halstead, T. K., 338
Haly, A. R., 349
Hamada, K., 43
Hamann, H., 139
Hamer, G. K., 22
Hamm, P., 23
Hammaker, R. M., 63
Hampson, P., 361
Hanabusa, M., 345
Hankin, D., 10
Hanlan, J. F., 15, 64
Hanna, A. W., 377
Hansen, P. E., 103
Hansen, R. S., 40, 276
Hanson, S. W., 379
Happey, F., 302
Haque, R., 321, 364
Hardy, P. M., 301
Hardy, W. A., 4
Hardy, W. N., 113
Harkness, A. L., 111
Harmon, J. F., 132, 134
Harris, A. B., 119
Harris, C. L., 23
Harris, J. M., 16
Harris, L., 234
Harris, R. K., 43, 72, 79, 85, 91, 110, 111, 125, 191, 193, 202, 210, 212, 213, 216, 231, 235, 250, 281, 389
Harrison, L. W., 75, 198, 357
Harrison, M., 108
Harrison, P. G., 47, 66, 274
Harrod, J. F., 240
Hart, B. T., 10
Hart, F. A., 386
Hartman, J. S., 43, 70, 71
Hartmann, H., 351
Hartmann, O., 194
Hartmann, S. R., 184
Hartnell, G. E., 128
Hashimoto, T., 331
Hatch, G. F., 41
Haubold, W., 45, 82

Haul, R., 163, 346
Haupt, J., 158, 159, 342
Hausser, K. H., 49, 140, 154, 168
Hawkins, B. L., 226, 234
Hawthorne, H. M., 162, 331
Hawthorne, M. F., 67
Hayama, M., 352
Hayamizu, K., 19, 49, 62, 87, 106
Hayashi, S., 348
Haylock, J. C., 301
Hazato, G., 291
Headley, L. C., 163, 347
Hecht, H. G., 351
Heckmann, G., 45, 261
Hedvig, P., 299
Heffernan, M. L., 100, 105, 206, 271, 272
Heide, K., 326
Heil, G., 29
Hélène, C., 319
Heller, P., 168
Hellier, D. G., 21, 97
Hemming, O. F. W., 306
Henderson, T. O., 311
Hennel, J. W., 329, 349
Henold, K. L., 248
Herak, J. N., 299
Herlem, M., 254, 372
Hermann, P., 179, 220
Hertz, F., 301
Hertz, H. G., 129, 130, 253, 370
Hess, R. E., 63, 64
Hess, S., 113
Hewson, M. J. C., 43, 72
Heybey, A., 295
Heyd, W. E., 87
Heyns, A. M., 330
Higashimura, T., 302
Higuchi, T., 366
Hijmans, J. P. A. M., 352
Hikita, T., 339, 340
Hilderbrand, R. L., 311
Hill, H. A. O., 43
Hill, H. D. W., 109, 136, 140, 175, 177, 179, 180, 186, 221, 283
Hill, J., 343
Hiller, J. M., 247
Hilt, R. L., 155
Himel, C. M., 246
Hinckley, C. C., 382, 384, 385
Hindermann, D. K., 5, 333
Hindman, J. C., 5
Hinton, J. F., 39
Hirai, A., 352
Hirai, R., 291
Hirano, S., 30
Hirano, T., 289
Hirao, K., 51
Hirotsu, T., 29
Hirvonen, M., 153
Hlubucek, J. R., 382
Ho, C., 307, 308
Ho, F. F.-L., 293, 309
Ho, P., 160

Hobbs, M. E., 7, 141
Hoboken, N. J., 63
Hoch, M. J. R., 296
Hoffman, R. A., 261
Hoffmann, R., 52
Hog, J. H., 59
Hoheisel, C., 139
Holland, R., 119
Hollister, C., 10
Holm, R. H., 389
Holmes, J. R., 10
Homer, J., 20, 24, 377
Honeybourne, C. L., 383
Hood, G. M., 162
Hoodless, I. M., 161
Hooper, H. O., 343
Hoppe, J. A., 313
Horii, T., 15
Horlick, G., 175
Horn, H.-G., 40, 69
Horn, R. R., 385
Horrocks, W. De W., jun., 381, 383, 387, 388
Horsley, W. J., 177
Horton, D., 76, 198
Hosonuma, K., 288
Howard, C. J., 158
Howard, J. C., 303
Howarth, O. W., 14
Howell, J. A. S., 71
Howell, W. C., 271
Howery, D. G., 348
Hsia, R. K. C., 106
Hsu, M., 180
Hubbard, P. S., 7, 132, 155
Huber, H., 380
Huber, L. M., 170
Huckerby, T. N., 78, 90
Hudec, J., 74
Huestis, W. H., 311
Hüther, H., 64
Hughes, D. G., 151, 353
Huiskamp, W. J., 351
Humphrey, S. A., 375
Hunt, E., 6
Hunt, J. P., 244
Hunter, M., 163
Huntress, W. T., 119, 120, 123
Hunziker, P., 43
Hursthouse, M., 40, 84
Huter, M., 347
Hutton, H. M., 104, 125, 128
Huysmans, W. G. B., 379
Hynes, T. V., 339

Ignatova, N. P., 41, 274
Ihrig, A. M., 35, 39, 68, 75, 89, 92, 105, 355, 357
Ikeda, T., 271
Imonishe, Y., 302
Inesi, G., 309
Inoue, Y., 298
Iodine, J. D., jun., 354
Ionin, B. I., 80
Irvin, K. J., 291
Irving, C. S., 253
Irwin, D. M., 338

Author Index

Irwin, M. A., 31
Ito, S. E., 294
Itoh, K., 95
Ivanov, V. T., 234
Iwadore, T., 106
Iwamura, H., 228
Izmest, E. V., 151
Izmestjev, I. V., 151

Jackman, L. M., 104, 204, 372
Jackson, C. H., 271
Jackson, C. J., 377
Jackson, R. L., 162
Jackson, W. R., 241
Jacobsen, J. P., 64
Jacques, M. St., 203
Jacquier, R., 106
Jaeckle, H., 167
Jagur-Grodzinski, J., 254
Jakobsen, H. J., 40, 62, 69, 274, 276
Jakubowski, A., 29
Jamsek-Vilfan, M., 331, 338
Janik, J. A., 147
Janik, J. M., 147
Jansen, T. R., 389, 390
Januszewski, H., 46, 253, 260
Janzen, W. R., 354
Jao, L., 248
Jardetzky, O., 285
Jasinski, A., 344
Jaureguiberry, C., 271
Jautelat, M., 40
Jeener, J., 113
Jefford, C. W., 97
Jenkins, J. M., 274
Jenkins, P. N., 172
Jennings, W. B., 238, 241
Jensen, F. R., 165
Jensen, H., 35, 55
Jensen, R. K., 41
Jentschura, U., 253
Jesson, J. P., 45, 76
Jeuell, C. L., 41, 105
Johannesen, R. B., 43, 77, 269
Johanson, R. G., 41
Johansson, A., 217
Johansson, B., 338
Johnson, A., 260, 388
Johnson, B. F. G., 381
Johnson, C. E., 27
Johnson, C. S., jun., 145, 156, 183, 342, 343
Johnson, D. L., 144, 388
Johnson, D. M., 94
Johnson, L. F., 68, 230, 269, 382
Johnson, R. N., 257
Johnson, T. W., 299
Johnson, W. R., 8
Johnston, M. D., jun., 62, 356, 357
Johnston, N. W., 288
Jokisaari, J., 86
Jolley, K. W., 366

Jonas, J., 7, 122, 123, 125, 129, 166, 167, 323
Jonas, V., 64, 219, 233, 372
Jones, A. J., 40
Jones, D. E., 175, 176, 178
Jones, D. W., 30, 76, 102, 105, 302
Jones, E. P., 159
Jones, E. W., 306
Jones, G. P., 158, 353
Jones, J., 123
Jones, R. G., 40, 76, 174
Jones, R. T., 308
Jordan, A. D., 121, 174, 252
Jordan, R. B., 121, 174, 244, 252
Joseph-Nathan, P., 202
Jost, R., 32, 96
Jouany, C., 66
Juan, C., 51
Jugie, G., 66
Jung, G., 41, 175, 286, 311
Jungnickel, B., 326
Jurga, J., 122
Jurga, K., 122

Kadaba, P. K., 153
Kaduk, B. A., 63
Kaeger, J., 134
Kaiser, R., 222
Kalinin, V. N., 218
Kambario, S. M., 290
Kamieńska-Trela, K., 46
Kamieński, B., 18
Kane, A. R., 45, 390
Kanert, O., 353
Kantschowska, I., 219
Kaplan, J. I., 222
Kappleman, A. H., 104, 105
Kaptein, R., 140, 177, 180, 283
Karger, J., 347
Kari, R. E., 10
Karplus, M., 6, 10, 57, 61, 345
Karraas, M., 354
Kasahara, M., 339, 340
Kase, T., 294
Kashin, A. N., 274
Kashiwagi, M., 294, 332
Kasugai, Y., 128, 256
Kato, H., 51, 52, 55, 62
Kato, T., 326
Kato, Y., 21
Katz, J. J., 111, 389
Kaufman, B., 38
Kaufmann, J., 47, 78, 255
Kawamori, A., 164
Kawamura, S., 15
Kawasaki, Y., 64
Kayne, F. J., 313
Kdanovich, V. I., 84
Kearns, D. R., 307
Keat, R., 88
Keiter, R. L., 73
Keller, G., 129
Keller, H. J., 371, 379

Keller, T., 41
Kellie, G. M., 76, 197
Kelly, D. P., 41
Kelsey, D. R., 380
Kenney, M. E., 390
Kent, P. W., 43, 314
Kermaas, L. F., 354
Kern, C. W., 9
Kessemeier, H., 111, 345
Kessenikh, A. V., 41, 78, 82, 250, 274
Khaddar, M. R., 83, 371
Khane, P. L., 128
Khanh, P. H., 289
Khanzada, A. W. K., 325
Khazanovich, T. N., 258
Kher, V. G., 128
Khmelnitski, L. I., 39, 64, 173, 253
Khoury, F., 117
Khutsishvili, G. R., 152
Kidd, R. G., 71
Kiichi, T., 151
Killough, J., 234
Kimland, B., 104
Kimmich, R., 317
Kimura, B. Y., 40, 70
King, A. R., 152, 350
King, N. L. R., 304
Kinsey, J. L., 114
Kintzinger, J. P., 123, 250, 251
Kishi, M., 382
Kisin, A. V., 107, 239
Kissin, Yu. V., 287
Kitching, W., 43, 78
Kitchlew, A., 123, 134, 171
Kivelson, D., 10, 124
Kivelson, M. G., 124
Klaeboe, P., 76
Kleier, D. A., 230
Klein, H.-F., 79
Klein, M. P., 177
Klemperer, W., 6, 144
Kleppner, D., 4
Kline, D., 340
Klose, G., 179, 220
Klose, H., 57, 76
Klotz, I. M., 299
Klumpp, G. W., 241
Knauss, L., 73
Knispel, R. R., 161
Knoeber, M. C., 40
Knottnerus, D. I. M., 334
Knox, K., 390
Kobayashi, R., 117
Kodama, T., 157
Koenig, S. H., 131, 308
Köhler, W., 364
Koeng, F. R., 87, 207, 221
Koermer, G. S., 385
Kohler, R. R., 309
Kohn, B. H., 348
Koide. T., 328
Kolb, H., 152
Kolinský, M., 287
Koller, J., 67, 337
Kollman, V. H., 68, 310
Koma, A., 353
Komenko, T., 382

Komoroski, R., 140, 172
Kondo, I., 44
Kondo, N. S., 87
Kopanev, V. D., 47
Kopf, P. W., 288
Kopp, J. P., 351
Kopple, K. D., 88, 259, 300
Koptyug, V. A., 241
Korchagina, D. V., 241
Korchemkin, M. A., 150
Korenevsky, V. A., 76, 107, 201, 239
Korringa, J., 133
Korver, P. K., 219, 224, 225, 366
Kosfeld, R., 121, 133, 322
Kost, D., 226
Kostelnik, R., 106
Kostka, A. G., 389
Kowalewski, J., 53
Kozerski, L. J., 64
Kozmin, A. S., 40
Krayuschkin, M. M., 39
Kreilick, R., 41
Kreishman, G. P., 320
Kreshkov, A. P., 255
Krishna, N. R., 193, 250, 280
Kriz, H. M., 339
Kronhaug, F. H., 18, 104
Kruger, G. J., 132
Krugh, T. R., 54
Krygowski, T. M., 18
Krynicki, K., 119, 166
Kubo, R., 113, 252, 259
Kubo, V., 246, 371
Kubota, H., 316, 348
Kuhlman, R., 70
Kuhlmann, K. F., 125, 172, 280, 281, 389
Kukolich, S. G., 49
Kula, R. J., 246
Kulek, T., 126
Kuliev, A. M., 363
Kulish, L. F., 218
Kumanova, M. D., 26
Kumar, A., 183, 193, 250
Kuntz, I. D., jun., 316, 348, 356
Kuppers, J. R., 364, 365
Kurilenko, O. D., 363
Kurkovskaya, L. N., 363
Kurland, R. J., 382, 387
Kuroda, N., 164
Kursanov, D. N., 84
Kushida, K., 19, 62, 87
Kushida, T., 345
Kydon, D. W., 125, 128

Laarhoven, W. H., 30
Ladd, J. A., 104
Lado, F., 221, 342
Lähteenmäki, U., 153
Lahajnar, E., 331
Lahajnar, G., 149
Lai, C. F., 167
Laisaar, S., 141, 174
Lalita, K., 114

Lalowicz, Z. T., 329, 349
La Mar, G. N., 110, 140, 282, 283, 389
Lambe, E. B. D., 4
Lambert, J. B., 68, 87, 207, 221
L'Amie, R., 76
Lammert, S. R., 329
Landa, B., 71
Landman, D., 105, 205, 227
Landsberger, F. R., 353
Lapidot, A., 253
Laposa, J. D., 39
Lappert, M. F., 47, 70
Larsen, D. W., 130, 131, 255, 256, 370
Larsson, E., 74, 76
Lassigne, C., 371
Laszlo, P., 355, 356, 374
Lauer, C., 154
Laurent, J.-P., 66, 80
Lauterbur, P. C., 33
Lavallee, D. K., 17
Lawlor, J. M., 104, 204, 271, 372
Lawrie, R. A., 349
Layton, B., 39
Lazarus, M. J., 354
Lazzaroni, R., 80
Lazzeretti, P., 27, 246
Leach, J. B., 66, 213
Lebedev, O. V., 39, 64, 173, 253
Lebel, G. J., 39
Leblond, J., 133
Lee, A. G., 311
Lee, D. M., 134
Lee, M., 342
Lee, R. E., 371
Lee, T.-C., 33
Lee, Y., 123, 129
Lefevre, F., 75, 87, 271
Lehman, J. W., 46
Lehmann, H., 307
Lehn, J. M., 123, 216, 250, 251
Leigh, J. S., 242
Lemanceau, B., 325, 329
Lemieux, R. U., 271
Lemius, B., 130
Lenkinski, R. E., 40, 111, 283, 386
Lentzner, H. L., 28
Lenz, R. W., 288
Lequan, R.-M., 41, 69, 79
Leroy, J. L., 341
Lessley, S. D., 43, 70
Letcher, J. H., 44
Letter, J. E., jun., 244
Levine, B. A., 382
Levine, Y. K., 311
Levinstone, A. R., 359
Levy, B., 53
Levy, G. C., 40, 41, 64, 139, 140, 178, 368
Lewis, D. W., 385
Lewis, J., 381
Leyden, D. E., 235, 236, 246

Li, N. C., 364
Lichter, R. L., 46, 68, 173, 367
Lichtman, W. M., 82
Liler, M., 65, 88, 234, 367
Lilja, H., 255
Lillya, C. P., 87
Lim, Y. Y., 370
Lincoln, S. F., 245
Lindblom, G., 131, 312
Lindemann, L. P., 39, 290
Linder, B., 361
Linderberg, J., 336
Lindman, B., 131, 217, 218, 254, 255
Lindon, J. C., 49, 108, 109, 172, 272
Lindqvist, I., 254
Lindsay, P. H., 78
Lindstrom, T. R., 307, 308
Liotta, C. L., 23
Lipofsky, J., 324
Lipowitz, J., 43
Lippert, E., 253
Lippmaa, E., 38, 39, 64, 139, 141, 173, 174, 253
Lipsicas, M., 117
Lipsky, S. R., 171, 382
Lissac, P., 290
Litzow, M. R., 47, 70
Liu, K. J., 297
Liu, N., 7
Llewellyn, J. P., 158
Lloyd, D., 87
Lochmuller, C. H., 346
Lockhart, N. C., 162, 331
Loewenstein, A., 218
Logan, N., 46
Loginova, E. I., 82
Lombardi, E., 287
Longuet-Higgins, H. C., 134
LoSurdo, A., 370
Louick, D. J., 219
Love, A. L., 233
Lovick, D. J., 233
Lowe, I. J., 342, 343
Lowe, J. P.. 53
Lowenstein, A., 130, 138
Lowman, P. W., 69
Lubas, B., 317
Lubensky, T. C., 144
Luber, J. R., 383
Lucas, M. M., 370
Lucken, E. A. C., 335
Lugtenburg, J., 86
Lukacs, G., 41
Lukevits, O. K., 198
Lunazzi, L., 88, 105
Lupton, E. C., jun., 16, 225
Lutz, O., 6
Luz, Z., 254
Luzikov, U. N., 238
Lyerla, J. R., 119, 120, 140
Lynch, G. F., 352
Lynch, L. J., 315, 349
Lynch, R. J., 14
Lynden-Bell, R. M., 52, 82, 258, 343

Author Index

Lyubimov, A. N., 331

Mabry, T. J., 31
McAndrew, B. A., 30
McArdle, P., 381
McBrierty, V. J., 299, 332
McCall, D. W., 296, 299
Macciantelli, D., 88
MacLean, C., 75
McClung, R. E. D., 139
McConnell, H. M., 61, 103, 309, 381
McCourt, F. R., 113
McCowan, J. D., 15, 64
Macdonald, C. J., 60
McDonald, G. N., 111
McDonald, I. R., 332, 343
MacDonald, R. L., 290
McDowell, C. A., 325
McElroy, R. G. C., 156
McErlane, K. M. J., 40
McEwen, G. K., 93
McFarland, C. W., 66
McFarlane, H. C. E., 72, 78, 221, 273, 274
McFarlane, W., 65, 72, 78, 82, 221, 273, 274, 278
McGarvey, B. R., 382, 387
McGrath, J. W., 326, 338
Maciel, G. E., 14, 38, 40, 55, 56, 62, 74, 103
McIntyre, H. M., 156, 343
McIver, J. W., 51, 62, 74
McIvor, M. C., 202
Mackenzie, R. K., 228
Mckinley, S. V., 228
McLaughlin, A., 318
MacLean, C., 93
McMillan, N. D., 159
McMurry, T. B. H., 41
McNeel, M. L., 237
MacNicol, D. D., 228
Macomber, R. S., 21
McRae, E. G., 358
McWeeny, R., 7, 25, 26
McWhan, D. B., 353
Maggio, M., 386
Magi, M., 39, 64, 173, 253
Mahajan, M., 333
Mahendroo, P. P., 152
Maher, J. P., 108
Mai, L. A., 198
Maia, H. L., 235
Maier, G., 153
Majerski, Z., 261, 383
Majid, Y. A., 325
Mak, C., 152
Mak, H. D., 291
Maklakov, A. I., 218
Maksić, Z. B., 21, 60
Malevanyi, B. A., 218
Mali, M., 164, 185, 341
Malinowski, E. R., 17, 63, 71, 78, 359, 360, 361
Mallard, D. J. H., 289
Mallion, R. B., 25, 26, 27, 103
Malm, S. M., 3
Malo, H., 86

Malrieu, J.-P., 54
Mamatyuk, V. I., 241
Mamayev, V. M., 42
Mamedov, F. N., 363
Manatt, S. L., 85
Mangini, A., 105
Mankowski-Favelier, R., 45, 80
Mann, B. E., 40, 45, 46, 69, 70, 74, 79, 80, 82, 193, 273
Mannschreck, A., 39
Mansfield, P., 185, 344, 345, 354
Mantione, R., 75, 271
Mantsch, H. H., 95, 310
Manzer, L. E., 40, 70, 80
Maraviglia, B., 118, 336
Marciacq-Rousselot, M.-M., 370
Margalit, Y., 138
Marini, G., 259
Markley, J. L., 177
Marks, T. J., 251
Marlborough, D. I., 301
Maroni, P., 76
Marsden, K. H., 349
Marshall, A. G., 380, 386
Marshall, J. L., 39, 68, 89, 100
Marsmann, H., 40, 69
Marthy, N. S. R. K., 301
Martin, G., 80
Martin, G. J., 87
Martin, J. S., 67, 189
Martin, M. L., 75, 87, 106, 271
Martin, R. B., 382
Martinelli, M., 261, 270
Martinez, D., 313, 350
Martins, A. F., 146
Marton, J., 227
Maruani, J., 344
Marx, G. S., 17
Marx, J., 324
Masamune, T., 87
Maskasky, J. E., 390
Maslov, P. G., 361
Mason, J., 36, 47
Masters, C., 45, 46, 74, 79, 80, 273
Mateescu, G. D., 41
Matthey, F., 45, 80
Mathieson, D. W., 20
Mathieu, A., 103
Matsui, T., 369
Matsukama, Y., 291
Matsumoto, K., 41
Matsuzaki, K., 288
Matthews, C. H., 333
Matthews, D. G., 43, 109
Matthews, R. S., 30, 102, 105
Matthews, R. W., 71
Matwiyoff, N. A., 68, 310
Maude, C., 128
Mauk, V. V., 363
Mayer, R. T., 246
Mayers, D. F., 305
Meakin, P., 76

Mehlkopf, A. F., 183
Mehrer, H., 160
Mehring, M., 49, 170, 182, 185, 345
Meić, Z., 60
Meister, R., 132, 133, 167
Mellon, E. K., 251
Melnikov, N. N., 41, 274
Memory, J. D., 221, 342
Mengenhauser, J. V., 346
Men'shova, L. M., 363
Mestdagh, M. M., 163, 346
Metcalf, B. W., 29
Metcalfe, J. C., 311, 319
Metcalfe, S. M., 319
Meyer, E. F., 156, 328
Meyer, H., 6, 118, 153, 336
Meyers, S. M., 153
Michel, C., 88
Michel, D., 162
Middleton, W. J., 235
Mikofajczyk, M., 45
Miles, H. T., 220
Miller, D. P., 42
Miller, E., 361
Miller, K., 372
Miller, M. A., 20
Miller, P. S., 87
Miller, S. I., 17, 366
Miller, W. B., 64
Miller, V. R., 67
Millett, F., 313, 314
Mills, H. H., 228
Mills, R. L., 336
Milne, G. W. A., 46, 64
Milone, L., 40, 84
Mims, W. B., 183
Mishima, K., 122
Miskow, M., 75
Miskow, M. H., 87
Missen, A. W., 388
Mitomo, N., 348
Mitsch, C. C., 45
Miyajima, G., 38, 64
Miyamoto, T., 47
Mizuno, O., 326, 353
Mo, Y. K., 41
Mochel, V. D., 298
Mock, N. H., 308
Modak, S. G., 128
Möller, U., 79
Monjol, P., 287
Montenay-Garestier, T., 319
Mooberry, E. S., 49, 351
Mooney, J. R., 390
Moores, B. M., 181
Moraga, L., 16, 107, 212
Morallee, K. G., 43, 314
Moreland, C. G., 45
Morgan, L. O., 245
Morgan, R. E., 122
Morgan, W. R., 235, 236
Morginangelli, A., 287
Moriarty, R. M., 219, 233
Morioka, M., 29
Morishima, I., 52, 55, 368, 369, 388
Moritz, A. G., 106, 172, 272, 365

Moroi, D. S., 145
Moroz, N. K., 218
Morris, B., 325
Moskowitz, J. W., 10
Moskvich, Yu. N., 330
Moss, G. P., 386
Moss, K. C., 71
Moss, R. E., 4
Mosser, S., 76, 212
Motrell, E. L., 14
Mottley, C., 343
Moulson, T., 53
Moulton, W. G., 351
Müller, D., 164
Muenter, J. S., 6
Muetterties, E. L., 45, 76
Muha, G. M., 347
Muir, A. R., 183
Mukherjee, R., 219, 233
Mukhomorov, V. K., 22
Mukouyama, Y., 290
Mulady, B., 159
Mulcahy, W., 156, 326
Muller, B. H., 132, 134, 337
Muller, D., 347
Muller, N., 60, 367
Muller-Warmuth, W., 158
Mundheim, O., 366
Muneyuki, R., 388
Muntz, R. H., 368
Murae, T., 271
Murata, I., 29
Murray, M., 43, 72, 80, 213
Murrell, J. N., 61
Murthy, D. S. N., 21
Musher, J. I., 23
Musso, J., 40
Mutter, M., 43
Myagi, M., 218
Myers, S. M., 118, 336
Myers, T. C., 311
Myint, T., 4

Nagal, Y., 64
Nagel, D. L., 236
Nagel, M., 162
Nageswara Rao, B. D., 280
Naider, F., 301
Nair, P. M., 61
Nakabayashi, M., 290
Nakagawa, T., 232
Nakamura, N., 151
Nakamura, Y., 333
Nakano, M., 366
Nakano, N. N., 366
Nakano, T., 64
Nakashima, T. T., 40
Nakatsuji, H., 51, 52, 55
Namikawa, K., 38
Nan-I Liu, 129
Narasimhan, P. T., 56, 57, 58, 59
Nathan, G., 110, 259
Natta, G., 287
Natterstadt, J. J., 38
Natusch, D. F. S., 282, 388
Naulet, N., 87

Navech, J., 76, 213
Navratil, M., 291
Nazaki, H., 30
Nechaev, Yu. D., 80
Neely, J. W., 244
Negita, H., 157
Negrebetskii, V. V., 41, 78, 82, 250, 274
Nelson, A. C., 49
Nelson, A. I., 354
Nelson, F. A., 272
Nelson, G. L., 368
Nelson, J., 256
Nelson, J. H., 307
Nelson, S. M., 256
Neuberg, R., 29
Neumann, R. M., 6, 144
Newman, R. C., jun., 64, 219, 233, 372
Newsoroff, G. P., 105, 205, 227
Newton, M. G., 76
Ng, P., 245, 371
Ng, S., 366, 374
Nichols, J. M., 85, 212
Nicholson, J. Y., 331
Nickerson, M. A., 146
Nieboer, E., 314
Niecke, E., 82
Niederberger, W., 102
Nielsen, J. T., 59, 64
Niemelä, L., 153
Niesen, L., 351
Nifant'ev, E. Ye., 40, 95
Nikonorova, L. K., 82
Nilles, G. P., 101
Nilsson, B., 227
Nishida, V. M., 371
Nishihama, T., 271
Nishihara, H., 352, 353
Nishioka, A., 20, 126, 128, 129, 298, 320
Nivard, R. J. F., 30
Nixon, J. F., 43, 66, 82
Noack, F., 113, 128, 133, 155, 165, 296
Noe, E. A., 80, 229
Noggle, J. H., 124, 166, 269, 389
Nolle, A. W., 152
Norberg, R. E., 343, 352
Norén, I. B. E., 307
North, A. C. T., 260, 306
Norton, J. R., 381
Novikov, S. S., 39, 64, 173, 253
Nuretdinov, I. A., 82

Odaka, T., 352
Odiot, S., 106
Odom, J. D., 47, 68, 69
Ödberg, L., 157
Ogawa, H., 29
Ogiwara, Y., 316, 348
Ohtsuki, M. A., 64
Oja, T., 339
Okabe, M., 330
Okada, K., 388
Okada, M., 288

Okajima, S., 294
Okamoto, Y., 105
Okawara, R., 64
Oki, M., 228
Okogun, J. I., 382
Olaf, H. G., 349
Olah, G. A., 41, 66, 105
Olavi, P., 246
Oldfield, E., 319
Olf, H. G., 293, 294, 316, 331, 332
Oliver, J. P., 77, 95, 248
Oliver, W. L., 92
Olivson, A., 139
Olofsson, G., 100
Olovsson, I., 329
Olsen, J. S., 308
Olsen, R. K., 233
Omelańczuk, J., 45
Onak, T., 88
Onak, T. P., 66, 213
Onaka, S., 47
O'Neil, J. W., 230, 236
Onsager, I., 356
Opie, M. C. A., 18
Oppenheim, I., 114, 118, 124
Oppermann, G., 163, 346
O'Reilly, D. E., 122, 137, 138, 153, 157, 329, 333, 337, 338
Ori, M., 80
Ormondroyd, S., 370
Orrell, K. G., 218, 235
Orwoll, R. D., 321
Oshenoff, D. D., 134
Osipenko, A. N., 151
Osredker, R., 185, 341
Ossman, G. W., 338
Ostlund, N. S., 51, 74
Oth, J. F. M., 29
Otnes, K., 147
Ottinger, R., 97
Ouellette, R. J., 88
Ovchinnikov, Yu. A., 234
Ozier, I., 115
Ozin, G. A., 43

Pachler, K. G. R., 140, 275, 282
Paci, M., 87, 199
Pack, G. R., 10
Packer, K. J., 346
Paddock, N. L., 43
Padilla, A. G., 94
Page, J. L., 154
Pajak, Z., 122
Pajari, L., 246
Pake, G. E., 324
Palmer, G., 217
Paolillo, L., 302, 310
Parekh, M., 70
Park, M. J., 339
Parker, G. W., 221, 342
Parker, J., 104
Parker, R. G., 122, 125, 151
Parkhurst, H. J., 123
Parry Jones, G., 182

Author Index

Parshall, G. W., 45
Parsons, I. W., 43, 80
Parsons, J. L., 318, 349
Partington, P., 76, 174, 311
Patel, D. J., 304, 306, 307
Patnaik, B. K., 289
Patterson, A., jun., 249
Paudler, W. W., 375
Paul, D. B., 106, 272
Paul, E. G., 280
Paul, I. C., 368
Paulis, N. J., 151
Pauwells, P. J. S., 180, 283
Pavlidou, E. G., 375
Pawliczek, J. B., 231
Peak, S., 246, 258, 371
Peake, S. C., 43, 72, 85, 212
Pearson, H., 226
Pearson, R. M., 346
Peat, I. R., 60, 298
Pecsok, R. L., 315, 348
Pedersen, B., 76
Pedley, J. B., 47, 70
Péguy, A., 83, 371
Pehk, T., 38, 39
Peierls, R. F., 338
Pellatt, M. G., 218
Pellizer, G., 43
Perdok, W. G., 334
Pereira, A. M. P., 31, 106
Perlin, A. S., 39
Perraud, R., 33, 363
Perry, A. J., 4
Pesek, J. J., 315, 348
Petch, H. E., 159
Peterlin, A., 293, 316, 331, 332, 349
Peternelj, J., 157
Petersen, I. B., 59
Petersen, T. E., 62
Peterson, E. M., 122, 138, 153, 157, 329, 333, 337, 338
Peterson, G. E., 340
Peterson, J., 164
Peterson, L. K., 70
Pethica, B. A., 370
Petit, M. A., 75, 385
Petke, J. D., 10
Petrakis, L., 41
Petrov, A. A., 80
Petrov, M. P., 350
Petrovskii, P. V., 84
Pfeifer, H., 162, 347
Pham-Quang-Tho, 290
Philipsborn, W. V., 205
Philips, L., 12, 43, 80, 86, 172
Phillips, W. D., 304
Pickett, H. M., 175
Pickett, J. H., 346
Pieranski, P., 126
Pierre, J., 33, 363
Pietropaolo, R., 40
Pihlaja, K., 76, 197
Pincus, P., 144
Pines, A., 49, 143, 168, 170, 184, 185, 334, 345
Pintar, M. M., 145, 147,
156, 157, 159, 161, 326, 329
Pirkle, W. H., 368
Pirs, J., 147
Pislewski, N., 126
Pitner, T. P., 305
Pittman, R. A., 327, 348
Pitzer, P. M., 53
Pivcova, H., 291
Plinke, G., 29
Poldy, F., 159, 343
Poleshchuk, N. S., 218
Pople, J. A., 1, 10, 25, 36, 51, 52, 59, 74, 92, 134, 195, 368
Popov, A. I., 254, 372
Poole, C. P., jun., 112, 323
Porter, L. J., 388
Porter, R. D., 41, 105
Portnova, S. L., 234
Poulter, C. D., 27, 100
Pound, R. V., 113, 324
Power, J. D., 53
Powles, J. G., 119, 121, 166, 342
Poynton, A. J., 32
Pratt, S. L., 300
Pregosin, P. S., 39, 41, 46, 64, 108
Preissing, G., 128, 133, 296
Prelesnik, A., 185, 341
Preston, P. N., 46, 105
Price, E., 295, 332
Price, K., 39, 41, 79
Prihodko, A. S., 39, 64, 173, 253
Primas, H., 194
Pritchard, D. E., 60
Provotorov, B. N., 342
Pruppacher, H. R., 134
Pryce, N. G., 317
Punkkinen, M., 151
Pupp, M., 49, 351
Purcell, E. M., 113, 324
Purdela, D., 41, 45
Purser, J. M., 47
Pynn, R., 147
Pyper, N. C., 110, 193, 250
Pyykko, P., 336

Quereshi, M. S., 370, 387
Quirt, A. R., 189

Raban, M., 226
Rabenstein, D. L., 246, 290
Rabinovitz, M., 30
Rabinowitz, J. R., 10
Raby, B. F., 74
Radeglia, R., 42, 59, 364
Rae, I. D., 32
Raftery, M. A., 311, 313, 314
Raghunathan, P., 325
Ragozzino, E., 128
Ragsdale, R. O., 43, 70
Rahkamaa, E., 65, 86, 246, 271, 354
Rakita, P. E., 239
Rakshys, J. W., jun., 228
Ramachandran, G. N., 88, 300
Ramadan, B., 171
Ramakanth, A., 151
Ramey, K. C., 219, 233, 358
Ramsey, N. F., 5, 50, 62, 114, 116, 143
Randall, E. W., 39, 40, 41, 46, 84, 108, 274
Randall, P. J., 344
Randall, R. F., 319
Randić, M., 60
Rao, B. D. N., 123, 134, 171, 193, 250, 333
Rao, K. V. S. R., 352
Rao, V. S., 61
Ratkovic, S., 156, 330
Rattle, H. W. E., 305
Raymond, M., 340
Rayner, L., 320
Raynes, W. T., 1, 7, 11, 20, 30, 31
Raza, M. A., 20, 30, 87
Read, S. H., 131
Rebane, T. K., 26
Redfield, A. G., 113, 159, 169, 185, 187, 270, 308, 345
Redwood, M. E., 43
Reed, G. H., 246
Reed, K., 151
Reedy, B. W., 174
Rees, R. G., 85, 212
Reeves, G. K., 343
Reeves, L. W., 87, 110, 219, 220
Reeves, P. C., 17
Reich, H. J., 40
Rein, R., 10
Reinheimer, J. D., 104, 105
Reisdorff, J., 29
Reisse, J., 97
Reiter, R. C., 367
Remeika, J. P., 353
Remin, M., 349
Renard, J. P., 352
Rennie, W. J., 76
Repko, E., 353
Resing, H. A., 159, 346
Retcofsky, H. L., 39
Reuben, G., 242
Reuben, J., 131, 313
Revel, M., 76, 213
Revitt, D. M., 46
Rewicki, D., 380
Reynolds, W. F., 22, 60, 86, 298
Rhee, C., 339, 341
Rhim, W. K., 49, 111, 143, 168, 185, 334, 345
Rhodes, M., 122
Rhodes, N. L., 181
Richards, R. E., 137, 314
Richardson, R. C., 134, 333
Riches, P. L., 306
Richer, J. C., 88
Richtering, H., 139

Rico, M., 103
Riddell, F. G., 76, 197
Riedel, K. H., 64
Riehl, J. W., 114
Rietz, R. R., 111, 178
Rigamonti, A., 164
Riggin, M. T., 161
Rigny, P., 271
Rigo, A., 75, 200
Rimmelin, P., 32, 96
Ripmeester, J. A., 153, 154, 155, 157, 327, 328, 333
Ritchie, G. L. D., 21
Rizvi, S. Q. A., 43, 109
Robb, I. D., 388
Robert, J. B., 64, 87, 93, 213
Roberts, B. W., 68
Roberts, H. G. Ff., 24, 25
Roberts, J., 82, 258
Roberts, J. D., 40, 41, 46, 68, 80, 106, 173, 226, 229, 234, 235, 298, 367, 387
Roberts, R. T., 353
Roberts, R. M. G., 76
Robertson, R. E., 381
Robinson, E. A., 43
Rochau, G., 156, 326
Rock, S. L., 63
Rodmar, S., 16, 107, 212
Roeder, S. B. W., 162, 181, 331
Röttele, H., 29
Rogers, D., 183
Rogers, L. B., 346
Rogers, M. D., 291
Rogers, M. T., 120, 121, 142, 183, 218, 225
Romers, C., 86
Ronayne, J., 373
Rondeau, R. E., 107, 174
Roos, B., 53
Roques, B., 271
Rose, D. G., 377
Roseberry, T., 111, 178
Rosen, J. F., 233
Rosén, U., 16, 107, 212
Rosenberg, D., 21, 64
Rosenberg, E., 40, 84
Rosenberg, H. M., 107
Rosenblatt, G., 352
Rosenblum, A., 88
Rosenthal, L., 370, 387
Ross, D. S., 88
Rosseel, T., 76
Rosser, M. J., 88
Rossotti, F. J. C., 376
Rossotti, H., 376
Roth, K., 380
Rothenberg, S., 10
Rothschild, W. G., 127, 321
Rowbotham, J. B., 104
Rowland, T. J., 353
Rowsell, D. G., 94
Royston, J., 337
Rubailo, A. I., 330
Rubin, I. D., 84

Rubinstein, M., 353
Rubčić, A., 60
Rudolph, R. W., 73
Rugheimer, J. H., 114, 130
Rummens, F. H. A., 85
Rupley, J. A., 304
Rupprecht, A., 312, 349
Rushworth, F. A., 154, 327
Rusnak, L. L., 244
Ryall, M., 354
Ryan, J. J., 301
Rydon, H. N., 235, 301

Sabourault, B., 377
Sacchi, C., 298
Sack, R. A., 252
Sadlej, A. J., 46
Saffar, Z. M. El., 156
Sahatjian, R. A., 87
Sahm, W., 47, 78, 255
Saika, A., 21
Saikachi, H., 29
St-Jacques, M., 85
Saito, T., 290
Saji, H., 352
Salie, D. L., 6
Salmeen, I., 217
Saluvere, T., 141, 174
Samek, Z., 31
Samuelson, G. L., 160
Samuelsson, E. G., 354
Samulski, E. T., 145, 173, 321
Sanchez, B., 246, 371
Sanders, J. K. M., 242, 379, 386
Santelli, M., 99
Santini, R. E., 173
Santry, D. P., 51, 52, 54
Santucci, S., 261, 270
Sarko, A., 287
Sarney, S. G., 313
Sartori, P., 91, 210
Sasaki, Y., 47
Sato, K., 126, 128, 129, 320
Satoh, S., 19, 62, 87
Sauer, J., 230
Saunders, J. K., 385
Saunders, L., 20
Sauvaitre, H., 40
Savitsky, G. B., 35, 38
Sawyer, D. W., 121
Sayer, B. G., 35, 39
Sayer, M., 352
Schaaf, T. F., 77, 95, 248
Schaefer, H. F., 10
Schaefer, T., 65, 102, 104, 106, 126, 208, 251, 358
Schaeffer, C. D., 63, 64
Schaeffer, J., 180
Schaeffer, R., 111, 178
Schakel, M., 241
Schamper, T. J., 259
Schastnev, P. V., 61
Schaumann, E., 219
Schaumburg, K., 35, 55, 59, 64, 80, 209
Schauwecker, P., 221
Scheidegger, U., 106

Scheie, C. E., 153, 157, 333, 337
Schell, F. M., 41
Schempp, E., 340
Schenck, G. E., 28, 378
Schenck, R., 63
Schenetti, L., 80, 106, 246
Scheraga, H. A., 302, 303
Schexnayder, D. A., 40
Schilling, G., 39
Schirmer, R. E., 269, 389
Schittenhelm, W., 111, 180, 283
Schlosberg, R. A., 41
Schmidbaur, H., 79, 94
Schmidt, C. F., 125, 140
Schmidt, G. W., 181
Schmidt, R. L., 356
Schmidt, V. H., 151
Schmiedel, H., 347
Schmulbach, C. D., 371
Schmutzler, R., 43, 72, 80, 85, 212, 213
Schnabel, B., 326, 344
Schneider, B., 220, 287, 291, 344
Schneider, H., 41
Schneider, W. G., 1, 195, 358
Schraml, J., 19
Schrobilgen, G. J., 43, 71
Schröder, G., 29
Schrumpf, G., 15, 86, 196
Schuerch, C., 287
Schuetz, R. D., 101
Schultz, C. W., 73
Schultz, P., 156, 326, 328
Schuster, R. E., 246, 258, 371, 383
Schutte, C. J. H., 330
Schwartz, M., 166
Schwarzhans, K. E., 327, 371, 379
Schweitzer, D., 49, 136, 140, 167, 168, 172
Schweizer, M. P., 306
Schwenk, A., 47, 78, 175, 255
Sciacovelli, O., 205
Scott, K. N., 105, 203
Scott, T. A., 329, 339
Scrowston, R. M., 105, 206, 366
Sears, R. E. J., 142
Seconi, D., 80
Seconi, G., 106
Sedlak, B., 353
Seeger, A., 160
Seelig, J., 380
Seers, R. E. J., 182
Seff, K., 383
Segal, G. A., 52
Segel, S. L., 352
Segne, A. L., 287
Seiter, C. H. A., 180
Sen, S. K., 6
Sentz, A., 159
Sepulchre, A., 41
Sergeev, N. M., 34, 40, 42, 43, 274, 327

Author Index

Sergeyev, N. M., 47, 65, 76, 95, 107, 201, 238, 239, 274
Serra, A. M., 261, 270
Servis, K. L., 248
Setkina, V. N., 84
Settine, J. M., 305
Sevostjanova, V. V., 39
Seyd, W., 347
Shaddick, R. C., 110, 219
Shafizadeh, F., 330
Shanan-Atidi, H., 226
Shanbhag, S., 354
Shanina, B. D., 218
Shapet'ko, N. N., 362, 363
Shapiro, B. L., 382
Sharp, A. R., 159, 329
Sharp, D. W. A., 88
Sharp, R. R., 21, 186
Shaw, B. L., 40, 45, 46, 74, 79, 80, 82, 273, 274
Shaw, D., 175, 180, 280, 282, 283
Shaw, K. N., 110, 219, 220
Shchori, E., 254
Sheetz, M. P., 309
Sheinin, E. B., 86
Sheldrick, G. M., 6, 43, 71
Sheline, R. K., 49, 351
Shen, M. C., 332
Sheppard, N., 52
Sheppard, W. A., 16
Shepperd, C. M., 126, 251
Sherwood, J. N., 162, 330, 331
Shevelev, S., 218
Shevlin, P. B., 386
Shigorin, D. N., 362
Shimizu, N., 17
Shimizu, T., 343
Shimjo, T., 352
Shimokawa, S., 106
Shimomura, K., 119, 157
Shimp, L. A., 251
Shingu, T., 382
Shinra, K., 218
Shiotani, A., 79, 94
Shiroki, H., 288
Shmyrev, I. K., 354
Shok, M., 382
Shono, T., 218
Shore, L., 348
Shortland, A., 80
Shortley, G. H., 191
Shoup, R. R., 80, 121, 135, 137, 140, 179, 216, 220, 237
Shporer, M. S., 254, 312, 313
Shubin, V. G., 241
Shulman, R. G., 307
Shvetsov-Shilovskii, N. I., 41, 274
Sibert, J. W., 246
Sichel, J. M., 36
Siegel, M., 117
Sievers, R. E., 43
Sillescu, H., 127, 218, 320
Silverman, D. N., 302

Silvidi, A. A., 164, 313, 338, 350
Simeral, L., 38
Simon, W., 221
Simonnin, M.-P., 41, 69, 79, 80
Sipe, J. P., tert., 383
Sisdo, M., 302
Sissons, D., 105
Skirrow, J. D., 220, 344
Skrunts, L. K., 218
Slade, R. M., 45, 46, 80
Slak, J., 164, 339
Slater, J. L., 49, 351
Sleezer, P. D., 78
Slessor, K. N., 84
Slichter, C. P., 112
Slichter, W. P., 296, 327
Slightom, E. L., 23
Slivnik, J., 149
Slonim, I. Ya., 331
Smidt, J., 183, 254, 293, 382
Smith, A. J., 43
Smith, B. A., 146
Smith, C. W., 146
Smith, D. W., 166
Smith, E. G., 297
Smith, G. V., 384, 385
Smith, G. W., 324, 328, 330
Smith, I. C. P., 95, 310
Smith, J. A. S., 337
Smith, J. D., 247
Smith, K. R., 241
Smith, L. A., 165
Smith, M., 118
Smith, M. R., 353
Smith, R., 388
Smith, S. L., 75, 92, 307, 355, 356, 357
Smith, W. B., 35, 105
Snobbart, A., 346
Snowden, B. S., jun., 348
Snyder, L. C., 54
Snyder, R. E., 151
Sobottka, J., 295
Soda, G., 336
Soest, J. F., 331
Sogn, J., 300
Sohma, J., 106
Soifer', G. B., 151
Solkan, V. N., 42
Soloman, I., 78
Sommer, J. M., 32, 96
Sondheimer, F., 29
Sone, T., 107
Soulati, J., 248
Sowerby, D. B., 46
Spaargaren, K., 219, 224, 225, 366
Spach, G., 301
Spalthof, W., 370
Spassov, S. L., 219, 232
Spear, R. J., 58
Speert, A., 356
Spegt, P., 373
Speight, P. A., 341
Spence, R. D., 352
Spencer, H. G., 289
Spielvogel, B. F., 47

Spiess, H. W., 49, 136, 140, 168, 172, 351
Spoerri, P. E., 17
Spragg, R. A., 235
Sprague, P. W., 41
Spratt, R., 256
Srivastava, V. C., 168
Stainbank, R. E., 40, 45, 46, 74, 80, 82
Staniforth, M. L., 386
Stassinopoulou, I., 76
Stec, W. J., 35, 65
Stedman, D. E., 76, 212
Steele, W. A., 122
Steener, P. R., 272
Stefaniak, L., 46, 253, 260
Stehlik, D., 154
Steigal, A., 230
Steinberg, M. P., 354
Stejskal, E. O., 246
Stephens, M. D., 104
Stephens, R. C., 43
Stephens, R. M., 302
Stephens, R. S., 70
Stepisnik, J., 338, 339
Sternheimer, R. M., 338
Sternhell, S., 58, 86, 105, 205
Sternlicht, H., 168, 175
Sternson, L. A., 76
Steur, R., 39
Stevens, K. W. H., 383
Stilbs, P., 65
Stocco, G. C., 94
Stock, L. M., 23
Stoddart, R. W., 319
Stoker, J., 287
Stolfo, J., 88
Stone, W. E., 163, 346
Storek, W., 364
Storey, H. T., 301
Story, H. S., 340
Stothers, J. B., 40, 70, 76, 100, 271
Strange, J. H., 122, 154, 157, 161, 162, 186, 328
Strauss, H. L., 175
Street, R., 343
Stricker, G., 6
Strouse, C. E., 68, 310
Studebaker, J. F., 304
Stungis, G. H., 130
Subbotin, O. A., 40
Sudmeier, J. L., 76
Sulima, L. V., 218
Sullivan, G. R., 382
Sullivan, J. F., 390
Sullivan, N., 354
Sumi, M., 290
Sung, C. C., 144
Sunko, D. E., 261, 383
Susic, M., 163, 347
Sutcliffe, L. H., 35, 46, 70, 101, 105, 355
Suzaki, T., 291
Svanholm, U., 64, 238
Svanson, S. E., 338
Swaelens, G., 76
Swain, C. G., 16, 225
Swain, J. R., 43, 66, 82

Swanenburg, T. J. B., 151
Swift, T. J., 217
Sykes, B. D., 300, 304
Sykora, S., 354
Symons, M. C. R., 370
Sze, S. N., 45

Tabata, Y., 164
Tabony, J. M., 109, 172, 272
Taddei, F., 18, 27, 80, 90, 106, 213, 246
Taft, R. W., 16
Takahashi, K., 38, 64, 107, 364
Takahashi, T., 271
Takats, J., 240
Takayanagi, S., 352
Takeda, M., 246
Takeuchi, Y., 378
Takeyami, Y., 291
Tanaka, K., 331
Tanaka, S., 353
Tanaka, Y., 122
Tancredi, T., 310
Tanner, J. E., 162, 297, 321
Tanno, T., 290
Taplick, T., 326
Tarasov, V. P., 47
Tarasova, A. I., 39
Tarpley, A. R., jun., 39, 40, 63, 77, 89, 298
Tarr, C. E., 146
Tatsuzaki, I., 339, 340
Taurins, A., 106
Tavernier, D., 87
Taylor, B., 35, 46, 70, 105
Taylor, G. T., 302
Taylor, J. S., 318
Taylor, K., 240
Taylor, N. F., 43
Taylor, P. C., 339
Taylor, P. W., 314
Tegenfeldt, J., 157, 329
Teichmann, H., 80
Temussi, P. A., 302, 310
Terenzi, M., 157
Terry, K. W., 158
Thoai, N., 231
Thomas, W. A., 86, 199
Thompson, B. C., 315
Thompson, K. H., 327
Thompson, R. C., 301
Thomson, J., 43
Thorpe, F. G., 78, 90
Tiddy, G. J., 217
Tiecco, M., 105
Tiezzi, E., 259
Timmons, C. J., 106, 207
Tjan, S. B., 381
Tobias, R. S., 94
Todo, I., 340
Todorova, R., 219
Tokuhiro, T., 135, 183
Tomchuk, E., 6, 125, 128
Tomić, L., 261, 383
Tomić, M., 261, 383
Tomita, K., 113, 259

Tompa, A. S., 295, 332
Tompa, K., 353
Topart, J., 78
Torchia, D. A., 301
Torgeson, D. R., 350, 351
Torgrimsen, T., 76
Tori, K., 382, 388
Torocheshnikov, V. N., 274
Torrey, H. C., 132, 159, 331
Torssell, K., 62
Tosolini, G., 76
Toth, F. I., 353
Towl, A. D. C., 55
T'Raa, J., 126, 251
Tracey, A. S., 84
Tribble, M. T., 20
Trill, T., 174
Tripp, V. W., 348
Trivellane, E., 302, 310
Tsang, T., 158
Tsau, T., 329
Tse, D., 342
Ts'o, P. O. P., 87
Tsuda, T., 352
Tsuno, Y., 17
Tsurugi, J., 15
Tsuruta, T., 289
Tsutsui, M., 80
Tsuyuki, T., 271
Tsuzuki, Y., 18
Tsvetkova, V. I., 287
Tuchagnes, J.-P., 80
Tukey, J. W., 176
Tuomikoski, P., 368
Tupčiauskas, A. P., 34, 47, 274
Turbitt, T. D., 28
Turnbull, A. G., 362
Turnbull, D., 161
Turner, D. W., 183
Turov, E. A., 350
Tutsch, R., 130
Tward, E., 171
Tweedale, A., 47, 70
Tyler, A., 134
Tyssee, D. A., 64

Uebersfeld, J., 133
Uetrecht, J. P., 29
Uhmann, R., 43
Ulmen, J., 231
Ulrich, S. E., 47, 66, 156, 274, 329
Ungermann, C. B., 66, 213
Uphaus, R. A., 389
Urry, D. W., 305
Uryu, T., 288
Ustynyuk, Yu. A., 34, 40, 42, 47, 65, 76, 95, 107, 201, 238, 239, 274
Utton, D. B., 158

Valentine, K. M., 152
Valic, M. I., 157, 326
Valle, M., 40, 84
van Brederode, H., 379

van Den Enden, J., 154, 328
van Den Heuvel, G. M., 151
van der Haak, P. J., 219, 224, 225, 366
Vanderhart, D. L., 137
Van der Hart, W. J., 61
Van der Lugt, W., 334
van der Werf, S., 234
Van Deursen, F. W., 15, 212
van de Ven, L. J. M., 39, 41
Vandewalle, M., 76
Van Dongen, J. P. C. M., 39
Van Geet, A. L., 237
van Putte, K., 154, 155, 328
van Schutz, J., 155
Van Steenwinkel, R., 342
Van Vleck, J. H., 324, 384
van Wazer, J. R., 35, 44, 65
Varvoglis, A. G., 375
Vasil'ev, A. F., 41, 274
Vasiléva, L. Yu., 317
Vasishth, R. C., 290
Vass, G., 41
Vasyanina, L. K., 362, 363
Vaughan, J., 18
Vaughan, R. W., 167
Vauthier, E., 40
Vaziri, C., 85, 203
Vega, S. P-., 114
Veksli, Z., 299
Veldhuis, R. G. M., 30
Veracini, C. A., 80
Verdini, A. S., 310
Verhegghe, G., 76
Verkade, J. G., 73, 93, 95
Verner, D., 74
Vernier, J., 93
Vernon, W. D., 110, 389
Versmold, H., 129, 130
Verzele, M., 271
Vestin, R., 53
Vickers, G. D., 183
Vigee, G. S., 245, 371
Viglino, P., 87, 199
Vikane, O., 214
Vikelsoe, J., 354
Vilfan, M., 147
Villatranca, J. J., 315
Vincent, E.-J., 40, 64
Virtanen, I., 246
Visser, F. R., 381
Vivarelli, P., 80, 106, 246
Vladimiroff, T., 71, 78
Voelter, W., 41, 175, 286, 311
Vogel, E., 29
Volarovich, M. P., 317
Vold, R. L., 80, 135, 136, 171, 179, 186, 216, 221, 321
Volkov, V. E., 330
Volz, H., 41
Volz-de Lecea, M., 41
Von Meerwall, E., 353

Author Index

Von Ostwalden, P. W., 106
von Philipsborn, W., 23
von Schütz, J. U., 156
Von Wijnsberghe, L., 96
Voss, J., 219
Vottero, P., 76
Vrscaj, S., 329
Vucelic, D., 163, 347

Waddington, T. C., 45, 334
Wade, C. G., 115, 145, 146, 147
Wade-Jardetzky, N. G., 285
Waegell, B., 97
Waldstein, P., 163, 347
Walker, I. M., 370, 387
Walker, N. A., 291
Wallace, W. E., jun., 163, 347
Wallach, D., 158
Walsh, H. C., 47, 68
Walter, W., 219
Walther, F. G., 4
Wander, J. D., 78, 382
Wandewalle, M., 96
Wang, C. H., 119, 120
Wang, R., 118, 184
Ward, G. A., 309
Ward, I. M., 294, 332
Warde, R. L., 313
Warren, J. P., 271
Wasylishen, R., 65, 104, 106, 208
Watanabe, H., 64
Watanabe, T., 352
Watkins, C. L., 145
Watson, B. M., 302
Watson, I. D., 366
Watt, I. C., 315, 349
Watton, A., 159
Watts, V. S., 355
Watts, W. E., 28
Waugh, J. S., 49, 114, 143, 168, 170, 182, 184, 185, 324, 334, 345
Wawzonek, S., 87
Weatherall, I., 45
Weaver, H. T., 160
Webb, G. A., 46, 253, 379
Weber, G., 376
Weber, P., 88
Wehrli, F. W., 41, 69
Wei, I. Y., 337
Weil, G., 373
Weiner, P. H., 359, 360, 361
Weinhaus, F., 153, 336
Weinstein, I. B., 307
Weisenthal, L., 4
Weissman, M., 30
Weissman, S. I., 384
Weithase, M., 155
Welti, D., 105
Wendisch, D., 85
Wenkert, E., 41
Wennerström, H., 131, 218, 254
Wert, C., 324

Wesslen, B., 288
West, B. O., 43
West, R., 88
West, R. J., 245
Westmoreland, T. D., 78
Wettstrom, R., 354
Whalley, W. B., 20
Wheeler, G. L., 85
Wheland, R. C., 229
Whidby, J. F., 246
Whipple, E. B., 84
White, A. H., 21
White, A. I., 39, 46, 108
White, A. M., 41, 105
White, D., 118, 184
White, D. M., 178, 299
White, D. W., 93, 95
White, J. P., 316
White, W. D., 337
Whitehead, M. A., 36, 39
Whitehurst, P. W., 219, 233
Whitesides, G. M., 80, 385
Whitney, P. M., 377
Whitten, J. L., 10
Widenlocher, G., 33
Wieder, M. J., 33, 46, 64
Wien, R., 304, 318
Wiersema, R. J., 67
Wilbur, D. J., 125, 167
Wilcox, C. F., 29
Wilczok, T., 317
Wileman, D. F., 43, 92, 211
Wiley, G. R., 17, 366
Wilkes, C. E., 286
Wilkinson, G., 80
Willcott, M. R., 40, 111, 283, 386
Williams, D. A. R., 228
Williams, D. H., 242, 373, 386
Williams, D. L., 114
Williams, J. M., 157, 337
Williams, K. L., 76, 212
Williams, M. K., 86, 199
Williams, R. J. P., 260, 274, 306, 314, 382
Willmorth, J. H., 326
Wilmshurst, T. H., 109, 172, 272
Wilson, G. L., 70
Wilson, G. V. H., 343
Winkler, H., 162
Winkler, P. F., 4
Winstein, S., 27, 29, 78
Winterhalter, K. H., 308
Wirth, H. E., 370
Wise, W. B., 219, 233
Witanowski, M., 46, 253, 260
Wölcke, U., 33
Woessner, D. E., 348
Wofsy, S. C., 6, 144
Wolf, D., 160
Wolf, F., 340
Wolf, H. C., 152, 156
Wolfe, J. P., 152, 350
Wolkowski, Z. W., 40, 231, 260
Wollan, D. S., 354

Woller, P. B., 236
Womble, L. R., 35
Wood, G., 87
Wood, R. J., 274
Woodbrey, J. C., 225
Woodman, C. M., 189
Woodward, C. K., 304
Woplin, J. R., 43, 72, 213
Workman, D. T., 164
Worvill, K. M., 216
Wray, V., 12, 43, 80, 86
Wright, G. E., 86
Wright, G. J., 18
Wu, A., 17
Wüthrich, K., 308
Wyatt, M., 380
Wylde, L. E., 161
Wyman, J., 376
Wyssbrod, H. R., 300

Xavier, A. V., 260, 274, 306, 314, 382

Yagi, T., 340
Yagshiev, B., 363
Yagupolskii, L. M., 43
Yajima, F., 258
Yamadaya, T., 352
Yamagata, K., 352
Yamagata, Y., 44
Yamamoto, G., 228
Yamamoto, H., 30
Yamamoto, K., 29
Yamamoto, O., 19, 49, 62, 87, 106
Yamane, T., 307
Yamasaki, A., 258, 330
Yamashita, J., 44
Yankelevich, A. Z., 82
Yannoni, C. S., 49, 185
Yasuoka, H., 353
Yi, P. N., 114
Yoder, C. H., 63, 64
Yonemitsu, T., 291
Yonezawa, T., 51, 52, 55, 368, 369, 388
Yoshida, M., 29
Yoshimura, Y., 388
Yoshioka, H., 31
Young, D. A. T., 67
Young, W. G., 78
Yukawa, Y., 17
Yuki, H., 105

Zachmann, H. G., 292, 295, 332
Zakrajsek, E., 337
Zambelli, A., 287, 298
Zane, L. I., 333
Zanger, M., 202
Zaucer, M., 337, 368
Zeidler, M. D., 118, 119, 120, 132
Zeltmann, A. H., 121, 245
Zetta, L., 41
Zhidomirov, G. M., 61
Ziao, N. K., 218

Ziessow, D., 41, 64, 253, 273
Zil'berman, B. D., 218
Zioudrou, C., 76
Zitserman, V. Yu., 258
Zuckerman, J. J., 47, 66, 274
Zuckermann, D. M., 168
Zumer, S., 164, 338
Zupancic, I., 147, 163, 185, 341, 347
Zviadaze, M. D., 152